Multidimensional Nanomaterials for Supercapacitors: Next Generation Energy Storage

Edited by

Sanjeev Verma

Department of Battery Manufacturing New Energy
Reliance Industries Limited, Navi Mumbai-400701
Maharashtra, India

Shivani Verma

Department of Chemistry, School of Physical Sciences
Doon University, Dehradun-248012
Uttarakhand, India

Saurabh Kumar

Department of Electronics and Communication Engineering
National Institute of Technology Hamirpur
Hamirpur-17700, Himachal Pradesh, India

&

Bhawna Verma

Department of Chemical Engineering and Technology
Indian Institute of Technology (Banaras Hindu University)
Varanasi-221005, Uttar Pradesh, India

Multidimensional Nanomaterials for Supercapacitors:

Next Generation Energy Storage

Editors: Sanjeev Verma, Shivani Verma, Saurabh Kumar and Bhawna Verma

ISBN (Online): 978-981-5223-40-8

ISBN (Print): 978-981-5223-41-5

ISBN (Paperback): 978-981-5223-42-2

First published in 2024.

need for a court order if at any point you breach any terms of this License Agreement. In no event will any delay or failure by Bentham Science Publishers in enforcing your compliance with this License Agreement constitute a waiver of any of its rights.

3. You acknowledge that you have read this License Agreement, and agree to be bound by its terms and conditions. To the extent that any other terms and conditions presented on any website of Bentham Science Publishers conflict with, or are inconsistent with, the terms and conditions set out in this License Agreement, you acknowledge that the terms and conditions set out in this License Agreement shall prevail.

Bentham Science Publishers Pte. Ltd.
80 Robinson Road #02-00
Singapore 068898
Singapore
Email: subscriptions@benthamscience.net

BENTHAM
SCIENCE

CONTENTS

PREFACE i

LIST OF CONTRIBUTORS iii

CHAPTER 1 INTRODUCTION OF NEXT-GENERATION MATERIALS 1
Neeraj Kumar, Shailendra Kumar Dwivedi, Om Prakash and *Shivani Verma*
INTRODUCTION 2
    Fundamental Theory of Supercapacitor 4
    Classifications of Supercapacitor 5
        Charge Storage Mechanism on Supercapacitors 6
    Classifications and Types of Nanomaterials 8
        Classification of Nanomaterials Based on Origin 8
        Classification Based on Dimensionality 9
        Classification Based on Material Used 11
    Multifunctional Future Materials, Their Properties, and Applications 11
        Carbon Based Materials 12
        Self-healing Polymers 14
        Metal-organic Frameworks (MOFs) 15
        Mxenes 16
        Composite Materials 17
        Nano-Inks and Quantum Dots 18
        Metamaterials 19
  SYNTHESIS TECHNIQUES 19
  FUTURE SCOPE OF NANOMATERIALS 21
  CONCLUSION 21
  REFERENCES 22

CHAPTER 2 SUPERCAPACITOR BASICS (EDLCS, PSEUDO, AND HYBRID) 29
Dinesh Bejjanki and *Sampath Kumar Puttapati*
  INTRODUCTION 29
  FARADAIC AND NON-FARADAIC 31
  CLASSIFICATION OF SUPERCAPACITOR 32
    Electric Double-layer Capacitors (EDLCs) 32
    Pseudocapacitor 33
    Hybrid Supercapacitor 35
    Electrode 36
    Carbon Materials 37
    Transition Metals 38
    Conducting Polymers 39
    Composite Materials 39
    Electrolyte 39
    Membrane 40
    Current Collectors 41
  CONCLUSION 41
  REFERENCES 42

CHAPTER 3 GRAPHENE AND ITS DERIVATIVES: CHEMISTRY, PROPERTIES, AND ENERGY STORAGE APPLICATION 49
Om Prakash, Vijay Kumar Juyal, Abhishek Pathak, Neeraj Kumar, Vivek Kumar. Shivani Verma, Akansha Agrwal and *Viveka Nand*
  INTRODUCTION 50

CHEMICAL EXFOLIATION 50
CHEMISTRY OF GRAPHENE 51
    Derivatives of Graphene 52
        Hydrogenated graphene (graphane) 53
        Fluorinated graphene (fluorographene) 53
        Oxidized Graphene (Graphene Oxide) 54
        Graphyne and Graphdiyne 56
    Other Miscellaneous Forms of Graphene 58
        Graphite 58
        Graphene quantum dots (GQDs) 59
        Carbon Nanotubes (CNTs) 60
        Fullerene 61
    Applications of Graphene and its darivatives 62
        Supercapacitors 62
        Lithium-ion battery 62
        Solar Cells 63
        Fuel Cells 63
        Water Filtration 64
CONCLUSION 64
FUTURE OUTLOOK 64
REFERENCES 65

CHAPTER 4 QUANTUM DOTS: CHEMISTRY, PROPERTIES, AND ENERGY STORAGE APPLICATIONS 71
Himadri Tanaya Das, T. Elango Balaji, Swapnamoy Dutta, Payaswini Das and Nigamananda Das
INTRODUCTION 71
    Synthesis and properties of QDs 73
        Laser ablation 73
        Electrochemical oxidation 74
        Chemical oxidation 75
        Microwave synthesis 75
        Thermal decomposition 76
PROPERTIES 76
DIFFERENT TYPES OF QUANTUM DOTS AND THEIR APPLICATIONS IN ENERGY STORAGE 77
    Quantum Dots applications in batteries 78
    QDs Applications in Supercapacitors 81
DISCUSSING THE PROS AND CONS OF QDS IN ENERGY STORAGE APPLICATIONS 82
CONCLUSION 83
REFERENCES 83

CHAPTER 5 METAL-ORGANIC FRAMEWORKS (MOFS): CHEMISTRY, PROPERTIES, AND ENERGY STORAGE APPLICATIONS 88
Nikhil Kumar, Nisha Gupta and Pallab Bhattacharya
INTRODUCTION 89
FUNDAMENTALS OF MOFS: CLASSIFICATION, SYNTHESIS AND PROPERTIES 90
    Transition metal-based MOFs 92
    Inner Transition Metal-based MOFs 94
    Mixed Metal-based MOFs 94

BRIEF DESCRIPTION OF ENERGY STORAGE SYSTEMS: CLASSI- FICATION, MECHANISM OF OPERATION, ADVANTAGE AND LIMI- TATIONS 95

BATTERIES 96

    Supercapacitors (SCs) 98

        Electric Double-layer Capacitor (EDLC) 100

        Pseudo Capacitor (PC) 100

        Hybrid SCs (HSCs) 101

RESPONSIBLE FACTORS IN MOFS FOR ENERGY STORAGE 102

ENERGY STORAGE PERFORMANCE OF VARIOUS MOF-BASED SYSTEMS 103

    Pristine MOFs 103

    MOF Composites 105

        MOF@rGO 106

        MOF@CNT 106

        MOF@NPs 107

    MOF Derived Materials 108

        MOF/C 109

        MOF/Metal-Carbon 110

        MOF/Metal Oxide 110

        MOF/Metal Hydroxide 111

        MOF/Metal Carbide or Sulphide or Nitride or Phosphide 112

SUMMARY AND FUTURE PERSPECTIVES 113

ACKNOWLEDGEMENTS 113

REFERENCES 113

CHAPTER 6 MXENE: CHEMISTRY, PROPERTIES, AND ENERGY STORAGE APPLICATIONS 120

Manisha Devi, Shipra Jaswal and *Swadesh Kumar*

INTRODUCTION 120

    Structure of MXene and MAX Phase 122

DIFFERENT APPROACHES FOR THE SYNTHESIS OF MXENES 122

    Top-down Synthetic Approach of MXene from MAX Precursor 123

    Wet Chemical Etching 123

        HF Etching 123

        Fluoride Salt Etching 124

        Alkali Etching 125

        Molten Salts Etching 126

        Electrochemical Method of Etching 126

        Intercalation/Delamination Method to Generate Delaminated MXenes (D-MXenes) 127

        Delamination of MXenes in Organic Solvents and Molecules 127

        Delamination of MXenes with Metal Ions 127

    Bottom-up Strategy 128

    Properties of MXenes 128

    Theoretical Capacity 128

    Electronic Band Structure 129

    Morphologies and Surface Chemistries 129

    Optoelectronic Properties 130

    Mechanical Properties 131

    Thermal Stability Properties 131

    Applications of MXenes in Energy Storage 132

    MXenes for Batteries 132

        Lithium-ion Batteries (LIBs) 132

Sodium Ion Battery (SIBs) 134
Potassium Ion Battery (PIBs) 134
MXenes for Supercapacitors (SCs) 135
CONCLUSION 137
REFERENCES 138

CHAPTER 7 DIFFERENT SUPERCAPACITORS' CHARACTERIZATIONS 145
Satendra Kumar, Hafsa Siddiqui, Netrapal Singh, Manoj Goswami, Lakshmikant
Atram, S. Rajveer, N. Sathish and Surender Kumar
INTRODUCTION OF SUPERCAPACITORS 145
Double-layer Formation and Faradaic Process 146
Electrochemical Characterizations 148
Scanning Electrochemical Microscopy 148
Cyclic Voltammetry and Potentiometry 149
Electrochemical Impedance Spectroscopy and Time Constant 151
Leakage Current and Self-discharge 152
Morphology Observation and Surface Analysis 153
Scanning Electron Microscopy 153
Transmission Electron Microscopy 153
X-ray Photoelectron Spectroscopy 153
Augur Electron Spectroscopy 155
Others 155
Phase, Structure, and Dynamics Observation 156
X-ray Diffraction 156
Raman Spectroscopy 158
Nuclear Magnetic Resonance Spectroscopy 159
Fourier Transform Infrared Spectroscopy 160
Others 161
CONCLUSION AND FUTURE OUTLOOK 162
REFERENCES 163

CHAPTER 8 ELECTROLYTES FOR ELECTROCHEMICAL ENERGY STORAGE
SUPERCAPACITORS 169
Priyanka A. Jha, Pardeep K. Jha and Prabhakar Singh
INTRODUCTION 169
TYPES OF ELECTROLYTES 173
Liquid electrolytes 173
Water-in-salt Electrolyte 173
Aqueous Electrolyte 174
Non-Aqueous Electrolyte 175
Solid-State 177
Redox-Active Electrolyte 178
Aqueous Electrolyte 178
INFLUENCE OF PORE SIZE ON PROPERTIES OF ELECTROLYTE 178
Liquid Electrolyte 178
Aqueous Electrolyte 178
Non-Aqueous Electrolyte 180
Redox-Active Electrolyte 181
PERFORMANCE OF ELECTROCHEMICAL SUPERCAPACITOR DEPENDING ON
ELECTROLYTE PERFORMANCE 181
CHALLENGES AND PERSPECTIVES OF ELECTROLYTES 182
SUMMARY 183

ACKNOWLEDGEMENT 183
REFERENCES 184

**CHAPTER 9 GRAPHENE-BASED FIBER SHAPE SUPERCAPACITORS FOR FLEXIBLE
ENERGY STORAGE APPLICATIONS** 190
Ankit Tyagi, Bhuvaneshwari Balasubramaniam and *Raju Kumar Gupta*
INTRODUCTION 191
    Evolution of Fiber Shape SCs 193
    Electrolyte used for Fiber Shape SCs 196
    Type of Fiber Shape SCs Device Structures 197
    Performance Evaluation of Fiber Shape SCs 198
    Wet-spinning of GO Fibers 201
    GO Fibers with Doping and Composite with Carbon and Polymer Materials 202
    GO Fibers Composite with Metal Oxides, Metal Sulfide, and MXene 204
CONCLUSION 206
REFERENCES 206

CHAPTER 10 QUANTUM DOTS-BASED NANOSTRUCTURES FOR SUPERCAPACITORS 212
Himadri Tanaya Das, Swapnamoy Dutta, T. Elango Balaji and *Niga* 212
INTRODUCTION 212
    Synthesis of Quantum Dots for Electrodes in Supercapacitors 214
    Performances or Reported Articles with Electrolyte 215
    Discussing the Pros and Cons of Supercapacitor 220
    Future Scope for Supercapacitor 221
CONCLUSION 221
REFERENCES 222

**CHAPTER 11 METAL-ORGANIC FRAMEWORKS (MOFS) BASED NANOMATERIALS
FOR SUPERCAPACITOR APPLICATIONS** 225
Pardeep K. Jha, Priyanka A. Jha and *Prabhakar Singh*
INTRODUCTION 225
SYNTHESIS OF METAL-ORGANIC FRAMEWORKS BASED NANOMATERIALS 228
    Synthesis Techniques 228
CLASSIFICATION 229
    Dimensional Morphology 230
    0D 230
    1D 231
    2D 232
    3D 232
    Compositional Classification 233
        Pristine-MOF 234
        MOF Composites 234
        MOF-CP 234
        MOF- Derivatives 234
        C-MOF 234
        NC-MOF 236
        Metal Oxides Composites 236
        Metal Hydroxides Composites 236
        Metal Sulfides Composites 236
        Other NC MOF Derivatives 237
        Hybrid MOF Derivatives 237
CHALLENGES 238

CONCLUDING REMARKS 238
ACKNOWLEDGEMENT 239
REFERENCES 239

**CHAPTER 12 MXENE-BASED NANOMATERIALS FOR HIGH-PERFORMANCE
SUPERCAPACITOR APPLICATIONS** 244
Zaheer Ud Din Babar, Ayesha Zaheer, Jahan Zeb Hassan, Ali Raza and Asif Mahmood
INTRODUCTION 245
    Charge Storage Pathways in the MXene-based Supercapacitors 247
        Aqueous Media 247
        Non-aqueous Media 251
    MXenes as Supercapacitor Electrodes 252
        Surface Chemistry 252
        Fabrication and Design 256
    Factors Affecting the Electrochemical Performance of Supercapacitors 257
        Synthesis 257
        Structure and Size of MXene 262
        Architecture of Electrodes 264
    Electrolyte 265
        Current Advances in the MXene-based High-performance Supercapacitors 266
CONCLUSION AND OUTLOOK 272
REFERENCES 273

**CHAPTER 13 RECENT DEVELOPMENTS IN THE FIELD OF SUPERCAPACITOR
MATERIALS** 284
Mani Jayakumar and Venkatesa Prabhu S.
INTRODUCTION 284
    Carbon composite-based Supercapacitor Electrode Materials 286
        Recent developments in Capacitor Materials of Metal-oxide and its Composites 287
        Capacitors using Conducting Polymers-carbon 290
        Composite Graphene Capacitors 291
        Composite Capacitors with CNTs 291
    Advancements in Micro-supercapacitors 292
    1. Supercapacitors' Benefits 293
    2. Standard uses for Supercapacitor Materials 293
        Electric and Hybrid Vehicles 293
        Electronic and Low-power Applications 294
        Military and Defense Applications 295
        Renewable Energy 295
        Industrial and Biomedical Applications 295
        Traction 295
SUMMARY AND OUTLOOK 295
REFERENCES: 296

**CHAPTER 14 SUPERCAPACITOR MATERIALS: FROM RESEARCH TO THE REAL
WORLD** 303
Ahmad Nawaz, Vikas Kumar Pandey and Pradeep Kumar
INTRODUCTION 303
COMPONENTS 305
    Electrodes 305
        Current Collector 305

Activated Carbon-based Materials 306
Binders 306
Additives used for Conductivity 307
Electrolyte 307
Electrolyte Degradation 308
Thermal Stability 308
Non-conventional Electrolytes 309
Separators 310
Supercapacitor Applications 310
Power Electronics 310
Memory Protection 311
Battery Enhancement 311
Portable Energy Sources 312
Adjustable Speed Drives (ASDs) 312
High Power Sensors and Actuators 313
Hybrid Electric Vehicles 314
Military and Aerospace Applications 315
CONCLUSION 315
REFERENCES 316

CHAPTER 15 FUTURE OUTLOOK AND CHALLENGES FOR SUPERCAPACITORS 321
Vikas Kumar Pandey and Bhawna Verma
INTRODUCTION 321
MARKET CHALLENGES 323
Challenges Based On Electrode Materials 324
Current collector challenges 324
Electrical Double-layer-based Materials 325
Pseudocapacitive Materials 326
Conducting Polymers 326
Transition Metal Oxides 327
Composite Based Materials 328
Electrolytes 330
Current Collectors 330
Computational Aspects 331
Protocols and strategies 332
CONCLUSION 333
REFERENCES 335

SUBJECT INDEX 340

PREFACE

Supercapacitors are a new class of superior energy storage devices that provide both high energy and power densities, bridging the gap between batteries and regular capacitors. The two primary charge storage processes of supercapacitors are the redox process and the electrochemical double layer. Considerable interest is being paid to strategies that would combine both mechanisms in a supercapacitor to improve its electrochemical characteristics. The energy storage capacity of supercapacitors can be greatly impacted by the electrode materials utilized to make these devices. For supercapacitors, a variety of materials are being used, including conducting polymers, carbon-based materials, layered structured materials, metal oxides, and sulfides. The energy and power density of supercapacitors might vary depending on the materials' shape and kind. This book discusses developments in next-generation supercapacitor materials such as Mxene, MOFs, Quantum dots, and graphene-based nanostructures. A brief history of nanostructural materials, chemistry and supercapacitors as energy storage devices is also provided. This technical book can be a very helpful reference for scientists, industrial practitioners, graduate and undergraduate students, and other professionals in the scientific and education domains.

This book attempts to present the most recent as well as future forming materials, and ground-breaking developments in nanostructured materials for supercapacitor applications. The numerous intriguing characteristics of nanoscale materials make them perfect for energy storage applications. Additionally, methods are used to improve their morphological, electronic, and electrical characteristics in order to improve their electrochemical performances. Numerous new nanocomposites based on Mxenes, MOFs, Quantum dots, and variants of graphene are discussed. In-depth descriptions of novel methods for synthesizing and customizing their electrochemical characteristics are provided. With thorough characterization, mechanistic techniques, and theoretical analysis, this book compiles information on the production and applications of nanomaterials for supercapacitors. Recent advances in cutting-edge technology, including flexible and wearable supercapacitors made of nanostructured materials, are discussed. The readers of this book are given both basic and specialized techniques for creating nanostructured materials for supercapacitors.

Sanjeev Verma
Department of Battery Manufacturing New Energy
Reliance Industries Limited, Navi Mumbai-400701
Maharashtra, India

Shivani Verma
Department of Chemistry, School of Physical Sciences
Doon University, Dehradun-248012
Uttarakhand, India

Saurabh Kumar
Department of Electronics and Communication Engineering
National Institute of Technology Hamirpur
Hamirpur-17700, Himachal Pradesh, India

&

Bhawna Verma
Department of Chemical Engineering and Technology
Indian Institute of Technology (Banaras Hindu University)
Varanasi-221005, Uttar Pradesh, India

List of Contributors

Ayesha Zaheer	Department of Physics "Ettore Pancini", University of Naples Federico II, Piazzale Tecchio, 80, 80125 Naples, Italy
Ali Raza	Department of Physics "Ettore Pancini", University of Naples Federico II, Piazzale Tecchio, 80, 80125 Naples, Italy
Asif Mahmood	School of Chemical and Biomolecular Engineering, The University of Sydney, Sydney, Australia Center for Clean Energy Technology, School of Mathematical and Physical Sciences, Faculty of Science, University of Technology Sydney, Sydney, Australia
Ahmad Nawaz	Center for Refining & Advanced Chemicals, King Fahd University of Petroleum and Minerals, Dhahran, 31261, Saudi Arabia Department of Chemical Engineering & Technology, Indian Institute of Technology (Banaras Hindu University), Varanasi-221005, India
Abhishek Pathak	Department of Chemistry, Govind Ballabh Pant Institute of Agriculture and Technology, 263145, India
Ankit Tyagi	Department of Chemical Engineering, Indian Institute of Technology Jammu, Jammu, 181221, J & K, India
Akansha Agrwal	Department of Applied Sciences, KIET Group of Institutions, Delhi-NCR, Meerut Road (NH-58), Ghaziabad 201206, India
Bhawna Verma	Department of Chemical Engineering & Technology, Indian Institute of Technology (Banaras Hindu University), Varanasi-221005, Uttar Pradesh, India
Bhuvaneshwari Balasubramaniam	Department of Material Science Programme, Indian Institute of Technology Kanpur, Kanpur 208016, UP, India
Dinesh Bejjanki	Department of Chemical Engineering, National Institute of Technology, Warangal 506004, India
Himadri Tanaya Das	Centre of Excellence for Advance Materials and Applications, Utkal University, Bhubaneswar 751004, Odisha, India
Hafsa Siddiqui	CSIR - Advanced Materials and Processes Research Institute (AMPRI), Bhopal-462026, India
Jahan Zeb Hassan	Department of Physics, Riphah Institute of Computing and Applied Sciences (RICAS), Riphah International University, 14 Ali Road, Lahore, Pakistan
Lakshmikant Atram	CSIR - Advanced Materials and Processes Research Institute (AMPRI), Bhopal-462026, India
Manisha Devi	Department of Chemistry, Gautam College Hamirpur, Himachal Pradesh, India
Manoj Goswami	Academy of Scientific and Innovative Research (AcSIR), Ghaziabad-201002, India CSIR - Advanced Materials and Processes Research Institute (AMPRI), Bhopal-462026, India

Mani Jayakumar Department of Chemical Engineering, Haramaya Institute of Technology, Haramaya University, Haramaya, Dire Dawa, Ethiopia

Neeraj Kumar School of Studies in Chemistry, Jiwaji University, Gwalior (M.P), India
IPS group of colleges, Shivpuri Link Road, Gwalior (M.P), India

Nigamananda Das Centre of Excellence for Advance Materials and Applications, Utkal University, Bhubaneswar 751004, Odisha, India
Department of Chemical Engineering, National Taiwan University of Science and Technology, Taipei, 10607, Taiwan

Nikhil Kumar Functional Materials Group, Advanced Materials & Processes (AMP) Division, CSIR-National Metallurgical Laboratory (NML), Burmamines, East Singhbhum, Jamshedpur, Jharkhand-831007, India

Nisha Gupta Functional Materials Group, Advanced Materials & Processes (AMP) Division, CSIR-National Metallurgical Laboratory (NML), Burmamines, East Singhbhum, Jamshedpur, Jharkhand-831007, India

Netrapal Singh Academy of Scientific and Innovative Research (AcSIR), Ghaziabad-201002, India
CSIR - Advanced Materials and Processes Research Institute (AMPRI), Bhopal-462026, India

N. Sathish Academy of Scientific and Innovative Research (AcSIR), Ghaziabad-201002, India
CSIR - Advanced Materials and Processes Research Institute (AMPRI), Bhopal-462026, India

Om Prakash Regional Ayurveda Research Institute, Ministry of Ayush, Gwalior, 474009, India

Payaswini Das CSIR-Institute of Minerals and Mining Technology, Bhubaneswar, Odisha, India

Pradeep Kumar Department of Chemical Engineering & Technology, Indian Institute of Technology (Banaras Hindu University), Varanasi-221005, India

Pallab Bhattacharya Functional Materials Group, Advanced Materials & Processes (AMP) Division, CSIR-National Metallurgical Laboratory (NML), Burmamines, East Singhbhum, Jamshedpur, Jharkhand-831007, India

Priyanka A. Jha Department of Physics, Indian Institute of Technology (Banaras Hindu University), Varanasi-221005, India

Pardeep K. Jha Department of Physics, Indian Institute of Technology (Banaras Hindu University), Varanasi-221005, India

Prabhakar Singh Department of Physics, Indian Institute of Technology (Banaras Hindu University), Varanasi-221005, India

Raju Kumar Gupta Department of Chemical Engineering, Indian Institute of Technology Kanpur, Kanpur-208016, UP, India
Center for Environmental Science and Engineering, Indian Institute of Technology Kanpur, Kanpur-208016, UP, India
Department of Sustainable Energy Engineering, Indian Institute of Technology Kanpur, Kanpur- 208016, UP, India

Shailendra Kumar Dwivedi School of Studies in Physics, Jiwaji University, Gwalior (M.P), India
Madhav Institute of Technology & Science, Gola ka Mandir, Gwalior-474005, India

Shivani Verma Department of Chemistry, School of Physical Sciences, Doon University, Dehradun-248012, Uttarakhand, India

Sampath Kumar Puttapati Department of Chemical Engineering, National Institute of Technology, Warangal 506004, India

Swapnamoy Dutta University of Tennessee, Bredesen Center for Interdisciplinary Research and Graduate Education, Knoxville, TN, 37996, USA

Shipra Jaswal Department of Chemistry, Gautam College Hamirpur, Himachal Pradesh, India

Swadesh Kumar Department of Chemistry, Gautam College Hamirpur, Himachal Pradesh, India

Satendra Kumar Academy of Scientific and Innovative Research (AcSIR), Ghaziabad-201002, India
CSIR - Advanced Materials and Processes Research Institute (AMPRI), Bhopal-462026, India

S. Rajveer Metallurgical and Materials Engineering, National Institute of Technology, Jamshedpur-831014, India

T. Elango Balaji Department of Chemical Engineering, National Taiwan University of Science and Technology, Taipei, 10607, Taiwan

Venkatesa Prabhu S. Center of Excellence for Bioprocess and Biotechnology, Department of Chemical Engineering, College of Biological and Chemical Engineering, Addis Ababa Science and Technology University, Addis Ababa, Ethiopia

Vikas Kumar Pandey Department of Chemical Engineering & Technology, Indian Institute of Technology (Banaras Hindu University), Varanasi-221005, India

Vijay Kumar Juyal Department of Chemistry, Govind Ballabh Pant Institute of Agriculture and Technology, Uttarkhand 263145, India

Vivek Kumar Regional Ayurveda Research Institute, Ministry of Ayush, Gwalior, 474009, India

Viveka Nand Department of Chemistry, Govind Ballabh Pant Institute of Agriculture and Technology, Uttarkhand 263145, India

Zaheer Ud Din Babar Scuola Superiore Meridionale (SSM), University of Naples Federico II, Largo S. Marcellino, 10, 80138, Italy
Department of Physics "Ettore Pancini", University of Naples Federico II, Piazzale Tecchio, 80, 80125 Naples, Italy

CHAPTER 1

Introduction of Next-Generation Materials

Neeraj Kumar[1,3], **Shailendra Kumar Dwivedi**[2,4,*], **Om Prakash**[5] and **Shivani Verma**[6]

[1] *School of Studies in Chemistry, Jiwaji University, Gwalior (M.P), India*

[2] *School of Studies in Physics, Jiwaji University, Gwalior (M.P), India*

[3] *IPS group of colleges, Shivpuri Link Road, Gwalior (M.P) India*

[4] *Madhav Institute of Technology & Science, Gola ka Mandir, Gwalior-474005, India*

[5] *Regional Ayurveda Research Institute, Ministry of Ayush, Gwalior, 474009, India*

[6] *Department of Chemistry, School of Physical Sciences, Doon University, Dehradun-248012, Uttarakhand, India*

Abstract: The "next-generation materials" are those materials that have high efficiency, high-performance structural stability, easy manufacturability, and multifunctional capabilities. These new materials can be classified based on dimension, shape, composition, and nanostructure like 0D, 1D, 2D, and 3D. These materials have unique enhanced properties *viz.* electronic, optical, mechanical, magnetic, optoelectronics, vitrification, thermal properties, *etc*. Due to these outstanding features, these smart materials could be a game changer for prospects. Tuning the properties of such advanced materials provides a wide variety of fascinating opportunities. This chapter aims to provide a comprehensive overview of materials used to fabricate supercapacitor point of view and several other latest applications. The nanomaterials, discussed in this chapter along with their properties are Graphene, nanotubes, nanocomposites, microwave-absorbing materials, nanoparticles, biomaterials, and self-healing polymers. It also discusses future directions for the development of advanced materials that perform well to anticipate future trends and highlight their relevance in real-world contexts. This chapter could become the torchbearer for new researchers working in the field of multifunctional advanced materials.

Keywords: Functional & smart materials, Flexible electronics, Multifunctional, Nanomaterials & nanofluids, Optoelectronics.

* **Corresponding author Shailendra Kumar Dwivedi:** School of Studies in Physics, Jiwaji University, Gwalior (M.P), India and Madhav Institute of Technology & Science, Gola ka Mandir, Gwalior-474005, India;
E-mail: shail.dew04@gmail.com

INTRODUCTION

The "Next-generation materials" or advanced materials are those materials that have the properties of high efficiency, high-performance structural stability, easy manufacturability, and multifunctional capabilities. The basic characteristics of these materials include being very light, intelligent, more durable, and active materials that can adjust appropriately to their surroundings. In recent years, the development of new materials and technologies has been associated with innovation, creativity, originality, and forward thinking. A self-assembly process is specifically designed to produce advanced materials comprising nanoscale structures [1, 2]. These materials are of great interest in scientific research efforts and industrial development because of their innovative potential applications in various fields. Advanced materials are future materials with improved properties that are consciously designed for superior performance. The major scientific contributions of the 21st century, and a new understanding of atomic and subatomic levels, laid the foundation for the creation of advanced materials. The development of such advanced future materials can even lead to the design of completely advanced products, such as portable supercomputers, mini electronic gazettes, flexible electronics and optoelectronic devices, automatic lightweight weapons, fire registrant materials, medical implantable devices, gas sensors, lightweight industrial equipment, intelligent robotics, *etc.*

Nowadays, the materials such as graphene, carbon nanotube, men, nanofluids, quantum dots, nanoparticles, metal-organic frameworks (MOFs), aerogel, nanocomposites, microwave absorbing materials, self-healing polymers, artificial spider silk, metal foam, synthetic fuel and lubricants, shrilk and many more have emerged as advanced materials for human beings. These materials have the potential to sort out human futuristic problems and are useful for the better advancement of human civilization. Advanced nanomaterials are very desirable in these domains because of their controllable production and beautiful design. Due to its vast applicability in a range of sectors, such as energy storage, electronics, optics, optoelectronics catalytic, absorption and separation, biomedical, luminescence, sensing, and environment, nanotechnology has gained a lot of attention in recent decades. The key features of advanced nanomaterials are their active surfaces, dimensions, and reaction conditions [3, 4].

The physical and chemical properties of advanced nanomaterials are greatly influenced by their dimensions and reaction conditions [3]. Thus, it is the right time to think about not only synthesizing materials but also tuning their physicochemical properties (Fig. **1a**) to develop next-generation materials. The beauty of advanced nanomaterials is their tunable properties; therefore, by changing the shape, size, and reaction conditions of the nanomaterials, one can change their functionality accordingly. So, to utilize these nanomaterials for the development of a new world, we need to develop advanced synthetic techniques so that more features of those materials can be explored in various fields for human beings.

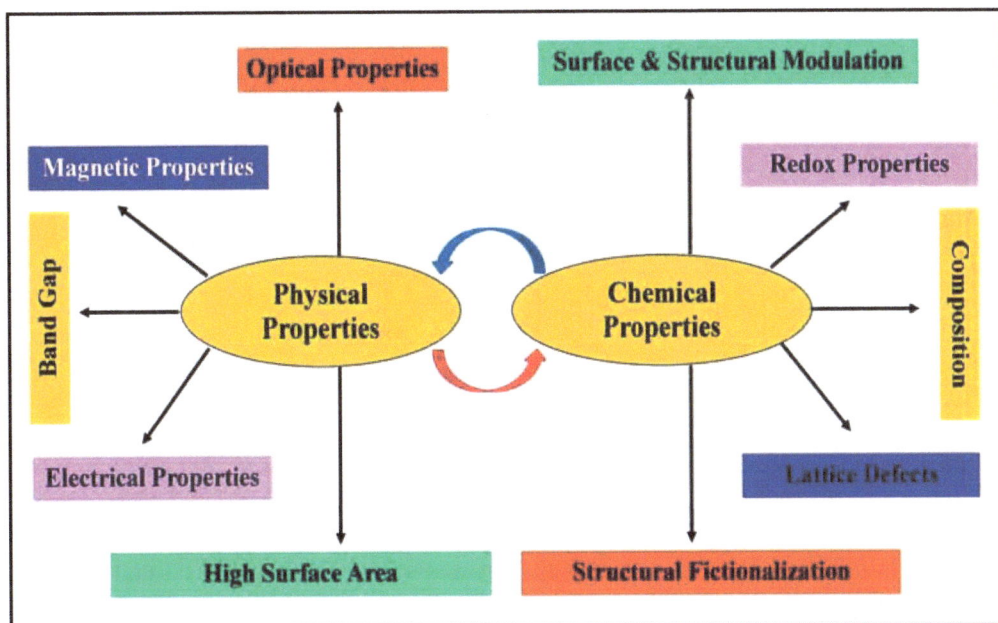

Fig. (1a). Physicochemical properties of nanomaterials.

Scientific legend; Andre Geim and Konstantin Novoselov in 2004 at Manchester University discovered a wonder material called "Graphene" by playing with a lump of graphite and Scotch tape. At that time, both did not know how to deal with and what to do. But nowadays, Graphene has become one of the extraordinary materials for the future world because of its properties like immensely strong, flexible, transparent, and conductivity. Shrilk could be another wonder material for the future world [4]. Shrilk is a biodegradable solution to plastic and is mainly made up of silk proteins and chitin developed by Javier Fernandez and Donald Ingber at the Wyss Institute of Biologically Inspired Engineering, Harvard University [5]. A material with huge absorption capability of electromagnetic radiations (microwave) was discovered named metamaterials. It is an advanced or a new class of materials that can have electromagnetic features including the negative value of permittivity, permeability, and refractive index that do not occur naturally. In 2011, another wonder material called Mxenes was discovered by two research groups led by Y.Gogotsi and M.Barsoum at Drexel University. Generally, Mxene is a 2D transition metal-based compound of carbides, carbonitrides, and nitrides used for wastewater treatment, energy storage, detection of various gases (gas sensors), and electronic applications [6, 7].

Nowadays, the continuous developments in high-performance energy storage devices have gained much attention from the scientific world and environmental security agencies of different countries. To fulfill the energy demand, various alternatives have come into existence but supercapacitor technologies could be the best alternative among all which can offer high power densities, large life cycles, quick charge and discharge response time as well as a clean and safe electrochemical energy storage [8 - 10]. This chapter aims to provide a comprehensive overview of several latest functional materials used for supercapacitors, storage mechanisms, criteria of formation and design fabrication, different electrodes and electrolyte materials, along with their properties, synthesis, applications, and future scopes.

Fundamental Theory of Supercapacitor

A supercapacitor (Fig. **1b**) is a device having a higher capacitance value than conventional capacitors at lower voltage limits. It fulfils the gap between rechargeable batteries and electrolytic capacitors and has 10-100 times more energy storage capacity per unit volume or mass than electrolytic capacitors [11]. A supercapacitor consists of a bi-electrode system that is separated from each other by an electrolyte separator. The supercapacitor device is composed of many parts like a current collector, two electrodes, a separator, and an electrolyte solution. Its characteristics are based on these constituents.

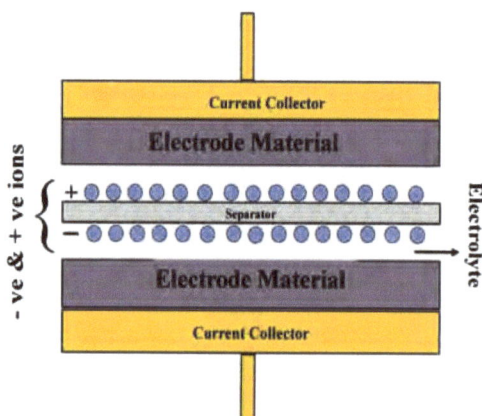

Fig. (1b). Basic device structure of supercapacitor.

The basic function of the separator is similar to that of the battery. It keeps apart the two electrodes to avoid a short circuit between the electrodes and allows ions to pass through. The basic principle of energy storage in supercapacitors is based on the charging-discharging cyclic process which happens at the electrode-electrolyte interface. In comparison with conventional capacitors, supercapacitor electrodes possess a highly effective surface area which leads to enhancement in capacitance value by a huge factor of 10000 than conventional capacitors [12, 13].

Classifications of Supercapacitor

There are two major classification standards for the supercapacitors. The first classification is based on the energy storage mechanisms of the different electrode materials, and the second classification is based on the different electrolytes. A supercapacitor based on different electrode materials can have different possible designs and leads to the production of symmetric supercapacitors, asymmetric supercapacitors, and hybrid supercapacitors. In case of a symmetric supercapacitor, the anode and cathodes are made of the same material (basically carbon materials) whereas, an asymmetric supercapacitor can have various possible combinations of electrode materials. If the combination of these two electrodes is in such a way that one is the capacitive type and the other is a capacitive Faradic (pseudocapacitive) or non-capacitive Faradic type, then that combination leads to the formation of hybrid capacitors [14]. Furthermore, the electrolyte-type supercapacitors are divided based on aqueous and organic electrolyte media. The aqueous electrolytes include acidic electrolytes (H_2SO_4 aqueous solution, 36%), basic electrolytes (strong bases KOH and NaOH), and the neutral electrolytes (KCl, NaCl, and other salts) which are mostly used in case of manganese oxide electrode material along with water as a solvent. The organic electrolyte commonly uses lithium salts, and quaternary amine salts along with solvents such as ACN, PC, GBL, THL, *etc* [15]. Based on the above discussion, supercapacitors can be classified based on charge-storage mechanism into three basic types as follows: (**i**) Pseudocapacitors (PCs), (**ii**) Hybrid capacitors (HCs), and (**iii**) Electric double-layer capacitors (EDLCs) [16 - 18] as shown in Fig. (**1c**).

H. Becker developed a "Low-voltage electrolytic capacitor in 1957 by using a porous carbon electrode system. At that time, he did not know the energy storage process and believed that a charge was stored in the carbon pores which provide energy. Later on, numerous scientists worked on this and finally electric-double layer mechanism came into existence. The first supercapacitor having low internal resistance was designed for military applications in 1982 by the Pinnacle Research Institute (PRI) and was commercialized under the brand name "Ultracapacitor- PRI" [19]. Supercapacitors have promising potential because of their excellent charge storage properties and high-power density for various energy storage applications. Supercapacitor devices can be widely used in photovoltaic, electric hybrid, wind power generation, *etc*. These are also used as power supplies in portable devices such as computers, digital cameras, mobile phones, and notebooks because of their lightweight and small size [20]. In addition, supercapacitors have many advantages compared to electrochemical batteries and fuel cells, such as short charging times, long cycle stability, and high-power density. Therefore, it is necessary to understand their energy storage mechanism.

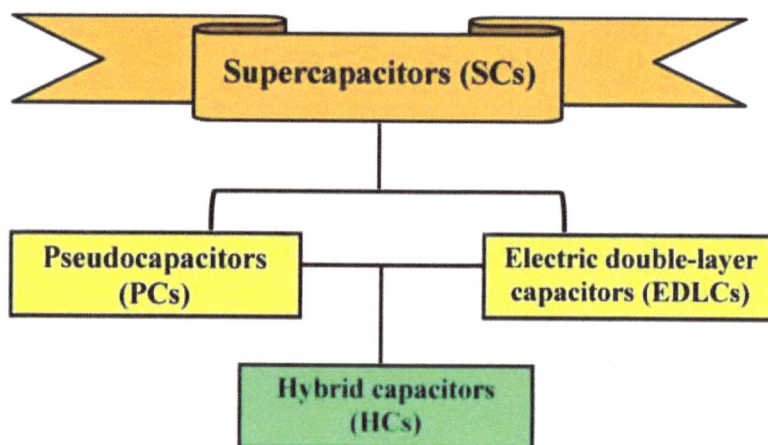

Fig. (1c). Types of supercapacitors.

Charge Storage Mechanism on Supercapacitors

Electrochemical Double-layer Capacitors

Electrostatic ion adsorption/desorption at the interface of electrode and electrolyte is the concept of energy storage used in EDLCs. When voltage is applied, there was no charge accumulation on the surface of the electrode, hence, opposite charges attract each other due to potential difference as a result of this the diffuse of electrolyte ions takes place over the separator as well as on the opposite charged electrode. A charge's double layer was formed to prevent ion recombination in electrodes. Thus, the charge storage takes place directly across the double layer of the electrode material without any charge transfer across the interface, and hence the capacitance value increases due to the capacitance effect [21].

Pseudocapacitors

Pseudocapacitors are completely non-electrostatic and obey the Faradaic redox mechanism for charge transfer between the electrode and the electrolyte. Commonly, transition metal oxide (MnO_2, RuO_2, Fe_3O_4, $MnFeO_2$ *etc.*) and conducting polymer (PANI–PANI, PPY/MWCNT, PANI/MWCNT, *etc.*) electrodes show high electrochemical pseudocapacitance behavior. When a potential is applied to the pseudocapacitor, redox reaction takes place on the electrode material interface and hence charge's passage across a double layer. In addition, due to this redox Faradic mechanism, pseudocapacitors possess high specific capacitance as well as energy densities in comparison to EDLCs [22].

Hybrid Supercapacitors

This type of supercapacitors was designed to achieve enhancement in energy density as compared to EDLC. HCs supercapacitors are based on the mechanism of double-layer ion adsorption/desorption and reversible Faradic reaction. The hybrid supercapacitor formation results from the coupling of different redox and EDLC materials like graphene or graphite, magnetic metal oxides, conducting polymers, and activated carbon [23]. Graphene and carbon nanotubes are very popular carbon-derived nanomaterials that are being used as efficient electrode materials in the design of supercapacitors. These materials have many outstanding features like high mechanical properties with great specific surface area and most importantly competent electrical properties [24]. Furthermore, carbon fiber, carbon derivatives, xerogel, activated carbon, and template carbon have been applied as efficient electrode materials in the design of supercapacitors. These materials possess durable power density, powerful lifecycles, lasting cycle durability, and desirable coulumbic reliability [25]. Nowadays, magnetic metal oxide nanoparticles have received great attention from energy storage devices like supercapacitors with high specific capacitance. Magnetic metal oxide nanoparticles are class of an attractive type of material because they are cheap and easy to prepare in large quantities [26].

Recently, the spinel ferrite which has nominal composition MFe_2O_4, where M is magnesium, copper, manganese, nickel, zinc, and cobalt. This has been successfully synthesized and exhibited a notable discharge of capacitance up to 1000 mA hg^{-1}, which is about three times higher than commercial anodes made from graphite [27]. Yao *et al.* [28] have successfully developed a carbon-coated Zn ferrite/graphene composite by a general multistep strategy. During the anodic process, one broad peak rises at ~ 1.50-2.10 V, exhibiting the oxidation of the base zinc ions (Zn^0 to Zn^{2+}) and iron ions (Fe^0 to Fe^{3+}). The electrochemical studies have revealed that the electrode offers a discharge capacity with a value of 1235 mA h g^{-1} and a loss of about 465 mA h g^{-1} over 150 cycles with good cycling performance. The nickel molybdate NiMoO$_4$ has been studied as one of the most popular candidates for supercapacitor electrodes. 2D nickel molybdate like-nanoflakes synthesized *via* rapid microwave-assisted, have achieved 1739 F g^{-1} of specific capacitance at 1 mV s^{-1} of scan rates. Huang *et al.* [29] have demonstrated that the three-dimensional interconnected nickel molybdate-lik--nanoplate arrays revealed a specific capacitance as high as 2138 F g^{-1} at a current density of 2 mA cm^{-2}.

Classifications and Types of Nanomaterials

Nature is a wonderful gift from God to human beings. The observation and examination of nature, natural processes, and evaluation of verities of elemental structures inspired us to solve human futuristic problems. Researchers aim to replicate something in nature that they find to be incredibly amazing (Biomimicry). Human inventions have been influenced and created by natural structures [8]. Controlled organization and properties with nano-scale precision have led to the creation of multi-functional advanced nanomaterials with miniaturization. Thus, nanomaterials are broadly classified based on three basic criteria; (1) Classification of nanomaterials based on origin [9]; (2) dimensionality, and (3) on the material used in the synthesis process. The existing classifications are based on research articles, textbooks, internet sources, and expert's knowledge of the various disciplines. According to the first approach called dominant bond type (in technical disciplines), the materials are divided into four categories Ceramics, Metals, Polymers, and Composites. According to the International Organization of Standardization ISO norms and the German Institute for Standardization (GIS) norms materials are classified as Materials of Glass (Ceramic, Metal, Stone, Paint and Color, Paper, Leather, Plastic, Textile, and Wood), Composite Materials and Raw Materials [30 - 33].

Classification of Nanomaterials Based on Origin

In everyday life, materials are important. Most of the time, people interface with these materials through products knowingly or unconsciously, voluntarily or involuntarily. Based on the origin of the materials, they are broadly classified into types; natural and synthetic. Natural materials (biotic) exist in nature and are produced by bio-geochemical or mechanical processes. They are not chemically changed as much. For example, a wooden table, its shape might be changed, but the material is still wood. Similarly, glass might be considered as a natural substance due to its origin from sand, which has been melted and then cooled. Synthetic materials are also made from natural resources and may or may not be chemically identical to a naturally occurring substance. These materials are produced by anthropogenic processes. The succession of chemical processes used to transform natural resources into synthetic goods is termed chemical synthesis. Fig. (**2a**) gives an idea of the classification of materials along with some examples.

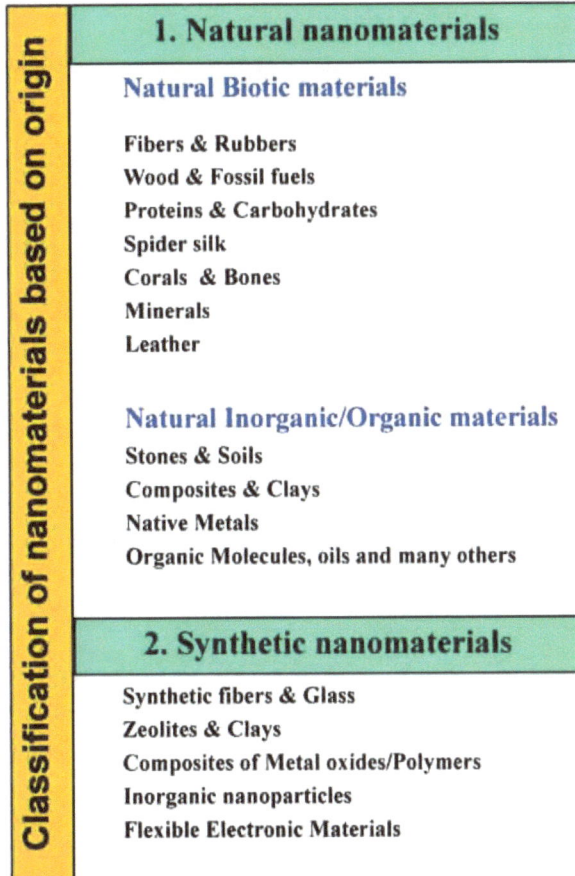

Fig. (2a). Classification of nanomaterial materials based on Origin.

Classification Based on Dimensionality

Dimensionality is another criterion for the classification of nanomaterials. The shape and size of nanomaterials, ranging from 1-100 nm, are the basis of their classification. Further, they can be divided into four classes, based on their dimensionality and shape, *i.e.*; 0D, 1D, 2D, and 3D (Fig. **2b**).

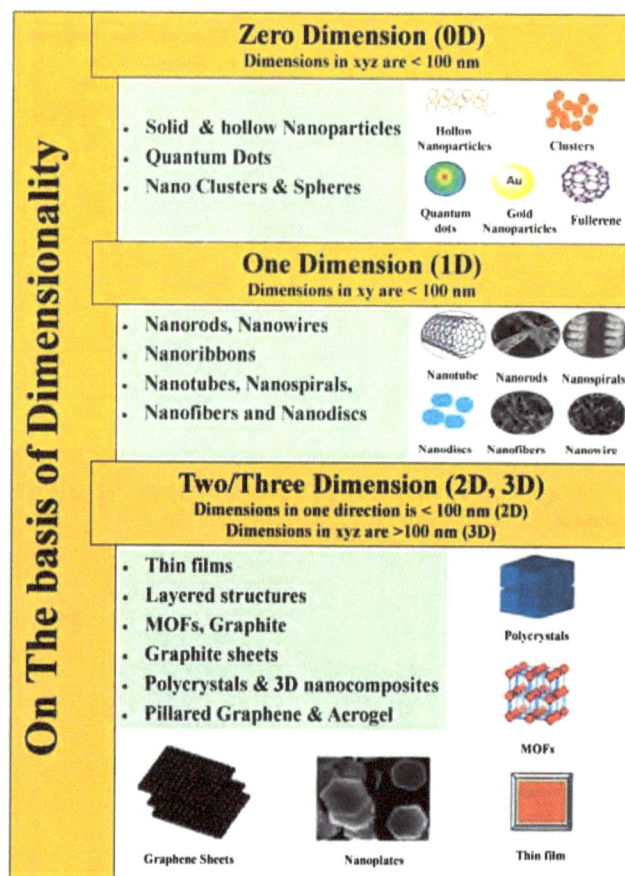

Fig. (2b). Schematic representation of the classification of nanomaterials based on dimensionality (adopted from refs. [34-37]).

Zero-dimensional (0D) nanomaterials are materials with all of their dimensions at nanoscale, or below 100 nm in size. Spherical NMs, nanorods, Cubes and polygons, hollow spheres, metal nanoparticles, fullerenes, and quantum dots are all included in 0D. Materials having only two dimensions in the nanoscale range are included in one-dimensional (1D) nanomaterials. The common examples are carbon nanotubes, ceramic, metallic nanodiscs, nanorods, nanofibers, and nanowires. Two-dimensional (2D) nanomaterials are those which have only one dimension in nanoscale while the other two are not. Some common examples of 2D materials are single and multi-layered structures, thin films, nanoplates, MOFs, *etc* [38]. Further, three-dimensional (3D) nanomaterials are those having dimensions in different directions with all of their dimensions beyond 100 nm.

Classification Based on Material Used

Intentionally created functional nanomaterials come in a wide range of varieties, and more are predicted to be developed in the future. Fig. (**2c**) displays some frequently used materials and it is expected that by tuning their basic properties, these materials have the potential to generate revolutionary advanced material for future generation.

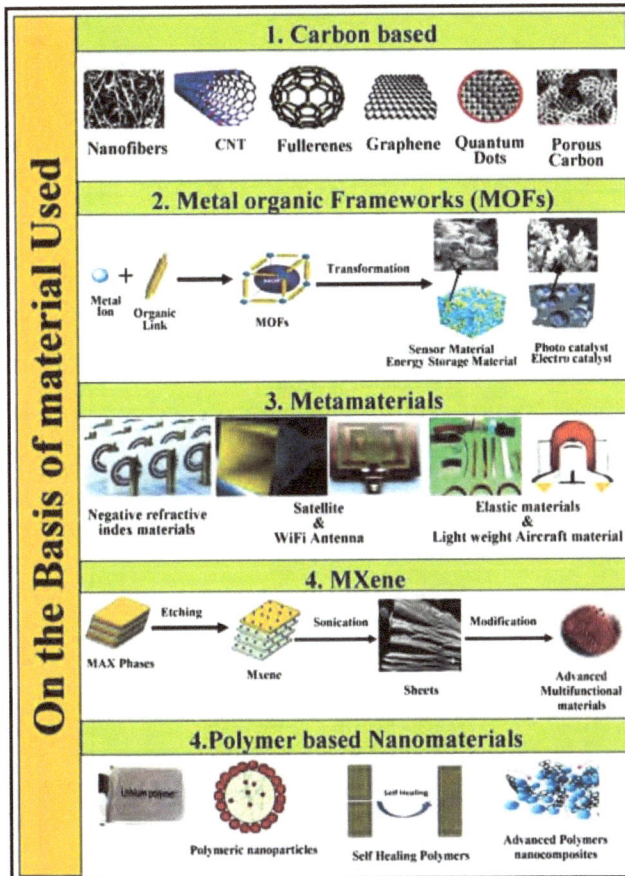

Fig. (2c). Classification of materials based on the route of synthesis (adapted from refs. [41-45, 63, 71, 77, 81-82, 91-93]).

Multifunctional Future Materials, Their Properties, and Applications

Various above-mentioned functional materials such as Graphene, Mxene, MOFs, perovskites, self-healing polymers, Shrilk, Metameterias, quantum dots, and advanced nanomaterials, have been developed for future generation. However, many properties of these materials still need to be explored for the development of a new world with new technologies. Here, various properties and applications (Fig. **3a**) of highly demanded materials with their functionality are discussed.

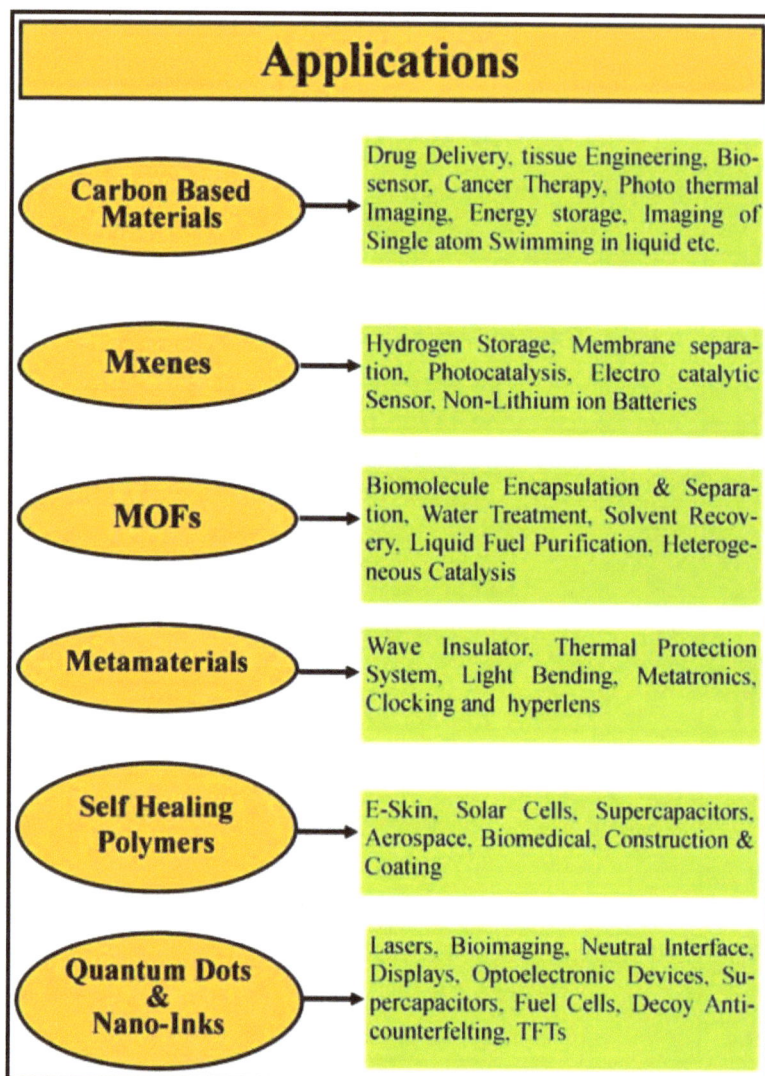

Fig. (3a). Application areas of multifunctional materials.

Carbon Based Materials

Carbon-based nanomaterials have outstanding properties such as high surface area, lightweight, high electrical conductivity, high thermal and chemical stability, corrosion resistive and non-oxidizing nature. Carbon materials provide a vast range of forms and textures, and they are simple to produce. Carbon is a solid-state allotrope that provides a wide variety of structures and is easy to process.

The carbon material is an ancient but a new substance with ongoing and continuous discoveries. Carbon materials include activated carbons, carbon black, graphite, carbon nanofibers, glassy carbons, fullerenes, carbon nanotubes, and wonder material graphene [39, 40]. Graphene, a monolayer of sp^2-bonded carbon atoms has attracted significant scientific interest due to its outstanding properties such as excellent enormous specific surface area (2620 m^2 g^{-1}), mechanical properties (Young's modulus of $1TP_a$ and intrinsic strength of $130GP_a$), high electronic conductivity (electron mobility of 2.5×10^5 cm^2 $V^{-1}s^{-1}$at room temperature), high thermal conductivity (above 3000 WmK^{-1}), along with many other properties [41 - 44]. It is easily prepared from graphite flakes. Many researchers investigate the dispersion behavior of graphene and its oxide in organic solvents such as NMP (N-methyl-2- pyrrolidone), THF (tetra hydro furan), Acetone, N-N-dimethylformamide, ethylene glycol, and many others to expand its processability. However, there is still a challenge to understand how graphene disperses in these liquids [45].

Graphene quantum dots (GQDs) are a very smart and latest zero-dimensional (0D) member of the carbon family consisting of single to few layers of graphene sheets with lateral dimensions of 10 nm [46]. GQDs have demonstrated extraordinary physiochemical properties including non-zero band gap, edge effect, and quantum confinement effect. GQDs have a lot of potential in the electronic and optical industries due to their exceptional characteristics. Pan *et al.* [47] have demonstrated a facile hydrothermal route for the synthesis of GQDs having blue luminescent features and analyzed their fluorescent properties for the first time. The applicability of GQDs in fluorescence imaging, magnetic resonance imaging, bioimaging, dual-modal imaging, and two-photon imaging can be explored using this research work. Recently, scientists from the University of Manchester have fabricated a novel nano-Petri dish by adopting graphene-decorated 2D MoS_2 materials to develop a new technique for observing how atoms swim in liquids [48]. Moreover, the team at the National Graphene Institute successfully captured the images of a single atom swimming in liquid for the first time. This finding could have a great impact on the future development of green technology such as hydrogen technology. Moreover, to translate the underlying information into practical applications, it is essential to integrate carbon elements, especially nanocarbons, with other components to build functional or structural materials. Carbon composites, for instance, are used in cylinders for high-pressure hydrogen storage that can operate at 50 MPa due to their high specific mechanical characteristics. Functional materials for energy storage that combine a carbon substance with a different element, such as a metal oxide or conductive polymer, are a hot topic of research. Future research in carbon-based materials will undoubtedly be vigorous due to the numerous challenging topics that exist today. The creation of novel carbons with various structures and textures, as well as the comprehension and customization of surface chemistry, are all incredibly significant and closely related to the creation of new applications.

Self-healing Polymers

The ability of a substance to repair physical loss is known as self-healing. To produce self-healing polymers, both physical and chemical methods have been chosen. These include covalent-bond reformation and rearranging heterogeneous self-healing systems, shape-memory effects, diffusion and flow, and supra-molecular chemistry dynamics [49]. Due to their tendency to repair scratch-damaged or maintain their original physico-chemical and mechanical characteristics, smart self-healing polymers have gained a lot of research interest in recent years [50, 51]. Self-healing polymers became a hot topic of the current scenario after the first international conference on "self-healing materials" held at Delft University of Technology, Netherland, in 2007 [52]. These self-healing polymers are a unique class of intelligent materials having automatic healing properties much similar to human skin, as these materials can repair internal flaws, cracks, or damage caused by any matrix and can rebuild the mechanical properties (such as tensile strength) of the damaged area. Carolyn Dryin 1996 reported the first autonomic healing polymer. Recently (in 2015), a group of NASA scientists named Scott R. Zavada and his co-workers developed a polymer that has bulletproof properties. Moreover, a bullet will rupture the polymer, but at the same time, the temperature impact causes the polymer to flow and rejoin, closing the gap once the bullet has passed through [53]. Such kind of polymers could be helpful for healing damage in satellites and spacecraft caused by high-speed debris. The self-healing properties of polymers allow them to repair cracked or damaged material parts, extending the durability and life of many materials, decreasing waste, and improving performance while using them in real-world applications like construction, automotive, aerospace, biomedical engineering, and defense. Recently, significant progress has been made in the design and development of new self-healing polymers using a variety of chemical techniques. These polymers can spontaneously mend cracks and damage in moderate circumstances, which is desirable in many applications [54 - 56]. Ginting M. *et al.* reported a self-healable polyacrylic acid/polypyrrole-Fe (PAA/PPy-Fe) composite utilized for antibacterial and electrical conductivity properties. The antibacterial activity was studied against *E. coli.* revealing a 1.26-1.56 cm zone of inhibition after 12 hours of incubation. The composite exhibited reversible restorability when applied in an electrical circuit powered by 3V batteries, consisting of an LED [57]. Self-healing polymers have excellent potential for applications to space suits. Habitats and inflatable structures as reported by Pernigoni*et al.* The data for hyperelastic and viscoelastic response and damage and healing time was recorded. An effective self-healing ability was shown by the polymer (ASTM 1708) with polyamide under pressurized conditions [58]. Recently, Gao H. *et al.* explored the mechanical and conductive properties of solid, stretchable, and self-healable poly (oxime-urethane) and graphene composite. The self-healing composite showed a tensile strength of 6 MPa, 1000% elongation, and 48 MJ m^{-3} toughness [59]. The self-healing polymers also demonstrate protective properties. Owing to these properties, a composite based on PDO-2,5 polymer and oxime-urethane, as a protective film on the inner wall of the tire was utilized by Liu X. *et al.* [60] The composite was found to be self-healing and puncture-resistant. Hence, these materials can be successfully used as protective coatings for automobiles, electronics, and diplomas. Wang S. *et al.* developed a dynamically cross-linked polyurethane hot melt adhesive (DPU-HMA) possessing superior solvent resistance, high bonding strength, fast curing speed, and excellent bonding effects on wood, plastic, metal, and composite substrates [61]. These polymers are of great interest in biomedicine due to their self-healing properties. Jiang C *et al.* designed a self-healing poly(oximeurethane) elastomer having biocompatible, biodegradable, and mechanically adjustable properties. It was used for in vivo repair of the tissues. The results were validated in three animal models for nerve coaptation, bone immobilization, and aortic aneurysm, providing a new perspective to biomedical engineering [62].

Metal-organic Frameworks (MOFs)

A class of porous materials with exceptional chemical and structural tunability composed of metal anodes and organic linkers belonging to metal-organic frameworks (MOFs). Because of their porosity, stability, long-range order, conductivity, particle morphology, and synthetic adaptability, MOFs could be excellent platforms for identifying design features for advanced functional materials for specific applications. For cost-effective technologies, MOFs are the worthiest candidates to replace materials such as ordered silica, zeolites, and highly porous materials in various fields like fuel cells, sensors, gas storage, catalysis, and purification [63]. Radhakrishnan S. *et al* studied the electrochemical applications of Cobalt phenylphosphonate (CP) - MOF, such as energy storage and electrocatalysis. The CP-MOF exhibited excellent catalytic performance toward the electro-oxidation of methanol with a good catalytic constant (7.79 x $10^5 cm^3 mol^{-1} s^{-1}$) and higher oxidation peak current (2.97 $\pm$ 0.11 mA cm^{-2}). A high specific capacity of 218 C g^{-1} at 0.25 A/g current density and 82% cyclic stability up to 8000 cycles was observed when used as electrode materials revealing its excellent energy storage properties [64]. Jamil *et al.* utilized Co and Ca- based MOF as catalysts for the production of biodiesel from waste cooking oil. The results demonstrated good agreement with the predicted results for the yield of biodiesel (84.5%) with a percentage error of less than $\pm$ 5%. The regenerated catalyst exhibits a notable biodiesel production drop of up to 7% after three cycles

[65]. MOFs have also been successfully used for sensing applications such as sensing antibiotics, pesticides [66], hydrogen peroxide [67], nitroaromatic compounds [68], *etc.* The application of MOFs in fuel cells has been of great utility. Wang H. *et al.* prepared Zn-MOF for H_2O_2 fuel cells [69] with a power density of 212 mW cm^{-2} and a current density of 630 mA cm^{-2}. Ziang Z.*et al.* reported the adsorbent properties of cage-based MOF for separation and purification of natural gas and C_2H_2 showing adsorption selectivities for C_2 hydrocarbons over CH_4 above 17.7, 5.0 for CO_2/CH_4 and 4.4 for C_2H_2/CO_2. The studies also revealed good hydrolytic stability of MOF under harsh chemical conditions, making it suitable for practical future applications [70].

Mxenes

Mxenes have been investigated as one of the potential materials for a wide range of applications. Mxenes have many outstanding features including high miscibility, availability of active sites, high surface area to volume ratio, high electrical conductivity, electron-rich density, surface charge state, enabling stable colloidal solutions in water, negative zeta-potential, mechanical properties of transition metal carbides/nitrides, effective absorption of electromagnetic waves and functionalized surfaces that make MXenes hydrophilic and easily bind to different species. Energy storage was the first MXene application that was investigated, and it still accounts for a sizable amount of MXene operations. MXenes' distinctive layered structures offer transition metal-active redox sites on the surface, while simultaneously improving electrolyte ion transport. Due to these characteristics, MXenes have become an attractive candidate for high-performance electrodes for electrochemical capacitors [71]. Yu. L *et al.* used a 2D screen printing technique for energy storage application of pure MXene-N ink with low viscosity having a capacitance value of 70.1 mF cm^{-2}. The supercapacitor exhibited the energy density and power density of 0.42 mWh cm^{-2} and 0.83 mWh cm^{-3} respectively [72]. Apart from this, Chen L. *et al* examined the electronic properties of $Ti_3C_2T_x$ MXene. A strong dispersion of more than 1 eV was shown by the electronic structure of $Ti_3C_2T_x$. Also, the work function measured for $Ti_3C_2T_x$ was found in the range of 3.9 to 4.8 eV [73]. MXenes bound to PVA showed an increased electrical conductivity of 7.25 $\times$ 10^{-3} Sm^{-1} as compared to pure PVA (1×10^{-13} Sm^{-1}) and the optical absorption coefficient was calculated to be in the range 4000-5000 cm^{-1} [74]. The mechanical properties of MXenes are also the focus of attention. Ti_3C_2/polyacrylamide nanocomposite hydrogels exhibited fracture strengths of 66.5 to 102.7 kPa, compressive strengths of 400.6 to 819.4 kPa, and elongations at break of 2158.6% to 3047.5%, revealing its impressive mechanical properties [75]. Yue Y. *et al* studied the magnetic properties of Zr_2N MXene. The studies revealed that the ground state of Zr_2N MXene is antiferromagnetic, but a magnetic state with applied strain greater than

(>) 4% tends to be ferromagnetic [76]. Based on the above properties, it can be concluded that MXene is a fascinating candidate for futuristic materials and can be applied for various applications.

Composite Materials

Composites are an important class of multifunctional materials consisting of more than one phase bonded together. These materials can be categorized into four classes based on matrix composition: metal, carbon, polymer, and ceramic matrix composites. These materials can be modified and utilized accordingly for various applications owing to their excellent physical and mechanical properties. Characteristics such as resistance to creep, creep rupture, wear, corrosion and fatigue, high modulus, high strength, low coefficient thermal expansion, and low density make composite materials reliable for countless applications such as aerospace, energy production, infrastructure, architecture, automotive, transportation, energy storage, marine, *etc.* Along with these applications, the composite materials also show good biological activity as reported by Abhilash M.R. *et al.* The antibacterial activity of Fe_2O_3/Cu_2O against *E. coli*, *P. aeruginosa*, *Staphaureus*, and *B. subtilis* was studied and the material was found to be less toxic against *Musmusculus* skin melanoma cells. The composite also exhibited a short time span for photocatalytic degradation of Rhodamine-B and Janus green dyes [77]. Similarly, a composite material based on chitosan, glutaraldehyde, reduced graphene oxide, and palladium was prepared by Ge L. *et al.* for catalytic degradation of organic pollutants [78]. Carbon-based composite materials are well-suited for energy storage properties with high cyclic stability as reported by Vidhya M.S. *et al.* [79]. A composite material of cobalt hydroxide with reduced graphene oxide was prepared as electrode material which delivered a high specific capacitance of 1100 Fg^{-1} at 0.5 Ag^{-1} current density and 98.1% cyclic stability after 2000 cycles. Nevertheless, carbon-based composites are also utilized as microwave absorbers by Feng A. *et al.* They prepared a hierarchical carbon fiber@cobaltferrite@manganesedioxide ($CF@CoFe_2O_4@MnO_2$) composite for microwave absorption, exhibiting a superior performance with minimum reflection loss value up to -34 dB [77]. Sankar S. *et al.* developed polymer-based composite materials for gas sensing and electrical properties. A composite of poly(aniline-co-indole) with varying contents of copper alumina exhibited excellent performance towards ammonia gas sensing and formation of p-n junction in the material. The composite revealed high electrical conductivity, gas sensing, and thermal stability making it a promising candidate for electronic and sensing applications [80].

Nano-Inks and Quantum Dots

Nanoparticle conductive inks and composites of nanomaterials are no longer a technology that is just used in academic labs; firms are now developing these formulations and putting them to use in real-world goods. Owing to their outstanding features such as high optical absorption coefficients (> 104 cm^{-1}) and tunable direct band gaps ranging from 1.1 to 1.5 eV [81 - 83], ternary and quaternary chalcogenides materials, Cu_2SnSe_3 (CTSe) and Cu_2ZnSnS_4 (CZTS), have received research attention to become an effective photovoltaic material [84 - 86]. Recently, ternary and quaternary semiconductor compounds CTSe and CZTS, have been used as efficient photoactive layers in heterojunction thin film solar cells. A combination of inorganic nanoparticles and conjugated polymers was used in the past to develop low-cost photovoltaic (PV), energy storage, and electrochemical sensors. Organic semiconductors that have undergone solution processing and inorganic/organic composite materials have the potential to lower the cost of solar energy devices and sensors significantly. The semiconductor nanocrystals and their nano-inks offer suitable energy band alignment for the fast exciton dissociation rate and charge transport as well as wide coverage of spectral region [87]. It has also been reported that the performances of solar cells are strongly dependent on electron/hole selective layers used. To create a buffer layer, various thin layers including ZnO, ZnS, CdS, and TiO_2 have been successfully incorporated into thin film solar cells. In thin films, the buffer layer is mainly used for the formation of a junction with an observer layer to allow the maximum number of photons in the absorber layer. Currently, rGO–CNF/Ce–TiO_2, Sr-CeTiO_2/CNF, and PANI/MOR emerged as an efficient charge carrier's selective layers [88]. Therefore, nano-inks based on these materials could be beneficial for multidisciplinary applications.

Quantum dots (QDs) are nanocrystals of semiconducting materials having a size resumed less than 10 nm. Currently, QDs have gained great research interest in a wide variety of different applications such as photovoltaic cells, photodetectors, biological imaging, and LEDs. Due to their minute size, a physically confined electron cloud is produced, called quantum confinement effect, and various properties such as optical, electronic, and chemical, can be enhanced to a great extent. The phrase "quantum dot" was first used in 1986. Legend Alexey Ekimov created QDs in a glass matrix in 1981, while Louis Brus created them in a colloidal solution in 1983. At the beginning of the twenty-first century, Prof. Xiaogang Peng, who worked at the Department of Chemistry, University of Arkansas developed a "Green Synthesis" method by replacing $CdMe_2$ with CdO with a non-coordinating solvent ODE. This allowed "QD synthesis" to be used in laboratories and industries all around the globe. In 2014, GaN was used to develop efficient blue light emitting diodes (LED), brilliant and energy-efficient white light sources, by employing high vacuum and temperature-controlled epitaxial growth of a multilayer semiconductor single crystal on a sapphire substrate. A Nobel Prize in Physics was awarded for the same. But the price is outrageous! It is anticipated that the benefits of both organic light-emitting diode and GaN-LED will be combined if the QDs-based devices (*i.e.*, QLED) can attain high performance comparable with GaN-LED (OLED). Recent research by Xiaogang Peng and his team supported this theory by producing a high-performing, low-cost QLED utilizing a "solution-processed synthesis approach". The photo luminescent QDs technology has been substantially commercialized in lighting and displays during the past ten years [89]. This was a revolutionary change in the field of ODs.

Metamaterials

The term "meta" means "beyond" in Greek. "Metamaterials" have unique qualities that go beyond those of natural materials. The properties of metamaterials are due to their structure rather than the construction materials. The most remarkable contribution in this field is V. G. Veselago's statement in 1968 that materials with both negative permittivity and negative permeability are theoretically feasible [90]. These materials have unique spatial alterations in their constituent components as they are man-made substances. These materials are extensively used for the modification of elastic, acoustic, or electromagnetic properties of materials. Metamaterials often used in microwave engineering, waveguides, dispersion compensation, smart antennas, and lenses also produce low/high-frequency band gaps to control wave propagations with varying wavelengths. For example, the metamaterial's permittivity and permeability can have positive or negative values. The selected frequency of surface-based metamaterials is exploited for wave guiding since the single-cell dimensions of these materials are less than their wavelength. Metamaterials have a wide variety of applications in different fields such as public safety, sensor identification, high-frequency combat communications, enhanced ultrasonic sensors, solar energy management for high-gain antennas, and distant aerospace applications [90, 91].

SYNTHESIS TECHNIQUES

Nanomaterials can be synthesized using two types of approaches; top down and bottom approach (Fig. **3b**). As the name suggests, the top-down approach uses bulk materials to produce nano-size particles. Various dispersion and aggregation techniques are employed to structure macroscopic particles to a nanometer scale. Physical methods are involved in this approach including mechano-chemical dispersion, plasmo-chemical method, and condensation from the gas phase. The original structure of the compact material is retained in this type of approach.

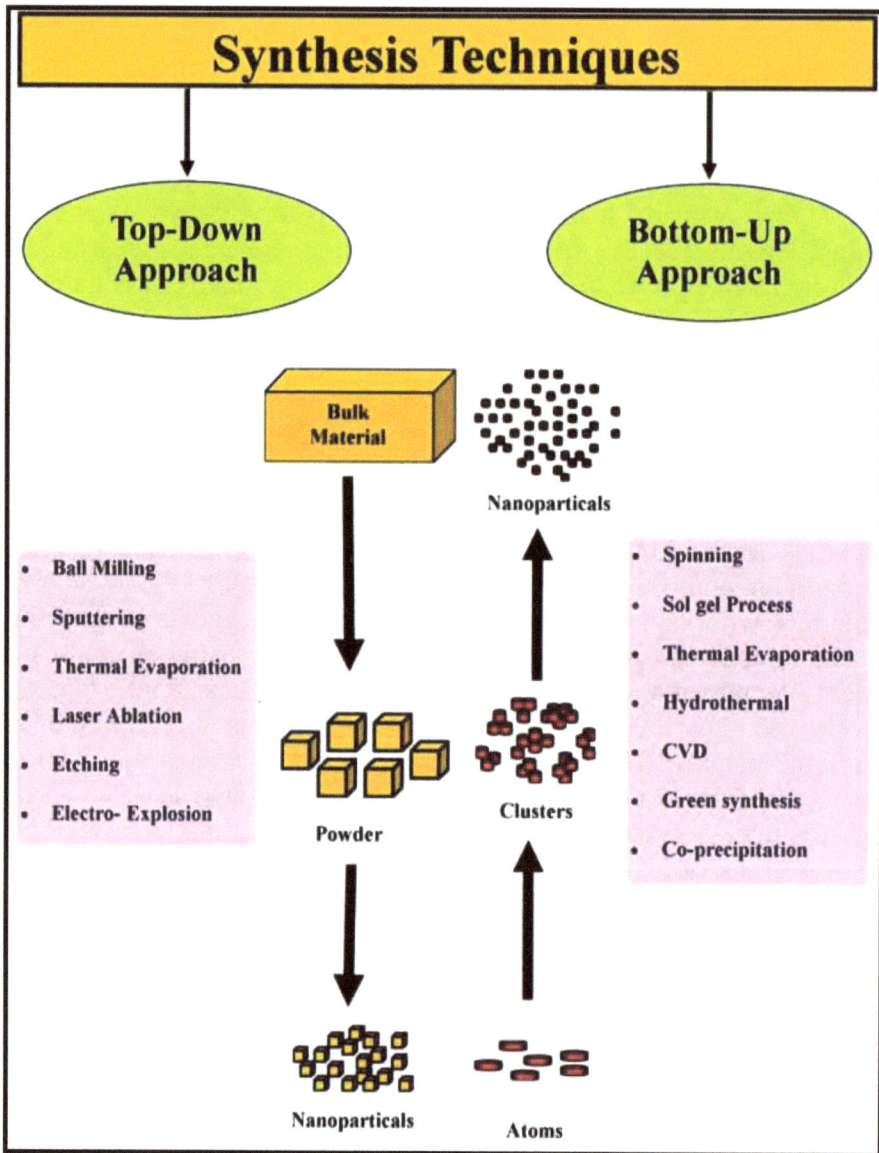

Fig. (3b). Synthesis techniques for synthesis of nanostructures.

In a bottom-up approach, basic building blocks *i.e.*; atoms and molecules are organized to build up nanostructures. This approach is largely about chemical methods for nanomaterial synthesis. Chemistry expertise is required in this method for assembling and structuring these nano-objects into nanomaterials. The electrical structure of nanoparticles may be affected due to variations in the position of atoms in nanoparticles created by aggregation of atoms. The objective of this approach is to prepare materials with excellent chemical, mechanical, optical, and magnetic properties, by using small-sized materials in the beginning [92 - 98].

FUTURE SCOPE OF NANOMATERIALS

Nanotechnology is considered to be the major technological advancement of the twenty-first century and has sparked interest worldwide. Today's nanotechnology is much more advanced than what is imagined in science fiction. Although the current focus of nanotechnology is primarily on new material composition, their prospective applications are extremely broad. The usage of nanotechnology has grown significantly, and it has the potential to become crucial shortly. It is anticipated to inspire a wide range of application domains across practically all industries and technological fields. It will open up new possibilities for producing life's necessities (such as electronics, medications, goods, cars, and homes) more effectively and affordably while utilizing fewer raw materials. There will be significant advancements in a variety of fields, including robotics, computing, energy, and food. Perhaps Richard Feynman's innovative lecture, which inspired others to investigate this technology and helped establish this industry, can be credited with making this possible. Due to its distinctive properties, future applications of nanotechnology are anticipated to be far more sophisticated. As nanomaterial-based engineering techniques progress, the goal of producing clean energy is becoming more and more attainable. Nanomaterials have produced promising outcomes, enabling the development of new types of solar and hydrogen fuel cell technologies, serving as effective catalysts for water splitting, and displaying outstanding hydrogen storage capabilities. Nanomedicine holds a bright future for nanomaterials. Therapeutic compounds can be delivered *via* nanocarriers. The future of advanced technology is linked with advancements in the field of nanotechnology. Although nanotechnology is maturing rapidly, it is still in a formative phase. Nobody can predict how quickly will these novel concepts finish the R&D stage and reach the market. However, with certainty, we can state that nanotechnology is here to stay and that it uses and applications can be produced in a way that is morally upright and beneficial to mankind.

CONCLUSION

Overall, the advanced and multifunctional materials field is growing exponentially across the scientific community owing to their fascinating flexible electronics and optoelectronic devices, lightweight weapons, fire registrant materials, gas sensors, energy storage and biomedical, *etc.*, which facilitate them to be qualified as the key materials for the next generation. These materials show high-level performance because of their unique properties. To achieve progress and growth in fundamental research and their practical applications, the exploration of the design of these advanced and multifunctional materials and their assemblies is facing more and more challenges. In this review, we summarized the basic properties, synthesis techniques, classifications, and recent applications in various fields. Finally, the continuous developments of elaborate design and assembly and enhancement in properties of such materials offer unprecedented opportunities for new applications as well as provide next-generation solutions.

REFERENCES

[1]   J.Y. Cheong, S.H. Cho, J. Lee, J.W. Jung, C. Kim, and I.D. Kim, "Multifunctional 1D nanostructures toward future batteries: A comprehensive review", *Adv. Funct. Mater.,* vol. 32, no. 49, p. 2208374, 2022.
[http://dx.doi.org/10.1002/adfm.202208374]

[2]   M. Hasan, J. Zhao, and Z. Jiang, "Micromanufacturing of composite materials: A review", *Int. J. Extreme Manuf.,* vol. 1, no. 1, p. 012004, 2019.
[http://dx.doi.org/10.1088/2631-7990/ab0f74]

[3]   M. Sajid, "Nanomaterials: Types, properties, recent advances, and toxicity concerns", *Curr. Opin. Environ. Sci. Health,* vol. 25, p. 100319, 2022.
[http://dx.doi.org/10.1016/j.coesh.2021.100319]

[4]   K.S. Novoselov, A.K. Geim, S.V. Morozov, D. Jiang, Y. Zhang, S.V. Dubonos, I.V. Grigorieva, and A.A. Firsov, "Electric field effect in atomically thin carbon films", *Science,* vol. 306, no. 5696, pp. 666-669, 2004.
[http://dx.doi.org/10.1126/science.1102896] [PMID: 15499015]

[5]   J.G. Fernandez, and D.E. Ingber, "Unexpected strength and toughness in chitosan-fibroin laminates inspired by insect cuticle", *Adv. Mater.,* vol. 24, no. 4, pp. 480-484, 2012.
[http://dx.doi.org/10.1002/adma.201104051] [PMID: 22162193]

[6]   J.A. Kumar, P. Prakash, T. Krithiga, D.J. Amarnath, J. Premkumar, N. Rajamohan, Y. Vasseghian, P. Saravanan, and M. Rajasimman, "Methods of synthesis, characteristics, and environmental applications of MXene: A comprehensive review", *Chemosphere,* vol. 286, no. Pt 1, p. 131607, 2022.
[http://dx.doi.org/10.1016/j.chemosphere.2021.131607] [PMID: 34311398]

[7]   A. Kirchon, L. Feng, H.F. Drake, E.A. Joseph, and H.C. Zhou, "From fundamentals to applications: A toolbox for robust and multifunctional MOF materials", *Chem. Soc. Rev.,* vol. 47, no. 23, pp. 8611-8638, 2018.
[http://dx.doi.org/10.1039/C8CS00688A] [PMID: 30234863]

[8]   B.E. Conway, *Electrochemical supercapacitors: Scientific fundamentals and technological applications.* Springer US, 1999.
[http://dx.doi.org/10.1007/978-1-4757-3058-6]

[9]   I.V. Barsukov, C.S. Johnson, J.E. Doninger, and V.Z. Barsukov, *New carbon based materials for electrochemical energy storage systems: Batteries, supercapacitors, and fuel cells.* Springer Netherlands: Dordrecht, 2006.
[http://dx.doi.org/10.1007/1-4020-4812-2]

[10]   C.K. Chan, H. Peng, G. Liu, K. McIlwrath, X.F. Zhang, R.A. Huggins, and Y. Cui, "High-performance lithium battery anodes using silicon nanowires", *Nat. Nanotechnol.,* vol. 3, no. 1, pp. 31-35, 2008.
[http://dx.doi.org/10.1038/nnano.2007.411] [PMID: 18654447]

[11]   F. Häggström, and J. Delsing, "IoT energy storage - A forecast", *Energy Harvesting and Systems,* vol. 5, no. 3-4, pp. 43-51, 2018.
[http://dx.doi.org/10.1515/ehs-2018-0010]

[12]   L.L. Zhang, and X.S. Zhao, *Carbon-based materials as supercapacitor electrodes.* vol. Vol. 38. Chemical Society Reviews, 2009, pp. 2520-2531.

[13]   J. Huang, B.G. Sumpter, and V. Meunier, "Theoretical model for nanoporous carbon supercapacitors", *Angew. Chem. Int. Ed.,* vol. 47, no. 3, pp. 520-524, 2008.
[http://dx.doi.org/10.1002/anie.200703864] [PMID: 18058966]

[14]   E. Hornbogen, G. Eggeler, and E. Werner, *Werkstoffe, "Aufbau und eigenschaften von keramik, metal, polymer und verbundwerkstoffen" (materials: Structure and properties of ceramics, metals, polymers and composite materials).* Springer: Berlin, 2017.

[15]   Z. Su, H. Guo, and C. Zhao, "Rational design of electrode–electrolyte interphase and electrolytes for rechargeable proton batteries", *Nano-Micro Lett.,* vol. 15, no. 1, p. 96, 2023.
[http://dx.doi.org/10.1007/s40820-023-01071-z] [PMID: 37037988]

[16]   L. Jiang, S. Shanmuganathan, G.W. Nelson, S.O. Han, H. Kim, I. Na Sim, and J.S. Foord, "Hybrid system of nickel–cobalt hydroxide on carbonised natural cellulose materials for supercapacitors", *J. Solid State Electrochem.,* vol. 22, no. 2, pp. 387-393, 2018.
[http://dx.doi.org/10.1007/s10008-017-3723-z]

[17]   B.E. Conway, *Electrochemical supercapacitors: Scientific fundamentals and technological applications.* Springer US, 1999.
[http://dx.doi.org/10.1007/978-1-4757-3058-6]

[18]   M. Zhi, C. Xiang, J. Li, M. Li, and N. Wu, "Nanostructured carbon–metal oxide composite electrodes for supercapacitors: A review", In: *Nanoscale* vol. 5. , 2013, pp. 72-88.
[http://dx.doi.org/10.1039/C2NR32040A]

[19]   H.I. Becker, *Low voltage electrolytic capacitor,* 1957.

[20]   S. Liu, L. Wei, and H. Wang, "Review on reliability of supercapacitors in energy storage applications", *Appl. Energy,* vol. 278, p. 115436, 2020.
[http://dx.doi.org/10.1016/j.apenergy.2020.115436]

[21]   N.I. Jalal, R.I. Ibrahim, and M.K. Oudah, "A review on Supercapacitors: Types and components", *J. Phys. Conf. Ser.,* vol. 012015, p. 2021, 1973.

[22]   A. Muzaffar, M.B. Ahamed, K. Deshmukh, and J. Thirumalai, "A review on recent advances in hybrid supercapacitors: Design, fabrication and applications", *Renew. Sustain. Energy Rev.,* vol. 101, pp. 123-145, 2019.
[http://dx.doi.org/10.1016/j.rser.2018.10.026]

[23]   D.P. Chatterjee, and A.K. Nandi, "A review on the recent advances in hybrid supercapacitors", *J. Mater. Chem. A Mater. Energy Sustain.,* vol. 9, no. 29, pp. 15880-15918, 2021.
[http://dx.doi.org/10.1039/D1TA02505H]

[24]   P. Chen, J-J. Yang, S-S. Li, Z. Wang, T-Y. Xiao, Y-H. Qian, and S-H. Yu, *Hydrothermal synthesis of macroscopic nitrogen doped graphene hydrogels for ultrafast supercapacitor.* vol. Vol. 2. Nano Energy, 2013, pp. 249-256.

[25]   J. Yin, D. Zhang, J. Zhao, X. Wang, H. Zhu, and C. Wang, *Meso-and micro-porous composite carbons derived from humic acid for supercapacitors.* vol. Vol. 136. Electrochimica Acta, 2014, pp. 504-512.

[26]   O. Masala, and R. Seshadri, "Synthesis routes for large volumes of nanoparticles", *Annu. Rev. Mater. Res.,* vol. 34, no. 1, pp. 41-81, 2004.
[http://dx.doi.org/10.1146/annurev.matsci.34.052803.090949]

[27]   S. Yuvaraj, R.K. Selvan, and Y.S. Lee, "An overview of AB 2 O 4 - and A 2 BO 4 -structured negative electrodes for advanced Li-ion batteries", *RSC Advances,* vol. 6, no. 26, pp. 21448-21474, 2016.

[http://dx.doi.org/10.1039/C5RA23503K]

[28]   L. Yao, Q. Su, Y. Xiao, M. Huang, H. Li, H. Deng, and G. Du, "Facial synthesis of carbon-coated ZnFe2O4/graphene and their enhanced lithium storage properties", *J. Nanopart. Res.,* vol. 19, no. 7, p. 261, 2017.
[http://dx.doi.org/10.1007/s11051-017-3935-2]

[29]   L. Huang, J. Xiang, W. Zhang, C. Chen, H. Xu, and Y. Huang, "3D interconnected porous NiMoO 4 nanoplate arrays on Ni foam as high-performance binder-free electrode for supercapacitors", *J. Mater. Chem. A Mater. Energy Sustain.,* vol. 3, no. 44, pp. 22081-22087, 2015.
[http://dx.doi.org/10.1039/C5TA05644F]

[30]   B. Mekuye, and B. Abera, "Nanomaterials: An overview of synthesis, classification, characterization, and applications", *Nano Select,* vol. 4, no. 8, pp. 486-501, 2023.
[http://dx.doi.org/10.1002/nano.202300038]

[31]   B.E. Marschallek, and T. Jacobsen, "Classification of material substances: Introducing a standards-based approach", *Mater. Des.,* vol. 193, p. 108784, 2020.
[http://dx.doi.org/10.1016/j.matdes.2020.108784]

[32]   V. Singh, P. Yadav, and V. Mishra, "Recent advances on classification, properties, synthesis, and characterization of nanomaterials", In: *Green synthesis of nanomaterials for bioenergy applications,* 2020, pp. 83-87.
[http://dx.doi.org/10.1002/9781119576785.ch3]

[33]   F. Asghari, Z. Jahanshiri, M. Imani, M. Shams-Ghahfarokhi, and M. Razzaghi-Abyaneh, "Antifungal nanomaterials: Synthesis, properties, and applications", In: *Nanobiomaterials in antimicrobial therapy.* William Andrew Publishing, 2016, pp. 343-383.
[http://dx.doi.org/10.1016/B978-0-323-42864-4.00010-5]

[34]   R. Aversa, M.H. Modarres, S. Cozzini, R. Ciancio, and A. Chiusole, "The first annotated set of scanning electron microscopy images for nanoscience", *Sci. Data,* vol. 5, no. 1, p. 180172, 2018.
[http://dx.doi.org/10.1038/sdata.2018.172] [PMID: 30152811]

[35]   B.W. Shiau, C.H. Lin, Y.Y. Liao, Y.R. Lee, S.H. Liu, W.C. Ding, and J.R. Lee, "The characteristics and mechanisms of Au nanoparticles processed by functional centrifugal procedures", *J. Phys. Chem. Solids,* vol. 116, pp. 161-167, 2018.
[http://dx.doi.org/10.1016/j.jpcs.2018.01.033]

[36]   S. Saha, S. Bansal, and M. Khanuja, *Book Chapter -2 "Classification of nanomaterials and their physical and chemical nature.* Nano-enabled Agrochemicals in Agriculture, 2022, pp. 7-34.

[37]   M. Inagaki, and F. Kang, *Materials science and engineering of carbon: Fundamentals Butterworth-Heinemann.,* 2014.

[38]   K. Khan, A.K. Tareen, M. Aslam, Y. Zhang, R. Wang, Z. Ouyang, Z. Gou, and H. Zhang, "Recent advances in two-dimensional materials and their nanocomposites in sustainable energy conversion applications", *Nanoscale,* vol. 11, no. 45, pp. 21622-21678, 2019.
[http://dx.doi.org/10.1039/C9NR05919A] [PMID: 31702753]

[39]   M. Inagaki, and L.R. Radovic, "Nanocarbons", In: *Letter to Editor/Carbon* vol. 40. , 2002, pp. 2263-2284.

[40]   Y. Hernandez, M. Lotya, D. Rickard, S.D. Bergin, and J.N. Coleman, "Measurement of multicomponent solubility parameters for graphene facilitates solvent discovery", *Langmuir,* vol. 26, no. 5, pp. 3208-3213, 2010.
[http://dx.doi.org/10.1021/la903188a] [PMID: 19883090]

[41]   G.K. Dimitrakakis, E. Tylianakis, and G.E. Froudakis, "Pillared graphene: A new 3-D network nanostructure for enhanced hydrogen storage", *Nano Lett.,* vol. 8, no. 10, pp. 3166-3170, 2008.
[http://dx.doi.org/10.1021/nl801417w] [PMID: 18800853]

[42]   M.D. Stoller, S. Park, Y. Zhu, J. An, and R.S. Ruoff, "Graphene-based ultracapacitors", *Nano Lett.,*

vol. 8, no. 10, pp. 3498-3502, 2008.
[http://dx.doi.org/10.1021/nl802558y] [PMID: 18788793]

[43] R.R. Nair, P. Blake, A.N. Grigorenko, K.S. Novoselov, T.J. Booth, T. Stauber, N.M.R. Peres, and A.K. Geim, "Fine structure constant defines visual transparency of graphene", *Science,* vol. 320, no. 5881, p. 1308, 2008.
[http://dx.doi.org/10.1126/science.1156965] [PMID: 18388259]

[44] J.I. Paredes, S. Villar-Rodil, A. Martínez-Alonso, and J.M.D. Tascón, "Graphene oxide dispersions in organic solvents", *Langmuir,* vol. 24, no. 19, pp. 10560-10564, 2008.
[http://dx.doi.org/10.1021/la801744a] [PMID: 18759411]

[45] R. Zhang, and Z. Ding, "Recent advances in graphene quantum dots as bioimaging probes", *J Anal Methods Chem ,* vol. 2, pp. 45-60, 2018.

[46] D. Pan, J. Zhang, Z. Li, and M. Wu, "Hydrothermal route for cutting graphene sheets into blue-luminescent graphene quantum dots", *Adv. Mater.,* vol. 22, no. 6, pp. 734-738, 2010.
[http://dx.doi.org/10.1002/adma.200902825] [PMID: 20217780]

[47] M.R. Younis, G. He, J. Lin, and P. Huang, "dots for bioimaging applications", *Front Chem.,* vol. 8, p. 424, 2020.
[http://dx.doi.org/10.3389/fchem.2020.00424] [PMID: 32582629]

[48] S. Wang, and M.W. Urban, "Self-healing polymers", *Nat. Rev. Mater.,* vol. 5, no. 8, pp. 562-583, 2020.
[http://dx.doi.org/10.1038/s41578-020-0202-4]

[49] H. Jiang, G. Zhang, X. Feng, H. Liu, F. Li, M. Wang, and H. Li, "Room-temperature self-healing tough nanocomposite hydrogel crosslinked by zirconium hydroxide nanoparticles", *Compos. Sci. Technol.,* vol. 140, pp. 54-62, 2017.
[http://dx.doi.org/10.1016/j.compscitech.2016.12.027]

[50] E. Su, and O. Okay, "Hybrid cross-linked poly(2-acrylamido-2-methyl-1-propanesulfonic acid) hydrogels with tunable viscoelastic, mechanical and self-healing properties", *React. Funct. Polym.,* vol. 123, pp. 70-79, 2018.
[http://dx.doi.org/10.1016/j.reactfunctpolym.2017.12.009]

[51] C. Dry, "Procedures developed for self-repair of polymer matrix composite materials", *Composite Structures,* vol. 35, pp. 263-269, 1996.

[52] S.R. Zavada, N.R. McHardy, K.L. Gordon, and T.F. Scott, "Rapid, puncture-initiated healing via oxygen-mediated polymerization", *ACS Macro Lett.,* vol. 4, no. 8, pp. 819-824, 2015.
[http://dx.doi.org/10.1021/acsmacrolett.5b00315] [PMID: 35596502]

[53] Y. Yang, H. Wang, L. Huang, M. Nishiura, Y. Higaki, and Z. Hou, "Terpolymerization of ethylene and two different methoxyaryl-substituted propylenes by scandium catalyst makes tough and fast self-healing elastomers", *Angew. Chem. Int. Ed.,* vol. 60, no. 50, pp. 26192-26198, 2021.
[http://dx.doi.org/10.1002/anie.202111161] [PMID: 34751988]

[54] S.R. White, J.S. Moore, N.R. Sottos, B.P. Krull, W.A. Santa Cruz, and R.C.R. Gergely, "Restoration of large damage volumes in polymers", *Science,* vol. 344, no. 6184, pp. 620-623, 2014.
[http://dx.doi.org/10.1126/science.1251135] [PMID: 24812399]

[55] Y.L. Rao, A. Chortos, R. Pfattner, F. Lissel, Y.C. Chiu, V. Feig, J. Xu, T. Kurosawa, X. Gu, C. Wang, M. He, J.W. Chung, and Z. Bao, "Stretchable self-healing polymeric dielectrics cross-linked through metal-ligand coordination", *J. Am. Chem. Soc.,* vol. 138, no. 18, pp. 6020-6027, 2016.
[http://dx.doi.org/10.1021/jacs.6b02428] [PMID: 27099162]

[56] M. Ginting, S.P. Pasaribu, I. Masmur, J. Kaban, and Hestina, "Self-healing composite hydrogel with antibacterial and reversible restorability conductive properties", *RSC Advances,* vol. 10, no. 9, pp. 5050-5057, 2020.
[http://dx.doi.org/10.1039/D0RA00089B] [PMID: 35498274]

[57] L. Pernigoni, and A.M. Grande, "Development of a supramolecular polymer based self-healing multilayer system for inflatable structures", *Acta Astronaut.,* vol. 177, pp. 697-706, 2020.
[http://dx.doi.org/10.1016/j.actaastro.2020.08.025]

[58] H. Gao, J. Xu, S. Liu, Z. Song, M. Zhou, S. Liu, F. Li, F. Li, X. Wang, Z. Wang, and Q. Zhang, "Stretchable, self-healable integrated conductor based on mechanical reinforced graphene/polyurethane composites", *J. Colloid Interface Sci.,* vol. 597, pp. 393-400, 2021.
[http://dx.doi.org/10.1016/j.jcis.2021.04.005] [PMID: 33892422]

[59] X. Liu, X. Liu, W. Li, Y. Ru, Y. Li, A. Sun, and L. Wei, "Engineered self-healable elastomer with giant strength and toughness via phase regulation and mechano-responsive self-reinforcing", *Chem. Eng. J.,* vol. 410, p. 128300, 2021.
[http://dx.doi.org/10.1016/j.cej.2020.128300]

[60] S. Wang, Z. Liu, L. Zhang, Y. Guo, J. Song, J. Lou, Q. Guan, C. He, and Z. You, "Strong, detachable, and self-healing dynamic crosslinked hot melt polyurethane adhesive", *Mater. Chem. Front.,* vol. 3, no. 9, pp. 1833-1839, 2019.
[http://dx.doi.org/10.1039/C9QM00233B]

[61] C. Jiang, L. Zhang, Q. Yang, S. Huang, H. Shi, Q. Long, B. Qian, Z. Liu, Q. Guan, M. Liu, R. Yang, Q. Zhao, Z. You, and X. Ye, "Self-healing polyurethane-elastomer with mechanical tunability for multiple biomedical applications in vivo", *Nat. Commun.,* vol. 12, no. 1, p. 4395, 2021.
[http://dx.doi.org/10.1038/s41467-021-24680-x] [PMID: 34285224]

[62] C.M. Oliva González, B.I. Kharisov, O.V. Kharissova, and T.E. Serrano Quezada, "Synthesis and applications of MOF-derived nanohybrids: A review", *Mater. Today Proc.,* vol. 46, pp. 3018-3029, 2021.
[http://dx.doi.org/10.1016/j.matpr.2020.12.1231]

[63] S. Radhakrishnan, S.C. Selvaraj, and B.S. Kim, "Morphology engineering of Co-MOF nanostructures to tune their electrochemical performances for electrocatalyst and energy-storage applications supported by DFT studies", *Appl. Surf. Sci.,* vol. 605, p. 154691, 2022.
[http://dx.doi.org/10.1016/j.apsusc.2022.154691]

[64] U. Jamil, A. Husain Khoja, R. Liaquat, S. Raza Naqvi, W. Nor Nadyaini Wan Omar, and N. Aishah Saidina Amin, "Copper and calcium-based metal organic framework (MOF) catalyst for biodiesel production from waste cooking oil: A process optimization study", *Energy Convers. Manage.,* vol. 215, p. 112934, 2020.
[http://dx.doi.org/10.1016/j.enconman.2020.112934]

[65] G.D. Wang, Y.Z. Li, W.J. Shi, B. Zhang, L. Hou, and Y.Y. Wang, "A robust cluster-based Eu-MOF as multi-functional fluorescence sensor for detection of antibiotics and pesticides in water", *Sens. Actuators B Chem.,* vol. 331, p. 129377, 2021.
[http://dx.doi.org/10.1016/j.snb.2020.129377]

[66] W. Dang, Y. Sun, H. Jiao, L. Xu, and M. Lin, "AuNPs-NH2/Cu-MOF modified glassy carbon electrode as enzyme-free electrochemical sensor detecting H2O2", *J. Electroanal. Chem.,* vol. 856, p. 113592, 2020.
[http://dx.doi.org/10.1016/j.jelechem.2019.113592]

[67] X.J. Zhang, F.Z. Su, D.M. Chen, Y. Peng, W.Y. Guo, C.S. Liu, and M. Du, "A water-stable Eu III - based MOF as a dual-emission luminescent sensor for discriminative detection of nitroaromatic pollutants", *Dalton Trans.,* vol. 48, no. 5, pp. 1843-1849, 2019.
[http://dx.doi.org/10.1039/C8DT04397C] [PMID: 30648716]

[68] H. Wang, Y. Zhao, Z. Shao, W. Xu, Q. Wu, X. Ding, and H. Hou, "Proton conduction of nafion hybrid membranes promoted by NH 3 -modified Zn-MOF with host–guest collaborative hydrogen bonds for H 2 /O 2 fuel cell applications", *ACS Appl. Mater. Interfaces,* vol. 13, no. 6, pp. 7485-7497, 2021.
[http://dx.doi.org/10.1021/acsami.0c21840] [PMID: 33543925]

[69] Z. Jiang, Y. Zou, T. Xu, L. Fan, P. Zhou, and Y. He, "A hydrostable cage-based MOF with open metal

sites and Lewis basic sites immobilized in the pore surface for efficient separation and purification of natural gas and C 2 H 2", *Dalton Trans.,* vol. 49, no. 11, pp. 3553-3561, 2020.
[http://dx.doi.org/10.1039/D0DT00402B] [PMID: 32118237]

[70]   Y. Chen, H. Yang, Z. Han, Z. Bo, J. Yan, K. Cen, and K.K. Ostrikov, "MXene-based electrodes for supercapacitor energy storage", *Energy Fuels,* vol. 36, no. 5, pp. 2390-2406, 2022.
[http://dx.doi.org/10.1021/acs.energyfuels.1c04104]

[71]   L. Yu, Z. Fan, Y. Shao, Z. Tian, J. Sun, and Z. Liu, "Versatile N-doped MXene ink for printed electrochemical energy storage application", *Adv. Energy Mater.,* vol. 9, no. 34, p. 1901839, 2019.
[http://dx.doi.org/10.1002/aenm.201901839]

[72]   T. Schultz, N.C. Frey, K. Hantanasirisakul, S. Park, S.J. May, V.B. Shenoy, Y. Gogotsi, and N. Koch, "Surface termination dependent work function and electronic properties of Ti3C2T x Mxene", *Chem. Mater.,* vol. 31, no. 17, pp. 6590-6597, 2019.
[http://dx.doi.org/10.1021/acs.chemmater.9b00414]

[73]   K.H. Tan, L. Samylingam, N. Aslfattahi, R. Saidur, and K. Kadirgama, "Optical and conductivity studies of polyvinyl alcohol-MXene (PVA-MXene) nanocomposite thin films for electronic applications", *Opt. Laser Technol.,* vol. 136, p. 106772, 2021.
[http://dx.doi.org/10.1016/j.optlastec.2020.106772]

[74]   P. Zhang, X.J. Yang, P. Li, Y. Zhao, and Q.J. Niu, "Fabrication of novel MXene (Ti 3 C 2)/polyacrylamide nanocomposite hydrogels with enhanced mechanical and drug release properties", *Soft Matter,* vol. 16, no. 1, pp. 162-169, 2020.
[http://dx.doi.org/10.1039/C9SM01985E] [PMID: 31774104]

[75]   Y. Yue, B. Wang, N. Miao, C. Jiang, H. Lu, B. Zhang, Y. Wu, J. Ren, and M. Wang, "Tuning the magnetic properties of Zr2N MXene by biaxial strain", *Ceram. Int.,* vol. 47, no. 2, pp. 2367-2373, 2021.
[http://dx.doi.org/10.1016/j.ceramint.2020.09.079]

[76]   M.R. Abhilash, G. Akshatha, and S. Srikantaswamy, "Photocatalytic dye degradation and biological activities of the Fe 2 O 3 /Cu 2 O nanocomposite", *RSC Advances,* vol. 9, no. 15, pp. 8557-8568, 2019.
[http://dx.doi.org/10.1039/C8RA09929D] [PMID: 35518681]

[77]   L. Ge, M. Zhang, R. Wang, N. Li, L. Zhang, S. Liu, and T. Jiao, "Fabrication of CS/GA/RGO/Pd composite hydrogels for highly efficient catalytic reduction of organic pollutants", *RSC Advances,* vol. 10, no. 26, pp. 15091-15097, 2020.
[http://dx.doi.org/10.1039/D0RA01884H] [PMID: 35495471]

[78]   M.S. Vidhya, G. Ravi, R. Yuvakkumar, D. Velauthapillai, M. Thambidurai, C. Dang, B. Saravanakumar, A. Syed, and T.M. Dawoud, "Functional reduced graphene oxide/cobalt hydroxide composite for energy storage applications", *Mater. Lett.,* vol. 276, p. 128193, 2020.
[http://dx.doi.org/10.1016/j.matlet.2020.128193]

[79]   A. Feng, T. Hou, Z. Jia, and G. Wu, "Synthesis of a hierarchical carbon fiber@cobalt ferrite@manganese dioxide composite and its application as a microwave absorber", *RSC Advances,* vol. 10, no. 18, pp. 10510-10518, 2020.
[http://dx.doi.org/10.1039/C9RA10327A] [PMID: 35492930]

[80]   S. Sankar, A. George, and M.T. Ramesan, "Copper alumina @ poly (aniline- co -indole) nanocomposites: synthesis, characterization, electrical properties and gas sensing applications", *RSC Advances,* vol. 12, no. 27, pp. 17637-17644, 2022.
[http://dx.doi.org/10.1039/D2RA02213C] [PMID: 35765439]

[81]   S.K. Dwivedi, D.C. Tiwari, S.K. Tripathi, P.K. Dwivedi, P. Dipak, T. Chandel, and N.E. Prasad, "Fabrication and properties of P3HT: PCBM/Cu2SnSe3 (CTSe) nanocrystals based inverted hybrid solar cells", *Sol. Energy,* vol. 187, pp. 167-174, 2019.
[http://dx.doi.org/10.1016/j.solener.2019.05.012]

[82]   S.K. Dwivedi, S.K. Tripathi, D.C. Tiwari, A.S. Chauhan, P.K. Dwivedi, and N. Eswara Prasad, "Low

cost copper zinc tin sulphide (CZTS) solar cells fabricated by sulphurizing sol-gel deposited precursor using 1,2-ethanedithiol (EDT)", In: *Solar Energy.* vol. 224. , 2021, pp. 210-217.

[83] S.K. Dwivedi, D.C. Tiwari, N. Kumar, S.K. Tripathi, and N.E. Prasad, "Electrochemical-based studies of Cu2SnSe3 nanocrystals and P3HT:PCBM for hybrid solar cells", *Bull. Mater. Sci.,* vol. 46, no. 3, p. 163, 2023.
[http://dx.doi.org/10.1007/s12034-023-03000-7]

[84] P. Uday Bhaskar, G. Suresh Babu, Y.B. Kishore Kumar, and V. Sundara Raja, "Investigations on co-evaporated Cu2SnSe3 and Cu2SnSe3–ZnSe thin films", *Appl. Surf. Sci.,* vol. 257, no. 20, pp. 8529-8534, 2011.
[http://dx.doi.org/10.1016/j.apsusc.2011.05.008]

[85] S.K. Dwivedi, S.K. Tripathi, D.C. Tiwari, A.S. Chauhan, P.K. Dwivedi, and N. Eswara Prasad, "Low cost copper zinc tin sulphide (CZTS) solar cells fabricated by sulphurizing sol-gel deposited precursor using 1,2-ethanedithiol (EDT)", *Sol. Energy,* vol. 224, pp. 210-217, 2021.
[http://dx.doi.org/10.1016/j.solener.2021.04.046]

[86] D.C. Tiwari, S.K. Dwivedi, P. Dipak, and T. Chandel, "PEDOT: PSS: RGO nanocomposite as a hole transport layer (HTLs) for P3HT: PCBM based organic solar cells", *AIP Conference Proceedings,* vol. 1953, p. 100065, 2018.

[87] N. Kumar, S. Kumar Dwivedi, D. Chandra Tiwari, and R. Tomar, "Study of rGO-CNF/Ce-TiO2 based hetrojunction for optoelectronic devices", *Mater. Lett.,* vol. 315, p. 131945, 2022.
[http://dx.doi.org/10.1016/j.matlet.2022.131945]

[88] M.G. Bawendi, M.L. Steigerwald, and L.E. Brus, "The quantum mechanics of larger semiconductor clusters ("quantum dots")", *Annu. Rev. Phys. Chem.,* vol. 41, no. 1, pp. 477-496, 1990.
[http://dx.doi.org/10.1146/annurev.pc.41.100190.002401]

[89] V.G. Veselago, "The electrodynamics of substances with simultaneously negative values of $\$\$$ and μ", *Sov. Phys. Usp.,* vol. 10, no. 4, pp. 509-514, 1968.
[http://dx.doi.org/10.1070/PU1968v010n04ABEH003699]

[90] X. Dai, Z. Zhang, Y. Jin, Y. Niu, H. Cao, X. Liang, L. Chen, J. Wang, and X. Peng, "Solution-processed, high-performance light-emitting diodes based on quantum dots", *Nature,* vol. 515, no. 7525, pp. 96-99, 2014.
[http://dx.doi.org/10.1038/nature13829] [PMID: 25363773]

[91] M.S. Zaini, J. Ying Chyi Liew, S.A. Alang Ahmad, A.R. Mohmad, and M.A. Kamarudin, "Quantum confinement effect and photoenhancement of photoluminescence of PbS and PbS/MnS quantum dots", *Appl. Sci.,* vol. 10, no. 18, p. 6282, 2020.
[http://dx.doi.org/10.3390/app10186282]

[92] W.K. Bae, Y.S. Park, J. Lim, D. Lee, L.A. Padilha, H. McDaniel, I. Robel, C. Lee, J.M. Pietryga, and V.I. Klimov, "Controlling the influence of Auger recombination on the performance of quantum-dot light-emitting diodes", *Nat. Commun.,* vol. 4, no. 1, p. 2661, 2013.
[http://dx.doi.org/10.1038/ncomms3661] [PMID: 24157692]

[93] Z. Vaseghi, and A. Nematollahzadeh, "Nanomaterials", In: *A book chapter-2, Green synthesis of nanomaterials for bioenergy applications.* John Wiley, 2020, pp. 23-82.
[http://dx.doi.org/10.1002/9781119576785.ch2]

[94] L.A. Kolahalam, I.V. Kasi Viswanath, B.S. Diwakar, B. Govindh, V. Reddy, and Y.L.N. Murthy, "Review on nanomaterials: Synthesis and applications", *Mater. Today Proc.,* vol. 18, pp. 2182-2190, 2019.
[http://dx.doi.org/10.1016/j.matpr.2019.07.371]

[95] S. Anu Mary Ealia, and M.P. Saravanakumar, "A review on the classification, characterisation, synthesis of nanoparticles and their application", *IOP Conf. Series Mater. Sci. Eng.,* vol. 263, p. 032019, 2017.
[http://dx.doi.org/10.1088/1757-899X/263/3/032019]

CHAPTER 2

Supercapacitor Basics (EDLCs, Pseudo, and Hybrid)

Dinesh Bejjanki[1] and **Sampath Kumar Puttapati**[1,*]

[1] Department of Chemical Engineering, National Institute of Technology, Warangal 506004, India

Abstract: Over the past few years, supercapacitors have been spotlighted because of the challenges faced by other energy storage systems. The supercapacitor possesses excellent power density and long-term durability with an eco-friendly nature. Due to their wide range of advantages, supercapacitors are applicable especially in electric vehicles, heavy-duty vehicles, telecommunication, electric aircraft, and consumer electronic products. As per the charge storage mechanism, supercapacitors are divided into three categories based on their charge-storing method: electric double-layer capacitors (EDLCs), pseudocapacitors, and hybrid capacitors. The electrode materials such as graphene, activated carbon, metal oxides, conducting polymers, *etc.*, were widely applied, for better performance. The electrolyte is a crucial component in the mechanism of the supercapacitor to run the system at a higher voltage and thus there are various electrolytes such as solid, inorganic, and organic based on the application of the materials, and the electrolytes are chosen. However, the supercapacitors suffer from low energy density. Currently, research is more focused on advanced materials and various synthesis methods to overcome the drawbacks. This chapter provides a detailed understanding of supercapacitors with redox and non-redox reactions -the broad classification of the supercapacitor -their charge storage mechanism -various electrode materials -electrolytes (aqueous, non-aqueous, and solid) and current collectors, *etc.* Finally, the parameters that help in estimating the performance of supercapacitors are (specific capacitance, energy density, and power density) included.

Keywords: Classification of supercapacitor, EDLC, Hybrid supercapacitor, Pseudo capacitors.

INTRODUCTION

Energy consumption has risen drastically over the past decades as well as the diminution of non-renewable fuel sources, leading to work on new approaches to solve the issue [1, 2]. In reality, the rise in energy consumption is brought about by the increasing use of electronic gadgets in society [3]. There are limited energy

---
*[*] **Corresponding author Sampath Kumar Puttapati:** Department of Chemical Engineering, National Institute of Technology, Warangal 506004, India; E-mail: pskr@nitw.ac.in

Sanjeev Verma, Shivani Verma, Saurabh Kumar & Bhawna Verma (Eds.)

resources; it is also essential to produce clean, sustainable, and renewable energy sources [4]. Renewable sources with eco-friendly nature are warmly welcomed for the production of power. Environmental-friendly renewable technologies include wind, solar, and tidal energy [5]. At the same time, burning fossil fuels is hugely hazardous because of the release of CO_2 and additional pollutants into the environment [6]. Renewable energy technologies, like solar or wind, only produce energy when the sun is shining or the wind is blowing. Thus energy-efficient gadgets are facing issues. We cannot effectively utilize renewable energy until appropriate energy storage technology is developed. Innovative electrical drive energy storage systems must be coupled with renewable energy storage conversion technology [7, 8]. Batteries and supercapacitors are two potential electrical energy storage solutions that can meet the fundamental needs of storing energy from renewable energy technologies [9]. The term "supercapacitor" also refers to electrochemical capacitors. Consequently, because they can store energy for a long time, electrical energy will be available all the time. Due to their ability to provide continuous power, batteries, and supercapacitors are in high demand in the current scenario [10, 11]. Supercapacitors are energetic devices because of their high-power densities, whereas batteries are powerful because of their high energy densities [12]. Combining renewable energy technologies with supercapacitors or batteries may create a unique hybrid system that can save energy and produce simultaneously. Such a hybrid energy system holds great promise for powering a future electronic device [13]. Heavy-duty vehicles need to have both high-power density and energy density simultaneously. Therefore, hybrid devices must emerge in combination with supercapacitors and batteries. The new module supercapacitor showed excellent performance and they are proper candidates for hybrid energy systems. In the year 1990, the United States Department of Energy actively promoted financing for research on supercapacitors and batteries. It increased global awareness of the supercapacitors' potential [14]. Ever since, tremendous work has been devoted to supercapacitor research and advancement in hybrid electrode materials, nanocomposites, and appropriate electrolytes to enhance performance at the minimum cost. Simultaneously, a fundamental understanding of the supercapacitor's design, operation, performance, and component optimization resulted in improved supercapacitor performance and a significant increase in energy density [15, 16]. To enhance energy density, advanced hybrid supercapacitors were created, in which the electrodes are composited with carbon materials and metal oxides. In the hybrid supercapacitor, the electrochemical process of the electroactive substance occurs at the interface of electrode and electrolyte either by oxidation-reduction or adsorption, and intercalation, mechanism. In this manner with the fundamentals of supercapacitors and electrode materials, the capacitance and the energy density of the electrode materials can be boosted [17].

This chapter will give a comprehensive understanding of supercapacitors, their broad classification, and their working mechanism. It also includes the main components of a supercapacitor such as electrode materials, electrolytes, and the current collector. Lastly, the parameters such as specific capacitance, energy density, and power density are being discussed.

FARADAIC AND NON-FARADAIC

In electrochemical measurement, faradaic and non-faradaic reactions are the main types of electrode characteristics (Fig. **1**). Charge transfer occurs at the electrode during the redox reaction in the faradaic process. However, the occurrence of electrochemical reactions at the electrode surface does not entail the presence of a faradaic reaction [18]. An electrical charge is introduced on both electrodes while an electrochemical reaction occurs and the ions are transported away from it because charge should not be retained on the electrode. The faradaic process is commonly found in fuel cells and lead-acid batteries. There is no charge transfer in the non-faradaic process. Similarly, in the adsorption and desorption method, the electric charge and ions stay at or on electrodes.

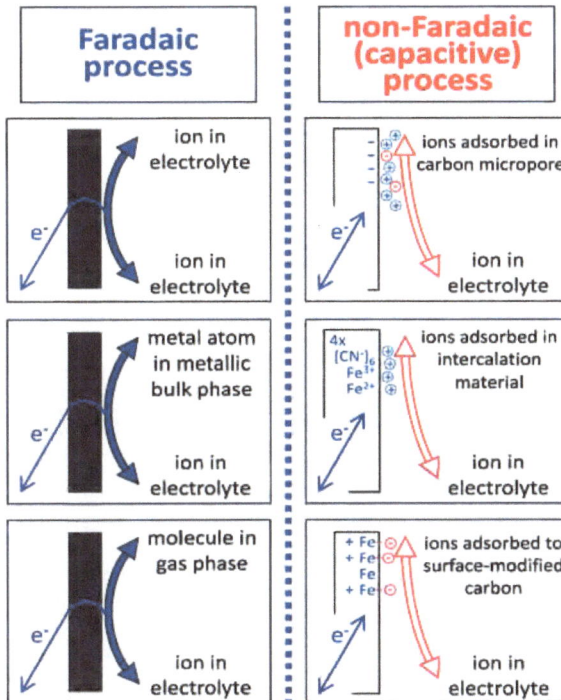

Fig. (1). Difference in charge storage vs charge transfer in faradaic and non-faradaic reactions. Adapted with permission from Reference [18], Copyright (2018), Chemical Physics.

CLASSIFICATION OF SUPERCAPACITOR

Supercapacitors are primarily divided into three groups based on the charge storage. Fig. (**2**) shows their classification as follows: (i) electric double-layer capacitors (EDLC), (ii) pseudocapacitors, and (iii) hybrid capacitors. In relation to energy storage, supercapacitors vary from conventional capacitors. Superca-pacitors store the charge chemically between the electrodes and the electrolyte, whereas dielectric capacitors store the charge electrostatically [19]. In supercapacitors, active electrolytes are between the electrodes; nevertheless, a dielectric membrane is used, whereas in the dielectric capacitor, no electrolyte is required [20].

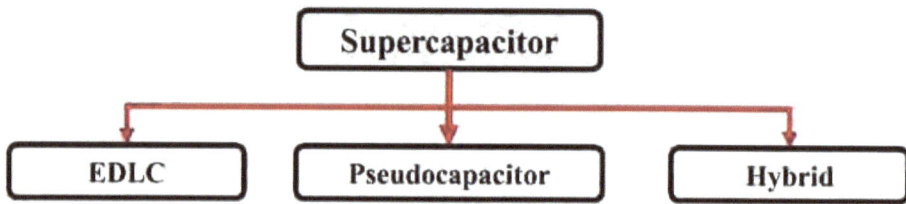

```
                    ┌─────────────────────┐
                    │   Supercapacitor    │
                    └─────────────────────┘
         ┌───────────────────┼───────────────────┐
         ▼                   ▼                   ▼
    ┌─────────┐      ┌─────────────────┐     ┌─────────┐
    │  EDLC   │      │ Pseudocapacitor │     │ Hybrid  │
    └─────────┘      └─────────────────┘     └─────────┘
```

Fig. (2). Classification of supercapacitors.

Electric Double-layer Capacitors (EDLCs)

In EDLCs, electrolyte ions attach to the electrode surface due to electrostatic attraction [21]. While charging, there is a growth of electric charges over the interface of electrolyte and electrode, leading to the development of an electric double layer [22]. Here, the charge is stored in an electric double-layer form and no charge transfer process occurs; for example, the redox process. In response to voltage applications, an electrical layer of charge forms over the electrode material's surface lattice assembly. To balance the charges further, ions in the electrolyte with the opposite polarity will get deposited on the other electrode surface [23]. The charge with two layers on the surface of electrodes is divided with solvent and forms a Helmholtz plane. The charge of the electrical layer on the electrode surface is larger due to the difference in the size of ions and electrons. To balance this additional electronic charge, ions of diffused layer come into the scene. Up to the outer Helmholtz layer, the electrode potential fluctuates linearly, but as it gets closer to the diffuse layer, it increases exponentially. EDLC offers a quick charge/discharge cycle with the electrostatic mechanism. EDLC has a longer cycle life than pseudocapacitors due to its non-Faradaic charge storage behavior *i.e.*, in EDLC, no chemical reactions are involved, therefore the electrode morphology remains fixed. As a result, this device has a sizable reversible capacitance and a long cycle life [24, 25]. Fig. (**3**) shows a schematic illustration of an EDLC charging and discharging operation. During charging, the ions in the electrolyte are adsorbed onto to electrode surface, and during discharge, the ions return to the electrolyte solution meanwhile electrons travel *via* an external circuit where a load is connected. The active material's surface area determines how many ions are adsorbed/desorbed during charging/discharging on the electrode surface. Previously, activated carbon materials were used in EDLC. However, in modern capacitor advances, nanoscale materials with high surface area, such as graphene [26] carbon nanopetals [27], biomass-derived activated carbon [28], carbon nanotubes, and quantum dots [29 - 31] are used.

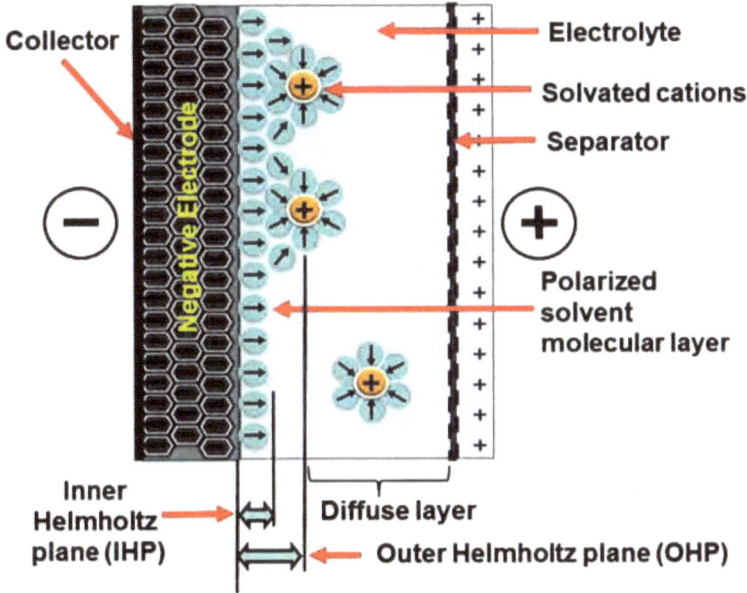

Fig. (3). Variations of the electric double layer at a negative electrode surface in the Helmholtz prototype, with the outer Helmholtz plane (OHP) and inner Helmholtz plane (IHP) highlighted. IHP denotes the range of closest distance of mainly adsorbed ions (usually negative ions), whereas the OHP denotes the length of non-specifically deposited ions. The OHP also serves as the plane from which the diffused layer emerges. Adapted with permission from Reference [32], Copyright (2022), Elsevier.

The following equations show the charge-discharge mechanism of EDLC,

$$E_n + E_p + I^- + I^+ \leftrightharpoons E_n^-//I^+ + E_p^+//I^- \qquad (1)$$

Where, E_n and E_p signify negative and positive electrodes, I^+ and I^- signify the cationic and anionic electrolyte, and // signifies the double layer on the interface.

Pseudocapacitor

In pseudocapacitors, energy is stored *via* charge transfer processes or faradaic reactions [33, 34]. Fig. (**4**) shows a schematic illustration of the pseudocapacitor.

The oxidation/reduction occurs at the capacitive electrode, and these processes must be reversible for a supercapacitor to have a long cycle life. In pseudocapacitors, electrodes experience continual chemical alterations due to quick charge/discharge cycles. Although charge transfer occurs in a pseudocapacitor, the adsorption of ions does not result in a chemical change of electrode material. In this scenario, the charge is transmitted *via* electrochemical redox reactions, *i.e.*, by adsorption and intercalation over the surface of the electrode. Along with EDLC, pseudo-capacitance behaviour has been seen in supercapacitor devices. Supercapacitors are produced using a variety of electrode materials, such as transition metals/metal-oxides, conductive polymers, metal organics frame (MOF), and mxene nanocomposites [35 - 39].

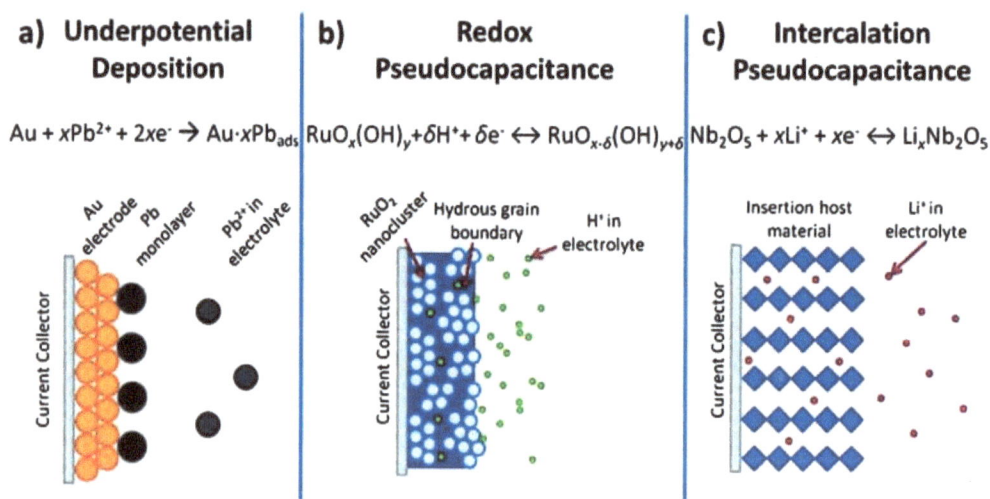

a) Underpotential Deposition

$$Au + xPb^{2+} + 2xe^- \rightarrow Au\cdot xPb_{ads}$$

b) Redox Pseudocapacitance

$$RuO_x(OH)_y + \delta H^+ + \delta e^- \leftrightarrow RuO_{x-\delta}(OH)_{y+\delta}$$

c) Intercalation Pseudocapacitance

$$Nb_2O_5 + xLi^+ + xe^- \leftrightarrow Li_xNb_2O_5$$

Fig. (4). Differentiation of pseudocapacitors mechanism types. Adapted with permission from Reference [43], Copyright (2020), Chinese Academy of Sciences.

These materials are subjected to reversible processes of oxidation and reduction reactions over a long time, these pseudo-capacitors often have very high energy densities with high capacitance. Nanomaterials with excellent porosity are highly favored electrodes because of the strong interaction between electrolyte ions and electrode material [40]. In comparison with EDLC, pseudocapacitors perform far better for high energy density but lack power density.

The three categories of electrochemical methods that contribute to charge transfer in pseudocapacitance are mentioned below.

Adsorption pseudocapacitance: Adsorption pseudocapacitance appears to be a function of electrolyte ion desorption and adsorption on the electrode surface. In the following equation (2.2), lead ions are adsorbed onto the gold surface [41].

$$xPb^2 + Au + 2xe^- \leftrightarrow xPb_{ads.} Au \qquad (2)$$

Intercalation pseudocapacitance: The van der Waals gaps as well as the grid of the electrode material are intercalated with electrolyte ions within intercalation pseudocapacitance. For illustration, the following equation (2.3), which describes how Li-ions from the electrolyte insert further within the grid of crystalline niobium oxide, could be used to describe the intercalation pseudocapacitance process [42].

$$xLi^+ + Nb_2O_5 + xe^- \leftrightarrow Nb_2Li_xO_5 \qquad (3)$$

Redox pseudocapacitance: In redox pseudocapacitance, as ions undergo electrochemical adsorption onto the electrode surface and with a charge transfer of faradaic reaction happens, thus redox pseudocapacitance occurs. For illustration, the following equation (2.4) is used to describe the redox reaction of RuO_2 in an $HClO_4$ electrolyte [43].

$$RuO_y(OH)_x + nH^+ + ne^- \rightarrow RuO_{x-n}(OH)_{y+n} \qquad (4)$$

Hybrid Supercapacitor

The term 'hybrid capacitors' suggests supercapacitors with the synergy of EDLC and pseudocapacitor materials. The energy-storing mechanism will follow both types of capacitors' behavior [44, 45]. In hybrid supercapacitors, to store energy asymmetric electrodes are utilized to reach a greater energy density. In the assessment of symmetric capacitors, hybrid supercapacitors hold a high voltage range with better specific capacitance [46]. Overall, three different forms of electrodes are often employed, including an asymmetric electrode, battery-type electrode, and composite electrodes. In an asymmetric capacitor, two electrodes show high capacity by nature, but one exhibits EDLC behavior and the other is made of a pseudocapacitive electrode [47]. In a battery, one electrode will be made of carbon, whereas the other is an electrode from the battery. In composite, pseudocapacitive electrode material is combined with carbon-based materials [48, 49]. The main advantages of using hybrid supercapacitors include negative self-rate, large voltage windows, high energy density, and large specific capacitance. Because of their tremendous capacity for charge storage, these kinds of supercapacitors have recently attracted the attention of several sectors. When compared to pseudocapacitors and EDLC, hybrid supercapacitors possess exceptional properties. This novel supercapacitor generation has much potential. The hybrid supercapacitors' ability to hold more charge by fusing redox and non-redox processes results in extremely high power and energy densities. When compared to EDLC and pseudo-capacitors, hybrid supercapacitors have large capacitance. A few hybrid supercapacitors operate at high potentials depending on the kind of electrolyte used during their fabrication, these supercapacitors are designed to provide electricity in a very effective manner to the electronic devices for the next generation. Hybrid supercapacitors filled the void between pseudo-capacitors and EDLC. For illustration of the energy-storing mechanism in a hybrid supercapacitor is shown in equation (2.5), where $Ni(OH)_2$ is employed as the electrode and aqueous KOH is used as the electrolyte [50].

$$Ni(OH)_2 + OH^- \leftrightharpoons NiOOH + H_2O + e^- \qquad (5)$$

Electrode

The electrode is the component of a supercapacitor that is actively employed because the sort of electrode-active specific material determines how much charge can be stored inside. The electrode should possess characteristics like high surface area, great electrical conductivity, strong redox activity, and porous structure. In EDLC, the electrode-active component doesn't always perform charge transfer processes, whereas pseudo-capacitors undergo such reactions. Pseudocapacitors are composed of conducting polymers with high electric conductivity and are more flexible than the compact electrode materials used in EDLC. For outstanding performance, the choice of electrode-active substances is of importance. A recent advancement, in the area of nanotechnology boosts supercapacitor electrode materials to synthesize novel nanomaterials. Due to their distinct characteristics as compared to macro and micro materials, nanostructured electrode materials are highly preferred. Nanomaterials have distinctive properties including high surface area, strong thermal stability, chemical stability, and excellent electric conductivity. Nanostructured materials make it simple to alter the electrode active substance with a highly porous structure. Because of their huge surface area, electrolyte ions may quickly diffuse *via* the pores, improving overall performance. As a result, materials with a greater surface area are selected. Carbon nanomaterials are commonly utilized for the same purpose including graphene oxide, carbon nanofibers, carbon nanotubes, reduced graphene oxide, mesoporous carbon, *etc.*

Supercapacitor electrode materials are made using nanocomposites. The exceptional act of nanocomposite electrodes has made these especially attractive alternatives for constructing large capacitance and energy-density supercapacitors.

One key issue grounded in electrode materials is cycle life, which occurs when electrode degradation occurs and performance decays after 1000 cycles. The electrode-active materials are divided into three categories, depending on how they store charge: (i) material with EDLC behavior (ii) material with redox or pseudocapacitance nature, and (iii) hybrid material with both EDLC and pseudocapacitance [51]. Fig. (**5**) provides a brief electrode material classification.

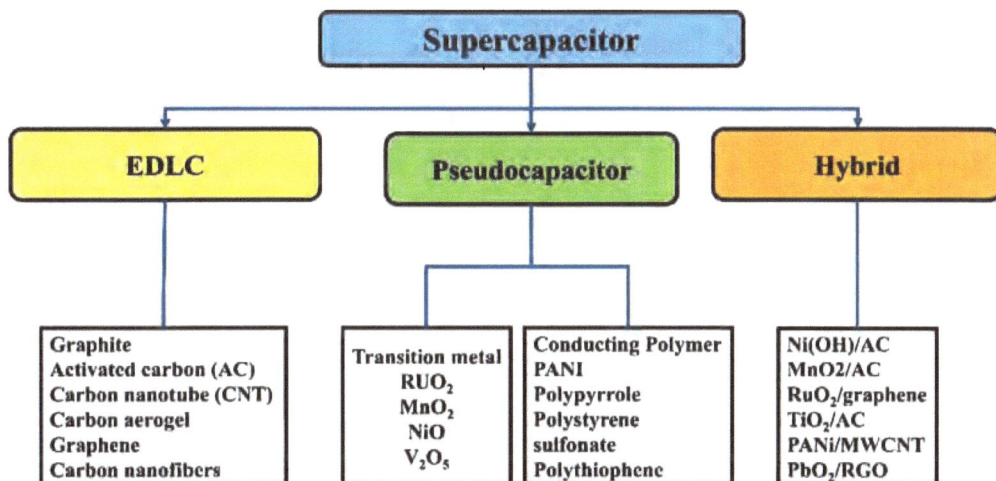

Fig. (5). Flow chart of different active materials for supercapacitors, Adapted with permission from Reference [77]. Copyright (2021), Springer Link.

A supercapacitor device's electrode materials can be made of a variety of materials, including carbon nanotubes, conducting polymers, and transition metal oxides. The following section provides an overview of several materials utilized as electrode materials for the development of a supercapacitor [52].

Carbon Materials

Carbon nanostructured materials have already been driven to supercapacitor studies because of their exclusive qualities like superior electrical conductivity, high surface area, high electrochemical and chemical stability, and eco-friendly nature [53, 54]. Carbon nanomaterials have always been utilized in EDLCs, pseudocapacitors, and hybrid capacitors. Carbon nanostructures store electrical energy through the development of electrochemical double layers. Carbon nanoparticles do not participate in redox reactions, which illustrates that carbon-based nanomaterials are EDLC in nature and have a much longer life than other forms of supercapacitors. A porous structure is required for the carbon nanomaterial to be employed in supercapacitor in case electrolyte ions could freely diffuse through the porous structures, and the availability of the electrolyte ions to the extreme internal pores might well take part in charge storage, thus improving the performance of supercapacitor.Nanoporous carbon-based electrode materials should execute reversible adsorption and intercalation to store large amounts of energy in supercapacitors for a prolonged period. That is, electrolyte ion adsorption at the electrode or electrolyte interface during charging must be reversibly removed from the surface during the discharging stage. If the ions are not desorbed completely or if they have any chemical interactions or chemical complex formation with the electrode, then the performance will suffer, and the electrode design will fail in such cases, the available active surface area for adsorption would decrease. In the application of carbon nanomaterial, porous structure is significant, in addition to pore size distribution and porosity for supercapacitor fabrication. Because the pore diameters are in the mesoporous region (2-50 nm), the efficiency will be remarkable because the mobility of electrolyte ions would be advantageous for generating larger capacitance [55]. Whereas, supercapacitor performance is shown to be weak in the situation of micropores (pore sizes < 2 nm) and macropores (pore sizes > 50 nm). Because there is minimal change in the structure, the carbon nanoparticles' good electrical conductivity benefits the flow of electrons even after 1000 cycles. Carbon nanofibers, carbon nano petals, carbon nanotubes (CNTs), graphene, mesoporous carbons, and other nanostructured carbon-based materials are employed as electrode active materials for supercapacitors. The application uses both single and multi-walled carbon nanotubes (MWCNT) [56]. In the cited articles, detailed characterizations are carried out for solar-reduced graphene oxide [57] and activated carbon [58].

Transition Metals

Transition metal oxide (TMO) materials draw attention because of the high specific capacitance in supercapacitors because they store charge from redox reactions. Because of their excellent electronic, mechanical, and redox characteristics, they are commonly employed in a variety of power generation and energy storage applications [59, 60]. Nano crystallites and porous TMO-based electrodes are greatly favored because electrolyte transport into the inner pores is high, resulting in improved performance. The redox capacitance is caused by its many valence state shifts. There are two types of transition metal-oxides: (i) noble metal-oxides such as IrO_2, RuO_2, *etc* [61]. and (ii) transition metal oxides (TMO) such as NiO_2, MnO_2, SnO_2, Co_3O_4, *etc*. Noble metal oxides are more expensive than base TMO. The low price and ecological nature of base TMO have drawn the attention of many for energy storage and conversion; nevertheless, efficiency is somewhat low compared with the noble metal oxides. A key disadvantage of supercapacitors' is low energy density, but it can be overcome by employing innovative TMO or hydroxides with unique electrode designs. The best technique for attaining the optimum supercapacitor performance is the nanostructured design of the electrode. The detailed properties of TMO are discussed. Electronically highly conductive materials have recently captured the attention of flexible energy storage and conversion systems. They showed successful results in those applications due to their strong electronic conductivities and ease of processing.

Conducting Polymers

Electronically conductive composite electrodes for supercapacitor applications have several advantages, including redox-type charge storage that acts as a capacity promoter, enabling the opportunity to fabricate flexible electrode materials, and thus initiating the prospect of flexible supercapacitors (FS). FS is required to be used in wearable and flexible electronics, which have grown in popularity in recent years. A doping method is used to store charge in electrically conducting polymers. Polyaniline/PANI [62, 49], polypyrrole [63 - 65], and poly(3,4-ethylenedioxythiophene)/PEDOT [66] are a few examples of electrical conducting polymers utilized in supercapacitor.

Composite Materials

As described in the preceding section, numerous strategies have been developed to overcome the technical issues associated with the fabrication of EDLC and pseudo-capacitive electrode materials. However, advanced electrode materials are needed for high-performance energy storage systems. Composting of carbon and pseudo-capacitive materials emerges as potential alternatives for achieving significantly improved energy storage capacities and successfully building the gap between standard capacitors and battery materials. As previously noted, conductive polymers lack poor cyclic stability and thus limits their use as independent electrode materials in energy storage application. Combining carbonaceous active materials with conductive polymers or with transition metals has led to enhanced performance [67]. Recently, binary ternary composite materials achieved a high capacitance with long-term cyclic stability, and here are a few examples of composite materials $Ni(OH)/AC$ [68], MnO_2/AC [69], $NiO/Mn_2O_3/RGO$ [70], MFG/PANI [71], and PANi/MWCNT *etc* [72].

Electrolyte

The electrolyte impacts the power and energy densities of the supercapacitor because electrolyte resistance is a critical factor. The cumulative resistances inside the device are mentioned as 'electrochemical series resistance'. The power density would be poor if such electrolyte's resistance is greater. Supercapacitors employ a variety of electrolytes, including ionic, organic, and aqueous. They are categorized into two types: solid and liquid. Fig. (6) depicts many forms of liquid and solid electrolytes. The stable voltage window that an electrolyte may work is used to select it. Aqueous electrolytes (such as alkali and acid) have relatively low resistances and are thus an excellent choice for the fabrication of supercapacitors [73, 74]. The aqueous electrolyte solution can have drawbacks, such as being unstable at high voltages, resulting in electrode degradation, and being ecologically toxic. On the other side, organic electrolytes are exceedingly combustible and poisonous, especially when operating at higher working voltage. The cause of why aqueous electrolytes have low resistance is because the protons are small in size and have high mobility, which reduces resistance. Organic electrolytes, on the other hand, have higher resistance because of their much bigger size. However, the choice of electrolyte in addition to the permeability of electrode design is demanding for optimal supercapacitor performance. As the device operates for more than 1000 cycles, electrolyte degradation increases resistance, resulting in a drop in capacitance and therefore, lower energy density. Because of this, organic and aqueous electrolytes are not commonly used in commercially available systems, however, ionic liquid electrolytes are used in industrial supercapacitors. These electrolytes have dissimilar properties such as strong conductivity and a wide electrochemical voltage window [75, 76]. Ionic electrolytes are safe to handle because they are naturally non-flammable. Depletion of electrolytes could be reduced by increasing the concentration of ionic contents. They are also chemically and environmentally stable, making them excellent applicants for supercapacitors. Because of the improved electrochemical properties, a subset of electrolytes known as 'gel polymeric electrolytes (GPE)' has added attention [77]. GPE could be used to create solid-state supercapacitors (SSS). GPE-based SSS are particularly favorable in the use of upcoming wearable and flexible electronics. As a result, a new form of electrolyte, titled 'gel polymer with ionic liquid-based electrolytes', is projected by integrating aspects of both gel polymer and ionic liquid electrolytes [78]. These include all of the benefits of their equivalents and disadvantages yet, the cost aspect remains a key issue.

Membrane

Electrodes separating membrane's functions are: (i) to enable electrolyte ions to permit through and (ii) to prevent short circuit of supercapacitor. The electrolyte membrane with high ionic and low electric conductivity is favored. A regular paper or commonly available Whatman filter paper would also serve. These electrolyte membranes are cheap and hence widely available nano-structured electrolytic membranes have also recently been developed [79, 80]. Nafion membrane is one example, because of their nanoscale characteristics, nafion membranes have excellent ionic conductivity and are frequently employed in industrial supercapacitors. However, Nafion is quite expensive, it is necessary to build innovative electrolyte membranes using low-cost polymeric materials; the membrane must have a porous shape to transmit ions from one side to another [81].

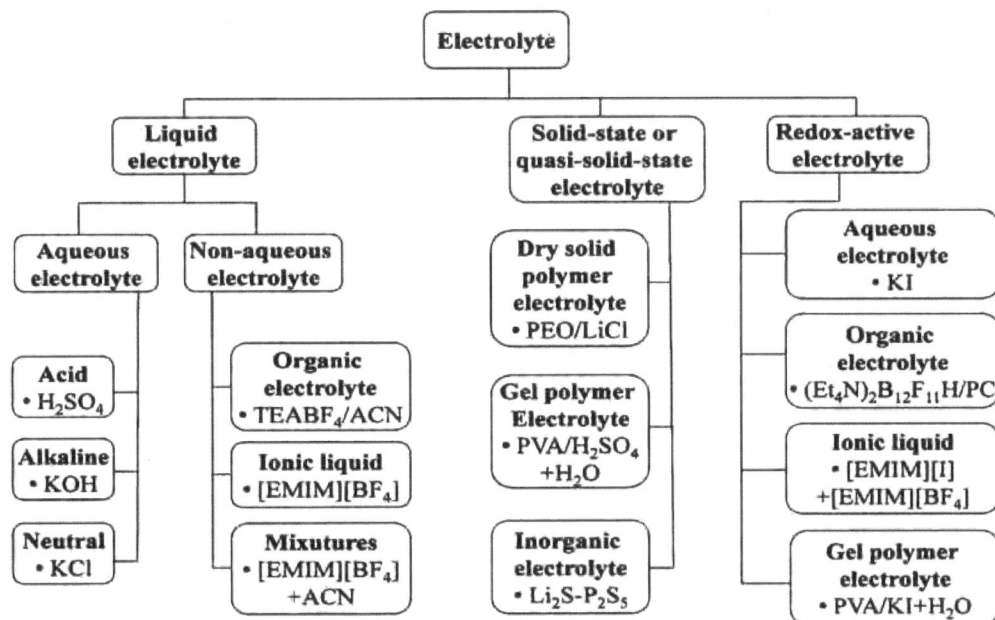

Fig. (6). Electrolyte classification for supercapacitors. Adapted with permission from Reference [69], Copyright (2015), Royal Society of Chemistry.

Current Collectors

The current collector's role is to accept electrons from active electrode material and carry electrons to external load. Metal plates, such as aluminum and copper, are utilized for this purpose. Steel plate is one example of an alloy utilized for the same function. In general, 2 main collectors are utilized on the anode and cathode surfaces of supercapacitors [82].

CONCLUSION

This chapter is a comprehensive overview of supercapacitors, including charge storage types, their merits, and demerits. Supercapacitors are classified as EDLC, pseudocapacitor, and hybrid capacitor. Because EDLC has non-redox energy storage, no electrode degradation occurs over a high number of cycles. However, when redox capacitors undergo quasi-reactions, chemical reactions occur to the active electrode materials, causing them to lower their performance after a few hundred cycles. To prevent this, hybrid capacitors have already been designed to overcome the drawbacks of EDLC capacitors and pseudocapacitors while retaining all of their benefits. When compared to batteries, supercapacitors have excellent power densities, but lower energy densities. To meet the requirements, future supercapacitors need to have a higher energy density. The biggest difficulty for the supercapacitor market is the high cost of manufacture, which makes them 'unaffordable' to the general public.

REFERENCES

[1]   F. Martins, C. Felgueiras, M. Smitkova, and N. Caetano, "Analysis of fossil fuel energy consumption and environmental impacts in European countries", *Energies,* vol. 12, no. 6, p. 964, 2019.
[http://dx.doi.org/10.3390/en12060964]

[2]   K. Whiting, L.G. Carmona, and T. Sousa, "A review of the use of exergy to evaluate the sustainability of fossil fuels and non-fuel mineral depletion", *Renew. Sustain. Energy Rev.,* vol. 76, pp. 202-211, 2017.
[http://dx.doi.org/10.1016/j.rser.2017.03.059]

[3]   S. Needhidasan, M. Samuel, and R. Chidambaram, "Electronic waste – an emerging threat to the environment of urban India", *J. Environ. Heal. Sci. Eng.,* pp. 1-9, 2014.
[http://dx.doi.org/10.1186/2052-336X-12-36]

[4]   A. Hussain, S.M. Arif, and M. Aslam, "Emerging renewable and sustainable energy technologies: State of the art", *Renew. Sustain. Energy Rev.,* vol. 71, pp. 12-28, 2017.
[http://dx.doi.org/10.1016/j.rser.2016.12.033]

[5]   K. Shivarama Krishna, and K. Sathish Kumar, "A review on hybrid renewable energy systems", *Renew. Sustain. Energy Rev.,* vol. 52, pp. 907-916, 2015.
[http://dx.doi.org/10.1016/j.rser.2015.07.187]

[6]   M.R. Lotfalipour, M.A. Falahi, and M. Ashena, "Economic growth, CO2 emissions, and fossil fuels consumption in Iran", *Energy,* vol. 35, no. 12, pp. 5115-5120, 2010.
[http://dx.doi.org/10.1016/j.energy.2010.08.004]

[7]   S. Sridhar, and S.R. Salkuti, "Development and future scope of renewable energy and energy storage systems", *Smart Cities,* vol. 5, no. 2, pp. 668-699, 2022.
[http://dx.doi.org/10.3390/smartcities5020035]

[8]   A.K. Rohit, K.P. Devi, and S. Rangnekar, "An overview of energy storage and its importance in Indian renewable energy sector", *J. Energy Storage,* vol. 13, pp. 10-23, 2017.
[http://dx.doi.org/10.1016/j.est.2017.06.005]

[9]   W. Zuo, R. Li, C. Zhou, Y. Li, J. Xia, and J. Liu, "Battery-supercapacitor hybrid devices: Recent progress and future prospects", *Adv. Sci.,* vol. 4, no. 7, p. 1600539, 2017.
[http://dx.doi.org/10.1002/advs.201600539] [PMID: 28725528]

[10]   E. Armelin, M.M. Pérez-Madrigal, C. Alemán, and D.D. Díaz, "Current status and challenges of biohydrogels for applications as supercapacitors and secondary batteries", *J. Mater. Chem. A Mater. Energy Sustain.,* vol. 4, no. 23, pp. 8952-8968, 2016.
[http://dx.doi.org/10.1039/C6TA01846G]

[11]   Y. Kumar, A. Gupta, A.K. Thakur, S.J. Uke, V. Khatri, A. Kumar, M. Gupta, and Y. Kumar, "Advancement and current scenario of engineering and design in transparent supercapacitors: electrodes and electrolyte", *J. Nanopart. Res.,* vol. 23, no. 5, p. 119, 2021.
[http://dx.doi.org/10.1007/s11051-021-05221-5]

[12]   J. Libich, J. Máca, J. Vondrák, O. Čech, and M. Sedlaříková, "Supercapacitors: Properties and applications", *J. Energy Storage,* vol. 17, pp. 224-227, 2018.

[http://dx.doi.org/10.1016/j.est.2018.03.012]

[13]   D. Lemian, and F. Bode, "Battery-supercapacitor energy storage systems for electrical vehicles: A review", *Energies,* vol. 15, no. 15, p. 5683, 2022.
[http://dx.doi.org/10.3390/en15155683]

[14]   M. Xu, J. Qu, and M. Li, "National policies, recent research hotspots, and application of sustainable energy: Case of china, usa, and european countries", *Sustain,* vol. 14, no. 16, 2022.
[http://dx.doi.org/10.3390/su141610014]

[15]   P. Lamba, P. Singh, P. Singh, P. Singh, A.K. Bharti, A. Kumar, M. Gupta, and Y. Kumar, "Recent advancements in supercapacitors based on different electrode materials: Classifications, synthesis methods and comparative performance", *J. Energy Storage,* vol. 48, p. 103871, 2022.
[http://dx.doi.org/10.1016/j.est.2021.103871]

[16]   A. Muzaffar, M.B. Ahamed, K. Deshmukh, and J. Thirumalai, "A review on recent advances in hybrid supercapacitors: Design, fabrication and applications", *Renew. Sustain. Energy Rev.,* vol. 101, pp. 123-145, 2019.
[http://dx.doi.org/10.1016/j.rser.2018.10.026]

[17]   A. Eftekhari, "The mechanism of ultrafast supercapacitors", *J. Mater. Chem. A Mater. Energy Sustain.,* vol. 6, no. 7, pp. 2866-2876, 2018.
[http://dx.doi.org/10.1039/C7TA10013B]

[18]   P.M. Biesheuvel, S. Porada, and J.E. Dykstra, "The difference between Faradaic and non-Faradaic electrode processes", *Phys. Chem-ph,* pp. 1-10, 2018.

[19]   A.G. Pandolfo, and A.F. Hollenkamp, "Carbon properties and their role in supercapacitors", *J. Power Sources,* vol. 157, no. 1, pp. 11-27, 2006.
[http://dx.doi.org/10.1016/j.jpowsour.2006.02.065]

[20]   L.E. Helseth, "Comparison of methods for finding the capacitance of a supercapacitor", *J. Energy Storage,* vol. 35, p. 102304, 2021.
[http://dx.doi.org/10.1016/j.est.2021.102304]

[21]   Y. Hirai, K. Okada, R. Kurokawa, S. Matsumoto, and Y. Sato, "Electrical property of EDLC and electrochemical interaction between separator and electrolyte", *J. Int. Counc. Electr. Eng.,* vol. 6, no. 1, pp. 72-77, 2016.
[http://dx.doi.org/10.1080/22348972.2016.1173781]

[22]   K.K. Kar, J.K.P. Sravendra, and R. Editors, *Polylactic Acid (PLA) Carbon Nanotube Nanocomposites.* vol. B. Springer, 2015, pp. 283-297.

[23]   Y. Zhang, H. Feng, X. Wu, L. Wang, A. Zhang, T. Xia, H. Dong, X. Li, and L. Zhang, "Progress of electrochemical capacitor electrode materials: A review", *Int. J. Hydrogen Energy,* vol. 34, no. 11, pp. 4889-4899, 2009.
[http://dx.doi.org/10.1016/j.ijhydene.2009.04.005]

[24]   L. Yin, S. Li, X. Liu, and T. Yan, "Ionic liquid electrolytes in electric double layer capacitors", *Sci. China Mater.,* vol. 62, no. 11, pp. 1537-1555, 2019.
[http://dx.doi.org/10.1007/s40843-019-9458-3]

[25]   P. Sharma, and T.S. Bhatti, "A review on electrochemical double-layer capacitors", *Energy Convers. Manage.,* vol. 51, no. 12, pp. 2901-2912, 2010.
[http://dx.doi.org/10.1016/j.enconman.2010.06.031]

[26]   P. Kamedulski, M. Skorupska, P. Binkowski, W. Arendarska, A. Ilnicka, and J.P. Lukaszewicz, "High surface area micro-mesoporous graphene for electrochemical applications", *Sci. Rep.,* vol. 11, no. 1, p. 22054, 2021.
[http://dx.doi.org/10.1038/s41598-021-01154-0] [PMID: 34764324]

[27]   J. Cherusseri, and K.K. Kar, "Hierarchically mesoporous carbon nanopetal based electrodes for flexible supercapacitors with super-long cyclic stability", *J. Mater. Chem. A Mater. Energy Sustain.,*

vol. 3, no. 43, pp. 21586-21598, 2015.
[http://dx.doi.org/10.1039/C5TA05603A]

[28] D. Bejjanki, P. Banothu, V. B. Kumar, and P. S. Kumar, "Biomass-derived N-doped activated carbon from eucalyptus leaves as an efficient supercapacitor electrode material", *C,* vol. 9, no. 1, p. 24, 2023.
[http://dx.doi.org/10.3390/c9010024]

[29] J. Cherusseri, and K.K. Kar, "Self-standing carbon nanotube forest electrodes for flexible supercapacitors", *RSC Advances,* vol. 5, no. 43, pp. 34335-34341, 2015.
[http://dx.doi.org/10.1039/C5RA04064G]

[30] J. Cherusseri, R. Sharma, and K.K. Kar, "Helically coiled carbon nanotube electrodes for flexible supercapacitors", *Carbon,* vol. 105, pp. 113-125, 2016.
[http://dx.doi.org/10.1016/j.carbon.2016.04.019]

[31] F.A. Permatasari, M.A. Irham, S.Z. Bisri, and F. Iskandar, "Carbon-based quantum dots for supercapacitors: Recent advances and future challenges", *Nanomaterials,* vol. 11, no. 1, p. 91, 2021.
[http://dx.doi.org/10.3390/nano11010091] [PMID: 33401630]

[32] D. Govindarajan, and K.K. Chinnakutti, *Oxide Free Nanomaterials for Energy Storage and Conversion Applications.* vol. 3. Elsevier, 2022, pp. 51-74.
[http://dx.doi.org/10.1016/B978-0-12-823936-0.00010-3]

[33] T. Das, V. K. Pandey, S. Verma, S. K. Pandey, and B. Verma, "Optimization of the ratio of aniline, ammonium persulfate, para -toluenesulfonic acid for the synthesis of conducting polyaniline and its use in energy storage devices", *Int J Energy Res,* vol. 46, no. 14, pp. 19914-19928, 2022.
[http://dx.doi.org/10.1002/er.8690]

[34] P. Bhojane, "Recent advances and fundamentals of Pseudocapacitors: Materials, mechanism, and its understanding", *J. Energy Storage,* vol. 45, p. 103654, 2022.
[http://dx.doi.org/10.1016/j.est.2021.103654]

[35] J. Cherusseri, and K.K. Kar, "Hierarchical carbon nanopetal/polypyrrole nanocomposite electrodes with brush-like architecture for supercapacitors", *Phys. Chem. Chem. Phys.,* vol. 18, no. 12, pp. 8587-8597, 2016.
[http://dx.doi.org/10.1039/C6CP00150E] [PMID: 26946975]

[36] J. Cherusseri, and K.K. Kar, "Ultra-flexible fibrous supercapacitors with carbon nanotube/polypyrrole brush-like electrodes", *J. Mater. Chem. A Mater. Energy Sustain.,* vol. 4, no. 25, pp. 9910-9922, 2016.
[http://dx.doi.org/10.1039/C6TA02690G]

[37] J. Cherusseri, and K.K. Kar, "Polypyrrole-decorated 2D carbon nanosheet electrodes for supercapacitors with high areal capacitance", *RSC Advances,* vol. 6, no. 65, pp. 60454-60466, 2016.
[http://dx.doi.org/10.1039/C6RA01402J]

[38] B. Xu, H. Zhang, H. Mei, and D. Sun, "Recent progress in metal-organic framework-based supercapacitor electrode materials", *Coord. Chem. Rev.,* vol. 420, p. 213438, 2020.
[http://dx.doi.org/10.1016/j.ccr.2020.213438]

[39] S. Panda, K. Deshmukh, S.K. Khadheer Pasha, J. Theerthagiri, S. Manickam, and M.Y. Choi, "MXene based emerging materials for supercapacitor applications: Recent advances, challenges, and future perspectives", *Coord. Chem. Rev.,* vol. 462, p. 214518, 2022.
[http://dx.doi.org/10.1016/j.ccr.2022.214518]

[40] Y. Jiang, and J. Liu, "Definitions of pseudocapacitive materials: A brief review", *Energy Environ. Mater.,* vol. 2, no. 1, pp. 30-37, 2019.
[http://dx.doi.org/10.1002/eem2.12028]

[41] C. Choi, D.S. Ashby, D.M. Butts, R.H. DeBlock, Q. Wei, J. Lau, and B. Dunn, "Achieving high energy density and high power density with pseudocapacitive materials", *Nat. Rev. Mater.,* vol. 5, no. 1, pp. 5-19, 2019.
[http://dx.doi.org/10.1038/s41578-019-0142-z]

[42]   C. Zhong, D. Sun, Y. Deng, W. Hu, J. Qiao, and J. Zhang, "Compatibility of electrolytes with inactive components of electrochemical supercapacitors", In: *Electrolytes for Electrochemical Supercapacitors.* Routledge Handbooks, 2016, pp. 255-274.
       [http://dx.doi.org/10.1201/b21497-4]

[43]   S. Zhao, P. Liu, G. Cheng, L. Yu, and H. Zeng, "Preparation and Pseudocapacitor Properties of Self-Supported Nickel Sulfides Electrode Materials", *Huaxue Jinzhan,* vol. 32, no. 10, pp. 1582-1591, 2020.

[44]   G. Ma, M. Dong, K. Sun, E. Feng, H. Peng, and Z. Lei, "A redox mediator doped gel polymer as an electrolyte and separator for a high performance solid state supercapacitor", *J. Mater. Chem. A Mater. Energy Sustain.,* vol. 3, no. 7, pp. 4035-4041, 2015.
       [http://dx.doi.org/10.1039/C4TA06322H]

[45]   N. Xu, X. Sun, X. Zhang, K. Wang, and Y. Ma, "A two-step method for preparing Li 4 Ti 5 O 12 –graphene as an anode material for lithium-ion hybrid capacitors", *RSC Advances,* vol. 5, no. 114, pp. 94361-94368, 2015.
       [http://dx.doi.org/10.1039/C5RA20168C]

[46]   S.W. Zhang, B.S. Yin, X.X. Liu, D.M. Gu, H. Gong, and Z.B. Wang, "A high energy density aqueous hybrid supercapacitor with widened potential window through multi approaches", *Nano Energy,* vol. 59, pp. 41-49, 2019.
       [http://dx.doi.org/10.1016/j.nanoen.2019.02.001]

[47]   Y. Shao, M.F. El-Kady, J. Sun, Y. Li, Q. Zhang, M. Zhu, H. Wang, B. Dunn, and R.B. Kaner, "Design and Mechanisms of Asymmetric Supercapacitors", *Chem. Rev.,* vol. 118, no. 18, pp. 9233-9280, 2018.
       [http://dx.doi.org/10.1021/acs.chemrev.8b00252] [PMID: 30204424]

[48]   D. Bejjanki, G. Uday Bhaskar Babu, K. Kumar, and S.K. Puttapati, "SnO2/RGOatPANi ternary composite via chemical oxidation polymerization and its synergetic effect for better performance of supercapacitor", *Mater. Today Proc.,* pp. 2-7, 2022.

[49]   G.V. Ramana, P.S. Kumar, V.V.S.S. Srikanth, B. Padya, and P.K. Jain, "Electrochemically active polyaniline (PANi) coated carbon nanopipes and PANi nanofibers containing composite", *J. Nanosci. Nanotechnol.,* vol. 15, no. 2, pp. 1338-1343, 2015.
       [http://dx.doi.org/10.1166/jnn.2015.9056] [PMID: 26353652]

[50]   N.R. Chodankar, H.D. Pham, A.K. Nanjundan, J.F.S. Fernando, K. Jayaramulu, D. Golberg, Y.K. Han, and D.P. Dubal, "True meaning of pseudocapacitors and their performance metrics: Asymmetric versus hybrid supercapacitors", *Small,* vol. 16, no. 37, p. 2002806, 2020.
       [http://dx.doi.org/10.1002/smll.202002806] [PMID: 32761793]

[51]   P. Forouzandeh, V. Kumaravel, and S.C. Pillai, "Electrode materials for supercapacitors: A review of recent advances", *Catalysts,* vol. 10, no. 9, p. 969, 2020.
       [http://dx.doi.org/10.3390/catal10090969]

[52]   M. Ates, A. Chebil, O. Yoruk, C. Dridi, and M. Turkyilmaz, "Reliability of electrode materials for supercapacitors and batteries in energy storage applications: A review", *Ionics,* vol. 28, no. 1, pp. 27-52, 2022.
       [http://dx.doi.org/10.1007/s11581-021-04296-3]

[53]   Z. Lu, Y. Chao, Y. Ge, J. Foroughi, Y. Zhao, C. Wang, H. Long, and G.G. Wallace, "High-performance hybrid carbon nanotube fibers for wearable energy storage", *Nanoscale,* vol. 9, no. 16, pp. 5063-5071, 2017.
       [http://dx.doi.org/10.1039/C7NR00408G] [PMID: 28265639]

[54]   L. Jiang, L. Sheng, X. Chen, T. Wei, and Z. Fan, "Construction of nitrogen-doped porous carbon buildings using interconnected ultra-small carbon nanosheets for ultra-high rate supercapacitors", *J. Mater. Chem. A Mater. Energy Sustain.,* vol. 4, no. 29, pp. 11388-11396, 2016.
       [http://dx.doi.org/10.1039/C6TA02570F]

[55]   P.H. Jampani, O. Velikokhatnyi, K. Kadakia, D.H. Hong, S.S. Damle, J.A. Poston, A. Manivannan, and P.N. Kumta, "High energy density titanium doped-vanadium oxide-vertically aligned CNT composite electrodes for supercapacitor applications", *J. Mater. Chem. A Mater. Energy Sustain.*, vol. 3, no. 16, pp. 8413-8432, 2015.
[http://dx.doi.org/10.1039/C4TA06777K]

[56]   F. Ran, X. Yang, and L. Shao, "Recent progress in carbon-based nanoarchitectures for advanced supercapacitors", *Adv. Compos. Hybrid Mater.*, vol. 1, no. 1, pp. 32-55, 2018.
[http://dx.doi.org/10.1007/s42114-017-0021-2]

[57]   V. Gedela, S.K. Puttapati, C. Nagavolu, and V.V.S.S. Srikanth, "A unique solar radiation exfoliated reduced graphene oxide/polyaniline nanofibers composite electrode material for supercapacitors", *Mater. Lett.*, vol. 152, pp. 177-180, 2015.
[http://dx.doi.org/10.1016/j.matlet.2015.03.113]

[58]   J. Wang, X. Zhang, Z. Li, Y. Ma, and L. Ma, "Recent progress of biomass-derived carbon materials for supercapacitors", *J. Power Sources*, vol. 451, p. 227794, 2020.
[http://dx.doi.org/10.1016/j.jpowsour.2020.227794]

[59]   S.R. Ede, S. Anantharaj, K.T. Kumaran, S. Mishra, and S. Kundu, "One step synthesis of Ni/Ni(OH) 2 nano sheets (NSs) and their application in asymmetric supercapacitors", *RSC Advances*, vol. 7, no. 10, pp. 5898-5911, 2017.
[http://dx.doi.org/10.1039/C6RA26584G]

[60]   L. Li, R. Li, S. Gai, F. He, and P. Yang, "Facile fabrication and electrochemical performance of flower-like Fe 3 O 4 @C@layered double hydroxide (LDH) composite", *J. Mater. Chem. A Mater. Energy Sustain.*, vol. 2, no. 23, pp. 8758-8765, 2014.
[http://dx.doi.org/10.1039/C4TA01186D]

[61]   D-Q. Liu, S-H. Yu, and S-K. Joo, "Electrochemical Performance of Iridium Oxide Thin Film for Supercapacitor Prepared by Radio Frequency Magnetron Sputtering Method", *ECS Meet. Abstr.*, vol. 2, no. 5, p. 525, 2008.

[62]   A. Eftekhari, L. Li, and Y. Yang, "Polyaniline supercapacitors", *J. Power Sources*, vol. 347, pp. 86-107, 2017.
[http://dx.doi.org/10.1016/j.jpowsour.2017.02.054]

[63]   Y. Huang, H. Li, Z. Wang, M. Zhu, Z. Pei, Q. Xue, Y. Huang, and C. Zhi, "Nanostructured Polypyrrole as a flexible electrode material of supercapacitor", *Nano Energy*, vol. 22, pp. 422-438, 2016.
[http://dx.doi.org/10.1016/j.nanoen.2016.02.047]

[64]   J. Zhao, J. Wu, B. Li, W. Du, Q. Huang, M. Zheng, H. Xue, and H. Pang, "Facile synthesis of polypyrrole nanowires for high-performance supercapacitor electrode materials", *Prog. Nat. Sci.*, vol. 26, no. 3, pp. 237-242, 2016.
[http://dx.doi.org/10.1016/j.pnsc.2016.05.015]

[65]   K. Puttapati, "A brief notes on metal oxide-carbon nanomaterial-polypyrrole/ polyaniline ternary nanocomposites as hybrid type supercapacitor electrode materials", *Nanosci. Nanotechnol.-Asia*, vol. 5, no. 7, pp. 130-136, 2015.

[66]   J. García-Torres, S. Colombi, L.P. Macor, and C. Alemán, "Multitasking smart hydrogels based on the combination of alginate and poly(3,4-ethylenedioxythiophene) properties: A review", *Int. J. Biol. Macromol.*, vol. 219, pp. 312-332, 2022.
[http://dx.doi.org/10.1016/j.ijbiomac.2022.08.008] [PMID: 35934076]

[67]   T.S. Bhat, P.S. Patil, and R.B. Rakhi, "Recent trends in electrolytes for supercapacitors", *J. Energy Storage*, vol. 50, no. January, p. 104222, 2022.
[http://dx.doi.org/10.1016/j.est.2022.104222]

[68]   L. Zhang, S. Yang, J. Chang, D. Zhao, J. Wang, C. Yang, and B. Cao, "A review of redox electrolytes

for supercapacitors", *Front Chem.*, vol. 8, p. 413, 2020.
[http://dx.doi.org/10.3389/fchem.2020.00413] [PMID: 32582626]

[69]   V.V.S.S. Srikanth, G.V. Ramana, and P.S. Kumar, "Perspectives on state-of-the-art carbon nanotube/polyaniline and graphene/polyaniline composites for hybrid supercapacitor electrodes", *J. Nanosci. Nanotechnol.*, vol. 16, no. 3, pp. 2418-2424, 2016.
[http://dx.doi.org/10.1166/jnn.2016.12471] [PMID: 27455650]

[70]   Y. Tian, J. Yan, L. Huang, R. Xue, L. Hao, and B. Yi, "Effects of single electrodes of Ni(OH)2 and activated carbon on electrochemical performance of Ni(OH)2–activated carbon asymmetric supercapacitor", *Mater. Chem. Phys.*, vol. 143, no. 3, pp. 1164-1170, 2014.
[http://dx.doi.org/10.1016/j.matchemphys.2013.11.017]

[71]   H. Shen, Y. Zhang, X. Song, Y. Liu, H. Wang, H. Duan, and X. Kong, "Facile hydrothermal synthesis of actiniaria-shaped α-MnO2/activated carbon and its electrochemical performances of supercapacitor", *J. Alloys Compd.*, vol. 770, pp. 926-933, 2019.
[http://dx.doi.org/10.1016/j.jallcom.2018.08.228]

[72]   D. Bejjanki, and S.K. Puttapati, "Easy Synthesis of NiO-Mn2O3@Reduced Graphene Oxide Ternary Composite as Electrode Material for Supercapacitor Application", *J. Electron. Mater.*, vol. 52, no. 7, pp. 4729-4737, 2023.
[http://dx.doi.org/10.1007/s11664-023-10436-4]

[73]   D. Bejjanki, S.K. Puttapati, H. Pant, and V.V.S.S. Srikanth, "Novel MgO/few-layer graphene-filled polyaniline ternary nanocomposite as efficient electrode material in aqueous supercapacitors", *Bull. Mater. Sci.*, vol. 46, no. 1, p. 48, 2023.
[http://dx.doi.org/10.1007/s12034-022-02885-0]

[74]   M. Ghasem Hosseini, and E. Shahryari, "A novel high-performance supercapacitor based on Chitosan/GO-MWCNT/PANI", *J. Colloid Interface Sci.*, vol. 496, pp. 371-381, 2017.
[http://dx.doi.org/10.1016/j.jcis.2017.02.027] [PMID: 28237755]

[75]   M. Deschamps, E. Gilbert, P. Azais, E. Raymundo-Piñero, M.R. Ammar, P. Simon, D. Massiot, and F. Béguin, "Exploring electrolyte organization in supercapacitor electrodes with solid-state NMR", *Nat. Mater.*, vol. 12, no. 4, pp. 351-358, 2013.
[http://dx.doi.org/10.1038/nmat3567] [PMID: 23416727]

[76]   W. Ye, H. Wang, J. Ning, Y. Zhong, and Y. Hu, "New types of hybrid electrolytes for supercapacitors", *Journal of Energy Chemistry*, vol. 57, pp. 219-232, 2021.
[http://dx.doi.org/10.1016/j.jechem.2020.09.016]

[77]   S. Alipoori, S. Mazinani, S.H. Aboutalebi, and F. Sharif, "Review of PVA-based gel polymer electrolytes in flexible solid-state supercapacitors: Opportunities and challenges", *J. Energy Storage*, vol. 27, p. 101072, 2020.
[http://dx.doi.org/10.1016/j.est.2019.101072]

[78]   R. Jamil, and D.S. Silvester, "Ionic liquid gel polymer electrolytes for flexible supercapacitors: Challenges and prospects", *Curr. Opin. Electrochem.*, vol. 35, p. 101046, 2022.
[http://dx.doi.org/10.1016/j.coelec.2022.101046]

[79]   C. Zhong, Y. Deng, W. Hu, J. Qiao, L. Zhang, and J. Zhang, "A review of electrolyte materials and compositions for electrochemical supercapacitors", *Chem. Soc. Rev.*, vol. 44, no. 21, pp. 7484-7539, 2015.
[http://dx.doi.org/10.1039/C5CS00303B] [PMID: 26050756]

[80]   S. Ahankari, D. Lasrado, and R. Subramaniam, "Advances in materials and fabrication of separators in supercapacitors", *Materials Advances*, vol. 3, no. 3, pp. 1472-1496, 2022.
[http://dx.doi.org/10.1039/D1MA00599E]

[81]   C.Y. Bon, L. Mohammed, S. Kim, M. Manasi, P. Isheunesu, K.S. Lee, and J.M. Ko, "Flexible poly(vinyl alcohol)-ceramic composite separators for supercapacitor applications", *J. Ind. Eng. Chem.*, vol. 68, pp. 173-179, 2018.

[http://dx.doi.org/10.1016/j.jiec.2018.07.043]

[82] A. Abdisattar, M. Yeleuov, C. Daulbayev, K. Askaruly, A. Tolynbekov, A. Taurbekov, and N. Prikhodko, "Recent advances and challenges of current collectors for supercapacitors", *Electrochem. Commun.,* vol. 142, p. 107373, 2022.
[http://dx.doi.org/10.1016/j.elecom.2022.107373]

CHAPTER 3

Graphene and its Derivatives: Chemistry, Properties, and Energy Storage Application

Om Prakash[1], Vijay Kumar Juyal[2], Abhishek Pathak[2], Neeraj Kumar[3,6], Vivek Kumar[1], Shivani Verma[5], Akansha Agrwal[4] and Viveka Nand[2,*]

[1] *Regional Ayurveda Research Institute, Ministry of Ayush, Gwalior, 474009, India*

[2] *Department of Chemistry, Govind Ballabh Pant Institute of Agriculture and Technology, Uttarakhand 263145, India*

[3] *School of Studies in Chemistry, Jiwaji University, Gwalior (M.P), India*

[4] *Department of Applied Sciences, KIET Group of Institutions, Delhi-NCR, Meerut Road (NH-58), Ghaziabad 201206, India*

[5] *Department of Chemistry, School of Physical Sciences, Doon University, Dehradun-248012, Uttarakhand, India*

[6] *IPS group of colleges, Shivpuri Link Road, Gwalior (M.P), India*

Abstract: Graphene has attracted a lot of attention in recent years since its discovery because of its unique structural, mechanical, optical, electric, and thermal properties, making it a viable candidate for a wide range of applications. Graphene, a 2-dimensional network of carbon atoms with high conductivity and surface area is a potential material for high-performance applications. For conceivably ground-breaking uses in lithium-ion batteries, solar cells, sensing, and photocatalytic applications, graphene is being used as a filler or composite material with polymers, metals, and metal oxides. Graphene's primary derivatives are graphene oxide (GO) and reduced-graphene oxide (rGO). Graphite can be oxidised to produce GO, and it can be reduced to produce rGO. There is a lot of interest in the application of energy storage in different industries because of the fascinating features of graphene and its derivatives. In the last decade, there has been a lot of interest in the energy storage applications of nanomaterials based on graphene, and numerous groups have started working in this area all over the world. Graphene is perfect for the manufacture of energy storage devices due to its exceptional compatibility, solubility, and selectivity. It is possible to do this, especially if they have been exposed to metal oxide, which causes only minor sheet restacking. The high conductivity of the interconnected networks of graphene is another factor influencing it as a material for energy storage applications.

Keywords: Graphene, Graphene derivatives, Graphene oxide, Reduced Graphene oxide.

* **Corresponding author Viveka Nand:** Department of Chemistry, Govind Ballabh Pant Institute of Agriculture and Technology, Uttarakhand 263145, India; E-mail: imvivekanand@gmail.com

INTRODUCTION

The graphene material and its derivatives are considered frontier materials for future technology [1]. Graphene was discovered in 2004 by Andre Geim and Konstantin Novoselov by the process of scotch tape peeling [2]. It is a hexagonal or honeycomb-like 2D sheet with sp^2 hybridized carbon atoms having a thickness in the order of atom diameter [3]. With its extended honeycomb network, it can be wrapped to form 0D fullerenes, rolled to form 1D nanotubes, and stacked to form 3D graphite. These display extraordinary electrical, mechanical, and thermal properties because of the long-range π-conjugate system which attracted theoretical studies and has become more exciting for experimentalists in recent years [4]. Strong chemical durability, highly ordered structure, high thermal conductivity, abundant surface areas, high Young's modulus, and high electron mobility are characteristics of the ideal graphene [5]. Its aromatic rings, reactive functional groups, and free electrons make graphene materials and their derivatives a fascinating field of study across multiple disciplines (Fig. **1**) [6].

Graphene can be synthesized in two main ways: top-down and bottom-up methods. In top-down graphene production, precursors such as graphite are structurally broken down, then the interlayers are separated and deposited on graphene sheets. Among these methods are oxidation-reduction of GO, arc discharge, liquid phase exfoliation, and mechanical exfoliation. Whereas, carbon source gases are used to synthesize graphene on a substrate in bottom-up methods like epitaxial growth, chemical vapour deposition, and total organic synthesis [1]. The graphene-based materials chemistry, properties, and energy storage application are discussed here. Due to its remarkable optical, electrical, thermal, and mechanical properties, there has been an increase in interest in using them in various biomedical applications during the past several years, including drug delivery systems, biosensors, and imaging systems. Because of this, research using nanomaterials from the graphene family has produced positive outcomes in a number of scientific fields. It is crucial to conduct more research in order to fully understand how these materials interact with biological systems [4]. This chapter has demonstrated the chemistry of graphene and its derivatives with its application in energy storage and other allied fields.

CHEMICAL EXFOLIATION

When compared to other popular methods like epitaxial growth, micromechanical cleavage, and Hummer's methods, chemical exfoliation is thought to be a very efficient and economical top-down synthesis process. Exfoliation of bulk graphite is used in the chemical exfoliation procedure to create graphene. The number of layers and lateral dimension of the resulting graphene may be greatly influenced by choosing an appropriate starting material of graphite. Using reducing solvents or oxidation, graphite layers are separated during chemical exfoliation. By widening the interlayer space between graphite flakes, oxidation or reduction solutions primarily aim to decrease the van der Waals force. As reducing agents, solvents like methanesulfonic acid and hydrazine hydrate N-methyl-2-pyrrolidon have been utilized often.

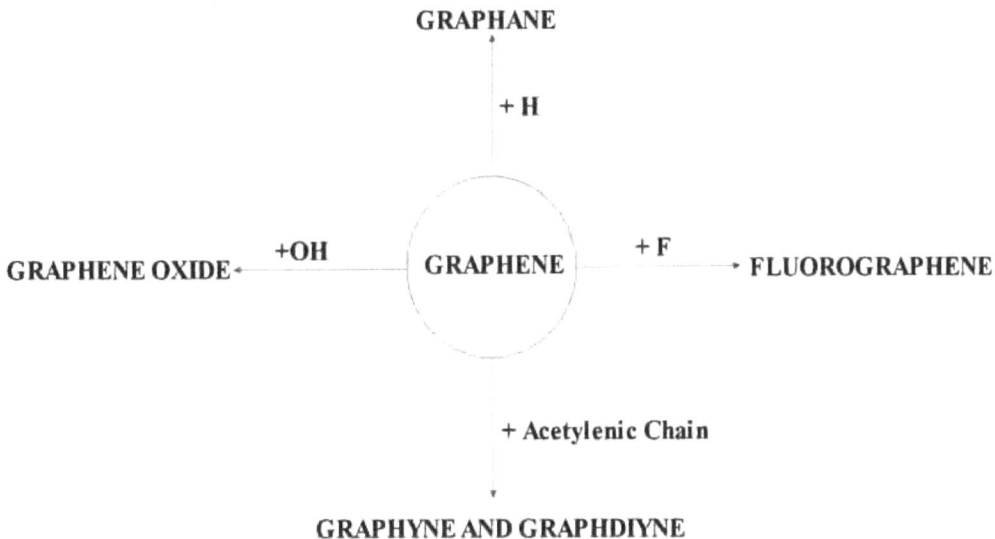

GRAPHANE

+ H

GRAPHENE OXIDE ◄   +OH   GRAPHENE   + F   ► FLUOROGRAPHENE

+ Acetylenic Chain

GRAPHYNE AND GRAPHDIYNE

Fig. (1). Schematic diagram of various derivatives of Graphene.

CHEMISTRY OF GRAPHENE

The chemistry of graphene is the cause of all of its incredible properties. The transparency of graphene is a significant feature. Only 2% of the light shining on it is absorbed by it. This is due to the fact that those atoms are spaced apart because of their thinness. Its strength, which is 200 times stronger than steel, is another significant attribute that may be described using chemistry [7]. Each carbon forms a covalent link with three additional carbon atoms, and these bonds are exceedingly strong. Since it is also incredibly flexible, this is unexpected. One of the best electrical conductors is graphene. This is so that electrons, the building blocks of electricity, can pass through the tightly bound carbon atoms. Because of covalent bonding, and the hexagonal form electrons can flow more quickly. The best heat conductor is known as graphene which is shown in Fig. (**2**).

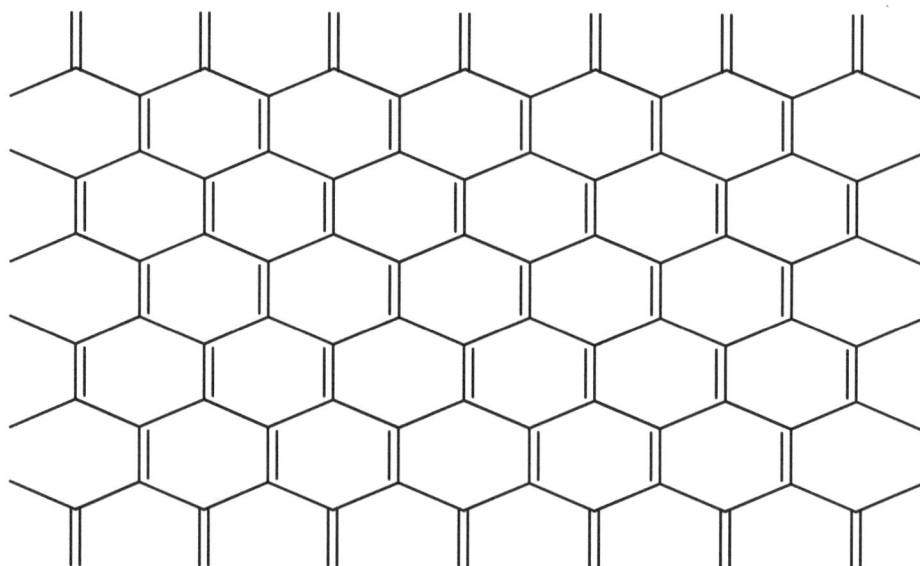

Fig. (2). Structure of Graphene and Shape.

Graphene contains a stable two-dimensional geometrical structure and includes delocalized p-conjugated groups. The distinctive structure and intensely anisotropic features of Graphene lead to a wide range of applications. It depends on how sheets are oriented and dislocated in an ensemble to produce characteristic properties.

Every orientation offers unique technical uses in the areas of sensors, energy-based storage, *etc*. Scientists are interested in it because of its easy synthesis, easily affordable, and handling practices in electronic devices. Its distinctive properties can be retained in various shapes, are highly compatible with industrial processes, and can be produced in large volumes within a reasonable time and effort. A graphene sheet's huge surface area enables p electrons to travel freely over the carbon skeleton, improving conductivity and enabling complete reactive. Typically, a mixture of potent oxidizing chemicals like H_2SO_4 and $KMnO_4$ can open nanotubes into a graphene sheet. Although delocalized p electrons on the surface of pure graphene make it chemically inert, but, band gap can be created by covalent chemistry, using doping, converting to GO, making fluoride or thiolate-based derivatives, converting it to nanoribbons and from its edges [8].

Derivatives of Graphene

Hydrogenated graphenes (graphane), Fuorinated graphenes (fuorographene) and oxidized graphenes (graphene oxide) are some derivatives of graphene which resembles the properties of graphenes. Moreover, acetylene or diacetylene chains between carbon hexagons show properties like graphene and are called graphyne and graphdiyne, respectively. These are not directly synthesized from graphene but resemble graphene with many similarities in structure and properties.

Hydrogenated graphene (graphane)

Hydrogenated graphene is known as graphane which, unlike graphene, consist of sp^3 C-C bond instead of sp^2 C-C bonds frameworks [9]. It was prepared by reacting graphene with H2 plasma at low pressure (~10 Pa) and low temperature (4 - 160 K) using Ar mixed with 10% H2, which is reversible by annealing and plasma irradiation [10]. This hydrogenation process converts conducting graphene to insulator graphane.

Graphane is found in two conformers: chain-like conformer and boat-like conformer. Chain-like conformers have H-atoms alternated on both sides of carbon atoms whereas in boat-like conformers, H-atoms alternating in pairs [9]. The calculated C-C bond length in chair-type conformer is found to be 0.152 nm which is greater than graphene (bond length 0.142) but similar to the diamond. Whereas, two C-C bond lengths (0.152 nm and 0.156 nm) are found in boat type conformer. Here, the short bond length resembles two C-atoms bonded to hydrogen on the opposite side whereas the other bond length is slightly high due to repulsion between two hydrogen atoms, which resemble two C-atoms bonded to the same side of hydrogen [11]. The chair type conformer is the most favoured as the binding energy of chair type conformer is 6.56 eV per atom which is more than boat form with 6.50 eV per atom which is also higher than analogous formula compounds like acetylene (5.90 eV per atom) and benzene (6.49 eV per atom) [9]. As a result, it is considered the most stable hydrocarbon compound with the form of a -CH framework with two-dimensional hydrocarbon units.

Partial or complete hydrogenation of graphene can change its band gap due to the formation of graphane. This is observed due to the hydrogenation of graphene. P-doped graphane can be used as high-temperature superconductors with critical temperatures above 90 °C [12]. It can also be used as hydrogen storage. Any disorder in the process of hydrogenation will lead to the contraction in the lattice constant by 2% [13].

Fluorinated graphene (fluorographene)

Fluorographene is considered a monolayer of graphite fluoride and is found in chair-like conformers (Fig. **3**). It is believed that the boat-type conformation is difficult to obtain because F- atoms alternate on the carbon atom layer in pairs, which gives high repulsion between the carbon atoms. The conversion of

Fluorographene to graphene can be done at high temperatures with potassium iodide [14]. Flurographene was first prepared by graphene on copper foil in the presence of xenon difluoride, the substance was first generated in 2010. Fluorographenes are prepared by reacting graphene with CF_4 [15] or XeF_2 [16] at room temperature and other methods include the chemical and mechanical exfoliation of graphite fluoride [17, 18]. A green method of producing fluorographene also involved the sonochemical exfoliation of graphite fluoride using N-methyl-2-pyrrolidone at ambient temperature [19] and the ultrasonication of GO with HF under hydrothermal conditions [20].

Fluorographene obtained by reaction of graphene with XeF_2 shows a band gap of 3.8 eV which shows luminesces in UV-Vis light and is similar to diamond in its optical properties [21]. Whereas, Fluorographeneprepared by graphene with XeF_2 at 70 0C shows insulating properties and is stable like Teflon up to 400 0C even in the presence of air [16]. A 25% fluorination (C_4F) was found when graphene was partially exposed to XeF_2 gas, whereas full fluorination (CF) was obtained when both sides were exposed to XeF_2 gas [22].

Fluorographene is a powerful insulator, just like other fluorocarbons (such as perfluorohexane). Although fluorographene is reactive in the presence of chemicals, it is thermally stable. The stable conformation of fluorographene consists of trans-linked cyclohexane chairs in an infinite number of covalent C-F bonds stacked in AB sequence. This conformation can be understood from the graphite monofluoride (CF)n structure, which is composed of fluorographene layers weakly bound to each other. Estimated C-F distance, C-C distance, and C-C-C angle are 136-138 pm, 157-158 pm, and 110°, respectively. Fluorographene is regarded as a semiconductor with a wide gap due to its substantially nonlinear I-V characteristics and nearly gate-independent resistance. Recent research has shown that the carrier mobility of Field Effect Transistors with a graphene channel dramatically increases when fluorographene is used as a passivation layer [23].

Oxidized Graphene (Graphene Oxide)

Graphene derivatives with wide bandgaps, like GOs, are semiconductors that potentially take the place of silicon in electronic applications [24]. Graphene sheets can typically be exfoliated by chemical oxidation to produce GO. These oxidizing substances, including HNO_3, H_2SO_4, H_2O_2, and $KMnO_4$ oxidized graphene by affixing epoxide, carboxylate, and hydroxyl groups to produce the material that is known as Graphene Oxide [25].

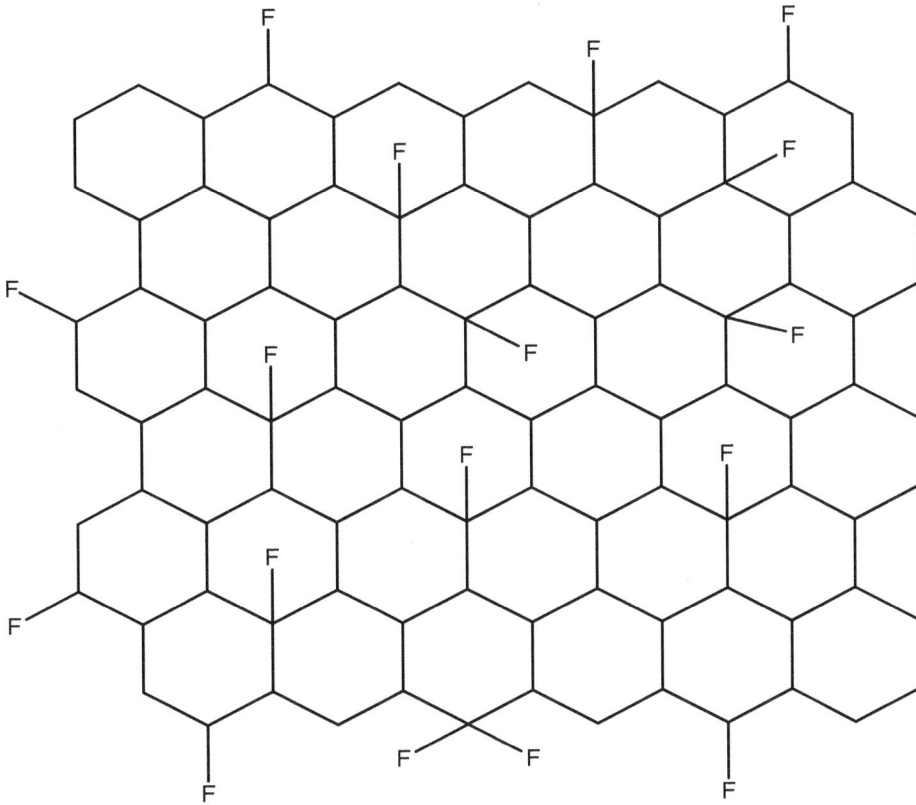

Fig. (3). Structure of fluro graphene.

Numerous oxygen functional groups in Graphene's fully oxidised state lessen the material's electrical conductivity, making it suitable for use as a conductance-based sensor. The removal of oxygen functional groups re-establishes the aromaticity of double-single bonds and can increase the conductivity by many orders of magnitude (Fig. **4**). Since some oxygen-containing groups are still available even after prolonged contact with the reductant, this technique is unable to restore the material's property to a pure Graphene sheet. Due to its dual properties of conductivity and chemical activity (reactive sites), rGO is a strong contender to serve as an active component in the development of sensors. In essence, when the chemical reduction of GO takes place, a large-scale generation of graphene monolayers is produced which shows a 2-3 times reduction in conductivity than that of pure graphene. Lattice vacancies that cannot be filled due to the reduction in GO are responsible for the decrease in conductivity.

On the base plane, in the centre, or at the edges, graphene is covalently connected to functional groups that contain oxygen [26]. GR (GO) is a prospective and economically advantageous choice for field-effect transistors, sensors, optical devices, biomedical applications, renewable energy devices, and energy storage because of the attachment of these functional groups. The various oxidic functional groups found in GO make it very useful for enhancing the efficiency of photoelectrochemical solar cells [27]. Due to its potential to produce graphene in mass quantities, it has attracted great attention [28]. According to the degree of oxidation, GO can either be a semiconductor or an insulator and have a variety of controllable electronic and optical properties and hence can be used in various fields of science.

GO is mainly prepared by Hummer's method. The process consists of oxidizing graphite with $KMnO_4$ and $NaNO_3$ in H_2SO_4, reducing excess $KMnO_4$ to $MnSO_4$ with H_2O_2, and washing with MeOH [29]. Other methods included Brodie method [30] in which fuming HNO_3 solution with $KClO_3$ was used to oxidize graphite. Another method proposed by using H_2SO_4–H_3PO_4 mixture which improves yields mild temperature conditions and no toxic gas excretion [31].

Graphite oxides formed can be of wide range of chemical compositions, such as C8O4.61H6.70, C8O2.54H3.91 [32], C8O3.78–5.05H2.9–4.4 [33], and C8O3.5–4.3H2.5–2.9 [34] depending on the reaction conditions and types of oxygen species attached such as hydroxyl, carboxylic, epoxy, and carbonyl groups.

Graphyne and Graphdiyne

In 1987, Baughman *et al.* theoretically put forward novel compounds such as graphyne and graphyne-family members, which are sp and sp^2 hybridized and made up of one-atom-thick forms of carbon (Fig. **5**) [35]. The members of the graphyne family have a distinctive atom arrangement which gives them several unusual features and makes them interesting materials for a variety of potential applications which include separation of gas, desalination of water, anode material of batteries, storage for H_2, and catalysis applications [36].

In graphyne structure, carbon hexagons linked to linear acetylenic chains whereas are two acetylenic chains are found between carbon hexagons in graphdiyne [37]. In graphyne, acetylenic bonds replace 1/3 of C-C bonds in graphene whereas in graphdiyne two acetylenic linkages are present. In graphyne and graphdiyne, the binding energy was found 7.95 eV and 7.78 eV per atom whereas the lattice parameter was found 0.686 nm and 0.944 nm, respectively [38]. Both graphyne and graphdiyne are semiconductive in nature with band gaps around 0.5-0.6 eV [39]. A different number of acetylenic links was assumed to result in a lower thermal conductivity for graphyne than graphene. Graphyne and graphdiyne thin layers were discussed widely for their mechanical properties as a function of their acetylenic linkages and their arrangements [40].

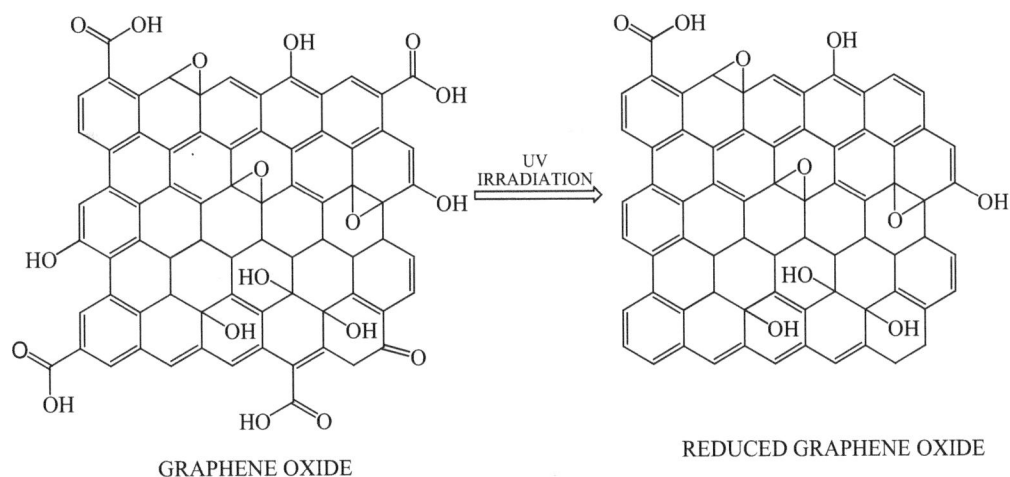

Fig. (4). Reduction of graphene oxide reduced graphene oxide.

The C-C single bonds of graphene can be entirely or partially replaced with acetylenic groups (C≡C) to create these structures. In 2010, Li *et al.* achieved the first graphdiyne film which is a member of the graphyne family, *via* a cross-coupling reaction on the surface of Cu [41]. On ZnO nanorod arrays, the fabrication of graphene films of varying thicknesses was accomplished [42]. Additionally, graphyne has been transformed into several morphologies, including nanotubes, nanowires, and nanowalls. A structure consisting of a hexagonal lattice and a rectangular lattice is among the suggested geometries for graphyne. The rectangular 6,6,12-graphyne lattice may have the greatest promise for future uses of all the theoretical configurations (Fig. 6).

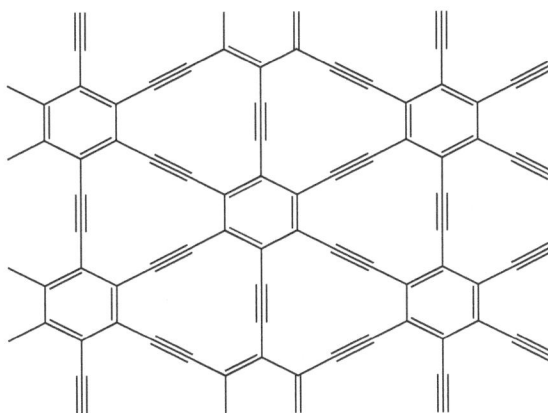

Fig. (5). Chemical structure of graphyne.

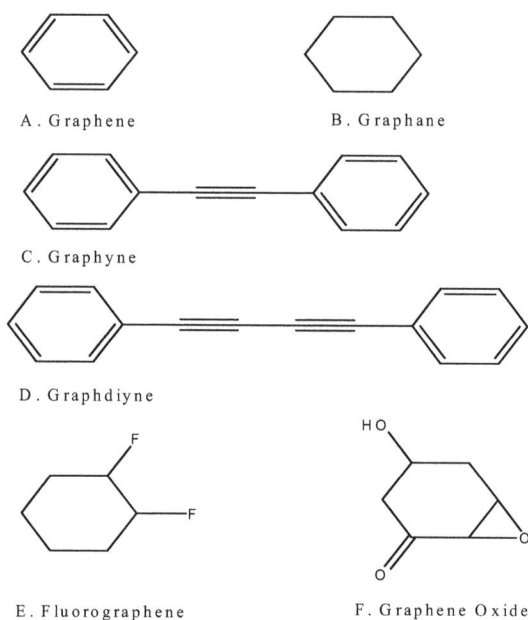

A. Graphene   B. Graphane

C. Graphyne

D. Graphdiyne

E. Fluorographene   F. Graphene Oxide

Fig. (6). Various monomer linkages of Graphene derivatives.

Other Miscellaneous Forms of Graphene

Graphite

Graphite, a crystalline form of carbon, is comprised of several layers of graphene. This natural substance is stable under normal conditions but does not efficiently conduct heat or electricity [43]. It is found in metamorphic rocks and associated with minerals such as quartz, calcite, micas, and tourmaline [44]. Graphite comes in different forms including crystalline, amorphous, lump, and pyrolytic graphite. Crystalline graphite consists of isolated, flat, plate-like particles with unbroken hexagonal edges, while amorphous graphite is made of very fine graphite flakes [45]. HOPG, or highly ordered pyrolytic graphite, has an angular spread of less than 1 degree between the graphite sheets [46]. The structure of graphite is made of trigonal planar carbon sheets arranged in a honeycomb lattice, with a bond length of 0.142 nm and a distance between planes of 0.335 nm [47]. The bonds between the layers are relatively weak van der Waals bonds, which can easily be occupied by gases, enabling the graphene-like layers to separate and glide past each other [48].

Graphite is divided into two main categories: alpha (hexagonal) and beta (rhombohedral) [49]. Despite having similar properties, they differ in the way their graphene layers are stacked. Alpha graphite has ABA stacking, while beta graphite, which is less common and less stable, has ABC stacking. Mechanical treatment can change alpha graphite into beta graphite, and heating above 1300°C can revert the beta form back to the alpha form [46]. In the production of batteries, both natural and synthetic graphite are utilized as anode materials to create electrodes [50]. The demand for graphite, particularly for nickel-metal hydride and lithium-ion batteries, rose in the late 1980s and early 1990s due to the popularity of portable electronics such as CD players and power tools. The growing use of laptops, mobile phones, tablets, and smartphones has further increased the demand for batteries and therefore, graphite. It is expected that the increasing demand for electric vehicle batteries will drive up the demand for graphite even more.

Graphene quantum dots (GQDs)

GQDs are tiny nanoparticles made up of graphene, with a size smaller than 100 nm [51]. They are regarded as a unique material for a variety of sectors including biology, optoelectronics, energy, and the environment due to their extraordinary qualities such as sustained photoluminescence, low toxicity, chemical stability, and noticeable quantum confinement effect [52]. Due to functional groups at their edges, GQDs, which are made up of one or more layers of graphene and are chemically and physically stable with a high surface-to-mass ratio, are simple to disperse in water [53]. The fluorescence emission of GQDs can span a wide range of spectra, from ultraviolet to visible to infrared. The electronic structure of GQDs is highly sensitive to the crystallographic orientation of their edges, with zigzag-edge GQDs displaying metallic behaviour if their diameter is around 7-8 nm [54]. In general, as the number of graphene layers or carbon atoms per graphene layer increases, the energy gap also increases [55].

Graphene quantum dots (GQDs) have been attracting significant attention as a new class of materials with a wide range of applications in various fields, such as energy storage, biomedical imaging, and optoelectronics. The properties of GQDs are determined by the size, shape, and chemical functionalization of the graphene sheets.

Top-down methods involve the reduction of bulk graphitic materials into smaller graphene sheets, which are then further cut into GQDs. These methods include cutting methods such as exfoliation, mechanical grinding, and sonication, as well as chemical reduction methods using potent mixed acids [56]. While these methods are straightforward, they typically require extensive purification due to the use of harsh chemicals.

Contrarily, bottom-up techniques create GQDs from tiny organic molecules like citric acid and glucose [57]. These methods involve the self-assembly of the organic molecules into graphene sheets, followed by cutting them into smaller GQDs. These methods are often more cost-effective and environmentally friendly than top-down methods but may result in a lower yield of GQDs.

GQDs offer a wide range of possible applications despite the many preparation techniques because of their distinct electrical, optical, spin, and photoelectric properties, which are brought on by the quantum confinement effect and edge effect [58]. These properties make GQDs ideal for use in bioimaging, temperature sensing, drug delivery, LED lighter converters, photodetectors, photoluminescent materials, and even in new energy storage systems, such as supercapacitors, lithium-ion batteries, and lithium-sulphur batteries [59].

Overall, GQDs have shown promising potential as a multifunctional material that could have a significant impact on various fields. Further research is needed to fully understand the properties and potential applications of GQDs [60].

Carbon Nanotubes (CNTs)

Carbon nanotubes (CNTs) are indeed one of the derivatives of graphene, and they have gained a lot of attention due to their unique properties and potential applications. CNTs are made of rolled-up graphene sheets and have a tubular structure with a diameter ranging from a few nanometers to several microns [61].

Single-wall carbon nanotubes (SWCNTs) are single-layer tubes with a diameter of less than 1 nanometer, whereas multi-wall carbon nanotubes (MWCNTs) are composed of multiple nested tubes with different diameters. The electrical conductivity of CNTs is highly dependent on the arrangement of atoms within the tubes, and both metallic and semiconducting CNTs can be synthesized [62, 63].

CNTs have high mechanical strength due to the strong covalent bonds between the carbon atoms, and they also have high thermal conductivity, making them useful in various applications such as electronics, energy storage, and thermal management. Additionally, CNTs have a high surface area, making them useful in catalytic reactions, and they can be functionalized with various chemical groups, opening up potential applications in areas such as biomedicine and environmental remediation [64, 65].

Carbon nanotubes (CNTs) are being widely studied for their potential applications in energy storage, especially in lithium-ion batteries and fuel cells. The CVD growth method is a popular method for synthesizing CNTs [66] because it allows for control over the diameter, length, and morphology of the tubes, which can have a significant impact on their properties and performance [67].

In lithium-ion batteries, CNTs are being explored as both an anode material and a conductive additive in composite electrodes. They have the potential to improve the performance of batteries by increasing the energy density and cycle life. The electrochemical properties of CNTs, such as their lower density, higher tensile strength, and higher rigidity, make them suitable for this application.

In fuel cells, CNTs are being studied as catalyst supporters due to their high surface area and excellent electrical conductivity. In fuel cells, a catalyst is required to improve the efficiency of the oxygen reduction reaction (ORR) and oxygen evolution reaction (OER), which are critical for fuel cell efficiency. Porous carbon, graphene, CNTs, and other carbon polymorphs are common materials used as catalyst supporters in fuel cells [68].

In conclusion, CNTs have a promising potential in energy storage applications, and ongoing research is focused on optimizing their performance and developing new materials and methods for their synthesis and utilization.

Fullerene

Fullerenes have unique properties and have been the subject of much research and study since their discovery. They have a high degree of stability, making them ideal for use as a stable structure for chemical and biological compounds. Additionally, fullerenes have unique electronic and optical properties, making them useful for applications in areas such as electronics, optics, and catalysis [69]. In the field of medicine, fullerenes have shown potential as a delivery system for drugs and as a protective agent against oxidative stress. In electronics, they have been used as electron acceptors in solar cells, and as a host material for the storage of hydrogen.

The production of fullerenes remains an active area of research, with scientists exploring new methods to improve the yields and purity of these fascinating materials [70]. As our understanding of fullerenes and their properties continues to grow, it is likely that new and exciting applications will emerge, making them a valuable area of study in materials science and beyond. In recent years, research has focused on the use of fullerenes in lithium-ion batteries, sodium-ion batteries, and supercapacitors [71]. For example, the addition of fullerene to the cathode of a lithium-ion battery has been shown to increase the battery's capacity and reduce its polarization, thus improving the battery's performance and stability. Additionally, fullerene derivatives have been used as cathode materials in sodium-ion batteries, showing promising results in terms of capacity and rate performance. Furthermore, the high electron-acceptance capacity of fullerene makes it a suitable candidate for use in supercapacitors, where it can store and release large amounts of energy rapidly [72, 73].

Despite the promising applications of fullerenes in rechargeable batteries and other energy storage devices, there are still some challenges that need to be overcome [74]. One of the major challenges is the relatively high cost of fullerene production, which makes it difficult to scale up its use for commercial applications. Additionally, further research is needed to fully understand the mechanisms behind the behaviour of fullerene in energy storage devices, in order to develop more efficient and durable energy storage systems based on fullerene technology [75].

In conclusion, fullerene holds great potential in the field of rechargeable batteries and other energy storage devices. Despite some challenges, the unique properties of fullerene, such as its solubility in organic solvents [76], high electron-acceptance capacity, and versatility for functionalization, make it a promising candidate for further research and development [77].

Applications of Graphene and its derivatives

Supercapacitors

Graphene is used in energy accumulator systems like cells and supercapacitors. It is frequently employed as the anode in Li-ion batteries and has a magnitude of roughly 1000 mAh g-1, which is threefold greater than the graphite electrode. Additionally, Graphene provides batteries with a longer lifespan and a recharge time of only a few seconds. Since it is flexible, it can be produced as a solid-state supercapacitor in textiles for ready-to-wear electronics [78]. This material shows good steadiness and a specific capacitance of roughly 100-264 F g^{-1}. The electrochemical stability and rate capabilities may be enhanced by adding functional species to the graphene matrix. Specific capacitance of roughly 300–1500 F g^{-1} is present in graphene-based organic hybrid materials with conducting polymers, for instance, polypyrrole, polyaniline (PANI), and poly(ethylenedioxythiophene) (PEDOT). By enhancing its electrical double-layer capacitance and pseudocapacitance, graphene-based supercapacitors could perform better. One efficient approach to reducing electrical double-layer capacitance is to maximise the surface area and materials' porosity. Moreover, the introduction of useful species with redox abilities might boost pseudocapacitance and even have synergistic effects on performance [79].

Lithium-ion battery

An electrochemical energy depository system based on the intercalation and deintercalation of Li ions, a Li-ion battery is typically composed of an anode, a cathode, and an electrolyte (separator). Porosity has a significant impact on how well the electrode materials operate. The vast surface area that graphene offers allows functional species to grow or deposit, which enhances the hybrid materials' porosity. Additionally, the porosity boosts the movement of electrolytes, which is advantageous for charging and discharging the battery's system. The electrochemical performances would also be affected by the electron exchange between the graphene and components [80]. The low energy density of lithium-ion batteries, the subpar cycle performance, and the increased cost are all becoming increasingly evident as science and technology advance. As a result, research has focused on creating graphene energy storage systems to address the numerous drawbacks of lithium-ion batteries [81].

Solar Cells

Due to its high electron movement, high thermal stability, and great flexibility, graphene is considered a next-generation conducting substance that may restore indium tin oxide and fluorine tin oxide in optoelectronic devices. The specialty of high electron mobility, which improves charge separation and hole transportation in photovoltaic systems is the main justification for the usage of graphene-based materials as electrodes in these devices [82]. In comparison to other materials, graphene demonstrates a large range of novel and occasionally enigmatic optical and electrical properties like zero band-gap semi-conductivity, great optical transparency, high carrier movement, and high tensile strength [83].

Fuel Cells

The catalyst has a significant impact on the price and longevity of fuel cells. Due to its great surface area and stability, carbon is frequently employed as the supporting material for catalyst nanoparticles in order to increase catalytic efficiency [84]. A catalyst can be supported on mutual sides of graphene sheets because it is a 2D structured carbon material with a huge surface area. Additionally, since its discovery, graphene has demonstrated remarkable performance as a support material because of its mechanical, structural, and electrical properties. One type of the most popular hybrid electrocatalysts for fuel cells is graphene-supported noble metal nanoparticles [85]. Applications of graphene derivatives as an active component of fuel cells have been shown to offer numerous benefits. The utility of graphene-based materials as electrocatalysts and fuel oxidation seems promising due to their more surface area and intense conductive characteristics. Additionally, strong tensile strength, minimal fuel permeability, and high ionic conductivity are all characteristics of polymer membranes that have been coupled with graphene. Bipolar plates' conductivity and corrosion resistance can be enhanced with graphene [86].

Water Filtration

Depending upon the creation size and the high pressure, graphene nanoporous membranes can be utilised for water filtration with a performance of 33% to 100%. A total of 97% of the NaCl in saltwater may be rejected by graphene membranes. Upon direct contact with cells, nanocarbon-based materials like graphene and CNTs can also stop the growth of germs [87]. Due to the extraordinary qualities demonstrated by the mixtures like greater absorptivity, specific conductance, tuneable optical behaviour, and lifespan, a number of researchers are working on the use of graphene/metal oxide for water purification [88].

CONCLUSION

Graphene and its derivatives (Graphene oxide, reduced Graphene oxide, Graphene, Graphyne, and Flurographene) have attracted considerable attention in fields of medicine, engineering, and other fields of science. Flurographene due to its wide band gap may be used in Field Effect Transistors (FET) whereas GO is useful in photoelectrochemical solar cells. Graphyne can be utilised as semiconductors due to low band gap and graphane can work as superconductors. Graphite is used as an anode material in battery electrodes. Graphene quantum dots are useful in solar cells. The excellent electrochemical characteristics of CNT, such as their lower density, higher tensile strength, and higher rigidity, are used in energy storage research. Fullerenes are helpful during battery cycling because of their excellent redox chemistry and electron-accepting capabilities. The lithium-ion battery. The applications of graphene and its derivatives in energy storage are gaining popularity due to their strong tensile strength, highly conductive characteristics, high ionic conductivity, and minimal fuel permeability. Other application comprises water purification, durable lightweight material synthesis, sensitive medical appliances, catalytic reactor, electrical sensors, and targeted drug delivery agents.

FUTURE OUTLOOK

Efficient electricity storage devices are essentials for the future due to increasing demands of energy and durable materials are needed for building appliances and infrastructures. Graphene and its derivatives are the solution that shows the feasible way which utilise its nano-scale properties for practical macro-scale applications. However, the cost-effectiveness and environmental risks are also associated with it but the researchers are ongoing to attain a cost-effective and environment-friendly high-quality single graphene layer so that drawbacks may be minimized. Still, more research is needed to establish graphene and its derivatives for practical applications in the future.

REFERENCES

[1]   X.J. Lee, B.Y.Z. Hiew, K.C. Lai, L.Y. Lee, S. Gan, S. Thangalazhy-Gopakumar, and S. Rigby, "Review on graphene and its derivatives: Synthesis methods and potential industrial implementation", *J. Taiwan Inst. Chem. Eng.,* vol. 98, pp. 163-180, 2019.
[http://dx.doi.org/10.1016/j.jtice.2018.10.028]

[2]   R. Rudrapati, "Graphene: Fabrication methods, properties, and applications in modern industries", In: *Graphene Production and Application.* IntechOpen: London, United Kingdom, 2020.
[http://dx.doi.org/10.5772/intechopen.92258]

[3]   K.S. Novoselov, A.K. Geim, S.V. Morozov, D.E. Jiang, Y. Zhang, S.V. Dubonos, I.V. Grigorieva, and A.A. Firsov, "Electric field effect in atomically thin carbon films", In: *science* vol. 306. , 2004, no. 5696, pp. 666-669.
[http://dx.doi.org/10.1126/science.1102896]

[4]   M.J. Allen, V.C. Tung, and R.B. Kaner, "Honeycomb carbon: A review of graphene", *Chem. Rev.,* vol. 110, no. 1, pp. 132-145, 2010.
[http://dx.doi.org/10.1021/cr900070d] [PMID: 19610631]

[5]   K.S. Novoselov, L. Colombo, P.R. Gellert, M.G. Schwab, and K. Kim, "A roadmap for graphene", In: *nature* vol. 490. , 2012, no. 7419, pp. 192-200.
[http://dx.doi.org/10.1038/nature11458]

[6]   J.C. Meyer, A.K. Geim, M.I. Katsnelson, K.S. Novoselov, T.J. Booth, and S. Roth, "The structure of suspended graphene sheets", *Nature,* vol. 446, no. 7131, pp. 60-63, 2007.
[http://dx.doi.org/10.1038/nature05545] [PMID: 17330039]

[7]   A. Sinitskii, and J.M. Tour, "Graphene electronics, unzipped", *IEEE Spectr.,* vol. 47, no. 11, pp. 28-33, 2010.
[http://dx.doi.org/10.1109/MSPEC.2010.5605889]

[8]   O.V. Prezhdo, P.V. Kamat, and G.C. Schatz, "Virtual issue: Graphene and functionalized graphene", *J. Phys. Chem. C,* vol. 115, no. 8, pp. 3195-3197, 2011.

[9]   M.H.F. Sluiter, and Y. Kawazoe, "Cluster expansion method for adsorption: Application to hydrogen chemisorption on graphene", *Phys. Rev. B Condens. Matter,* vol. 68, no. 8, p. 085410, 2003.
[http://dx.doi.org/10.1103/PhysRevB.68.085410]

[10]   D.C. Elias, R.R. Nair, T.M.G. Mohiuddin, S.V. Morozov, P. Blake, M.P. Halsall, A.C. Ferrari, D.W. Boukhvalov, M.I. Katsnelson, A.K. Geim, and K.S. Novoselov, "Control of graphene's properties by reversible hydrogenation: evidence for graphane", *Science,* vol. 323, no. 5914, pp. 610-613, 2009.
[http://dx.doi.org/10.1126/science.1167130] [PMID: 19179524]

[11]   J.O. Sofo, A.S. Chaudhari, and G.D. Barber, "Graphane: A two-dimensional hydrocarbon", *Phys. Rev. B Condens. Matter Mater. Phys.,* vol. 75, no. 15, p. 153401, 2007.
[http://dx.doi.org/10.1103/PhysRevB.75.153401]

[12]   G. Savini, A.C. Ferrari, and F. Giustino, "First-principles prediction of doped graphane as a high-temperature electron-phonon superconductor", *Phys. Rev. Lett.,* vol. 105, no. 3, p. 037002, 2010.
[http://dx.doi.org/10.1103/PhysRevLett.105.037002] [PMID: 20867792]

[13]   L. Feng Huang, and Z. Zeng, "Lattice dynamics and disorder-induced contraction in functionalized graphene", *J. Appl. Phys.,* vol. 113, no. 8, p. 083524, 2013.
[http://dx.doi.org/10.1063/1.4793790]

[14]   Villamanca, D., Colin, M., Ching, K., Rawal, A., Wu, Y., Kim, D.J., Dubois, M. and Chen, S., "Preparation and properties of graphene oxyfluoride films", *Applied Surface Science,* vol. 646, p. 158822, 2024.

[15]   W. Feng, P. Long, Y. Feng, and Y. Li, "Two-dimensional fluorinated graphene: Synthesis, structures, properties and applications", *Adv. Sci. (Weinh.),* vol. 3, no. 7, p. 1500413, 2016.
[http://dx.doi.org/10.1002/advs.201500413] [PMID: 27981018]

[16]   R.R. Nair, W. Ren, R. Jalil, I. Riaz, V.G. Kravets, L. Britnell, P. Blake, F. Schedin, A.S. Mayorov, S. Yuan, and M.I. Katsnelson, "Fluorographene: A two-dimensional counterpart of Teflon", In: *small* vol. 6. , 2010, no. 4, pp. 2877-2884.

[17]   R. Zbořil, F. Karlický, A.B. Bourlinos, T.A. Steriotis, A.K. Stubos, V. Georgakilas, K. Safarova, D. Jancik, C. Trapalis, and M. Otyepka, "Graphene fluoride: A stable stoichiometric graphene derivative and its chemical conversion to graphene", In: *small* vol. 6. , 2010, no. 24, pp. 2885-2891.

[18]   H. Chang, J. Cheng, X. Liu, J. Gao, M. Li, J. Li, X. Tao, F. Ding, and Z. Zheng, "Facile synthesis of wide-bandgap fluorinated graphene semiconductors", *Chemistry,* vol. 17, no. 32, pp. 8896-8903, 2011. [http://dx.doi.org/10.1002/chem.201100699] [PMID: 21714019]

[19]   P. Gong, Z. Wang, J. Wang, H. Wang, Z. Li, Z. Fan, Y. Xu, X. Han, and S. Yang, "One-pot sonochemical preparation of fluorographene and selective tuning of its fluorine coverage", *J. Mater. Chem.,* vol. 22, no. 33, pp. 16950-16956, 2012. [http://dx.doi.org/10.1039/c2jm32294c]

[20]   Z. Wang, J. Wang, Z. Li, P. Gong, X. Liu, L. Zhang, J. Ren, H. Wang, and S. Yang, "Synthesis of fluorinated graphene with tunable degree of fluorination", *Carbon,* vol. 50, no. 15, pp. 5403-5410, 2012. [http://dx.doi.org/10.1016/j.carbon.2012.07.026]

[21]   K.J. Jeon, Z. Lee, E. Pollak, L. Moreschini, A. Bostwick, C.M. Park, R. Mendelsberg, V. Radmilovic, R. Kostecki, T.J. Richardson, and E. Rotenberg, "Fluorographene: A wide bandgap semiconductor with ultraviolet luminescence", *ACS Nano,* vol. 5, no. 2, pp. 1042-1046, 2011. [http://dx.doi.org/10.1021/nn1025274] [PMID: 21204572]

[22]   J.T. Robinson, J.S. Burgess, C.E. Junkermeier, S.C. Badescu, T.L. Reinecke, F.K. Perkins, M.K. Zalalutdniov, J.W. Baldwin, J.C. Culbertson, P.E. Sheehan, and E.S. Snow, "Properties of fluorinated graphene films", *Nano Lett.,* vol. 10, no. 8, pp. 3001-3005, 2010. [http://dx.doi.org/10.1021/nl101437p] [PMID: 20698613]

[23]   K.I. Ho, M. Boutchich, C.Y. Su, R. Moreddu, E.S.R. Marianathan, L. Montes, and C.S. Lai, "A self-aligned high-mobility graphene transistor: decoupling the channel with fluorographene to reduce scattering", *Adv. Mater.,* vol. 27, no. 41, pp. 6519-6525, 2015. [http://dx.doi.org/10.1002/adma.201502544] [PMID: 26398725]

[24]   N. Lu, Z. Li, and J. Yang, "Electronic structure engineering via on-plane chemical functionalization: A comparison study on two-dimensional polysilane and graphane", *J. Phys. Chem. C,* vol. 113, no. 38, pp. 16741-16747, 2009. [http://dx.doi.org/10.1021/jp904208g]

[25]   M.M. Haley, S.C. Brand, and J.J. Pak, "Carbon networks based on dehydrobenzoannulenes: Synthesis of graphdiyne substructures", *Angew. Chem. Int. Ed. Engl.,* vol. 36, no. 8, pp. 836-838, 1997. [http://dx.doi.org/10.1002/anie.199708361]

[26]   M.M. Haley, "Synthesis and properties of annulenic subunits of graphyne and graphdiyne nanoarchitectures", *Pure Appl. Chem.,* vol. 80, no. 3, pp. 519-532, 2008. [http://dx.doi.org/10.1351/pac200880030519]

[27]   A.N. Banerjee, "Graphene and its derivatives as biomedical materials: Future prospects and challenges", *Interface Focus,* vol. 8, no. 3, p. 20170056, 2018. [http://dx.doi.org/10.1098/rsfs.2017.0056] [PMID: 29696088]

[28]   P. Blake, E.W. Hill, A.H. Castro Neto, K.S. Novoselov, D. Jiang, R. Yang, T.J. Booth, and A.K. Geim, "Making graphene visible", *Appl. Phys. Lett.,* vol. 91, no. 6, p. 063124, 2007. [http://dx.doi.org/10.1063/1.2768624]

[29]   W.S. Hummers Jr, and R.E. Offeman, "Preparation of graphitic oxide", *J. Am. Chem. Soc.,* vol. 80, no. 6, pp. 1339-1339, 1958. [http://dx.doi.org/10.1021/ja01539a017]

[30]   B.C. Brodie, "XIII. On the atomic weight of graphite", *Philos. Trans. R. Soc. Lond.,* vol. 149, pp. 249-259, 1859.
        [http://dx.doi.org/10.1098/rstl.1859.0013]

[31]   D.C. Marcano, D.V. Kosynkin, J.M. Berlin, A. Sinitskii, Z. Sun, A. Slesarev, L.B. Alemany, W. Lu, and J.M. Tour, "Improved synthesis of graphene oxide", *ACS Nano,* vol. 4, no. 8, pp. 4806-4814, 2010.
        [http://dx.doi.org/10.1021/nn1006368] [PMID: 20731455]

[32]   T. Nakajima, and Y. Matsuo, "Formation process and structure of graphite oxide", *Carbon,* vol. 32, no. 3, pp. 469-475, 1994.
        [http://dx.doi.org/10.1016/0008-6223(94)90168-6]

[33]   R. Yazami, P. Touzain, Y. Chabre, D. Berger, and M. Coulon, "Lithium-graphite oxide cells. I. Characteristics and dehydration study of graphite oxide,", In: *J. Chem Miner.* vol. 22. , 1985, no. 3, pp. 398-411.

[34]   S.K. Srivastava, and J. Pionteck, "Recent advances in preparation, structure, properties and applications of graphite oxide", *J. Nanosci. Nanotechnol.,* vol. 15, no. 3, pp. 1984-2000, 2015.
        [http://dx.doi.org/10.1166/jnn.2015.10047] [PMID: 26413611]

[35]   R.H. Baughman, H. Eckhardt, and M. Kertesz, "Structure-property predictions for new planar forms of carbon: Layered phases containing s p 2 and s p atoms", *J. Chem. Phys.,* vol. 87, no. 11, pp. 6687-6699, 1987.
        [http://dx.doi.org/10.1063/1.453405]

[36]   B.G. Kim, and H.J. Choi, "Graphyne: Hexagonal network of carbon with versatile Dirac cones", *Phys. Rev. B Condens. Matter Mater. Phys.,* vol. 86, no. 11, p. 115435, 2012.
        [http://dx.doi.org/10.1103/PhysRevB.86.115435]

[37]   M.J. McAllister, J.L. Li, D.H. Adamson, H.C. Schniepp, A.A. Abdala, J. Liu, M. Herrera-Alonso, D.L. Milius, R. Car, R.K. Prud'homme, and I.A. Aksay, "Single sheet functionalized graphene by oxidation and thermal expansion of graphite", *Chem. Mater.,* vol. 19, no. 18, pp. 4396-4404, 2007.
        [http://dx.doi.org/10.1021/cm0630800]

[38]   K.P. Loh, Q. Bao, G. Eda, and M. Chhowalla, "Graphene oxide as a chemically tunable platform for optical applications", *Nat. Chem.,* vol. 2, no. 12, pp. 1015-1024, 2010.
        [http://dx.doi.org/10.1038/nchem.907] [PMID: 21107364]

[39]   Y. Zhu, S. Murali, W. Cai, X. Li, J.W. Suk, J.R. Potts, and R.S. Ruoff, "Graphene and graphene oxide: Synthesis, properties, and applications", *Adv. Mater.,* vol. 22, no. 35, pp. 3906-3924, 2010.
        [http://dx.doi.org/10.1002/adma.201001068] [PMID: 20706983]

[40]   S.W. Cranford, D.B. Brommer, and M.J. Buehler, "Extended graphynes: Simple scaling laws for stiffness, strength and fracture", *Nanoscale,* vol. 4, no. 24, pp. 7797-7809, 2012.
        [http://dx.doi.org/10.1039/c2nr31644g] [PMID: 23142928]

[41]   G. Li, Y. Li, H. Liu, Y. Guo, Y. Li, and D. Zhu, "Architecture of graphdiyne nanoscale films", *Chem. Commun.,* vol. 46, no. 19, pp. 3256-3258, 2010.
        [http://dx.doi.org/10.1039/b922733d] [PMID: 20442882]

[42]   X. Qian, H. Liu, C. Huang, S. Chen, L. Zhang, Y. Li, J. Wang, and Y. Li, "Self-catalyzed growth of large-area nanofilms of two-dimensional carbon", *Sci. Rep.,* vol. 5, no. 1, p. 7756, 2015.
        [http://dx.doi.org/10.1038/srep07756] [PMID: 25583680]

[43]   N.N. Greenwood, and A. Earnshaw, *Chemistry of the Elements.* Elsevier, 2012.

[44]   O. Yenigun, and M. Barisik, "Local heat transfer control using liquid dielectrophoresis at graphene/water interfaces", *Int. J. Heat Mass Transf.,* vol. 166, p. 120801, 2021.
        [http://dx.doi.org/10.1016/j.ijheatmasstransfer.2020.120801]

[45]   D.M. Sutphin, and J.D. Bliss, "Disseminated flake graphite and amorphous graphite deposit types. An

analysis using grade and tonnage models", *CIM Bull.,* vol. 83, no. 940, pp. 85-89, 1990.

[46]   A.D. McNaught, and A. Wilkinson, *Compendium of chemical terminology.* vol. 1669. Blackwell Science Oxford, 1997.

[47]   P. Delhaes, *Graphite and precursors.* CRC Press, 2000.
[http://dx.doi.org/10.1201/9781482296921]

[48]   D.D.L. Chung, "Review graphite", *J. Mater. Sci.,* vol. 37, no. 8, pp. 1475-1489, 2002.
[http://dx.doi.org/10.1023/A:1014915307738]

[49]   T. Latychevskaia, S-K. Son, Y. Yang, D. Chancellor, M. Brown, S. Ozdemir, I. Madan, G. Berruto, F. Carbone, A. Mishchenko, and K.S. Novoselov, "Stacking transition in rhombohedral graphite", *Front. Phys.,* vol. 14, no. 1, p. 13608, 2019.
[http://dx.doi.org/10.1007/s11467-018-0867-y]

[50]   M. Gutierrez, M. Morcrette, L. Monconduit, Y. Oudart, P. Lemaire, C. Davoisne, N. Louvain, and R. Janot, "Towards a better understanding of the degradation mechanisms of Li-ion full cells using Si/C composites as anode", *J. Power Sources,* vol. 533, p. 231408, 2022.
[http://dx.doi.org/10.1016/j.jpowsour.2022.231408]

[51]   S. Ghosh, B. Sachdeva, P. Sachdeva, V. Chaudhary, G.M. Rani, and J.K. Sinha, "Graphene quantum dots as a potential diagnostic and therapeutic tool for the management of Alzheimer's disease", *Carbon Letters,* vol. 32, no. 6, pp. 1381-1394, 2022.
[http://dx.doi.org/10.1007/s42823-022-00397-9]

[52]   T.K. Henna, and K. Pramod, "Graphene quantum dots redefine nanobiomedicine", *Mater. Sci. Eng. C,* vol. 110, p. 110651, 2020.
[http://dx.doi.org/10.1016/j.msec.2020.110651] [PMID: 32204078]

[53]   P. Tian, L. Tang, K.S. Teng, and S.P. Lau, "Graphene quantum dots from chemistry to applications", *Mater. Today Chem.,* vol. 10, pp. 221-258, 2018.
[http://dx.doi.org/10.1016/j.mtchem.2018.09.007]

[54]   K.A. Ritter, and J.W. Lyding, "The influence of edge structure on the electronic properties of graphene quantum dots and nanoribbons", *Nat. Mater.,* vol. 8, no. 3, pp. 235-242, 2009.
[http://dx.doi.org/10.1038/nmat2378] [PMID: 19219032]

[55]   C. Wimmenauer, J. Scheller, S. Fasbender, and T. Heinzel, "Single-particle energy – and optical absorption – spectra of multilayer graphene quantum dots", *Superlattices Microstruct.,* vol. 132, p. 106171, 2019.
[http://dx.doi.org/10.1016/j.spmi.2019.106171]

[56]   L. Tang, R. Ji, X. Cao, J. Lin, H. Jiang, X. Li, K.S. Teng, C.M. Luk, S. Zeng, J. Hao, and S.P. Lau, "Deep ultraviolet photoluminescence of water-soluble self-passivated graphene quantum dots", *ACS Nano,* vol. 6, no. 6, pp. 5102-5110, 2012.
[http://dx.doi.org/10.1021/nn300760g] [PMID: 22559247]

[57]   S. Wang, Z.G. Chen, I. Cole, and Q. Li, "Structural evolution of graphene quantum dots during thermal decomposition of citric acid and the corresponding photoluminescence", *Carbon,* vol. 82, pp. 304-313, 2015.
[http://dx.doi.org/10.1016/j.carbon.2014.10.075]

[58]   A.D. Güçlü, P. Potasz, and P. Hawrylak, "Electric-field controlled spin in bilayer triangular graphene quantum dots", *Phys. Rev. B Condens. Matter Mater. Phys.,* vol. 84, no. 3, p. 035425, 2011.
[http://dx.doi.org/10.1103/PhysRevB.84.035425]

[59]   M.K. Kumawat, M. Thakur, R.B. Gurung, and R. Srivastava, "Graphene quantum dots from mangifera indica: application in near-infrared bioimaging and intracellular nanothermometry", *ACS Sustain. Chem.& Eng.,* vol. 5, no. 2, pp. 1382-1391, 2017.
[http://dx.doi.org/10.1021/acssuschemeng.6b01893]

[60]   F. Shi, and Q. Liu, "Recent advances on the application of graphene quantum dots in energy storage",

Recent Pat. Nanotechnol., vol. 15, no. 4, pp. 298-309, 2021.
[http://dx.doi.org/10.2174/1872210515666210120115159] [PMID: 33494687]

[61]   S. Iijima, "Helical microtubules of graphitic carbon", In: *Nature* vol. 354. , 1991, no. 6348, pp. 56-58.
[http://dx.doi.org/10.1038/354056a0]

[62]   J.W. Mintmire, B.I. Dunlap, and C.T. White, "Are fullerene tubules metallic?", *Phys. Rev. Lett.,* vol.
68, no. 5, pp. 631-634, 1992.
[http://dx.doi.org/10.1103/PhysRevLett.68.631] [PMID: 10045950]

[63]   J.W.G. Wilder, L.C. Venema, A.G. Rinzler, R.E. Smalley, and C. Dekker, "Electronic structure of
atomically resolved carbon nanotubes,", In: *Nature* vol. 391. , 1998, no. 6662, pp. 59-62.
[http://dx.doi.org/10.1038/34139]

[64]   M-F. Yu, O. Lourie, M.J. Dyer, K. Moloni, T.F. Kelly, and R.S. Ruoff, "Strength and breaking
mechanism of multiwalled carbon nanotubes under tensile load", In: *Science* vol. 287. , 2000, no.
5453, pp. 637-640.
[http://dx.doi.org/10.1126/science.287.5453.637]

[65]   Y-K. Kwon, and P. Kim, *Unusually high thermal conductivity in carbon nanotubes.* High Therm.
Conduct. Mater, 2006, pp. 227-265.
[http://dx.doi.org/10.1007/0-387-25100-6_8]

[66]   P. Nikolaev, "Gas-phase production of single-walled carbon nanotubes from carbon monoxide: A
review of the hipco process", *J. Nanosci. Nanotechnol.,* vol. 4, no. 4, pp. 307-316, 2004.
[http://dx.doi.org/10.1166/jnn.2004.066] [PMID: 15296221]

[67]   R. Carter, L. Oakes, A.P. Cohn, J. Holzgrafe, H.F. Zarick, S. Chatterjee, R. Bardhan, and C.L. Pint,
"Solution assembled single-walled carbon nanotube foams: Superior performance in supercapacitors,
lithium-ion, and lithium–air batteries", *J. Phys. Chem. C,* vol. 118, no. 35, pp. 20137-20151, 2014.
[http://dx.doi.org/10.1021/jp5054502]

[68]   R. Amin, P.R. Kumar, and I. Belharouak, "Carbon nanotubes: Applications to energy storage devices",
Carbon Nanotub. World Electron., vol. 10, pp. 5772-94155, 2020.

[69]   A.F. Hebard, "Buckminsterfullerene", *Annu. Rev. Mater. Sci.,* vol. 23, no. 1, pp. 159-191, 1993.
[http://dx.doi.org/10.1146/annurev.ms.23.080193.001111]

[70]   P.W. Dunk, J.J. Adjizian, N.K. Kaiser, J.P. Quinn, G.T. Blakney, C.P. Ewels, A.G. Marshall, and
H.W. Kroto, "Metallofullerene and fullerene formation from condensing carbon gas under conditions
of stellar outflows and implication to stardust", *Proc. Natl. Acad. Sci. USA,* vol. 110, no. 45, pp.
18081-18086, 2013.
[http://dx.doi.org/10.1073/pnas.1315928110] [PMID: 24145444]

[71]   Q. Dong, C.H.Y. Ho, H. Yu, A. Salehi, and F. So, "Defect passivation by fullerene derivative in
perovskite solar cells with aluminum-doped zinc oxide as electron transporting layer", *Chem. Mater.,*
vol. 31, no. 17, pp. 6833-6840, 2019.
[http://dx.doi.org/10.1021/acs.chemmater.9b01292]

[72]   Z. Du, N. Jannatun, D. Yu, J. Ren, W. Huang, and X. Lu, "C 60-Decorated nickel–cobalt phosphide as
an efficient and robust electrocatalyst for hydrogen evolution reaction", In: *Nanoscale* vol. 10. , 2018,
no. 48, pp. 23070-23079.

[73]   M. Di Giosia, P.H.H. Bomans, A. Bottoni, A. Cantelli, G. Falini, P. Franchi, G. Guarracino, H.
Friedrich, M. Lucarini, F. Paolucci, S. Rapino, N.A.J.M. Sommerdijk, A. Soldà, F. Valle, F. Zerbetto,
and M. Calvaresi, "Proteins as supramolecular hosts for C 60 : A true solution of C 60 in water",
Nanoscale, vol. 10, no. 21, pp. 9908-9916, 2018.
[http://dx.doi.org/10.1039/C8NR02220H] [PMID: 29790558]

[74]   Z. Jiang, Y. Zhao, X. Lu, and J. Xie, "Fullerenes for rechargeable battery applications: Recent
developments and future perspectives", *Journal of Energy Chemistry,* vol. 55, pp. 70-79, 2021.
[http://dx.doi.org/10.1016/j.jechem.2020.06.065]

[75] M. Chen, R. Guan, and S. Yang, "Hybrids of fullerenes and 2D nanomaterials", *Adv. Sci.,* vol. 6, no. 1, p. 1800941, 2019.
[http://dx.doi.org/10.1002/advs.201800941] [PMID: 30643712]

[76] R.S. Ruoff, D.S. Tse, R. Malhotra, and D.C. Lorents, "Solubility of fullerene (C60) in a variety of solvents", *J. Phys. Chem.,* vol. 97, no. 13, pp. 3379-3383, 1993.
[http://dx.doi.org/10.1021/j100115a049]

[77] A. Palkar, A. Kumbhar, A.J. Athans, and L. Echegoyen, "Pyridyl-Functionalized and water-soluble carbon nano onions: First supramolecular complexes of carbon nano onions", *Chem. Mater.,* vol. 20, no. 5, pp. 1685-1687, 2008.
[http://dx.doi.org/10.1021/cm7035508]

[78] V.B. Mbayachi, E. Ndayiragije, T. Sammani, S. Taj, E.R. Mbuta, and A. khan, "Graphene synthesis, characterization and its applications: A review", *Results in Chemistry,* vol. 3, p. 100163, 2021.
[http://dx.doi.org/10.1016/j.rechem.2021.100163]

[79] L.L. Zhang, and X.S. Zhao, "Carbon-based materials as supercapacitor electrodes", *Chem. Soc. Rev.,* vol. 38, no. 9, pp. 2520-2531, 2009.
[http://dx.doi.org/10.1039/b813846j] [PMID: 19690733]

[80] Y. Wang, and G. Cao, "Developments in nanostructured cathode materials for high-performance lithium-ion batteries", *Adv. Mater.,* vol. 20, no. 12, pp. 2251-2269, 2008.
[http://dx.doi.org/10.1002/adma.200702242]

[81] A. Iwan, and A. Chuchmała, "Perspectives of applied graphene: Polymer solar cells", *Prog. Polym. Sci.,* vol. 37, no. 12, pp. 1805-1828, 2012.
[http://dx.doi.org/10.1016/j.progpolymsci.2012.08.001]

[82] D. Wei, and Y. Liu, "Controllable synthesis of graphene and its applications", *Adv. Mater.,* vol. 22, no. 30, pp. 3225-3241, 2010.
[http://dx.doi.org/10.1002/adma.200904144] [PMID: 20574948]

[83] H.K. Bisoyi, and S. Kumar, "Carbon-based liquid crystals: Art and science", *Liq. Cryst.,* vol. 38, no. 11-12, pp. 1427-1449, 2011.
[http://dx.doi.org/10.1080/02678292.2011.597882]

[84] Y. Sun, Q. Wu, and G. Shi, "Graphene based new energy materials", *Energy Environ. Sci.,* vol. 4, no. 4, pp. 1113-1132, 2011.
[http://dx.doi.org/10.1039/c0ee00683a]

[85] D.R. Kauffman, and A. Star, "Graphene versus carbon nanotubes for chemical sensor and fuel cell applications", *Analyst,* vol. 135, no. 11, pp. 2790-2797, 2010.
[http://dx.doi.org/10.1039/c0an00262c]

[86] H. Su, and Y.H. Hu, "Recent advances in graphene-based materials for fuel cell applications", *Energy Sci. Eng.,* vol. 9, no. 7, pp. 958-983, 2021.
[http://dx.doi.org/10.1002/ese3.833]

[87] H. Wang, X. Mi, Y. Li, and S. Zhan, "3D graphene-based macrostructures for water treatment", *Adv. Mater.,* vol. 32, no. 3, p. 1806843, 2020.
[http://dx.doi.org/10.1002/adma.201806843] [PMID: 31074916]

[88] R.K. Upadhyay, N. Soin, and S.S. Roy, "Role of graphene/metal oxide composites as photocatalysts, adsorbents and disinfectants in water treatment: A review", *RSC Advances,* vol. 4, no. 8, pp. 3823-3851, 2014.
[http://dx.doi.org/10.1039/C3RA45013A]

CHAPTER 4

Quantum Dots: Chemistry, Properties, and Energy Storage Applications

Himadri Tanaya Das[1,*], **T. Elango Balaji**[2], **Swapnamoy Dutta**[3], **Payaswini Das**[4] and **Nigamananda Das**[1,2]

[1] *Centre of Excellence for Advance Materials and Applications, Utkal University, Bhubaneswar 751004, Odisha, India*

[2] *Department of Chemical Engineering, National Taiwan University of Science and Technology, Taipei, 10607, Taiwan*

[3] *University of Tennessee, Bredesen Center for Interdisciplinary Research and Graduate Education, Knoxville, TN, 37996, USA*

[4] *CSIR-Institute of Minerals and Mining Technology, Bhubaneswar, Odisha, India*

Abstract: Currently, Quantum dot nanomaterials have received a lot of attention due to their intriguing features. Most intriguing is how they can be used as electrodes to create safe chemical-free supercapacitor parts and produce clean energy. Due to their high charge storage capacity and stability, quantum dot electrodes are increasingly in demand for high-tech hybrid supercapacitors. This chapter covers the electrochemical performance, physiochemical characteristics, and synthesis of numerous quantum dots. They are also provided with information about the electrochemical characteristics of various supercapacitors. To show readers the potential of this field of study, the best operational factors are highlighted.

Keywords: Quantum dots, Nanostructured, Electrochemical, Energy storage devices, Batteries, Supercapacitors.

INTRODUCTION

Exploration of materials of different dimensions and their versatile applications have opened up several routes of advancement. In today's research, nanostructured materials have been widely used. Their dimensions as well as morphologies have exhibited a large scope that can be employed in upcoming technologies and quantum dots (QDs) are one of them. In general, QDs can be defined as nanocrystal-semiconducting materials having the smallest diameters

---

[*] **Corresponding author Himadri Tanaya Das:** Centre of Excellence for Advance Materials and Applications, Utkal University, Bhubaneswar 751004, Odisha, India; E-mail: himadridas@utkaluniversity.ac.in

Sanjeev Verma, Shivani Verma, Saurabh Kumar & Bhawna Verma (Eds.)

which vary in between the 2–10 nm nanoscale range. Owing to the smaller particle size, tunable characteristics, and high quantum yield-like significant features, QDs became an attractive choice of material to study in diverse fields of science and engineering such as energy storage (battery, SCs, *etc.*), medicine, and catalysis, solar cells, and sensing. Their adjustable surface area (SA), electro-optical characteristics, and unique quantum confinement effects have made them more impactful. Reduced dimensions of such materials have impacted electronic and optical properties in a great way which further affected applicative purposes. Excitons are restricted to a significantly smaller volume of semiconductor materials in QDs, which is by their exciton Bohr radius. Less energy band splitting occurs as a result, creating an area of quantum confinement. The electronic energy bands are connected to such a region of discrete and quantized electron-hole pairs in various dimensions within a material. To precisely tune the energy bandgap, the size and composition of the QDs can be modified which favors the applications. Specifically, by varying the diameter of QDs typically between 2 to 10 nm, the bandgap of QDs can be adjusted which ultimately results in the improvement of their electrical and optical characteristics. Similar to an atom, the energy levels in QDs are quantized, but compared to the atoms, the distinct factor is that in a QD, the energy level in between separation is lesser; besides, this separation varies based on the size of the QDs having a larger energy level in between separation. The interval between a QD's energy levels must not exceed kT(*the quantity of heat required to enhance the thermodynamic entropy of a system*)-otherwise, electrons could be excited to higher energy levels by thermal energy [1].

In recent research, QD sizes are precisely controlled and modified which has led to several visions of investigating the physical and chemical properties for different applications. Based on the core type, shape, structure, size like characteristics, QDs are typically sorted. In general, QD cores are prepared with different elements such as cadmium, indium, and carbon encapsulated by chalcogenides [2]. Nevertheless, recent studies have attempted to develop shell structures to enhance the structural stability as well as applicative perspective [3]. Formation of core–shell structure has also been proved useful in lessening the toxicity owing to the existence of an exterior shell layer surrounding the core QDs. One of the recent studies explained that *via* combining two nanocrystal semiconductors having separate bandgap energies, alloyed QDs are developed which have displayed distinguishable characteristics that are different from their parent semiconductor properties [4]. Gradient and homogeneous internal structures are observed in alloyed-QDs, where it was minutely noticed that without tweaking the size of the nanocrystals, their internal structural arrangements and composition vary and so, due to the approach of tuning the optical and electrical characteristics, the result from alloyed QDs of CdSxSe1-

x/ZnS formed with different composition emits a wide range of light wavelengths [5]. Several reports have also focused on their shapes such as cylindrical QDs, spherical QDs, tetrahedral QDs, conical QDs, pyramidal QDs, *etc.*, and their resultant experimental impact on different kinds of applications [3d, 6].

Synthesis and properties of QDs

The selection of the synthesis method is useful as it has a profound impact on the size, morphology as well as on the characteristics of the prepared product. QDs are found to be prepared in different ways in reports. Top-down synthesis is a technique where QDs are produced by thinning down bulk semiconductors. Laser ablation, chemical and electrochemical oxidation, ultrasonic, arc discharge, *etc.* are the commonly utilized top-down techniques for fabricating QDs. Bottom-up approaches are also popular techniques and are considered very impactful in the fabrication of QDs. These techniques are mainly categorized into two important segments which are vapor-phase and wet-chemical techniques; and are subdivided into different techniques such as wet chemical techniques like sol-gel, hot-solution decomposition, microemulsion, competitive reaction chemistry, and electrochemical method whereas molecular beam epitaxy, ion sputtering, *etc.* techniques are considered as vapor-phase techniques [7]. In addition, techniques like microwave synthesis, hydrothermal, and thermal decomposition techniques can also be used in the bottom-up approach.

Laser ablation

Owing to easy morphology control and better efficiency, laser ablation has been utilized in several studies to form QDs [8]. Importantly, the laser ablation of solids in liquid state is a simple as well as a versatile route to develop nanostructures. Sun *et al.* used a laser ablation technique to synthesize CQDs using graphite and water vapor under inert atmosphere conditions [9]. From several reports, it was confirmed that *via* adjusting laser pulse width, the QD size can be controlled as well and the adjustment affects QD's nucleation and growth [10]. Owing to the interaction between the graphite flakes and the laser beams, an immediate rise in pressure vapor/plasma plume and temperature was noticed at the interface of the liquid medium and the graphite flake. Afterward, the liquid confinement leads to the formation of a bubble at the laser focus which then swiftly expands and spreads out to the highest radius. Subsequently, the completion of a laser pulse width results in the start of the shrinking of the bubble along with the pressure of the surrounding liquid, which leads to the cooling of the inner region and forms a cluster of nuclei. The bubbles containing various cluster densities can be obtained by tuning the width of the laser pulse, thereby obtaining the CQDs with tunable sizes. In another study, Lihe Yan and the group employed a simple, fast, and low-cost, femtosecond laser ablation technique to synthesize high-quality TMDs QDs (MoS2 and WS2), which exhibited a layered structure and good optical (photoluminescent) properties [11]. Firstly, femtosecond laser ablation was used to cut bulk MoS2 and WS2 into small nanoparticles, and therefore ultrasonic process was employed to exfoliate the prepared nanoparticles into MoS2 and WS2 QDs. Importantly, during the injection of femtosecond pulses into the targets, multiphoton-adsorption ionization happens, and a plasma plume is formed at elevated pressure and temperature. In these conditions, the nanoparticles can be obtained by surface functionalization, and the Coulombic explosion of nanoparticles happens spontaneously. A particular report which has explored the laser ablation preparation technique to analyze the CdSe QDs utilized a nanosecond pulsed laser [12]. From TEM, it was seen that the prepared CdSe QDs have particle sizes (average) of around ~5.0-5.1 nm. Similarly, Manish Kumar Singh and the group explored the control synthesis of SnO_2 QDs *via* laser ablation in liquid where the range of diameter was between 1 to 5 nm [13].

Electrochemical oxidation

Due to the cost-effectiveness, high purity, adequate yield, simple adjustment of size, and good reproducibility, electrochemical oxidation is considered beneficial to synthesize QDs. However, in studies, most CQDs have been developed using this technique. Zhou *et al.* mentioned the preparation of CQDs from MWCNT *via* electrochemical synthesis [14]. In another study, Kang *et al.* approached an alkali-supported electrochemical technique to synthesize tunable CQDs which were found to have excellent electronic and optical characteristics [15]. The ultrasmall particles can be obtained by careful cutting *via* electrochemical oxidation of graphite honeycomb. The as-prepared ultrasmall particles can turn into minuscule graphite fragments. Thus, such an approach is easily adaptable and a simple direct approach for synthesizing high-quality CQDs. Utilizing graphite rods as both negative and positive electrodes and NaOH/EtOH as electrolytes, the CQDs were prepared by varying current densities ranging between 10 to 200 mA cm^{-2} along with various size ranges. High crystalline CQDs were prepared by oxidation of graphite electrochemically under different alkaline alcohols by Liu *et al.* [16]. Notably, a colourless dispersion of CQDs was formed but under ambient conditions, and with time, it slowly changed to bright yellow color, which indicates that with time, surface species got oxygenated. It is also well-known that certain employed potentials and pH significantly influence the generation of CQDs. In a different study, Yuwu Chi and the group demonstrated the synthesis of ZnO QDs through electrochemical oxidation where QDs seemed to have uniformly distributed particles with a diameter of about 5.0 nm, and demonstrated excellent photoluminescent and electro-chemiluminescent behavior [17].

Chemical oxidation

For large-scale production, chemical oxidation is an effective and convenient technique to produce QDs. In addition, for requirements in complicated devices for QDs, chemical oxidation is used [18 - 20]. One of the large-scale approaches was demonstrated by Qiao *et al.* where CQDs were developed using excitation-wavelength (λex) dependent PL [18]. They use nitric acid etching with activated carbon materials such as coal, wood, and coconut. The three different types of CQDs generated have diameters between 2 and 6 nm and are evenly spread after being subjected to a passivation procedure using amine-terminated substances such as 4,7,10-trioxa-1,13-tridecanediamine (TTDDA).

Microwave synthesis

The substrate's chemical bonds are broken using intense energy supplied by an electromagnetic wave or microwave with a higher wavelength between 1 mm and 1 m in this method of synthesis and as a result, homogeneous heating is provided simultaneously and the reaction time is effectively shortened. Typically, in this process, a resultant uniform size distribution of QDs has been observed [19]. At first, Zhu *et al.* introduced an easy and economical microwave pyrolysis technique for developing CQDs exhibiting stable, bright luminescence and remarkable water dispersion [19c]. In this approach, polyethylene glycol and saccharide with different concentrations were mixed along with Di water (distilled) to make a transparent solution and afterward, kept under thermal treatment for 2-10 minutes in a microwave oven. The colorless solution changes to yellow with increasing reaction time and lastly turned into a dark brown-colored solution, which implies that CQDs were finally formed. In another study, glucose-derived water-soluble crystalline GQDs were prepared by Tang and the group where a size-controlled microwave-assisted hydrothermal technique was utilized [19d]. The monodispersed characteristics of the GQDs were noticed mostly because of the rapid, spontaneous, uniform heating of the water solvent and glucose *via* irradiating microwaves, which facilitates homogenous nucleation and growth of the GQDs. With persisting microwave-heating starting from 1- 9 minutes, the size of the GQDs can be varied from 1.65- 21 nm. Shuit-Tong Lee and the group used a one-pot microwave-assisted approach to prepare silicon QDs by using glutaric acid and silicon nanowires as precursors which showed superior aqueous dispersibility, short reaction times (5 min), favorable sizes (~4 nm), robust pH- and photostability and also the QDs exhibited strong fluorescence properties (~15%) [20]. In another study, Wang *et al.* utilized irradiation of microwaves in aqueous phase arrangement for the formation of CdTe/CdS/ZnS core-shell-shell QDs which displayed aqueous dispersibility, notable spectral properties, and photostability [21]. In the work, CdS shell, having lattice constant and band gap energy in between those of ZnS and CdTe, was utilized as a buffer layer between the ZnS shell and CdTe core and such structural experimentation led to the step-by-step conversion of lattice spacing from the emitting CdTe core to the encircling ZnS shell, that shrinks the strain within the QD and also gave better photo luminance and reduced non-radiative surface defects. Other than these, graphitic carbon nitride and chromium (Cr)-doped CdS QDs preparation was also achieved through microwave synthesis where better optical properties and size-controlled synthesis were observed in lesser reaction times [22].

Thermal decomposition

It is a popular strategy to develop diverse nanomaterials with semiconducting and magnetic properties. In recent studies, it was observed that the heat employed externally can dehydrate and lead to carbonization of organics which eventually leads to the formation of CQDs. Simple operation, no solvent involvement, broad precursor tolerance, less reaction time, and economical and scalable production are the several advantages that thermal decomposition can offer [19d, 23]. In a study, Ma *et al.* used a sand bath to perform a simple and direct carbonization reaction of ethylene diamine tetra acetic acid (EDTA) and maintain a temperature range of 260-280 ^{0}C to prepare N-doped GQDs [23a]. During EDTA decarboxylation, the obtained species might constantly blend collectively to form graphite-like structural arrangements under the solid-state reaction environments, which is due to the conversion of N-containing compounds into different graphitic carbon nitrides at elevated pyrolysis temperatures. Martindale *et al.* synthesized CQDs with a high yield of 45% by straightforward pyrolysis of citric acid at 180 oC for 40 h [24]. The forming CQDs are in size ranging from 6.8±2.3 nm relatively distributed all over. In another study, AgInS/ZnS QDs were prepared through thermal decomposition [25]. The unique arrangement utilized a temperature of around 170 °C to heat a reaction mixture of Ag/ In-acetate, indium, and oleic acid in dodecanethiol to develop AIS QDs which resulted in a quantum yield (QY) of almost 13%. After ZnS shell growth, the final development of AIS/ZnS QDs resulted in the formation of almost 41% QY. Yang *et al.* utilized thermal decomposition at a temperature of 200 °C to synthesize ZnO QDs (hexagonal wurtzite structure) where a slight introduction of sodium dodecyl sulfate was done [26].

PROPERTIES

In addition to their remarkable applications in energy storage, Quantum dots (QDs) exhibit intriguing chemical and physical properties that underpin their utility in various fields. These nanoscale semiconductor materials possess size-dependent electronic structures, resulting in tunable optical and electronic properties. The quantum confinement effect, arising from the confinement of charge carriers within the QD volume, imparts unique characteristics to these nanomaterials [27].

Chemically, quantum dots' surface properties play a pivotal role in determining their stability, reactivity, and interactions with their surroundings. The ratio of surface atoms to core atoms is significantly higher in quantum dots compared to bulk materials, leading to an increased surface area-to-volume ratio. This high surface area enhances surface reactivity and makes QDs sensitive to surface ligands and passivation. Proper surface functionalization is essential to control aggregation, improve stability, and enable targeted applications [28].

Physically, quantum dots exhibit size-dependent fluorescence, a phenomenon commonly referred to as the "quantum size effect." As the size of the QD changes, so does its bandgap, resulting in a shift in the emitted light's color. This tunability of emission wavelengths makes quantum dots valuable in applications such as displays, sensors, and biological imaging [29].

Furthermore, quantum dots' ability to efficiently capture, separate, and transport charge carriers is a significant advantage in energy-related applications. Their small size and quantum confinement effect enable efficient charge transfer, making them ideal candidates for use as electrodes in supercapacitors. The ability to tailor their band structure and energy levels through the precise control of size and composition further enhances their suitability for energy storage devices [29].

In summary, the chemical and physical properties of quantum dots are pivotal in defining their versatility and functionality. Their tunable electronic and optical characteristics, surface reactivity, and charge transport capabilities contribute to their role as electrodes for energy storage applications and underscore their potential for revolutionizing various technological domains. This chapter not only delves into their electrochemical performance and synthesis but also aims to highlight the fundamental properties that make quantum dots a promising avenue for scientific exploration and practical innovation [30].

DIFFERENT TYPES OF QUANTUM DOTS AND THEIR APPLICATIONS IN ENERGY STORAGE

Recently, reports on carbonaceous materials, conducting polymer, and transition metal oxide/hydroxides were frequently reported. Among them, carbonaceous QDs grabbed consideration for their broad applicative range due to their economical synthesis, abandoned accessible SA, and eco-safety. Carbonaceous materials exhibit sufficient conductivity, long-term stability, and rapid charge-discharge characteristics which make them an idle candidate in energy storage devices (ESDs). Widely explored carbonaceous materials are graphene, activated carbon, and carbon nanotubes. Conducting polymers have rapid redox characteristics, conductivity, and good structural stability which ensure their competence in efficient usage for batteries, SCs, and fuel cells like ESDs. Some of the widely reported conducting polymers are polypyrrole, polyaniline, polythiophene, polyindole poly 3,4-ethylenedioxythiophene, *etc*. Due to their unique properties, these are not only used as electrode materials but also used in separators and membranes. Transition metal oxide/hydroxides offer minimal resistance and high charge storage capabilities with long cycle life. Sufficient space between the layers in metal hydroxides facilitates the redox reaction through intercalation and deintercalation.

The nanosizing of electrode materials has been verified to be a thriving strategy to increase the electrochemical activity of ESDs. Scaling down these materials to quantum dots resulted in improved accessibility to the electrolytes, which increases the efficiency of the reactions happening at the electrode-electrolyte interface. Among the various quantum dots studied, the reports on carbon-based quantum dots (CQDs) were frequent. Shortly after the introduction of CQDs, graphene quantum dots (GQDs) gained much attention, even though they both have similar characteristics, GQDs show enhanced crystalline structure. By taking advantage of these properties, GQDs can be composited with other materials to achieve superior electrochemical activities.

Quantum Dots applications in batteries

QDs are tested broadly for battery, and SCs like ESDs. In this section, we will mention a few vital studies of different battery systems where QDs are being used. For instance, Chao *et al.* used the solvothermal process to prepare GQDs coated onto the VO_2 and used as an electrode to enhance the electrochemical activity of batteries (Li/ Na-ion battery) [32]. In the study, it was observed that the as-synthesized electrodes provide a Na-storage capacity of 306 mAh g^{-1} at 100 mA g^{-1}, and a capacity of more than 110 mA h g^{-1} after 1500 cycles at 18 A g^{-1}. Moreover, the lithiation demonstrated a capacity of more than 420 mA h g^{-1} with good retention (94% after 1500 cycles). Importantly, the layer of surface functionalized GQDs on the VO_2 surface assists in the infiltration of electrolyte, ion-transfer, and hence improves the reaction kinetics. In another study, Daugherty *et al.* used an infrared-supported pyrolysis approach to react with the initial compounds such as glucose, urea, and ammonia sulfate to develop S/N co-doped GQDs at 280 °C [33]. The catalytic action, equivalent series resistance, stability, and voltage efficiency were considerably enhanced due to the S/N co-doping. The impact of GQDs, which are composed of lattice N atoms, O functionalities, and S dopants, which facilitate surface catalytic movement and promote movement of charges around the anode/anolyte interface for the vanadium redox couples (V(II)/V(III)), can also be linked to the increased activity. Similarly, CQDs are also used for different batteries as electrodes. Perumal Elumalai and the group prepared a CQD-Bi_2O_3 composite by utilizing spoiled (denatured) milk-derived CQDs and used as anode material in lithium-ion battery (LiBs), which demonstrated electrochemical performance and exhibited a discharge capacity as high as 1500 mA h g^{-1} at 0.2 C rate [34]. Due to the existing conductive carbon network, the as-synthesized composite anode material has delivered a discharge capacity of 1200 mA h g^{-1} after a couple of cycles (around 3 cycles). Like the $GQDs/VO_2$, the $CQDs/VO_2$ combination is also explored by researchers. Balogun *et al.*, used free-standing CQDs coated VO_2 interwoven nanowires as cathode material for LIB and SIBs where a solution of 1 M LiPF6 in 1:1 by volume of ethylene carbonate/dimethyl carbonate is used as an electrolyte [34]. The as-prepared electrode demonstrated capacities of 420 and 328 mA h g^{-1} at a current density rate of 0.3 C for Li and Na storage, respectively. In another study, CQDs derived from the chemical oxidation of D-(+)-glucose are employed as anode material for both LIB and SIBs [35]. The as-synthesized CQDs exhibited stable and excellent cycling activity after 500 cycles (864.9 mA h g^{-1} at 0.5 C) with the capacity retention of 91.6% and a good rate performance (340.2 mA h g^{-1} at 20 C) for LIBs. On the other hand, the CQDs demonstrated a specific capacity of 323.9 mA h g^{-1} at 0.5 C and capacity retention of 72.4% after 500 charge/discharge cycles, indicating excellent cycling stability for the SIB anode. For the LIB cell experimentation, the electrolyte used was 1 M lithium hexafluorophosphate (LiPF6), dissolved in a 1:1 vol% mixture of ethylene carbonate (EC) and diethyl carbonate (DEC), and for the SIB cell, M sodium perchlorate (NaClO4) in a 1:1 vol% mixture of the EC and propylene carbonate (PC) electrolytes. CQDs have also been used for potassium-ion batteries where a high specific capacity of 195.3 mA h g^{-1} (Fig. **1a**) and 67.43% capacity retention after 150 cycles at a current density of 0.1 A g^{-1} showed a high rate capability of up to 3.2 A g^{-1} (Fig. **1b**) [31a]. Apart from CQDs, and GQDs, other QDs like SnO2, ZnS, SnS, *etc.* have also been studied by different groups. For instance, Gao *et al.*, employed a hydrothermal method to prepare 3D SnO_2 QDS@graphene frameworks which exhibited enhanced specific area and promoted ion transport [36]. The fabricated LIB using the prepared anode following excellent electrochemical performances are being observed with a high reversible capacity (1300 mA h g^{-1} at 100 mA g^{-1}), excellent rate performance (642 mA h g^{-1} at 2000 mA g^{-1}), and super long cycle stability (when the current density is 10 A g^{-1}, the capacity loss is less than 2% after 5000 cycles). In another study, molten Na infused into SnO_2 QD-covered commercial carbon cloth (Na/SnO_2-CC) was developed with a combined method of wet-chemical synthesis and colloidal solution approach [31b]. The prepared composite is used as an anode in sodium metal battery setup (Na-Na3V2(PO4)3), which exhibited high initial specific capacity (Fig. **1d**) and stable electrochemical activity and exhibited an ultrahigh current density of 20 mA cm^{-2} in Na symmetric battery for almost 400 cycles (Fig. **1c**) and delivers superior cycling activity as well as reversible rate capability. Previously, the same research group investigated SnO_2 QDs on carbon nanotubes which demonstrated a stable discharge capacity of 845 mA h g^{-1} at 100 mA g^{-1} after 90 cycles when the composite (SnO_2 QD-CNT) was employed as anode material for LiBs [37]. Similarly, the SnO_2 QDs@ graphene oxide is also used as an anode for LIB which demonstrated potential electrochemical performances [38]. The good dispersion and high mass loading of SnO_2 QDs on the graphene sheet can offer the formation of several active sites which resulted in improved lithium storage capacity. Other modified SnO_2 QDs or composites such as SnO_2 QD modified N-doped carbon spheres, SnO_2 QD anchored on amorphous carbon coated MWCNT, *etc.*, have also been used as anode in LIBs which demonstrated promising performances such as cycle life and stability, high-rate capability, and excellent capacitive contribution properties [39]. Other than these, ZnS, MoP, and SnS are used in different studies for LIB as well as SIBs [40].

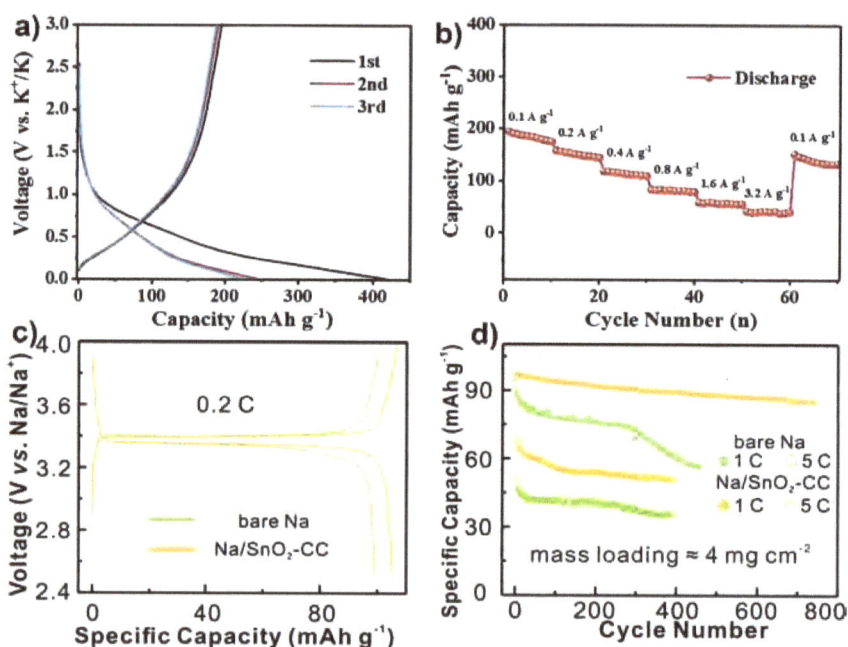

Fig. (1). a & b) Electrochemical performances of HHC as anode for PIBs: first three charge/ discharge curves at a current density of 0.1 A g^{-1}, rate capability at various current densities from 0.1 to 3.2 A g^{-1}, **c**) The cycle stability at 1 and 5 C, **d**) voltage profiles at 0.2 C of Na-Na3V2(PO4)3 full cells with bare Na anode and Na/SnO2-CC anode. Copyrights (Elsevier, 2019) (ACS, 2021) Reproduced from Ref [31].

QDs Applications in Supercapacitors

Quantum dots are used as electrode materials for supercapacitors due to less cost, facile synthesis method, and excellent electrochemical properties. Chen *et al.* reported CQDs assembled on a layered carbon. The fabricated device exhibited a high volumetric capacitance of 157.4 F cm^{-3}. Also the material retained good stability of 137.8 F cm^{-3} after 4000 cycles [42]. Athika *et al.* reported denatured milk-derived CQDs which delivered high discharge time (Fig. **2a**) and a Cs of 95 F g^{-1} with 100% coulombic efficiency even after 1000 cycles (Fig. **2b**). [41a] Jian *et al.* reported a conducting polymer, a polypyrrole composited with CQDs as all-solid-state supercapacitors which supplied a good Cs of 315 mF cm^{-2} and enhanced charge-discharge capabilities as shown in Fig. (**2c**) with capacitance retention of 85.7% after 2000 cycles (Fig. **2d**). [41b] Kakaei *et al.* reported polyaniline functionalized carbon dots having core-shell morphology. The as-synthesized composite exhibited a high capacity of 264.6 F g^{-1} and the composite retains high stability even after 5000 cycles [43]. Naushad *et al.* reported Co$_3$O$_4$ composited with nitrogen-doped CQDs. The composite revealed excellent specific capacitance (Cs) of 1782 F g^{-1}. Also, the device showed high energy densities (Ed) and power densities (Pd) of 36.9 W h g^{-1} and 480 W kg^{-1}, respectively [44].

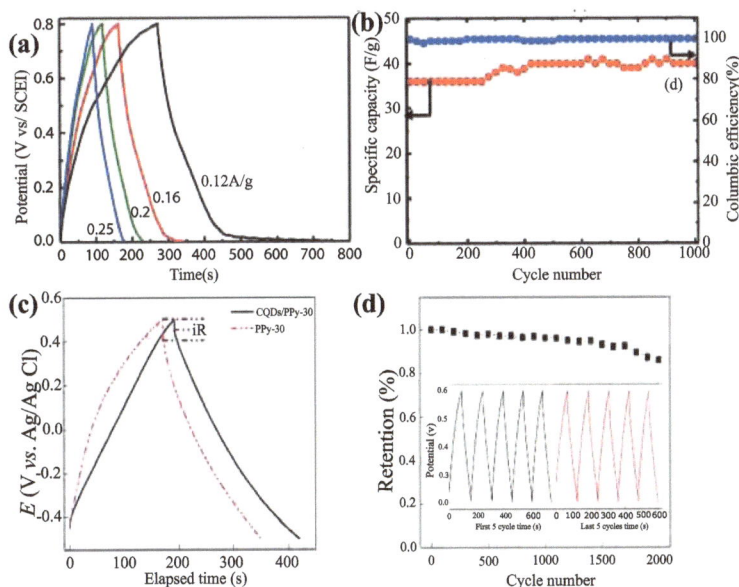

Fig. (2). a) Galvanostatic charge-discharge profiles at different current densities, **b)** Cycle-life data along with the Columbic efficiency at 0.12 Ag^{-1}, **c)** Comparison of GCD curves of CQDs/PPy-30 and PPy-30 electrodes at a current density of 0.5 mA cm^{-2}, **d)** Cycle stability at 2 mA cm^{-2}, inset showing the GCD curves of the first and the last five cycles. (Copyrights, 2019) (Copyrights, 2017). Reproduced from Ref [41].

GQDs exhibited better performance than CQDs due to their superior crystalline nature, Hu *et al.* stated that GQD deposited on carbon nanotubes electrochemically and the hybrid composite exhibited a Cs of 44 mF cm^{-2} which is 200% higher than the bare CNT electrode [45]. Liu *et al.* reported GQDs as negative electrodes for asymmetric micro-supercapacitors having the positive electrode as polyaniline. The as-fabricated device exhibits Cs of 210 μF cm^{-2} with a high power density of 7.46 μW cm^{-2} [46]. Ganganboina *et al.* reported GQDs decorated on Fe$_3$O$_4$-halloysite nanotubes which exhibit a Cs of 418 F g^{-1} with high Ed and Pd of 29 W h kg^{-1} and 5.2 kW kg^{-1} [47].

Metal oxide quantum dots have also been reported recently due to the good redox activity and pseudocapacitive characteristics. Geng *et al.* reported SnO$_2$ quantum dots which possessed high SA of 285.5 m^2 g^{-1} with a Cs of 315 F g^{-1} and the as-fabricated asymmetric solid-state supercapacitor exhibited 62 W h kg^{-1} at a power density of 1 kW kg^{-1} [48].

DISCUSSING THE PROS AND CONS OF QDS IN ENERGY STORAGE APPLICATIONS

Electrodes are very crucial segments of electrochemical energy storage system, as it precisely impacts the overall performance. The presence of the heteroatomic functional groups on the surface of QDs is useful in offering a broad range of active sites. QDs and their composites can be utilized as both active electrodes [49] as well as current collectors [50]. Owing to their small particle size, large SA, and adjustable surface function, they have exhibited large capacity [51], superior ionic conductivity [52], adequate rate capability [53], and excellent cycle stability [54], considerably enhancing the electrochemical activity of SCs and battery like tools. Importantly, decreasing the size of cathode or anode materials to QD level is very favorable in lessening the volume change stress, enhancing electrode dynamics, and reducing the movement distance of Li/Na-ions, or other ions within the battery. In addition, embedding QDs in carbon electrodes amplifies the distance between carbon atom layers, decreases the degree of order, forms heterojunctions, as well as improves the diffusion rate and storage capacity of Li-ions. *Via* increasing specific SA and doping N on carbon materials, electrochemical characteristics can be enhanced. Thus, the storage of electrolyte ions in deep holes is highly favored, as a result, the rate and capacitance activities are considerately enriched. However, a few aspects still need improvements. For instance, preparation strategies of QDs or doped QDs still require more upgradation, as issues like synthetic complexity, high cost, and low yield were observed which would restrict the large-scale development and application of QDs. Moreover, the reproducibility and polydispersity difficulties of QD's development led to failure in ensuring the electrode material stability during the period of cycling activity [55]. So, developing facile, economical, and efficient approaches is required to be illustrated broadly in reports. Secondly, it is vital to understand through both experiments and theoretical calculations the interface mechanism of QDs with other materials, and its effect on the characteristics of developed composite compounds. Proper interpretation of the impacts of the components of batteries will be useful to comprehend the effect of crucial parameters of QDs on the electrochemical outcomes of the battery or for other energy storage and conversion devices.

CONCLUSION

QDs are a novel category of nanomaterials that are also defined as zero-dimension particles. The utilization of QDs has been expanded over the past few years rapidly in various applications. In this chapter, we discussed different QDs and their workability for SCs as well as batteries. One of the favorable aspects of QDs is that the synthesis of QDs is explored in various ways and some of them were low-cost and simple processes. Studies have also shown that feedstocks like waste biomass, fruit peel, discarded portions of vegetables, and agricultural waste can be utilized to prepare QDS which can contribute in the future for turning waste into energy . In addition, in recent years, studies have attempted to develop several composite and hybrid materials which in turn enhanced the physical, chemical, and optical properties and as a result, the application prospects were potentially influenced. Even the ease in the tunability of QDs makes them more attractive for energy applications. However, it is required to minutely explore simple and efficient purification techniques to eradicate residual reagents. Such novel techniques will be useful in obtaining excellent yields that do not involve the eradication of initial constituents at all and achieving this will be favorable in not only doing cost-effective lab-scale research but also will be approachable for large-scale production and usage. Additionally, it is very important to relate their characteristics with the application mechanisms which need more consideration as more growth in understanding the complete operation will bring great help in energy application scenarios.

REFERENCES

[1]   D.C. Agrawal, *Introduction to nanoscience and nanomaterials.* World Scientific Publishing Company, 2013.
       [http://dx.doi.org/10.1142/8433]

[2]   H. Li, X. Jiang, A. Wang, X. Chu, and Z. Du, *Frontiers in chemistry* vol. 8. , 2020, p. 669.bD. Vasudevan, R. R. Gaddam, A. Trinchi, and I. Cole, *J. Alloys Compd.* vol. 636. , 2015, p. 395p. 404.cS. Yu, X. Zhang, L. Li, J. Xu, Y. Song, X. Liu, S. Wu, and J. Zhang, *Materials Research Express* vol. 6. , 2019, p. 0850p. 0857.

[3]   J. Bailes, *Nanoparticles in Biology and Medicine* Springer, 2020, pp. 343-349.bZ. Wang, and M. Tang, *Environ. Res.* vol. 194. , 2021, p. 110593.cM. A. Boles, D. Ling, T. Hyeon, and D. V. Talapin, *Nature materials* vol. 15. , 2016, pp. 141-153.dS. Hong, and C. Lee, *The plant pathology journal* vol.

34. , 2018, p. 85.

[4]   W.A.A. Mohamed, H. Abd El-Gawad, S. Mekkey, H. Galal, H. Handal, H. Mousa, and A. Labib, "Quantum dots synthetization and future prospect applications", *Nanotechnol. Rev.,* vol. 10, no. 1, pp. 1926-1940, 2021.
[http://dx.doi.org/10.1515/ntrev-2021-0118]

[5]   G. Vastola, Y.W. Zhang, and V.B. Shenoy, "Experiments and modeling of alloying in self-assembled quantum dots", *Curr. Opin. Solid State Mater. Sci.,* vol. 16, no. 2, pp. 64-70, 2012.
[http://dx.doi.org/10.1016/j.cossms.2011.10.004]

[6]   A. Nair, J. T. Haponiuk, S. Thomas, and S. Gopi, *Biomed. Pharmacother.* vol. 132. , 2020, p. 110834.bG. S. Selopal, H. Zhao, Z. M. Wang, and F. Rosei, *Adv. Funct. Mater.* vol. 30. , 2020, p. 1908762.cH. T. Das, P. Barai, S. Dutta, N. Das, P. Das, M. Roy, M. Alauddin, and H. R. Barai, *Polymers* vol. 14. , 2022, p. 1053.dH. T. Das, S. Dutta, P. Das, and N. Das, *Quantum Dots and Polymer Nanocomposites* CRC Press, 2022, pp. 395-411.

[7]   P.K. Tiwari, M. Sahu, G. Kumar, and M. Ashourian, *Comput. Intell. Neurosci.,* vol. 2021, p. 2096208, 2021.
[http://dx.doi.org/10.1155/2021/2096208] [PMID: 34413883]

[8]   L. Cao, X. Wang, M. J. Meziani, F. Lu, H. Wang, P. G. Luo, Y. Lin, B. A. Harruff, L. M. Veca, and D. Murray, *Journal of the American Chemical Society* vol. 129. , 2007, pp. 11318-11319.bY.-P. Sun, X. Wang, F. Lu, L. Cao, M. J. Meziani, P. G. Luo, L. Gu, and L. M. Veca, *The Journal of Physical Chemistry C* vol. 112. , 2008, pp. 18295-18298.cX. Li, H. Wang, Y. Shimizu, A. Pyatenko, K. Kawaguchi, and N. Koshizaki, *Chem. Commun.* vol. 47. , 2010, pp. 932-934.

[9]   Y.P. Sun, B. Zhou, Y. Lin, W. Wang, K.A.S. Fernando, P. Pathak, M.J. Meziani, B.A. Harruff, X. Wang, H. Wang, P.G. Luo, H. Yang, M.E. Kose, B. Chen, L.M. Veca, and S.Y. Xie, "Quantum-sized carbon dots for bright and colorful photoluminescence", *J. Am. Chem. Soc.,* vol. 128, no. 24, pp. 7756-7757, 2006.
[http://dx.doi.org/10.1021/ja062677d] [PMID: 16771487]

[10]   S. Hu, J. Liu, J. Yang, Y. Wang, and S. Cao, "Laser synthesis and size tailor of carbon quantum dots", *J. Nanopart. Res.,* vol. 13, no. 12, pp. 7247-7252, 2011.
[http://dx.doi.org/10.1007/s11051-011-0638-y]

[11]   Y. Xu, L. Yan, X. Li, and H. Xu, "Fabrication of transition metal dichalcogenides quantum dots based on femtosecond laser ablation", *Sci. Rep.,* vol. 9, no. 1, p. 2931, 2019.
[http://dx.doi.org/10.1038/s41598-019-38929-5] [PMID: 30814552]

[12]   S. Horoz, L. Lu, Q. Dai, J. Chen, B. Yakami, J.M. Pikal, W. Wang, and J. Tang, "CdSe quantum dots synthesized by laser ablation in water and their photovoltaic applications", *Appl. Phys. Lett.,* vol. 101, no. 22, p. 223902, 2012.
[http://dx.doi.org/10.1063/1.4768706]

[13]   M.K. Singh, M.C. Mathpal, and A. Agarwal, "Optical properties of SnO_2 quantum dots synthesized by laser ablation in liquid", *Chem. Phys. Lett.,* vol. 536, pp. 87-91, 2012.
[http://dx.doi.org/10.1016/j.cplett.2012.03.084]

[14]   J. Zhou, C. Booker, R. Li, X. Zhou, T.K. Sham, X. Sun, and Z. Ding, "An electrochemical avenue to blue luminescent nanocrystals from multiwalled carbon nanotubes (MWCNTs)", *J. Am. Chem. Soc.,* vol. 129, no. 4, pp. 744-745, 2007.
[http://dx.doi.org/10.1021/ja0669070] [PMID: 17243794]

[15]   H. Li, X. He, Z. Kang, H. Huang, Y. Liu, J. Liu, S. Lian, C.H.A. Tsang, X. Yang, and S.T. Lee, "Water-Soluble Fluorescent Carbon Quantum Dots and Photocatalyst Design", *Angew. Chem. Int. Ed.,* vol. 49, no. 26, pp. 4430-4434, 2010.
[http://dx.doi.org/10.1002/anie.200906154]

[16]   M. Liu, Y. Xu, F. Niu, J.J. Gooding, and J. Liu, "Carbon quantum dots directly generated from electrochemical oxidation of graphite electrodes in alkaline alcohols and the applications for specific

ferric ion detection and cell imaging", *Analyst (Lond.)*, vol. 141, no. 9, pp. 2657-2664, 2016.
[http://dx.doi.org/10.1039/C5AN02231B]

[17]   L. Chen, N. Xu, H. Yang, C. Zhou, and Y. Chi, "Zinc oxide quantum dots synthesized by electrochemical etching of metallic zinc in organic electrolyte and their electrochemiluminescent properties", *Electrochim. Acta*, vol. 56, no. 3, pp. 1387-1391, 2011.
[http://dx.doi.org/10.1016/j.electacta.2010.10.050]

[18]   Z.A. Qiao, Y. Wang, Y. Gao, H. Li, T. Dai, Y. Liu, and Q. Huo, "Commercially activated carbon as the source for producing multicolor photoluminescent carbon dots by chemical oxidation", *Chem. Commun. (Camb.)*, vol. 46, no. 46, pp. 8812-8814, 2010.
[http://dx.doi.org/10.1039/c0cc02724c] [PMID: 20953494]

[19]   Q. Wang, X. Liu, L. Zhang, and Y. Lv, *Analyst* vol. 137. , 2012, pp. 5392-5397.bH. Zhu, X. Wang, Y. Li, Z. Wang, F. Yang, and X. Yang, *Chem. Commun.*, 2009, pp. 5118-5120.cL. Tang, R. Ji, X. Cao, J. Lin, H. Jiang, X. Li, K. S. Teng, C. M. Luk, S. Zeng, and J. Hao, *ACS nano* vol. 6. , 2012, pp. 5102-5110.ddX. Zhai, P. Zhang, C. Liu, T. Bai, W. Li, L. Dai, and W. Liu, *Chem. Commun. (Camb.)* vol. 48. , 2012, pp. 7955-7957.

[20]   Y. He, Y. Zhong, F. Peng, X. Wei, Y. Su, Y. Lu, S. Su, W. Gu, L. Liao, and S.T. Lee, "One-pot microwave synthesis of water-dispersible, ultraphoto- and pH-stable, and highly fluorescent silicon quantum dots", *J. Am. Chem. Soc.*, vol. 133, no. 36, pp. 14192-14195, 2011.
[http://dx.doi.org/10.1021/ja2048804] [PMID: 21848339]

[21]   Y. He, H.T. Lu, L.M. Sai, Y.Y. Su, M. Hu, C.H. Fan, W. Huang, and L.H. Wang, "Microwave Synthesis of Water-Dispersed CdTe/CdS/ZnS Core-Shell-Shell Quantum Dots with Excellent Photostability and Biocompatibility", *Adv. Mater.*, vol. 20, no. 18, pp. 3416-3421, 2008.
[http://dx.doi.org/10.1002/adma.200701166]

[22]   M. Shkir, Z. R. Khan, K. V. Chandekar, T. Alshahrani, A. Kumar, and S. AlFaify, *Applied Nanoscience* vol. 10. , 2020, pp. 3973-3985.bH. Li, F.-Q. Shao, H. Huang, J.-J. Feng, and A.-J. Wang, *Sens. Actuators B Chem.* vol. 226. , 2016, pp. 506-511.

[23]   C.-B. Ma, Z.-T. Zhu, H.-X. Wang, X. Huang, X. Zhang, X. Qi, H.-L. Zhang, Y. Zhu, X. Deng, and Y. Peng, *Nanoscale* vol. 7. , 2015, pp. 10162-10169.bB. Chen, F. Li, S. Li, W. Weng, H. Guo, T. Guo, X. Zhang, Y. Chen, T. Huang, and X. Hong, *Nanoscale* vol. 5. , 2013, pp. 1967-1971.

[24]   B.C.M. Martindale, G.A.M. Hutton, C.A. Caputo, and E. Reisner, "Solar hydrogen production using carbon quantum dots and a molecular nickel catalyst", *J. Am. Chem. Soc.*, vol. 137, no. 18, pp. 6018-6025, 2015.
[http://dx.doi.org/10.1021/jacs.5b01650] [PMID: 25864839]

[25]   S. Chen, M. Ahmadiantehrani, N.G. Publicover, K.W. Hunter Jr, and X. Zhu, "Thermal decomposition based synthesis of Ag-In-S/ZnS quantum dots and their chlorotoxin-modified micelles for brain tumor cell targeting", *RSC Advances*, vol. 5, no. 74, pp. 60612-60620, 2015.
[http://dx.doi.org/10.1039/C5RA11250H] [PMID: 26236473]

[26]   L. Yang, J. Yang, X. Liu, Y. Zhang, Y. Wang, H. Fan, D. Wang, and J. Lang, "Low-temperature synthesis and characterization of ZnO quantum dots", *J. Alloys Compd.*, vol. 463, no. 1-2, pp. 92-95, 2008.
[http://dx.doi.org/10.1016/j.jallcom.2007.12.006]

[27]   R. Wang, K.Q. Lu, Z.R. Tang, and Y.J. Xu, "Recent progress in carbon quantum dots: synthesis, properties and applications in photocatalysis", *J. Mater. Chem. A Mater. Energy Sustain.*, vol. 5, no. 8, pp. 3717-3734, 2017.
[http://dx.doi.org/10.1039/C6TA08660H]

[28]   X. Wang, Y. Feng, P. Dong, and J. Huang, "A Mini Review on Carbon Quantum Dots: Preparation, Properties, and Electrocatalytic Application", *Front Chem.*, vol. 7, p. 671, 2019.
[http://dx.doi.org/10.3389/fchem.2019.00671] [PMID: 31637234]

[29]   X. T. Zheng, A. Ananthanarayanan, K. Q. Luo, and P. Chen, *small* vol. 11. , 2015, pp. 1620-1636.

[30] R.M. El-Shabasy, M. Farouk Elsadek, B. Mohamed Ahmed, M. Fawzy Farahat, K.N. Mosleh, and M.M. Taher, "Recent Developments in Carbon Quantum Dots: Properties, Fabrication Techniques, and Bio-Applications", *Processes (Basel)*, vol. 9, no. 2, p. 388, 2021.
[http://dx.doi.org/10.3390/pr9020388]

[31] Y. Zhang, L. Yang, Y. Tian, L. Li, J. Li, T. Qiu, G. Zou, H. Hou, and X. Ji, *Materials Chemistry and Physics* vol. 229. , 2019, pp. 303-309.bY. Xu, E. Matios, J. Luo, T. Li, X. Lu, S. Jiang, Q. Yue, W. Li, and Y. Kang, *Nano Lett.* vol. 21. , 2021, pp. 816-822.

[32] D. Chao, C. Zhu, X. Xia, J. Liu, X. Zhang, J. Wang, P. Liang, J. Lin, H. Zhang, Z.X. Shen, and H.J. Fan, "Graphene quantum dots coated VO2 arrays for highly durable electrodes for Li and Na ion batteries", *Nano Lett.*, vol. 15, no. 1, pp. 565-573, 2015.
[http://dx.doi.org/10.1021/nl504038s] [PMID: 25531798]

[33] M.C. Daugherty, S. Gu, D.S. Aaron, B. Chandra Mallick, Y.A. Gandomi, and C.T. Hsieh, "Decorating sulfur and nitrogen co-doped graphene quantum dots on graphite felt as high-performance electrodes for vanadium redox flow batteries", *J. Power Sources*, vol. 477, p. 228709, 2020.
[http://dx.doi.org/10.1016/j.jpowsour.2020.228709]

[34] A. Prasath, M. Athika, E. Duraisamy, A. Selva Sharma, V. Sankar Devi, and P. Elumalai, "Carbon Quantum Dot-Anchored Bismuth Oxide Composites as Potential Electrode for Lithium-Ion Battery and Supercapacitor Applications", *ACS Omega*, vol. 4, no. 3, pp. 4943-4954, 2019.
[http://dx.doi.org/10.1021/acsomega.8b03490] [PMID: 31459678]

[35] M. Javed, A.N.S. Saqib, R. Ata ur, B. Ali, M. Faizan, D. A. Anang, Z. Iqbal, and S. M. Abbas, *Electrochim. Acta* vol. 297. , 2019, pp. 250-257.
[http://dx.doi.org/10.1016/j.electacta.2018.11.167]

[36] L. Gao, G. Wu, J. Ma, T. Jiang, B. Chang, Y. Huang, and S. Han, "SnO 2 Quantum Dots@Graphene Framework as a High-Performance Flexible Anode Electrode for Lithium-Ion Batteries", *ACS Appl. Mater. Interfaces*, vol. 12, no. 11, pp. 12982-12989, 2020.
[http://dx.doi.org/10.1021/acsami.9b22679] [PMID: 32078288]

[37] X. Lu, H. Wang, Z. Wang, Y. Jiang, D. Cao, and G. Yang, "Room-temperature synthesis of colloidal SnO2 quantum dot solution and ex-situ deposition on carbon nanotubes as anode materials for lithium ion batteries", *J. Alloys Compd.*, vol. 680, pp. 109-115, 2016.
[http://dx.doi.org/10.1016/j.jallcom.2016.04.128]

[38] K. Zhao, L. Zhang, R. Xia, Y. Dong, W. Xu, C. Niu, L. He, M. Yan, L. Qu, and L. Mai, "SnO 2 Quantum Dots@Graphene Oxide as a High-Rate and Long-Life Anode Material for Lithium-Ion Batteries", *Small*, vol. 12, no. 5, pp. 588-594, 2016.
[http://dx.doi.org/10.1002/smll.201502183] [PMID: 26680110]

[39] R. Jin, Y. Meng, and G. Li, *Appl. Surf. Sci.* vol. 423. , 2017, pp. 476-483.bC.-P. Wu, K.-X. Xie, J.-P. He, Q.-P. Wang, J.-M. Ma, S. Yang, and Q.-H. Wang, *Rare Met.* vol. 40. , 2021, pp. 48-56.

[40] S. Lee, S. Kim, J. Gim, M. H. Alfaruqi, S. Kim, V. Mathew, B. Sambandam, J. Hwang, and J. Kim, *Composites Part B: Engineering* vol. 231. , 2022, p. 109548.bM. Wei, B. Li, C. Jin, Y. Ni, C. Li, X. Pan, J. Sun, C. Yang, and R. Yang, *Energy Storage Materials* vol. 17. , 2019, pp. 226-233.cG. K. Veerasubramani, M.-S. Park, J.-Y. Choi, and D.-W. Kim, *ACS Applied Materials & Interfaces* vol. 12. , 2020, pp. 7114-7124.dH. Wang, K. Xie, Y. You, Q. Hou, K. Zhang, N. Li, W. Yu, K. P. Loh, C. Shen, and B. Wei, *Adv. Energy Mater.* vol. 9. , 2019, p. 1901806.

[41] M. Athika, A. Prasath, E. Duraisamy, V. Sankar Devi, A. Selva Sharma, and P. Elumalai, *Mater. Lett.* vol. 241. , 2019, pp. 156-159.bX. Jian, H.-m. Yang, J.-g. Li, E.-h. Zhang, L.-l. Cao, and Z.-h. Liang, *Electrochim. Acta* vol. 228. , 2017, pp. 483-493.

[42] G. Chen, S. Wu, L. Hui, Y. Zhao, J. Ye, Z. Tan, W. Zeng, Z. Tao, L. Yang, and Y. Zhu, "Assembling carbon quantum dots to a layered carbon for high-density supercapacitor electrodes", *Sci. Rep.*, vol. 6, no. 1, p. 19028, 2016.
[http://dx.doi.org/10.1038/srep19028] [PMID: 26754463]

[43]   K. Kakaei, S. Khodadoost, M. Gholipour, and N. Shouraei, "Core-shell polyaniline functionalized carbon quantum dots for supercapacitor", *J. Phys. Chem. Solids,* vol. 148, p. 109753, 2021.
[http://dx.doi.org/10.1016/j.jpcs.2020.109753]

[44]   M. Naushad, T. Ahamad, M. Ubaidullah, J. Ahmed, A.A. Ghafar, K.M. Al-Sheetan, and P. Arunachalam, "Nitrogen-doped carbon quantum dots (N-CQDs)/Co3O4 nanocomposite for high performance supercapacitor", *J. King Saud Univ. Sci.,* vol. 33, no. 1, p. 101252, 2021.
[http://dx.doi.org/10.1016/j.jksus.2020.101252]

[45]   Y. Hu, Y. Zhao, G. Lu, N. Chen, Z. Zhang, H. Li, H. Shao, and L. Qu, "Graphene quantum dots–carbon nanotube hybrid arrays for supercapacitors", *Nanotechnology,* vol. 24, no. 19, p. 195401, 2013.
[http://dx.doi.org/10.1088/0957-4484/24/19/195401] [PMID: 23579638]

[46]   W. Liu, X. Yan, J. Chen, Y. Feng, and Q. Xue, "Novel and high-performance asymmetric micro-supercapacitors based on graphene quantum dots and polyaniline nanofibers", *Nanoscale,* vol. 5, no. 13, pp. 6053-6062, 2013.
[http://dx.doi.org/10.1039/c3nr01139a] [PMID: 23720009]

[47]   A.B. Ganganboina, A.D. Chowdhury, and R. Doong, "Nano assembly of N-doped graphene quantum dots anchored Fe3O4/halloysite nanotubes for high performance supercapacitor", *Electrochim. Acta,* vol. 245, pp. 912-923, 2017.
[http://dx.doi.org/10.1016/j.electacta.2017.06.002]

[48]   J. Geng, C. Ma, D. Zhang, and X. Ning, "Facile and fast synthesis of SnO2 quantum dots for high performance solid-state asymmetric supercapacitor", *J. Alloys Compd.,* vol. 825, p. 153850, 2020.
[http://dx.doi.org/10.1016/j.jallcom.2020.153850]

[49]   J. Li, X. Yun, Z. Hu, L. Xi, N. Li, H. Tang, P. Lu, and Y. Zhu, *Journal of Materials Chemistry A,* vol. 7, pp. 26311-26325, 2019.bZ. Li, F. Bu, J. Wei, W. Yao, L. Wang, Z. Chen, D. Pan, and M. Wu, *Nanoscale,* vol. 10, pp. 22871-22883, 2018.

[50]   Z. Zhao, and Y. Xie, *J. Power Sources,* vol. 337, pp. 54-64, 2017.bW.-W. Liu, Y.-Q. Feng, X.-B. Yan, J.-T. Chen, and Q.-J. Xue, *Adv. Funct. Mater.,* vol. 23, pp. 4111-4122, 2013.

[51]   S. Huang, M. Wang, P. Jia, B. Wang, J. Zhang, and Y. Zhao, "N-graphene motivated SnO2@SnS2 heterostructure quantum dots for high performance lithium/sodium storage", *Energy Storage Mater.,* vol. 20, pp. 225-233, 2019.
[http://dx.doi.org/10.1016/j.ensm.2018.11.024]

[52]   H. Tan, H.W. Cho, and J.J. Wu, "Binder-free ZnO@ZnSnO3 quantum dots core-shell nanorod array anodes for lithium-ion batteries", *J. Power Sources,* vol. 388, pp. 11-18, 2018.
[http://dx.doi.org/10.1016/j.jpowsour.2018.03.066]

[53]   B. Wang, Y. Xie, T. Liu, H. Luo, B. Wang, C. Wang, L. Wang, D. Wang, S. Dou, and Y. Zhou, "LiFePO4 quantum-dots composite synthesized by a general microreactor strategy for ultra-high-rate lithium ion batteries", *Nano Energy,* vol. 42, pp. 363-372, 2017.
[http://dx.doi.org/10.1016/j.nanoen.2017.11.040]

[54]   X. Li, K. Hu, R. Tang, K. Zhao, and Y. Ding, "CuS quantum dot modified carbon aerogel as an immobilizer for lithium polysulfides for high-performance lithium–sulfur batteries", *RSC Advances,* vol. 6, no. 75, pp. 71319-71327, 2016.
[http://dx.doi.org/10.1039/C6RA11990E]

[55]   Q. Xu, Y. Niu, J. Li, Z. Yang, J. Gao, L. Ding, H. Ni, P. Zhu, Y. Liu, Y. Tang, Z.P. Lv, B. Peng, T.S. Hu, H. Zhou, and C. Xu, "Recent progress of quantum dots for energy storage applications", *Carbon Neutrality,* vol. 1, no. 1, p. 13, 2022.
[http://dx.doi.org/10.1007/s43979-022-00002-y]

Metal-Organic Frameworks (MOFs): Chemistry, Properties, and Energy Storage Applications

Nikhil Kumar[1], **Nisha Gupta**[1] and **Pallab Bhattacharya**[1,*]

[1] *Functional Materials Group, Advanced Materials & Processes (AMP) Division, CSIR-National Metallurgical Laboratory (NML), Burmamines, East Singhbhum, Jamshedpur, Jharkhand-831007, India*

Abstract: The scarcity of natural stocks of fossil fuels and the rising pollutant ions evolved from the burning of carbon-containing fuels, has triggered the necessity for clean, renewable, and sustainable energies to be generated and its subsequent storage in portable form to meet the on-demand consumption. However, the performances of storage materials are still limited for extensive real-world applications due to their sluggish ion diffusion kinetics, lack of efficiency in extreme weather conditions, poor chemical stability, and many more. Therefore, it is highly requisite to discuss the development and assess the performances of new advanced energy storage materials. In this chapter, we are specifically keen to discuss the design, synthesis, chemistry, and properties of various MOFs based electrode materials for energy storage devices like batteries and supercapacitors, which can necessarily store electrical energies by implementing the use of suitable electrode and electrolyte materials through an upright fabrication technique. Generally, MOFs contain both inorganic metal ions and organic ligands/linkers which enable great control over their structural and compositional modifications to optimize the properties like porosity, stability, surface area, redox activity, and electrical conductivity and show great promise to generate high energy storage performances, in the recent past. However, despite the current success, MOFs based electrode materials have faced a lot of challenges in terms of the choice of suitable metals and organic ligand moieties, rich host-guest interactions, preparation of composites with desired morphology and properties, control over composite composition, scalability of the process and many more which needs to be addressed for its full-proof use in the real-world application as energy storage materials and thusly, this chapter is important to discuss.

Keywords: Active Sites, Batteries, Compositional flexibility, Composites, Conductivity, Electrochemical performances, Efficiency, Durability, Energy Storage, Functionality, Hydrothermal, MOFs, Metal nodes, Organic ligands,

---

** **Corresponding author Pallab Bhattacharya:** Functional Materials Group, Advanced Materials & Processes (AMP) Division, CSIR-National Metallurgical Laboratory (NML), Burmamines, East Singhbhum, Jamshedpur, Jharkhand-831007, India; E-mail: pallab.b@nmlindia.org*

Sanjeev Verma, Shivani Verma, Saurabh Kumar & Bhawna Verma (Eds.)

Porosity, Physicochemical properties, Power density, Secondary building units, Surface area, Supercapacitors.

INTRODUCTION

Given the widespread global demand for clean energy consumption in today's modern world, the development of such clean electrochemical energy storage devices remains a challenge. To address the need for upcoming energy storage devices, MOFs are gaining the attention of researchers due to their porous crystalline nature. Using inorganic metal ions or clusters and multidentate organic ligands, a promising class of porous solid-state materials has been developed over the last two decades with innovative chemistry. MOFs also known as metal-organic materials (MOMs), porous coordination network materials (PCN), porous coordination polymers (PCPs), and metal-organic coordination network (MOCN) constructed with inorganic metal ions or clusters and multidentate organic ligands, have emerged as the most promising class of solid-state porous materials over the last two decades with innovative chemistry. The term "MOF" came into existence after the pioneering publications of mainly three scientists Omar M. Yaghi [1], G. Ferey [2], and S. Kitagawa [3]. These multifunctional materials have contributed extensively to the interests and benefits the society by diminishing health, energy, and environmental issues. MOFs are composed of two different kinds of components: metal inorganic clusters and organic linkers, which act as "joints" and "struts", respectively [4]. The possible connectivity of organic ligands with different metal nodes, secondary building units (SBUs), or supra-molecular building blocks (SBBs) leads to the creation of predictable MOFs with versatile architectures by reticular chemistry. SBUs and the topology of the target framework are the only criteria to synthesize application-specific framework materials with desired characteristics. The excellent features of MOFs such as ultrahigh porosity (till approx. 90% free volume), huge internal surface area (beyond almost 10000 m^2g^{-1}), and an infinite number of possibilities with a degree of variability for inorganic and organic components, make MOFs suitable candidates for numerous potential applications in adsorption, catalysis, clean energy, gas storage and gas/vapor separation, thin-film and membrane devices, drug delivery and biomedical imaging, luminescence, sensors, conductors, pharmaceuticals, spintronic, energy generation, energy conversion and energy storage devices [4, 5].

Importantly, the flexibility and geometry of organic building blocks play a key role in the structural engineering and functionality of targeted framework materials with desired topologies. These kinds of frameworks exhibit high mechanical and thermal strength and can sustain the same porosity even after the removal of guest solvent molecules. In this rationale, severe benzene di-, tri, tetra-

and hexa-carboxylates, azolates, and their related derivatives are frequently utilized as organic building units and resulted in a series of advantageous MOFs with en*via*ble features [6]. The versatile coordination modes and adaptable conformations of organic ligands with multi-donor sites have been widely used in developing polynuclear MOFs. In addition to this, several chiral organic molecules namely peptides, amino acids, and their related derivatives can construct chiral centers based on MOFs [7]. Nevertheless, MOFs demonstrate a few weak points like poor chemical stability which impede their applicability with full potential. Hence, it is highly desirable to further improve activities, enhance properties, and incorporate new functionalities. In this context, a variety of active functional materials like graphene, carbon black, and carbon nanotubes have been projected to overcome the shortcomings of pristine MOFs. This new approach opens a lot of opportunities for chemists and researchers to do innovations with creative expression and amazing multidimensional properties. The multifunctional character of novel MOF composites will surely stimulate the appearance of inventive, ground-breaking, diversified technologically important industrial applications. These composites are directly used as precursors for inorganic solids or pioneer advanced materials with potential practical applications in sensing, catalysis, protective coatings, separation, bio-medical as well as energy generation, conversion, and storage [8 - 11]. This chapter focused on the recent noteworthy progress in the advancement of MOFs or MOFs-based composites, their classification, synthesis, and properties along with their performance in energy storage applications. The selectivity of ligand and metal ions leads to the formation of flexible MOFs with dynamic features towards different kinds of electrochemical measurements (Fig. **1**).

The responsible key factors, synergistic effects of functional metal nodes, and active ligand sites for energy storage applications have also been discussed in detail. The authors believed that the present chapter would inspire and motivate the readers to adopt a sustainable way of making industrially active MOF composites employed for decorating new energy storage devices for future generations.

FUNDAMENTALS OF MOFS: CLASSIFICATION, SYNTHESIS AND PROPERTIES

MOFs are a unique class of highly porous crystalline polymeric materials with metal ions or clusters linked mutually by bridging organic ligands. Usually, the ligands are an organic molecule that possesses functional groups (amine, pyridine, carboxylic acid, and others) in their structure where a lone pair of electrons can be easily donated to the next metal, normally termed Lewis Base [12]. The porous crystalline MOF framework depends on the dimensionality and length of the ligand. Contrastingly, metal ions come from alkali, alkaline earth metals (s-block), p-block, d-block or transition metals, and f-block or rare earth metals and inner transition metals. The variable oxidation state and types of geometries of metal decide the structural transformation of the framework., MOFs can be classified as one-dimensional (1D), two-dimensional (2D), and three-dimensional (3D) infinite networks [13] based on dimensionality and crystal structure topology. Lastly, the most prominent problem that needs to be addressed is the method or different techniques used to produce the MOFs that engage various factors like a molar ratio of solvent or solvent mixture, pH value, temperature, pressure, and many others that hamper the formation of desired targeted materials. Over the past two decades, numerous synthetic methods have been used to synthesize MOFs such as hydrothermal, solvothermal, room temperature layering (diffusion) or slow evaporation, microwave synthesis, electrochemical, sonochemical, and mechanochemical [14]. Moreover, solvothermal and hydrothermal reactions are carried out in specific closed vessels under autogenous pressure above the boiling point of the solvent used. The majority of the MOFs are generally synthesized by solvo- or hydrothermal reactions employing polar solvents (dimethyl formamide, dimethyl acetamide, acetonitrile, methanol, ethanol, water, or their mixtures) with reaction temperatures between 50–260 °C and reaction time of some hours or some days also. These particular reactions require specific Teflon-lined autoclaves to perform reactions at high temperatures (~400 °C). In the slow diffusion method, a solution of metal ions and organic ligands is purged into the diffusion tubes to make separate layers of metal ions and organic linkers with a sandwiched buffer layer. These layers start to diffuse towards the buffer layer and the formation of crystallization starts over some time. The microwave-assisted synthesis of metal clusters or MOFs have its advantages like high yield, short reaction time, and low cost. In this technique, nucleation and shape, size and morphology-controlled crystal formation take place by applying microwaves. The electrochemical synthesis of MOFs involves anodic dissolution of metal ions into a reaction mixture of organic ligands and electrolytes in an electrochemical cell. Similarly, high-energy ultrasonic radiations have been used for the chemical transformations of molecules to the crystallization nuclei. The mechanochemical method involves mechanical forces to form coordination bonds by using ball mills or manual grinding. This particular method applies at room temperature and it is solvent-free. In some exceptional cases, a negligible amount of solvent can be added to initiate the reaction between two solid reagents (metal ions and organic ligands). It is an environmentally benign approach that reduces particle size, facilitates mass transfer, locally melts the reagent, and turns into a product with high purity, less reaction time, and excellent yield [14]. Some of the studies and literature also reveal high-throughput methods and conventional step-by-step methods. The composites, morphology-dependent MOFs, membranes, thin films, and other shapes require other specific different synthesis methods.

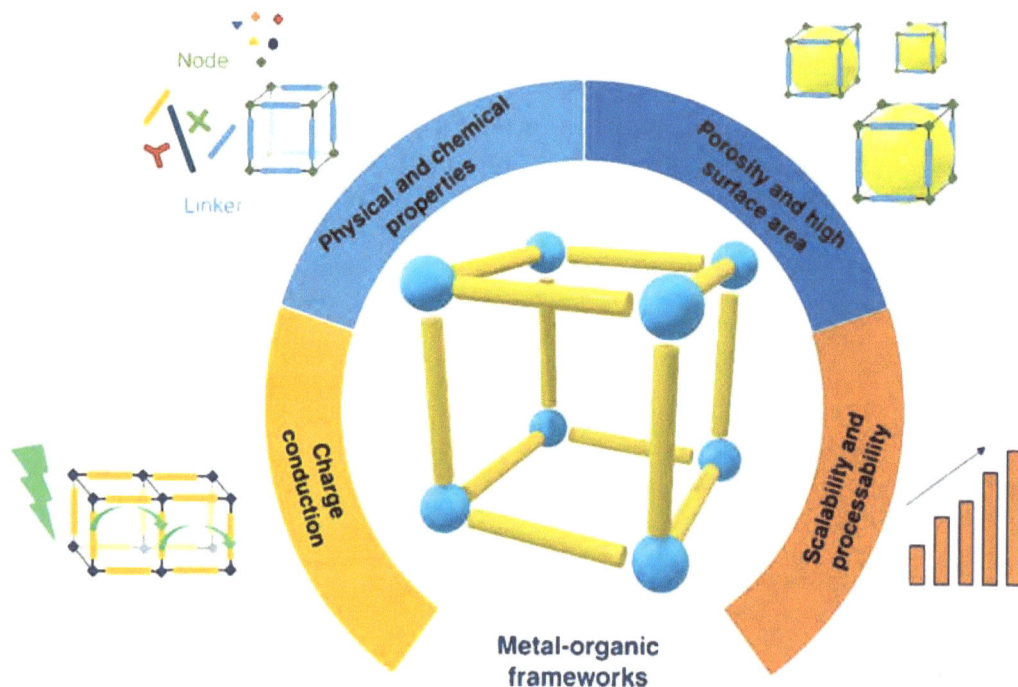

Fig. (1). Characteristics of tunable MOF applicable for electrochemical measurements. The scalability, processability, and functionality with physicochemical properties, adjustable porosity, and high charge conductivity are responsible features of MOFs utilized for electrochemical applications. Adapted with permission from Reference [6], Copyright (2019), Nature.

Transition metal-based MOFs

MOFs based on s-block and p-block metals have been reported in the literature but have received little attention in both basic, conceptual, and applied chemistry research due to their low stability and inflexible chemical properties. These MOFs have limited crystal structures, and the complexity of structure formation limits their applicability in a broader context. On the other hand, *d*-block or transition metals-based MOFs have high stability and are widely reported in the literature. Scandium-based MOFs are employed for highly selective hydrogen storage and carbon dioxide (CO_2) capture, sorption, catalysis, and more [15]. Titanium-based MOFs are well-known photoactive materials for photocatalytic hydrogen evolution reaction (HER). MUV-11 (MUV = Materials of Universidad de Valencia) is a hydroxamate titanium organic framework-based highly porous crystalline material with distorted octahedral geometry possessing excellent chemical stability and photoactivity in the acidic medium [16]. Less progress has been attained in the case of tri- and tetravalent metals-based MOFs as compared to divalent (Zn^{2+} and Cd^{2+}) metal ions. However, vanadium-based MOFs are rarely found but are mostly applicable in catalysis, magnetism, and separation of CO_2, CH_4, N_2, and electrode materials for batteries and supercapacitors (SCs) [17]. Chromium-based MOFs are less investigated but extensively used as carrier systems, gas sorption, catalysis, and environmental remediation [18]. Manganese (II) ions/clusters have been utilized as SBUs to synthesize new MOFs for magnetism and gas adsorption. The Lewis acidic nature of Mn^{2+} catalyzed various catalytic transformations, ozone decomposition, toluene oxidation, and many more conversions. The high stability, robustness, and large porosity enhance the applicability of Mn-based MOFs for energy generation, conversion, and storage devices through electrocatalysis [19]. Iron-based MOFs are well-known candidates for biomedical applications because of their low toxicity towards human beings, high stability, and easiness of synthesis. In addition to this, Fe-MOFs are also applied to other areas such as adsorption and degradation of volatile compounds, environment remediation, biosensing of glucose, catalysis, and drug delivery [20]. Copious of reports are available in the literature on cobalt-based MOFs, which have been broadly investigated due to the cheap and cost-effectiveness of cobalt salts. The penta- and hexa-coordinate geometries of cobalt lead to an increase in the possibilities of different coordination modes making Co-MOFs suitable candidates for electrocatalysis, oxygen and hydrogen evolution, oxygen reduction, catalysis, superconductivity, and magnetism [21]. Nickel-based MOFs are the center of attraction for researchers because of the presence of porosity with coordinatively unsaturated Ni^{2+} sites. The Lewis acidic nature and chemical and thermal stability of Ni-based MOFs enhance their applicability in chemistry i.e. catalytic transformation and conversions such as oligomerization of ethylene, gas separation, methanol oxidation, electrocatalysis towards energy generation, conversion, and storage [22]. A large number of reports are available on copper-containing MOFs which include crystal growth, orientation, modification and volatile organic compounds adsorption, fluorescent and colorimetric detection, sensors, and gas separation (CO_2, CH_4, N_2 *etc.*) [23]. Zinc-based MOFs have their importance and applicability for the separation of gases, materials, and compounds. Hybrid MOFs and composites are extensively used in catalysis for functional group transformation and organic synthesis, detection and removal of antibiotics in an aqueous medium (water), magnetism, photocatalysis, conductivity, and many more [24, 25]. A large number of Cd, Ag, and In-based MOFs are also reported as superior candidates for catalysis, dye adsorption, and SCs.

Inner Transition Metal-based MOFs

At present, Inner transition metal (f-block elements) or rare earth (RE) metals are recognized as an alternate choice for future research and next-generation technologies due to their unique electronic properties stated by their $4f$ and $5f$ sub-shell electronic configuration. It is well-known chemistry that high energy $4f$ orbitals are strongly shielded by less energetic $5s$ and $5p$ orbitals [26]. The electronic and magnetic properties of inner transition metals are not remarkably altered by flexible coordinating ligands. After carefully tuning metal node and organic ligand units, inner transition metal-based MOFs can be accumulated into fascinating network structures having diverse and multifaceted topologies, owing infinite possibilities for the advancement and development of new functional materials. These characteristics of RE-MOFs (La^{3+} to Lu^{3+}) extend their applicability in different kinds of sensing such as cations, anions, small molecules, nitro-aromatic or explosives, gases, vapor, solvents, pH, temperature, bio-sensing and so on. Apart from this, these frameworks also have large-scale applications in white light emission, medical/biological markers, drug delivery, near Infrared (NIR) ray emission, CO_2 valorization, and catalyzes like carbon-carbon bond formation, gas adsorption, and separation [27]. Substantially, Actinide (An)-based MOFs (An-MOFs) comprising actinide nodes and organic molecules or linkers represent an underexplored class of framework materials or coordination polymers, on account of facing challenges in their synthesis and further characterization. The unparalleled structural and coordination chemistry of An-series elements bestows a vast opportunity to explore the structural engineering, chemical reactivity, rational design, and versatile properties of An-MOFs considered one of the most interesting classes of MOFs. Significant progress and advancements have been noticed in the area of An-MOFs since 2003 in the synthesis, properties measurements, and applications. The topology, modularity, and porosity of An-MOFs are highly specified to emphasize a huge potential to refrain their electronic structures and resulting properties. Eventually, some important applications of An-MOFs especially thorium and uranium as scintillators, selective adsorbents, luminescent sensors, heterogeneous catalysts, conductors and semiconductors, and nuclear targets are highlighted.

Mixed Metal-based MOFs

Mixed-metal MOFs (MM-MOFs) came into existence with the synergism and advantage of complexity derived by commencing different kinds of metal ions in the same structure. This resulted in high performance with more stability concerning single-metal-based MOFs. A very simple definition of MM-MOFs as MOFs containing two different metal ions as nodes present anywhere in the same structure has been given by Chen and co-workers [27]. The occurrence of two or more metal nodes in the structure or metal ions immobilization metal complexes, metal nanoparticles as well as organometallic compounds present as guests inside the MOF cavities is known for heterogeneity in MOFs. More than one metal in the structure offers a large number of opportunities regarding multi-functionality, fine-tuning, and changes in the properties of framework material towards a targeted application. Therefore, the existence of two or more metals gives rise to a synergistic effect which leads to better performance as compared to single-meta--based MOFs predominantly in gas separation and storage, catalysis, photoactive materials, thermochromic thermometers, and sensing [28]. MM-MOFs contain two metal centers of dissimilar catalytic performance in close spatial proximity and particularly promote cascade or tandem reactions. These materials have better thermal, chemical, and water stability, which extends the robustness of crystal structure and makes it a suitable candidate for widespread applications. There are several advantages of bimetallic or mixed metallic MOFs due to the synergism of different metals: (1) they demonstrate improved adsorption capacity and electrical conductivity *via* substitution of the second metal site by regulating the molar ratio of bimetallic or mixed metal; (2) they show signs of selective adsorption capacity or good catalytic activity through incorporating metal ions with distinctive open metal sites; (3) they exhibit intriguing luminescence properties by following energy charge transfer mechanisms namely metal to ligand charge transfer, ligand to metal charge transfer, d–d transitions and specifically metal-to-metal charge transfer (MMCT) amid the bimetals [29].

BRIEF DESCRIPTION OF ENERGY STORAGE SYSTEMS: CLASSIFICATION, MECHANISM OF OPERATION, ADVANTAGE AND LIMITATIONS

Energy storage and related technologies are considered the backbone of the advancement and energy consumption as they have a major part in our day-to-day usage, *e.g.* smart watches, headphones, mobile phones, e-vehicles, power banks, batteries, supercapacitors, *etc* [30]. In the last couple of decades, extensive research on various electrode materials has been carried out, and notable advancements have been made. Electrochemical energy storage technologies are mainly classified as batteries and capacitors. These technologies include battery technologies, which work upon different electrochemical reactions to store charge/electricity explicitly lithium-ion (Li-ion) batteries, lead-acid (LA) batteries, sodium-sulfur (SS) batteries, Zn-air batteries, redox flow batteries, and SCs [31]. In electrochemical energy storage (EES) devices, the capacity to store energy and energy output is facilitated by the grouping and parting of ions and free electrons throughout the electrodes. These types of phenomena can ensue either in the bulk of the battery or near/at surfaces in the case of SCs. Already reported electrode materials for such systems are any inorganic or organic nonporous solids made up of simple constituent or porous materials (metal oxides, conductive polymers, and porous carbons) with greater compositional, chemical, and structural inaccuracy. Hence, crystalline porous materials (MOFs, COFs, or hybrid frameworks) have a good combination of high porosity, tenability, structural precision, and versatility [32]. Different types of new electrode materials for SCs and batteries provide enormous scope not only in terms of greater electrochemical performance but also for investigating storage and novel charge transfer mechanisms. A very limited number of metals have desired targeted properties for electrode materials such as attainable potential and molar mass. Based on these standpoints, inorganic materials act as cathodes and face a lot of challenges. As compared to this, organic materials emerged as promising candidates for advanced next-generation electrode materials in recent years since they have notable characteristics like the multi-redox nature of organic groups and their diverse molecular structures. Still, organic materials are also facing some more challenges such as low conductivity, less redox potential, and high solubility in electrolytes. Therefore, a large-scale practical application requires tackling these issues. Alternatively, researchers find a way to resolve these key issues by forming ordered networks-based structures (MOFs) by assembling metal sites and small organic molecules. The designed channels, abundant, and pores create the utmost of organic/metallic redox sites. Electrochemical energy storage systems are briefly discussed below.

BATTERIES

A battery is defined as a combined pack of one or more cells, containing a positively charged electrode termed a cathode, and a negatively charged electrode called anode, separated by a separator and an electrolyte (solid or liquid). In current years, batteries have become the most promising and emerging energy storage devices because of their high energy and power density, stable output voltage, improved safety performance, easily movable, and long cycle life which all allow mature modern industrialized technology and large-scale applicability [33]. So far, a large variety of materials have been tried, optimized, and employed to enhance energy storage and conversion efficiency. Nowadays, researchers have developed various metal ions-based batteries to replace old conventional batteries and upgrade energy storage technologies as per the necessity of the present generation. Here, brief classifications of some main batteries are discussed below:

1. Lithium Batteries (LIBs, LSBs, LABs)
2. Sodium-ion Batteries (SIBs)
3. Zinc Batteries (ZABs, ZN/CBs, ZIBs)
4. Ni−Fe Batteries (NFBs)

In this context, the very first is a lithium-based Li-ion battery (LIB) that works on the back-and-forth movement of Li-ions from the positive electrode (Li source) to the negative electrode (host for Li) through the electrolyte. LIBs are considered "workhorses" for electric vehicles and portable electronics. On the other hand, they still face several challenges, *e.g.*, large volume variation in the ion intercalation and deintercalation resulting in capacity fading, low power density due to sluggish kinetics of ion diffusion, and transmission of reactive sites [34]. In lithium-sulfur batteries (LSBs), sulphur is used as a cathode because of the abundance of low-cost sulphur resources in addition to its exceptionally high theoretical energy density (approx. 2600 WhKg^{-1}). LSBs are widely studied as alternative promising energy storage systems but some challenges limit their commercialization. LIBs have (i) severe polarization and sluggish electrochemical kinetics at the sulphur-based cathode, (ii) poor cycle stability due to the solubility of intermediate polysulfides (LiPSs) in several organic electrolytes, (iii) capacity fading because of a huge change in the volume of cathode throughout charging/discharging process, and (iv) shuttling of polysulfides causes damaging issues [35]. In lithium-air batteries (LABs), the combination of metal Li anode and oxygen cathode shows excellent performance because of their extremely high theoretical energy density (11140 WhKg^{-1}). Nonetheless, the applicability of LABs is hampered due to some challenges: (i) extra product (like Li$_2$CO$_3$) formed by side reactions may obstruct the active sites of O$_2$catalysts and hold back the diffusion of electrolyte ions, (ii) poor rate performance and large over potential caused by sluggish kinetics of electrochemical processes at oxygen cathode (oxygen evolution reaction (OER) and oxygen reduction reaction (ORR), and (iii) low power densities due to less electron transport and sluggish ion diffusion. Although Li-ion-based batteries are excellent commercial energy storage devices at present time, it is expected to reach an energy limit in upcoming years due to the high cost, criticality, and scarcity of lithium [36].

Sodium-ion batteries (SIBs) become the center of attraction and interested candidates for researchers, as sodium is an abundant and very low-cost metal with similar physicochemical properties to Li. Therefore, SIBs turn out to be an alternate potential contender to LIBs for commercial-scale industrial energy storage. Since Na$^+$ possesses a larger size and ionic radius than Li$^+$ and thus a fallout in slow and sluggish kinetics with large volume change at the time of charging and discharging processes. This further leads to poor cycle life and lesser rate capabilities [37]. As a consequence, a good number of LIB electrode materials could not apply to SIBs. This particular problem is restricting the production and development of similar types of SIBs. In continuation, Zn-based batteries are widely investigated due to their relatively low cost, the natural abundance of Zn source, high safety, and outstanding compatibility of aqueous electrolytes. Divalent Zn-ion (Zn^{2+}/Zn) couples gone through a two-electron redox, offers high energy density and theoretical capacity. Zinc-air batteries (ZABs), involving an Ag wire (cathode) were first reported in 1878, and after that, primary ZABs were applied for commercial purposes in the 1930s [38]. Conversely, primary batteries are not able to fulfill the energy demands and requirements through rechargeable power supplies nowadays. So, in the last few years, material science advancement successfully developed rechargeable ZABs. On the contrary, they are facing several challenges such as low power density and high polarization produced by inadequate reactant diffusion and slow-moving kinetics. The rechargeable ZABs involve air cathodes followed by electrochemical OER and ORR at the time of the charging/discharging process. In comparison to ZABs, zinc-nickel/cobalt batteries (Zn-Ni/Co) or (ZN/CBs) can deliver superior discharging voltages with lesser polarization. A ZN/CB system composed of Nickel or Cobalt-based oxides or hydroxides as cathodes attains the energy storage *via* reversible reaction in the presence of alkaline electrolytes. The combination of electrochemical properties of Ni/Co and metallic Zn in alkaline electrolytes determined their high-performance energy storage battery applications [39]. A zinc-ion battery (ZIB) prefers weak acid or neutral aqueous solutions as electrolytes while ZABs and ZN/CBs adopt alkaline electrolytes. There are very less chances to spoil the electrolyte by corroding the packaging material or current collectors and also avoiding injuries or accidents from electrolyte leakage. Additionally, the Zn electrode (anode) demonstrates a great theoretical capacity (820 mAhg^{-1}), low overpotential, and lesser toxicity which attract researchers for further production of ZIBs [40]. Substantially, Ni–Fe batteries (NFBs) have emerged and secured a prominent place in a group of aqueous alkaline (OH$^-$) batteries due to the abundance of Ni and Fe sources, stable output voltage, and excellent electrochemical properties [41]. The sluggish electron and ion transport leads to a poor life cycle and low rate capability which further limits the production of NFBs.

Supercapacitors (SCs)

SCs also termed ultracapacitors, belong to a class of energy storage devices but are different from conventional capacitors and batteries owing to their fast charge/discharge rates, longer cycle life, excellent power, and high energy density. Conversely, the comparatively low energy density turns out to be a significant hindrance towards large-scale wide applications. There are three main types of SCs based on their charge storage mechanisms. The first one is an electric double-layer capacitor (EDLC) and the second is pseudocapacitors (PCs). A new third type is known as hybrid SCs (HSCs). It shows the combined properties of EDLC and PC. The preparation of electrodes by using different techniques to enhance the energy densities along with excellent power densities is an important criterion for SCs. Some more general qualities and characteristics of promising electrodes must enclose abundant redox-active sites, high porosity, large surface area, and good ability for electron transfer. High conductivity with a greater number of active metal sites is recognized as a basic requisite feature of a MOF for electrochemical applications. A descriptive year-wise advancement of MOF-based electrode materials extensively utilized for different kinds of batteries and supercapacitors is portrayed in Fig. (**2**) [4].

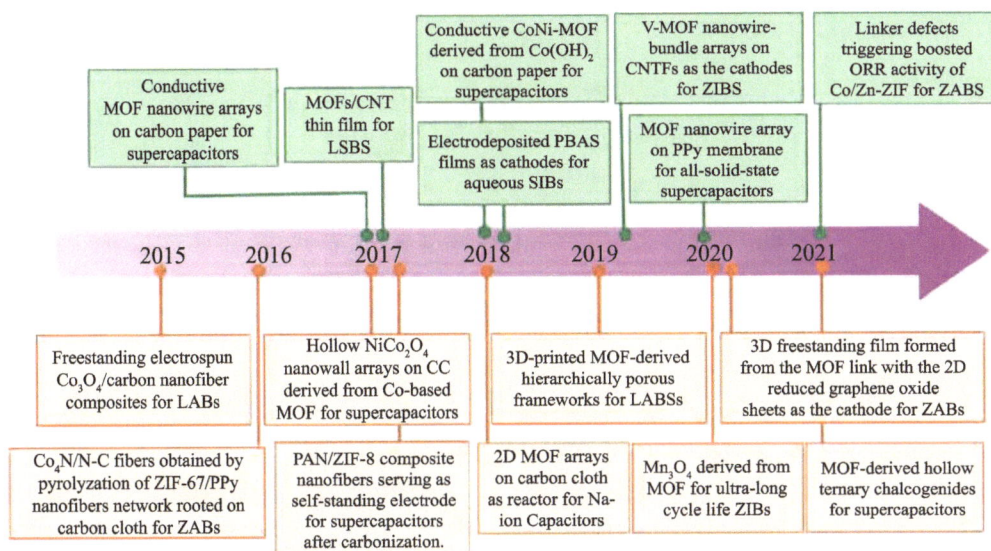

Fig. (2). Year-wise timeline of advancement in freestanding MOF-based/derived electrodes for numerous energy storage systems. Adapted with permission from Reference [4], Copyright (2022), ACS.

All three types of SCs (EDLCs, PCs, and HSCs) are briefly discussed here. Following these characteristics, porous materials (MOFs and their related derivatives) have been confirmed as idyllic and promising electrode materials for present and next-generation SCs as depicted in Table **1** [42 - 51].

Table 1. Some recent MOFs-based and related electrode materials for SCs.

MOF Based Materials	Electrolyte	Electrochemical Performance	Refs.
Ni/Co-N	1 M KOH	361.93 Cg^{-1} @ 2 $mAcm^{-2}$	42
$NiCo_2O_4$	2 M KOH	1055.3 Fg^{-1} @ 2.5 $mAcm^{-2}$	43
$Cu-Co_9S_8$	6 M KOH	2636 Fg^{-1} @ 2 Ag^{-1}	44
NiCo-LDH	2 M KOH	756 Cg^{-1} at 0.5 Ag^{-1}	45
Co-MOF	1 M LiOH	803 Fg^{-1} @ 0.5 $mAcm^{-2}$	46
Mn-MOF	2 M KOH	567.5 $mAh\ g^{-2}$ @ 1 Ag^{-1}	47

(Table 1) cont.....

MOF Based Materials	Electrolyte	Electrochemical Performance	Refs.
NiCoS	1 M KOH	2815.4 Fg^{-1} @ 1 $mAcm^{-2}$	48
Co_3O_4	1 M KOH	226.1 Cg^{-1} @ 1.3 Ag^{-1}	49
CoS_2	6 M KOH	2185 Fg^{-1} @ 1 $mAcm^{-2}$	50
Zn-Co-P	6 M KOH	2115.5 Fg^{-1} @ 1 Ag^{-1}	51

Electric Double-layer Capacitor (EDLC)

EDLCs are the systems that store electrical energy generated by intercalating charges produced at the interface of electrode and electrolyte shaping the double layer of charges. This physical deposition of charges through electrostatic attraction, is the main reason for fast charge and discharge kinetics, durable-life cycling ability (10^6 cycles or more), high power density, and finally negligible degradation because no chemical reaction or side reaction takes place in the process [52]. The most important feature of EDLC is its highly reversible charging-discharging cycles because of its non-faradaic electrical mechanism. The limited number of choices for the selection of electrode materials is the main drawback of EDLCs as EDLC-based devices are required to employ extremely high conductive material-based electrodes. Nowadays, the development and utilization of ionic conductive electrolytes overcome this particular shortcoming. So far, carbon-based materials are the superior candidates for the development of conductive electrodes for EDLCs. A family of carbonaceous materials such as graphene, MXenes, carbon nanotubes (CNTs), activated carbon, graphite, and many more have emerged as the most studied charge storage electrode materials [52, 53]. Furthermore, EDLCs can be divided into three types based on carbon content and properties of carbonaceous materials like structural engineering, morphology, and hybridization: (1) CNTs and graphene; (2) Carbon foams (microporous), carbon aerogels (nanoporous) and carbon derived materials; and (3) Activated carbon.

Pseudo Capacitor (PC)

PCs also known as faradaic SCs were first developed in 1997 and follow faradaic mechanisms to store electric energy by undergoing fast and reversible redox reactions such as charge transfer between the electrode and electrolyte or at/near the electrode surface. The electrode preparation for PCs is based on different methods available in the literature namely electrospinning, intercalation, and redox processes. Although the electrochemical signature of PCs is similar to EDLCs due to the faradaic process, these PCs show higher energy densities than EDLCs [53]. Pseudocapacitance occurs in those cases, where extents of charge (q) passage across the double layer and through the SCs cell during the faradaic charge-transfer process. For the preparation of electrodes, different methods include electrospinning, redox, and intercalation processes. The power density, cycle life stability, and durability of PCs are lower than EDLCs because of the involvement of chemical reactions. Most of the electrode materials of PCs include conductive polymers, metal-doped carbon, and metal oxides. Ruthenium oxide (RuO_2), manganese oxide (MnO_2), and electronically conductive polymers like polypyrroles, polythiophenes, and polyanilines are renowned pseudocapacitive materials having excellent specific capacitance as shown in their theoretical limit [54]. Metal-oxide-based PCs display constant capacitance values with changing voltage.

Hybrid SCs (HSCs)

HSCs are a new type of SCs composed of carbon-based EDLC-type electrode materials and non-polarizable electrode materials (such as metal or conducting polymer) to store the charge through both non-faradaic and faradaic processes. This characteristic makes them suitable candidates for high energy storage capacity by the combination of both battery-type and capacitor-type electrodes resulting in superior long cycling life and cost-efficiency as compared to EDLCs. HSCs exhibit higher operating voltage, superior energy density (up to ten times), and much higher capacitance than conventional symmetric SCs. Their self-discharge capacity is negligible and their standby current is lower. One important characteristic of HSCs is their discharging ability which is not zero (fully discharged) while traditional SCs discharge to zero volts. Although, both are very similar in construction, HSCs make use of graphite laced with a lithium-based anode and a different electrolyte. HSCs are one of the most promising candidates for EES devices due to their flexible and tunable performance mainly to achieve a high energy density and high power density. According to their architecture, optimized high-energy-density devices for batteries and high-power-density devices for SCs are basic key segments for the overall energy storage field [55]. The past achievement and new routes with relevant mechanisms unveil that different kinds of innumerable materials are employed for sodium-ion, lithium-ion, and zinc-ion capacitors but the same trend has not been observed in the case of calcium-ion, magnesium-ion, aluminum-ion, and potassium-ion-based capacitors. After these investigations and studies, these later hybrid ion capacitors are unable to reach the level of industrial and commercial scale as Na-ion, Zn-ion, and Li-ion capacitors. A large variety of electrolytes such as organic, aqueous, ionic liquids, solid, gel, and redox-active are illustrated in different parts of hybrid ion capacitors. So far, the commercial success of HSCs is under development and further divided into three main categories: 1. Asymmetric; 2. Composite; and 3. Battery-type. In general, SCs are proficient systems to deliver high-power density with faster kinetics than batteries, but the energy density of SCs is still far behind batteries.

RESPONSIBLE FACTORS IN MOFS FOR ENERGY STORAGE

Framework materials (MOFs, COFs, or open frameworks) are extensively employed for the development of energy storage systems due to their parallel characteristics as required for making efficient energy storage devices and future energy technologies. Researchers have continuously put efforts into exploring the chemistry behind potential framework materials utilized as electrode materials for energy storage devices (batteries and SCs). The main features of MOFs as novel electrode material are: 1. Porous framework is capable of accommodating more number of ions through a multi-electron redox reaction process, contributing towards high capacity; 2. The ordering of open channels adopted to allow rapid ionic transport in case of bulky anions and multivalent cations; 3. A large variation in the main parameters of an electrode (capacity and voltage) at the molecular level due to the vast structural adaptability, flexibility, and functional tenability of MOFs [56]. In addition to this, some more remarkable factors also responsible for energy storage through MOFs are fast electronic transmission, abundant active sites, large specific surface area, seamless contact, and more tunable porosities. Improving the electrode kinetics and reaction pathway is an important parameter in electrochemical-mediated processes employed in batteries and SCs. The sluggish electrode kinetics at the electrode surface or poor charge transfer limits energy efficiency, capacity, and electrode passivation related to energy applications. Among all emerging porous materials families, the MOF family has publicized significant achievements towards electrode development: (1) MOFs can present strong adsorbability and tunable affinity to the reactant; (2) active catalytic sites and metal centers are well-dispersed throughout the porous framework for synergistic catalysis; (3) deposition of the inert and insoluble product stated by MOF's nanopores compartments reminding the product growth into nanostructures; (4) component variations and structures of MOFs direct to analyze structure-property relationship for making the rational design and optimization possible. To date, numerous multifunctional porous MOFs have been applied as cathode materials to batteries (potassium ion, sodium ion, lithium ions, and zinc ion) [57]. Most of the pristine MOFs are poor conductors of electricity, MOF-related derivatives or MOF-based composites with improved conductivity were employed initially as good electrode materials for the storage of energy. In continuation, some freestanding MOFs namely ZIF-9 derived Co_3O_4 with carbon nanofiber-based composites. ZIF-L associated with other conductive substrates (NF, CNTF, CC, *etc.*) were derived as carbon-based composites for energy storage applications [58].

ENERGY STORAGE PERFORMANCE OF VARIOUS MOF-BASED SYSTEMS

The different characteristics of flexible organic ligands and metal clusters form porous MOFs with emerging structure-property relationships, large surface area, and high conductivity. Overall, as compared to pristine MOFs, a different type of MOF-based composites proved themselves as better candidates for energy storage applications due to the synergistic interaction of active species. In this context, allotropes of carbon (fullerenes, graphene, diamond, and nanotubes) shape up into different micro textures, dimensionality (0D to 3D), and forms (fabric, thin film, composite, powder, and foam) which are efficient and attractive candidates for various energy-related applications [59]. Among them, nanocarbons (CNTs and graphene) have become tremendous materials and are gaining attention due to their outstanding properties and large surface area. Graphene and its derivatized products namely graphene oxide (GO), surface-modified GO, and reduced graphene oxides (rGO) have a layer structure arranged with carbon atoms forming a two-dimensional honeycomb lattice. In contrast, CNTs are ordered allotropes of carbon with a high aspect ratio. These are further divided into two main categories: 1. Single-walled CNTs (0.4–2 nm diameter) and multi-walled CNTs (2–100 nm diameter). Both types of CNTs have excellent thermal and chemical stability, high tensile strength, and are ultra-lightweight. The outstanding thermal, mechanical, and electrical properties of CNTs and graphene claimed them as important and valuable nanostructured fillers in MO-based composites [60]. The combination of inorganic functional materials and nanocarbons results in an enhancement in their individual properties and is further predetermined for sustainable energy and environmental applications. To date, innumerable MOF–MOF-nanocarbon composites have been effectively prepared with CNTs, carbon monoliths, GO, and activated carbons and widely investigated for miscellaneous applications. The merits of MOF-based materials for electrochemical energy storage applications are demonstrated in Fig. (3). The correlation between different kinds of pristine MOFs and derivatized MOF composites with their crystal structure and physicochemical properties is shown. The substitution and addition of different kinds of conductive materials to make efficient MOF-based electrodes for ECS applications are also discussed.

Pristine MOFs

The design of pristine MOFs is accomplished by careful selection, modification, and variation of metal sites and organic ligands/linkers. The physicochemical properties of MOFs, such as electrical conductivity, electronic structures, hydrophilicity, separation efficiency of electron and hole, stability, and durability can be adjusted by a component design along with satisfactory advantages for excellent efficiency with durable energy storage applications. Activity and selectivity of pristine MOFs depend on the component design for metal sites by different strategies such as metal site defects, doping of heterometallic elements, coordination-based unsaturated metal sites, multiple valences, and oxidation state of metal centers along with external coordination ions.

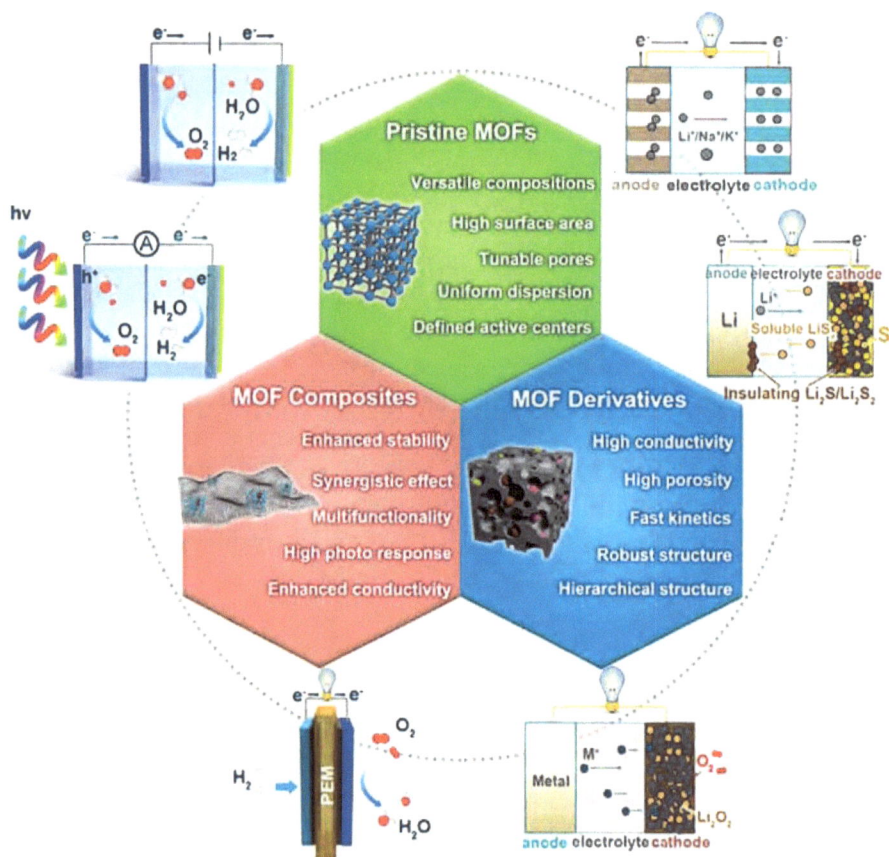

Fig. (3). Merits and advantages of MOF-based materials for electrochemical storage applications. Adapted with permission from Reference [14], Copyright (2020), ACS.

In addition to this, component designing of organic ligands is done*via* metal site adsorption, functional group modification, organic ligand defects, and selection of extensively large π-conjugated organic linkers/ligands. The electrochemical energy storage activity of pristine MOFs depends on the metal ions that are present in the basic frames of the as-prepared MOFs, coordinated organic ligand moieties, and guest solvent molecules. In most of the synthesized MOFs, the exclusion of the solvent molecules may create coordinatively unsaturated metal sites, act as Lewis acid centers with electron deficiency, and be extensively ready to accept electrons by reactant species [61]. Similar to metal nodes, the electrochemical energy storage capacity of pristine MOFs also depends on organic ligands, which provide various opportunities to design tunable functionalities. The organic ligand functionalization by introducing electron-donating functional groups ($-NH_2$ and $-SH$) can effectively improve the electrochemical and photochemical properties of pristine MOFs. The carboxylate ($COO-$) groups, nitrogen-containing (N-) groups, and planar and non-planar aromatic rings associated with organic ligands of pristine MOFs act as lithium insertion sites for further Li storage [62]. In the field of energy storage, the conductive MOF may be used without any conductive additive or binder. In a recent report, Li *et al.* [63] proposed a conductive Cu-MOF-based additive or binder-free high-performance solid-state supercapacitor. The resulting solid-state supercapacitor shows a very large surface area-normalized capacitance of ≈ 22 μFcm^{-2}. Therefore, MOF-based electrode materials could be promising for futuristic self-standing energy storage systems having higher energy density, and durable cycling performance.

MOF Composites

Several reports reveal that pristine MOFs are unable to offer desirable energy storage performance due to the inherent deficiencies in conventional MOFs such as low conductivity and restricted functionality. It has been demonstrated that mixing MOFs and high-functionality materials to produce MOF-based composites may overwhelm the limitations of the conventional MOFs along with keeping the initial recompenses of MOFs. A variety of functional materials like reduced graphene oxide (rGO), carbon nanotubes (CNTs), metal nanoparticles (MNPs), porous supports, complex molecules, and highly conductive substrates are recognized to couple with MOFs for the synthesis of MOF composites designed to boost the electrochemical energy storage applications. There are two general ways to introduce functionality by incorporating functional material in any conventional MOFs: (1) Active functionalized materials act as substrates to sustain MOFs. (2) The functionalized material is fully encapsulated inside the MOF crystals. A combination of highly sophisticated and advanced surface-sensitive characterization techniques namely ex-situ and *in-situ* high-resolution transmission electron microscopy (HRTEM), synchrotron X-ray spectroscopy, Raman spectroscopy, and Photo-/electrochemical properties measurements provide meticulous information and deep understanding of tailored pristine MOFs. The identification of the most active species, and the proposed mechanistic pathway lead to the creation of robust and durable MOFs for energy storage applications.

MOF@rGO

In recent years, the MOF and graphene composite emerged as a noteworthy materials for diverse applications, especially energy storage devices (batteries and SCs). Generally, the synthetic approaches for the formation of MOF/graphene materials are divided into main three categories based on MOFs and graphene combination: (1) Physical mixing; (2) *In-situ* growth; (3) excess metal-ion supported *in-situ* growth. In these routes, GO is initially employed and then *in-situ* reduced (rGO) to attain functional MOF/graphene-based nanocomposites. Graphene or reduced graphene or chemically modified graphene is extensively used for electrochemical energy-related applications owing to its extremely high specific surface area with abundant active functional groups. This further provides a model template for the controllable and appropriate growth of MOF particles. These MOFs and graphene assembly have several advantages: (1) the ultrathin-layered graphene sheets acquire stability in the presence of MOFs, confirming a large accessible surface with numerous active sites. (2) The coupling of MOFs and graphene promotes electrode transport over the whole electrode which further overcomes the poor conductivity of MOFs. (3) MOFs and graphene combination endow fast electrochemical processes with the durable life cycle of the electrode. (4) The composite has hierarchical pores which provide a perfect space for electrolyte entrance and reduce mass transfer resistance resulting in fast reaction kinetics [64, 65]. The shape and size of both the materials used for composites affect the overall performance of the composites. The coordination, electrostatic interactions, and strong chemical interface amid rGO and MOF nanocrystals lead to increased ion transport kinetics and adsorption energy for excellent energy storage performance. For example, Zn-MOF, rGO, and conductive polymer polyaniline (PANI) composites were successfully achieved by a step co-assembly method. The rGO was used to provide the porous conductive support for the crystalline structure of Zn-based MOF. The accumulation of PANI into an as-synthesized composite facilitates the smooth charge transfer between the electrode and electrolyte solution, which results in rGO/Zn-MOF@PANI possessing a high specific capacitance of 372 Fg^{-1} at a current density of 0.1 Ag^{-1} due to its large surface with high pores sizes, crystallinity and structure stability [64].

MOF@CNT

To upgrade existing and conventional electronic technology, the necessity to develop new energy and power supply devices is in demand. The fast-growing electronics market demands investigation of advanced energy storage devices. In this context, MOFs/CNT-based composites have attracted remarkable attention, due to their matchless electrochemical performances, mechanical flexibility, and long durable cyclability. CNTs are recognized as excellent candidates for energy-related applications because of their tough mechanical strength (due to the sp^2 carbon-carbon bond) and good elastic strain limit (up to 20%) [65]. In addition to this, CNTs possess high electrical conductivity, excellent mechanical properties, good chemical stability, low mass density, large specific surface area, and flexible dimensionality (1D fibers, 2D films, and 3D sponges). The adaptability in these assemblies from 1D to 2D to 3D is a unique and attractive feature of CNTs. Recent studies have verified that porous structure and active sites of CNTs form adaptable conditions for the transport of high charge during electrochemical reactions and notably enhance the functionality of batteries and SCs. Some additional features like ease of functionalization, good interaction with electrolyte ions and large aspect ratio make them a remarkable material for MOF-based composites [66]. Therefore, MOFs/CNT composites show good compressibility and str*etc*hability towards energy storage applications in flexible supercapacitors. For example, Ni-MOF/CNT composites were successfully derived through the solvothermal method by directly growing Ni-MOF onto the CNT surface. The Ni-MOF/CNT composite shows a superior capacitance of 1765 Fg^{-1} at a current density of 0.5 Ag^{-1}. The enhanced electrochemical performance indicates the specific structural effects of the Ni-MOF and the high conductivity of CNTs plays a major role in boosting energy storage performance. Further, this composite can be used to fabricate asymmetric devices, where rGO/g-C$_3$N$_4$ can be used as the negative electrode and Ni-MOF/CNT can be used as a positive electrode to achieve a superior energy density of 36.6 Whkg^{-1} and power density of 480 Wkg^{-1} [66].

MOF@NPs

The combination of nanoparticles (NPs) and porous MOFs develops MOF/nanoparticle composites, which are recognized as tremendous composite materials for energy storage technologies. There are two main strategies to encapsulate nanoparticles into the MOF's structure. The first method demonstrates that a MOF acts as a template to grasp guest nanoparticles inside their channels while in the second method, there is an encapsulation of pre-synthesized nanoparticles to MOFs [67]. The former method includes several well-known techniques like solid grinding, chemical vapor deposition (CVD), double solvent methods, and liquid impregnation whereas the second method involves some new techniques namely the self-sacrificing template technique. These MOFs/NP composites find potential applications in different areas. In particular, MOF layers built on NPs can be used as an excellent catalyst for heterogeneous catalysis, photocatalysis, and electrocatalysis. In comparison to normal MOFs, nanosized MOFs fabricate NPs in a better way and further enhance the applicability of conventional MOFs by reducing the size (approx. 50 nm) of MOF crystals. The strong interactions and synergistic effects among MOFs and

NPs contribute vastly to the electrochemical energy storage performances of MOF composites. There are several other functional materials like quantum dots (QDs), polyoxometalates (POMs), and metal complexes which are recognized as efficient materials for making MOF composites [68]. The physical and chemical properties of these functional materials are perfectly matched and enhance the stability and active sites of the composite for energy storage performance in batteries and SCs. However, most of the pristine MOFs have poor conductivity which restricts their performance in the field of energy storage. To overcome the poor conductivity of MOFs and produce a promising storage performance of the same, Wang's [68] group synthesized a composite of the Co-based zeolite imidazolate framework (ZIF-67) nanoparticles of MOFs crystal with the electrochemically deposited PANI onto carbon cloth to give a flexible conductive porous electrode. The composite PANI-ZIF-67-CC exhibits an outstanding areal capacitance of 2146 $mFcm^{-2}$ at 10 mVs^{-1} and is suitable for symmetric solid-state supercapacitor devices.

MOF Derived Materials

Presently, MOFs and MOF-based composites are extensively applied as precursors to fabricate different kinds of nanomaterials for energy generation, energy conversion, and energy storage applications. The high diversity and functionality of organic ligands and metal ions make MOFs idyllic podiums to fabricate a variety of functional compounds such as metal oxides, hydroxides, carbides, sulphides, nitrides, phosphides, carbons, and metal-carbon, along with their composites as displayed in Fig. (4). The design of a component of MOF-derived systems by using carbon, metal-carbon, metal oxides, hydroxides, nitrides, carbides, sulfides, and phosphides is shown. The surface modification of pristine MOFs with different types of materials leads to enhanced efficiency in the electrochemical energy storage system (Fig. 4).

All these kinds of MOF-derived materials are not only limited to the advantages of MOFs such as extensive surface area with high porosity but also propose innumerable prospects to decorate and design several structural/different compositional features through a controlled conversion process (for example ion-assisted solvothermal or pyrolysis) of already designed MOF precursors. Metal oxide, hydroxide, and other related nanomaterials possess controllable size, shape, functionality, and crystallinity and are commonly used in a variety of applications namely optics, catalysis, electronics, solar energy harvesting, electrochemical energy conversion and storage, and many more. Metal oxides and core-shell nanostructures are used to improve physicochemical properties like electric conductivity, semiconducting, and magnetic properties. They are considered as emerging composites for storage applications. The method of preparation for

MOF-derived materials is almost similar to the designed MOF-based composites (MOF@NPs). A familiar method is the generation of metal oxides inside the cavities of MOFs by following decomposition or oxidative annealing of already loaded precursors.

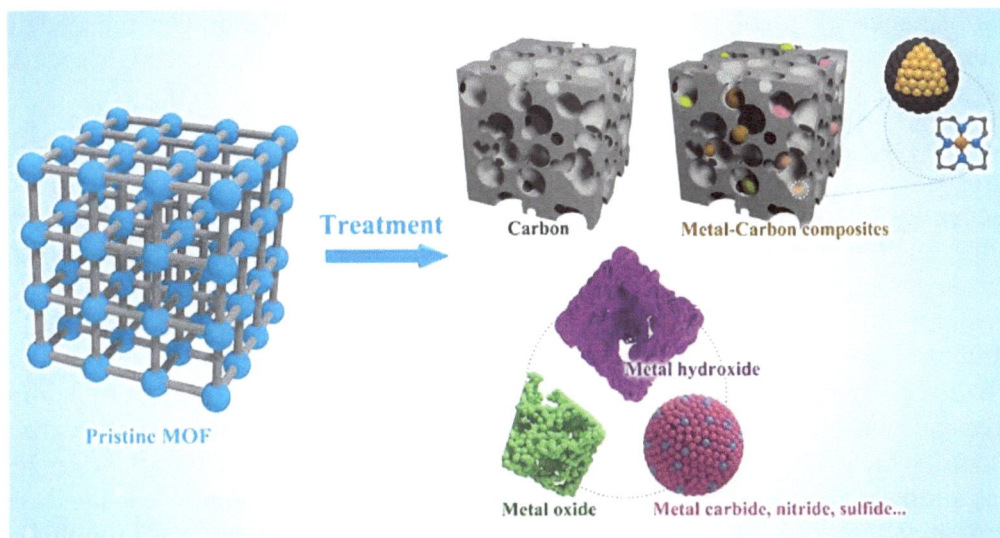

Fig. (4). The design of component of MOF-derived materials. Adapted with permission from Reference [14], Copyright (2020), ACS.

MOF/C

Carbon-based MOF composites contain a large surface area with high porosity and are extensively utilized for adsorption, as detoxifying agents in biological systems, and in electrochemical energy storage applications. A family of traditional adsorbents such as zeolites, carbon cloth, activated carbon black, carbon fiber, activated alumina, and silica emerges as potential candidates for various separation technologies in pharmaceutical, chemical, and petrochemical industries with high efficiency, low cost, and easy operation. Although, MOFs are born adsorbents and extraordinarily employed for storage, separation, and gas/liquid adsorption. MOFs and carbonaceous materials (MOF/C) and related composites have revealed tremendous physicochemical properties and are considered benchmark materials for industrial usage. Moreover, different loadings of charcoal in Ln–MOF/AC composites have remarkable applications for the removal of toxic pesticides like aldicarb or organochlorine. Even more, MOF/C composites are recognized as solid-phase adsorbents for selective hemoglobin extraction from human blood. However, this chapter only concentrates on the energy storage application of MOF-based materials, and thus performances of MOF and carbon-based composite electrodes have also been reviewed. The versatile functionality and active sites of MOF/C composites can demonstrate excellent energy storage performance in batteries and SCs. For example, a composite made of Cu-MOF, reduced graphene oxide, and a conducting polymer PANI synthesized through a chemical route includes an in situ polymerization process. The incorporation of PANI in composite boosted the pseudocapacitance performance by enabling high electrical conductivity and abundant presence of nitrogenous redox active sites. Finally, this composite showed a high specific capacitance of 276 Fg^{-1} operated at a current density of 0.5 Ag^{-1} [65].

MOF/Metal-Carbon

Generally, the metals used in metal-carbon are heteroatoms such as boron (B), nitrogen (N), phosphorus (P), or sulfur (S). All these metals are attuned to the carbon structure and can be able to efficiently regulate its inherent properties such as superficial and local chemical characteristics, and electronic, mechanical, and structural properties. An effective doping of the nitrogen atom is considered an idyllic choice due to its similar atomic size to the carbon atom and the remaining five valence electrons of a conductive nature form strong bonds with carbon atoms, which further demonstrates the broad range of applications. The easiest and simplest method to obtain *N*-doped MOF/metal-carbon composite is the direct carbonization of MOF precursors having *N*-containing ligands. Through this process, the active sites of the MOF/N-C composite are consistently distributed within the framework. As a result, N-doped MOF/metal-carbon composite material shows excellent electrochemical performance for energy generation and storage applications. In a report, an N-doped carbon composite was synthesized by using Tetracyanoquinodimethane (TCNQ) as a ligand in MOF. The resulting material acts as a promising electrode for SCs application, exhibiting a superior capacitance of 223.7 Fg^{-1} at 1 Ag^{-1}, where the capacitance retention of the prepared material is 73.4% after 3000 cycles at 1 Ag^{-1} [65]. Likewise, other heteroatoms (P, B, or S) may also be doped through a secondary carbon source by submerging MOFs in an organic solution restraining these elements. Remarkably, the heteroatom P or S doping may bring on redistribution of charges of carbon atom and result in the weakening of O–O bonds, which further enhance the electrochemical activities (OER, ORR, and energy storage) of MOF/metal-carbon nanocatalysts with durable long-term stability.

MOF/Metal Oxide

Metal oxides or derived nanomaterials are well-known remarkable catalysts with controllable size, shape, crystallinity, and functionality that are regularly applied in different fields. Their exceptional characteristics like high theoretical specific capacitance, great reversibility, and low cost make them ideal candidates for pseudocapacitive electrode materials, but contrary to this, their high surface energies and aggregation, result in loss of the pseudocapacitive performance. Whereas, metal oxides usually exhibit only small specific surface areas, which have chiefly restricted their applicability as electrode materials for other electrochemical energy storage systems. A cost-effective approach to increase the surface areas of metal oxides is developing an MOF-based composite, which further leads to achieving high pseudocapacitive performance. Numerous strategies are adopted to integrate the advantages and synergistic interactions of MOFs and metal oxides contributing effectively to electrochemical reactions and energy storage during the charging/discharging process. Mainly, the metal oxides efficiently increase the redox-active sites and MOFs provide a specific surface area for a variety of catalytic activities. Liu *et al.* [68] prepared a P-Co_3O_4@PNC from a Co-MOF through a pyrolysis–oxidation–phosphorization process. Heteroatom P doping creates structural defects, increasing the conductivity and also creating more electroactive sites. The resultant P-Co_3O_4@PNC composite has a superior specific capacity of 614 $mCcm^{-2}$ at 1 $mAcm^{-2}$. Furthermore, P-Co_3O_4@PNC and PNC are designed for a solid-state ASC device having an extraordinary energy density of 69.6 $Whkg^{-1}$ and power density of 750 W kg^{-1} with durable cycling stability of 96.8% after 10000 cycles at 20 Ag^{-1}.

Transition-metal oxide-based MOF derivatives have attracted the most attention due to their superior capacities and easiness of handling and fabrication for batteries and SCs. Their unique design and features are: (1) the calcinations of MOFs converted into the self-template formation of metal oxides possess convenient particle size, shape, and morphology while nano-size or hollow porous structures have more active sites, large volume expansion, and enhanced ion transport by distance shortening. (2) More than one metal or multimetallic oxides are synthesized by calcination of heterometallic MOF precursors and have shown better electrochemical activities than single metal oxide-based MOF composites.

MOF/Metal Hydroxide

Metal hydroxides have similar characteristics, features, and results to metal oxides. Layer double hydroxides (LDHs) are emerging materials with high visual contrast, fast response time, charge density, a large variety in elemental substitution, effective intermolecular interactions amid active site inter-layers and guest molecules, host-guest interactions ensuing in functionalization and immobilization capability, and also found their potential applications in different fields like drug delivery, catalysis, separation, photo- and electrochemical cells, smart luminescent systems and sensors. Presently, MOF/metal hydroxide or LDHs or 2D nanosheet-based composites have attracted huge interest from researchers and chemists owing to their proficient performance in electrochemical measurements and photocatalysis. The most important interesting features of LDHs are their physicochemical stability, non-toxicity, low cost, simple synthesis, and high electronic level due to the intercalation of large bulky anions inside the inner structure. Zhu *et al.* [69] synthesized $NiFe_2O_4$@NiCo-LDH nanocube as an anode for lithium-ion batteries. The as-prepared $NiFe_2O_4$@NiCo-LDH shows a high specific capacity of 636.9 mAhg^{-1} after 100 cycles, high Coulombic efficiency, and excellent rate performance. The inter-layers of LDHs contain aqueous water molecules which further help to improve their photo- as well as electrocatalytic activities towards energy storage systems like batteries and SCs.

MOF/Metal Carbide or Sulphide or Nitride or Phosphide

Metal carbides (MCs), metal sulphides, (MSs) metal nitrides (MNs), and metal phosphides (MPs) have been exciting active materials for electrodes preparation because of their exceptional physical properties, specifically high melting points (for example, tantalum carbide (TaC) and hafnium carbide (HfC) exhibit the highest known melting points). Besides this, they display good chemical stability and high electrical conductivity. In addition to this, these materials have exceptional mechanical and chemical properties for industrial applications initially as rotors within gas turbines, cutting tools, and protective layer coating in fusion reactors and later on in electrochemical workstations. Three main different interactions exist between the atoms of these compounds: ionic bond, covalent bond, and metallic bond. A combination of excellent electronic properties and good ionic conductivity makes them tremendous materials for electrochemical energy storage applications in batteries and capacitors. However, MOF-based composites of carbides, sulphides, nitrides, and phosphides undergo large volume expansion and are almost twice as high compared to metal oxides, potentially causing mechanical deterioration of the electrode all through cell reaction. As compared to Pt-group metals, several more advantages such as natural abundance and cost of the raw transition metals, exceptional thermal stability, and constant tolerance to ordinary catalyst poisons, make carbides, nitrides, sulphides, and phosphides different and alternative choice to the noble metals for electrochemical energy storage applications in recent years. Han *et al.* proposed a highly porous cobalt sulfide nanosheet array based on Ni foam substrate (Co_9S_8-NSA/NF) through a facile synthesis method. Co_9S_8-NSA/NF composite was prepared by solvothermal sulfurization, where at the first stage, 2D leaf-like MOF was grown on NF. The resultant composite is suitable as a binder-free electrode for supercapacitor applications having a high capacitance of 1098.8 Fg^{-1} at 0.5 Ag^{-1} [70].

SUMMARY AND FUTURE PERSPECTIVES

In summary, the ongoing research progress in the synthesis, design, and electrochemical energy storage applications of MOFs and MOF-based composites are thoroughly discussed. The synthetic strategies, advantages, and factors responsible for developing such electrodes to achieve high-performance energy storage devices like batteries and SCs are highlighted. In comparison to other conventional electrodes, MOF-based/derived electrodes become more efficient due to their conductive substrates, high mechanical stability, and fast ions and electron transport pathway. The poor conductivity limitation of pristine MOFs has also been overcome by using MOF composites-based electrodes. In combination with a flexible crystal structure, high specific surface area, a greater number of active sites and high porosity with tenability, MOF-based/derived electrodes possess numerous unique advantages for energy storage applications for batteries like sodium-ion batteries, lithium-ion batteries, lead-acetate batteries (SIBs, LIBs, and LABs), Ni-Fe batteries, zinc-ion batteries (ZABs, ZN/CBs, ZIBs), and SCs (EDLCs, PCs, and HSCs). After all these achievements, the power density, efficiency, durability, and life cycle of MOF-based electrodes still face a lot of challenges for large-scale commercial and industrial applications in the future. The existing problems, challenges, and difficulties in designing MOF-based electrodes are meticulously analyzed, discussed, and summarized from the perception of mechanism, performance, standard, multifunctionality, fabrication, scalability, and durability. The present chapter will provide a clear viewpoint about the design, properties, and applicability of different MOF-based electrode materials that are considered candidates for the next generation of sustainable energy storage technologies.

ACKNOWLEDGEMENTS

The corresponding author (DST-INSPIRE faculty registration No.: IFA17-MS135) is thankful to the DST-INSPIRE-FACULTY program for supporting the work by providing a suitable research grant. NK is thankful to CSIR-NML for the research support grant through the OLP-0426 project. The authors are also thankful to the CSIR, India for the support given through MLP 3119 project. The authors are thankful to the Director of CSIR-NML for his support and encouragement to conduct this work.

REFERENCES

[1]   H. Li, M. Eddaoudi, M. O'Keeffe, and O.M. Yaghi, "Design and synthesis of an exceptionally stable and highly porous metal-organic framework", *Nature,* vol. 402, no. 6759, pp. 276-279, 1999. [http://dx.doi.org/10.1038/46248]

[2]   A.K. Cheetham, G. Férey, and T. Loiseau, "Open-framework inorganic materials", *Angew. Chem. Int. Ed.,* vol. 38, no. 22, pp. 3268-3292, 1999.

[http://dx.doi.org/10.1002/(SICI)1521-3773(19991115)38:22<3268::AID-ANIE3268>3.0.CO;2-U] [PMID: 10602176]

[3]   S. Kitagawa, and M. Kondo, "Functional micropore chemistry of crystalline metal complex-assembled compounds", *Bull. Chem. Soc. Jpn.,* vol. 71, no. 8, pp. 1739-1753, 1998. [http://dx.doi.org/10.1246/bcsj.71.1739]

[4]   B. He, Q. Zhang, Z. Pan, L. Li, C. Li, Y. Ling, Z. Wang, M. Chen, Z. Wang, Y. Yao, Q. Li, L. Sun, J. Wang, and L. Wei, "Freestanding metal–organic frameworks and their derivatives: An emerging platform for electrochemical energy storage and conversion", *Chem. Rev.,* vol. 122, no. 11, pp. 10087-10125, 2022. [http://dx.doi.org/10.1021/acs.chemrev.1c00978] [PMID: 35446541]

[5]   A. Bavykina, N. Kolobov, I.S. Khan, J.A. Bau, A. Ramirez, and J. Gascon, "Metal-organic frameworks in heterogeneous catalysis: Recent progress, new trends, and future perspectives", *Chem. Rev.,* vol. 120, no. 16, pp. 8468-8535, 2020. [http://dx.doi.org/10.1021/acs.chemrev.9b00685] [PMID: 32223183]

[6]   A.E. Baumann, D.A. Burns, B. Liu, and V.S. Thoi, "Metal-organic framework functionalization and design strategies for advanced electrochemical energy storage devices", *Commun. Chem.,* vol. 2, no. 1, pp. 86-99, 2019. [http://dx.doi.org/10.1038/s42004-019-0184-6]

[7]   S. Dhibar, P. Bhattacharya, G. Hatui, S. Sahoo, and C.K. Das, "Transition metal-doped polyaniline/single-walled carbon nanotubes nanocomposites: efficient electrode material for high performance supercapacitors", *ACS Sustain. Chem. Eng.,* vol. 2, no. 5, pp. 1114-1127, 2014. [http://dx.doi.org/10.1021/sc5000072]

[8]   N. Kumar, M. Kumar, T.C. Nagaiah, V. Siruguri, S. Rayaprol, A.K. Yadav, S.N. Jha, D. Bhattacharyya, and A.K. Paul, "Investigation of new B -site-disordered perovskite oxide CaLaScRuO 6+δ : An Efficient oxygen bifunctional electrocatalyst in a highly alkaline medium", *ACS Appl. Mater. Interfaces,* vol. 12, no. 8, pp. 9190-9200, 2020. [http://dx.doi.org/10.1021/acsami.9b20199] [PMID: 32045211]

[9]   P. Bhattacharya, J.H. Lee, K.K. Kar, and H.S. Park, "Carambola-shaped SnO2 wrapped in carbon nanotube network for high volumetric capacity and improved rate and cycle stability of lithium ion battery", *Chem. Eng. J.,* vol. 369, pp. 422-431, 2019. [http://dx.doi.org/10.1016/j.cej.2019.03.022]

[10] J.W. Gittins, C.J. Balhatchet, S.M. Fairclough, and A.C. Forse, "Enhancing the energy storage performances of metal–organic frameworks by controlling microstructure", *Chem. Sci.,* vol. 13, no. 32, pp. 9210-9219, 2022. [http://dx.doi.org/10.1039/D2SC03389E] [PMID: 36092998]

[11] I. Hussain, S. Iqbal, C. Lamiel, A. Alfantazi, and K. Zhang, "Recent advances in oriented metal-organic frameworks for supercapacitive energy storage", *J. Mater. Chem. A Mater. Energy Sustain.,* vol. 10, no. 9, pp. 4475-4488, 2022. [http://dx.doi.org/10.1039/D1TA10213C]

[12] N. Kumar, K. Naveen, M. Kumar, T.C. Nagaiah, R. Sakla, A. Ghosh, V. Siruguri, S. Sadhukhan, S. Kanungo, and A.K. Paul, "Multifunctionality exploration of Ca 2 FeRuO 6 : An efficient trifunctional electrocatalyst toward oer/orr/her and photocatalyst for water splitting", *ACS Appl. Energy Mater.,* vol. 4, no. 2, pp. 1323-1334, 2021. [http://dx.doi.org/10.1021/acsaem.0c02579]

[13] Y. Tang, H. Zhang, Y. Jin, J. Shi, and R. Zou, "Boosting the electrochemical energy storage and conversion performance by structural distortion in metal-organic frameworks", *Chem. Eng. J.,* vol. 443, p. 136269, 2022. [http://dx.doi.org/10.1016/j.cej.2022.136269]

[14] T. Qiu, Z. Liang, W. Guo, H. Tabassum, S. Gao, and R. Zou, "Metal-organic framework-based

materials for energy conversion and storage", *ACS Energy Lett.,* vol. 5, no. 2, pp. 520-532, 2020.
[http://dx.doi.org/10.1021/acsenergylett.9b02625]

[15]   N.M. Padial, J. Castells-Gil, N. Almora-Barrios, M. Romero-Angel, I. da Silva, M. Barawi, A. García-Sánchez, V.A. de la Peña O'Shea, and C. Martí-Gastaldo, "Hydroxamate titanium–organic frameworks and the effect of siderophore-type linkers over their photocatalytic activity", *J. Am. Chem. Soc.,* vol. 141, no. 33, pp. 13124-13133, 2019.
[http://dx.doi.org/10.1021/jacs.9b04915] [PMID: 31319033]

[16]   Y. Yan, Y. Luo, J. Ma, B. Li, H. Xue, and H. Pang, "Facile synthesis of vanadium metal-organic frameworks for high- performance supercapacitors", *Small,* vol. 14, no. 33, p. 1801815, 2018.
[http://dx.doi.org/10.1002/smll.201801815] [PMID: 30028570]

[17]   J. Park, D. Feng, and H.C. Zhou, "Dual exchange in PCN-333: A facile strategy to chemically robust mesoporous chromium metal–organic framework with functional groups", *J. Am. Chem. Soc.,* vol. 137, no. 36, pp. 11801-11809, 2015.
[http://dx.doi.org/10.1021/jacs.5b07373] [PMID: 26317830]

[18]   S. Sangeetha, G. Krishnamurthy, S. Foro, and K. Raj, "Energy storage applications of cobalt and manganese metal–organic frameworks", *J. Inorg. Organomet. Polym. Mater.,* vol. 30, no. 11, pp. 4792-4802, 2020.
[http://dx.doi.org/10.1007/s10904-020-01593-8]

[19]   D. Fan, C. Chen, S. Lu, X. Li, M. Jiang, and X. Hu, "M. Jiang andX. Hu,"Highly stable two-dimensional iron monocarbide with planar hypercoordinate moiety and superior Li-ion storage performance,"", *ACS Appl. Mater. Interfaces,* vol. 12, no. 27, pp. 30297-30303, 2020.
[http://dx.doi.org/10.1021/acsami.0c03764] [PMID: 32396323]

[20]   P. Zhou, J. Wan, X. Wang, K. Xu, Y. Gong, and L. Chen, "Nickel and cobalt metal-organic-frameworks-derived hollow microspheres porous carbon assembled from nanorods and nanospheres for outstanding supercapacitors", *J. Colloid Interface Sci.,* vol. 575, pp. 96-107, 2020.
[http://dx.doi.org/10.1016/j.jcis.2020.04.083] [PMID: 32361050]

[21]   Y. Zhou, Z. Mao, W. Wang, Z. Yang, and X. Liu, "In-situ fabrication of graphene oxide hybrid Ni-based metal–organic framework (Ni–MOFs@ GO) with ultrahigh capacitance as electrochemical pseudocapacitor materials", *ACS Appl. Mater. Interfaces,* vol. 8, no. 42, pp. 28904-28916, 2016.
[http://dx.doi.org/10.1021/acsami.6b10640] [PMID: 27696813]

[22]   N. Kumar, T. Rom, V. Singh, and A.K. Paul, "Transition metal ions regulated structural and catalytic behaviors of coordination polymers", *Cryst. Growth Des.,* vol. 20, no. 8, pp. 5277-5288, 2020.
[http://dx.doi.org/10.1021/acs.cgd.0c00465]

[23]   T. Rom, N. Kumar, A. Agrawal, A. Gaur, and A.K. Paul, "Syntheses, crystal structures, topology and dual electronic behaviors of a family of amine-templated three- dimensional zinc-organophosphonate hybrid solids", *J. Mol. Struct.,* vol. 1263, p. 133087, 2022.
[http://dx.doi.org/10.1016/j.molstruc.2022.133087]

[24]   N. Kumar, and A.K. Paul, "Triggering Lewis acidic nature through the variation of coordination environment of Cd-centers in 2D-coordination polymers", *Inorg. Chem.,* vol. 59, no. 2, pp. 1284-1294, 2020.
[http://dx.doi.org/10.1021/acs.inorgchem.9b02997] [PMID: 31916441]

[25]   A.K. Paul, K. Naveen, N. Kumar, R. Kanagaraj, V.M. Vidya, and T. Rom, "First example of a nonananuclear silver sulfate hybrid cluster: green approach for synthesis of Lewis acid catalyst", *Cryst. Growth Des.,* vol. 18, no. 11, pp. 6411-6416, 2018.
[http://dx.doi.org/10.1021/acs.cgd.8b01258]

[26]   F. Saraci, V. Quezada-Novoa, P.R. Donnarumma, and A.J. Howarth, "Rare-earth metal–organic frameworks: From structure to applications", *Chem. Soc. Rev.,* vol. 49, no. 22, pp. 7949-7977, 2020.
[http://dx.doi.org/10.1039/D0CS00292E] [PMID: 32658241]

[27]   N. Kumar, R. Bhowal, D. Chopra, and A.K. Paul, "Structural enhancement under X-ray irradiation in

an octanuclear uranium-based 3D metal–organic framework", *Cryst. Growth Des.*, vol. 21, no. 10, pp. 5503-5507, 2021.
[http://dx.doi.org/10.1021/acs.cgd.1c00700]

[28]   S. Liu, Y. Qiu, Y. Liu, W. Zhang, Z. Dai, D. Srivastava, A. Kumar, Y. Pan, and J. Liu, "Recent advances in bimetallic metal–organic frameworks (BMOFs): Synthesis, applications and challenges", *New J. Chem.*, vol. 46, no. 29, pp. 13818-13837, 2022.
[http://dx.doi.org/10.1039/D2NJ01994A]

[29]   S. Abednatanzi, P. Gohari Derakhshandeh, H. Depauw, F.X. Coudert, H. Vrielinck, P. Van Der Voort, and K. Leus, "Mixed-metal metal-organic frameworks", *Chem. Soc. Rev.*, vol. 48, no. 9, pp. 2535-2565, 2019.
[http://dx.doi.org/10.1039/C8CS00337H] [PMID: 30989162]

[30]   P. Bhattacharya, S. Dhibar, M.K. Kundu, G. Hatui, and C.K. Das, "Graphene and MWCNT based bi-functional polymer nanocomposites with enhanced microwave absorption and supercapacitor property", *Mater. Res. Bull.*, vol. 66, pp. 200-212, 2015.
[http://dx.doi.org/10.1016/j.materresbull.2015.02.040]

[31]   N. Kumar, T. Rom, M. Kumar, T.C. Nagaiah, E. Lee, H.C. Ham, S.H. Choi, S. Rayaprol, V. Siruguri, T.K. Mandal, B.J. Kennedy, and A.K. Paul, "Unraveling the effect of A-site sr-doping in double perovskites Ca $2-x$ Sr x ScRuO 6 ($x = 0$ and 1): Structural interpretation and mechanistic investigations of trifunctional electrocatalytic effects", *ACS Appl. Energy Mater.*, vol. 5, no. 9, pp. 11632-11645, 2022.
[http://dx.doi.org/10.1021/acsaem.2c02101]

[32]   X. Yu, S. Yun, J.S. Yeon, P. Bhattacharya, L. Wang, S.W. Lee, X. Hu, and H.S. Park, "Pseudocapacitance: Emergent pseudocapacitance of 2D nanomaterials", *Adv. Energy Mater.*, vol. 8, no. 13, p. 1870058, 2018.
[http://dx.doi.org/10.1002/aenm.201870058]

[33]   S. Dhibar, P. Bhattacharya, G. Hatui, and C.K. Das, "Transition metal doped poly(aniline-c-pyrrole)/multi-walled carbon nanotubes nanocomposite for high performance supercapacitor electrode materials", *J. Alloys Compd.*, vol. 625, pp. 64-75, 2015.
[http://dx.doi.org/10.1016/j.jallcom.2014.11.108]

[34]   P. Bhattacharya, D.H. Suh, P. Nakhanivej, Y. Kang, and H.S. Park, "Iron oxide nanoparticle-encapsulated cnt branches grown on 3d ozonated cnt internetworks for lithium-ion battery anodes", *Adv. Funct. Mater.*, vol. 28, no. 29, p. 1801746, 2018.
[http://dx.doi.org/10.1002/adfm.201801746]

[35]   R. Mehek, N. Iqbal, T. Noor, M.Z.B. Amjad, G. Ali, K. Vignarooban, and M.A. Khan, "Metal–organic framework based electrode materials for lithium-ion batteries: a review", *RSC Advances,* vol. 11, no. 47, pp. 29247-29266, 2021.
[http://dx.doi.org/10.1039/D1RA05073G] [PMID: 35479575]

[36]   C. Yin, L. Xu, Y. Pan, and C. Pan, "Metal–organic framework as anode materials for lithium-ion batteries with high capacity and rate performance", *ACS Appl. Energy Mater.*, vol. 3, no. 11, pp. 10776-10786, 2020.
[http://dx.doi.org/10.1021/acsaem.0c01822]

[37]   Q. Yang, Y. Liu, H. Ou, X. Li, X. Lin, A. Zeb, and L. Hu, "Fe-Based metal–organic frameworks as functional materials for battery applications", *Inorg. Chem. Front.*, vol. 9, no. 5, pp. 827-844, 2022.
[http://dx.doi.org/10.1039/D1QI01396C]

[38]   Y. Li, and H. Dai, "Recent advances in zinc–air batteries", *Chem. Soc. Rev.*, vol. 43, no. 15, pp. 5257-5275, 2014.
[http://dx.doi.org/10.1039/C4CS00015C] [PMID: 24926965]

[39]   A.K. Thakur, M. Majumder, S.P. Patole, K. Zaghib, and M.V. Reddy, "Metal–organic framework-based materials: advances, exploits, and challenges in promoting post Li-ion battery technologies",

Materials Advances, vol. 2, no. 8, pp. 2457-2482, 2021.
[http://dx.doi.org/10.1039/D0MA01019G]

[40]   T. Chen, S. Chen, Y. Chen, M. Zhao, D. Losic, and S. Zhang, "Metal-organic frameworks containing solid-state electrolytes for lithium metal batteries and beyond", *Mater. Chem. Front.,* vol. 5, no. 4, pp. 1771-1794, 2021.
[http://dx.doi.org/10.1039/D0QM00856G]

[41]   J. Yang, J. Chen, Z. Wang, Z. Wang, Q. Zhang, B. He, T. Zhang, W. Gong, M. Chen, M. Qi, P. Coquet, P. Shum, and L. Wei, "High-capacity iron-based anodes for aqueous secondary nickel–iron batteries: recent progress and prospects", *ChemElectroChem,* vol. 8, no. 2, pp. 274-290, 2021.
[http://dx.doi.org/10.1002/celc.202001251]

[42]   L. Han, P. Tang, Á. Reyes-Carmona, B. Rodríguez-García, M. Torréns, J.R. Morante, J. Arbiol, and J.R. Galan-Mascaros, "Enhanced activity and acid pH stability of prussian blue-type oxygen evolution electrocatalysts processed by chemical etching", *J. Am. Chem. Soc.,* vol. 138, no. 49, pp. 16037-16045, 2016.
[http://dx.doi.org/10.1021/jacs.6b09778] [PMID: 27960335]

[43]   F.L. Li, Q. Shao, X. Huang, and J.P. Lang, "Nanoscale trimetallic metal–organic frameworks enable efficient oxygen evolution electrocatalysis", *Angew. Chem. Int. Ed.,* vol. 57, no. 7, pp. 1888-1892, 2018.
[http://dx.doi.org/10.1002/anie.201711376] [PMID: 29155461]

[44]   J. Yun, F.A. Schiegg, Y. Liang, D. Scieszka, B. Garlyyev, and A. Kwiatkowski, "Electrochemically formed NaxMn[Mn(CN)6] thin film anodes demonstrate sodium intercalation and deintercalation at extremely negative electrode potentials in aqueous media", *ACS Appl. Energy Mater.,* vol. 1, pp. 123-128, 2018.
[http://dx.doi.org/10.1021/acsaem.7b00022]

[45]   Z. Li, J. Cui, Y. Liu, J. Li, K. Liu, and M. Shao, "Electrosynthesis of well-defined metal–organic framework films and the carbon nanotube network derived from them toward electrocatalytic applications", *ACS Appl. Mater. Interfaces,* vol. 10, no. 40, pp. 34494-34501, 2018.
[http://dx.doi.org/10.1021/acsami.8b12854] [PMID: 30226043]

[46]   B. He, P. Man, Q. Zhang, H. Fu, Z. Zhou, C. Li, Q. Li, L. Wei, and Y. Yao, "All binder-free electrodes for high-performance wearable aqueous rechargeable sodium-ion batteries", *Nano-Micro Lett.,* vol. 11, no. 1, p. 101, 2019.
[http://dx.doi.org/10.1007/s40820-019-0332-7] [PMID: 34138024]

[47]   Y. Chen, W. Zhang, Z. Zhu, L. Zhang, J. Yang, H. Chen, B. Zheng, S. Li, W. Zhang, J. Wu, and F. Huo, "Co nanoparticles combined with nitrogen-doped graphitic carbon anchored on carbon fibers as a self-standing air electrode for flexible zinc–air batteries", *J. Mater. Chem. A Mater. Energy Sustain.,* vol. 8, no. 15, pp. 7184-7191, 2020.
[http://dx.doi.org/10.1039/D0TA00793E]

[48]   Y. Huang, C. Fang, R. Zeng, Y. Liu, W. Zhang, Y. Wang, Q. Liu, and Y. Huang, "*In situ*-formed hierarchical metal–organic flexible cathode for high-energy sodium-ion batteries", *ChemSusChem,* vol. 10, no. 23, pp. 4704-4708, 2017.
[http://dx.doi.org/10.1002/cssc.201701484] [PMID: 28891155]

[49]   Y. Mao, G. Li, Y. Guo, Z. Li, C. Liang, X. Peng, and Z. Lin, "Foldable interpenetrated metal-organic frameworks/carbon nanotubes thin film for lithium–sulfur batteries", *Nat. Commun.,* vol. 8, no. 1, p. 14628, 2017.
[http://dx.doi.org/10.1038/ncomms14628] [PMID: 28262801]

[50]   Y. Liu, G. Li, Z. Chen, and X. Peng, "CNT-threaded N-doped porous carbon film as binder-free electrode for high-capacity supercapacitor and Li–S battery", *J. Mater. Chem. A Mater. Energy Sustain.,* vol. 5, no. 20, pp. 9775-9784, 2017.
[http://dx.doi.org/10.1039/C7TA01526G]

[51]   M. Du, K. Rui, Y. Chang, Y. Zhang, Z. Ma, W. Sun, Q. Yan, J. Zhu, and W. Huang, "Carbon necklace incorporated electroactive reservoir constructing flexible papers for advanced lithium–ion batteries", *Small,* vol. 14, no. 2, p. 1702770, 2018.
[http://dx.doi.org/10.1002/smll.201702770]

[52]   S.J. Kwon, T. Kim, B.M. Jung, S.B. Lee, and U.H. Choi, "Multifunctional epoxy-based solid polymer electrolytes for solid-state supercapacitors", *ACS Appl. Mater. Interfaces,* vol. 10, no. 41, pp. 35108-35117, 2018.
[http://dx.doi.org/10.1021/acsami.8b11016] [PMID: 30230315]

[53]   N. Gupta, R.K. Sahu, T. Mishra, and P. Bhattacharya, "Microwave-assisted rapid synthesis of titanium phosphate free phosphorus doped Ti 3 C 2 MXene with boosted pseudocapacitance", *J. Mater. Chem. A Mater. Energy Sustain.,* vol. 10, no. 29, pp. 15794-15810, 2022.
[http://dx.doi.org/10.1039/D2TA04061A]

[54]   A. Sahoo, and Y. Sharma, "Synthesis and characterization of nanostructured ternary zinc manganese oxide as novel supercapacitor materialMater", *Chem. Phys.,* vol. 149, pp. 721-727, 2015.

[55]   M.Y. Ho, P.S. Khiew, D. Isa, T.K. Tan, W.S. Chiu, and C.H. Chia, "Overview on conducting polymer in energy storage and energy conversion system", *Nano,* vol. 09, p. 1430002, 2014.
[http://dx.doi.org/10.1142/S1793292014300023]

[56]   H.D. Yoo, S.D. Han, R.D. Bayliss, A.A. Gewirth, B. Genorio, N.N. Rajput, K.A. Persson, A.K. Burrell, and J. Cabana, "Rocking-chair-type metal hybrid supercapacitors", *ACS Appl. Mater. Interfaces,* vol. 8, no. 45, pp. 30853-30862, 2016.
[http://dx.doi.org/10.1021/acsami.6b08367] [PMID: 27775318]

[57]   M.L. Aubrey, and J.R. Long, "A dual−ion battery cathode via oxidative insertion of anions in a metal−organic framework", *J. Am. Chem. Soc.,* vol. 137, no. 42, pp. 13594-13602, 2015.
[http://dx.doi.org/10.1021/jacs.5b08022] [PMID: 26436465]

[58]   G. Nagaraju, S.C. Sekhar, B. Ramulu, and J.S. Yu, "High-performance hybrid supercapacitors based on MOF-derived hollow ternary chalcogenides", *Energy Storage Mater.,* vol. 35, pp. 750-760, 2021.
[http://dx.doi.org/10.1016/j.ensm.2020.12.005]

[59]   F. Meng, H. Zhong, D. Bao, J. Yan, and X. Zhang, "In situ coupling of strung Co 4 N and intertwined N–C fibers toward free-standing bifunctional cathode for robust, efficient, and flexible Zn–Air batteries", *J. Am. Chem. Soc.,* vol. 138, no. 32, pp. 10226-10231, 2016.
[http://dx.doi.org/10.1021/jacs.6b05046] [PMID: 27463122]

[60]   K. Liu, C. Li, L. Yan, M. Fan, Y. Wu, X. Meng, and T. Ma, "MOFs and their derivatives as Sn-based anode materials for lithium/sodium ion batteries", *J. Mater. Chem. A Mater. Energy Sustain.,* vol. 9, no. 48, pp. 27234-27251, 2021.
[http://dx.doi.org/10.1039/D1TA06996A]

[61]   J.D. Xiao, and H.L. Jiang, "Metal–Organic Frameworks for Photocatalysis and Photothermal Catalysis", *Acc. Chem. Res.,* vol. 52, no. 2, pp. 356-366, 2019.
[http://dx.doi.org/10.1021/acs.accounts.8b00521] [PMID: 30571078]

[62]   W.H. Li, K. Ding, H.R. Tian, M.S. Yao, B. Nath, W.H. Deng, Y. Wang, and G. Xu, "Conductive metal–organic framework nanowire array electrodes for high-performance solid-state supercapacitors", *Adv. Funct. Mater.,* vol. 27, no. 27, p. 1702067, 2017.
[http://dx.doi.org/10.1002/adfm.201702067]

[63]   L. Quoc Bao, T-H. Nguyen, H. Fei, I. Sapurina, F.A. Ngwabebhoh, C. Bubulinca, L. Munster, E.D. Bergerová, A. Lengalova, H. Jiang, T. Trong Dao, N. Bugarova, M. Omastova, N.E. Kazantseva, and P. Saha, "Electrochemical performance of composites made of rGO with Zn-MOF and PANI as electrodes for supercapacitors", *Electrochim. Acta,* vol. 367, p. 137563, 2021.
[http://dx.doi.org/10.1016/j.electacta.2020.137563]

[64]   Q.B. Le, T.H. Nguyen, H. Fei, C. Bubulinca, L. Munster, N. Bugarova, M. Micusik, R. Kiefer, T.T.

Dao, M. Omastova, N.E. Kazantseva, and P. Saha, "Electrochemical performance of composite electrodes based on rGO, Mn/Cu metal–organic frameworks, and PANI", *Sci. Rep.*, vol. 12, no. 1, pp. 664-676, 2022.
[http://dx.doi.org/10.1038/s41598-021-04409-y] [PMID: 35027598]

[65]   Y. Tong, D. Ji, P. Wang, H. Zhou, K. Akhtar, X. Shen, J. Zhang, and A. Yuan, "Nitrogen-doped carbon composites derived from 7,7,8,8-tetracyanoquinodimethane-based metal–organic frameworks for supercapacitors and lithium-ion batteries", *RSC Advances*, vol. 7, no. 40, pp. 25182-25190, 2017.
[http://dx.doi.org/10.1039/C7RA02543B]

[66]   P. Wen, P. Gong, J. Sun, J. Wang, and S. Yang, "Design and synthesis of Ni-MOF/CNT composites and rGO/carbon nitride composites for an asymmetric supercapacitor with high energy and power density", *J. Mater. Chem. A Mater. Energy Sustain.*, vol. 3, no. 26, pp. 13874-13883, 2015.
[http://dx.doi.org/10.1039/C5TA02461G]

[67]   L. Wang, X. Feng, L. Ren, Q. Piao, J. Zhong, Y. Wang, H. Li, Y. Chen, and B. Wang, "Flexible solid-state supercapacitor based on a metal–organic framework interwoven by electrochemically-deposited PANI", *J. Am. Chem. Soc.*, vol. 137, no. 15, pp. 4920-4923, 2015.
[http://dx.doi.org/10.1021/jacs.5b01613] [PMID: 25864960]

[68]   S. Liu, L. Kang, J. Zhang, E. Jung, S. Lee, and S.C. Jun, "Structural engineering and surface modification of MOF-derived cobalt-based hybrid nanosheets for flexible solid-state supercapacitors", *Energy Storage Mater.*, vol. 32, pp. 167-177, 2020.
[http://dx.doi.org/10.1016/j.ensm.2020.07.017]

[69]   L. Zhu, T. Han, Y. Ding, J. Long, X. Lin, and J. Liu, "A metal–organic-framework derived NiFe2O4@NiCo-LDH nanocube as high-performance lithium-ion battery anode under different temperatures", *Appl. Surf. Sci.*, vol. 599, p. 153953, 2022.
[http://dx.doi.org/10.1016/j.apsusc.2022.153953]

[70]   X. Han, K. Tao, D. Wang, and L. Han, "Design of a porous cobalt sulfide nanosheet array on Ni foam from zeolitic imidazolate frameworks as an advanced electrode for supercapacitors", *Nanoscale*, vol. 10, no. 6, pp. 2735-2741, 2018.
[http://dx.doi.org/10.1039/C7NR07931A] [PMID: 29296991]

CHAPTER 6

MXene: Chemistry, Properties, and Energy Storage Applications

Manisha Devi[1,*], Shipra Jaswal[1] and Swadesh Kumar[1]

[1] Department of Chemistry, Gautam College Hamirpur, Himachal Pradesh, India

Abstract: The growing interest and demand for energy storage applications have significantly encouraged the development of a broad range of functional 2D materials. Owing to the extraordinary properties including 2D lamellar structure, larger interlayer space, mechanical strength, high thermal and electrical conductivity, and negative zeta-potential, carbides/nitrides/carbonitrides of transition metal, usually known as MXenes have received the interest of researchers in the development of environmentally friendly materials for storage and conversion of energy. In this chapter, we focused on the MXene, their methods of preparation, the progress of development of various MXenes, and their modification for the storage of energy. Here, we have discussed the various storage devices for energy including batteries and superconductors. This chapter offers scientific inspiration and literature for the rational design and synthesis of high-capacity MXenes and their composites that can fulfill the increased demand for next-generation energy storage devices.

Keywords: Batteries, Energy storage, MXene, MXene-based nanocomposites, Supercapacitors.

INTRODUCTION

With the increased world population and industries, energy-related issues are of major concern for today's generation [1]. In this regard, the development of high-performance energy storage systems is essential to integrate various renewable energy resources into high-power energy devices. Up to now, various electrical energy storage devices (ESDs) and technologies have been developed for example batteries, capacitors, and supercapacitors [2, 3]. ESDs are very important in developing smart electronics and wearable textiles. Metal ions-based rechargeable batteries and supercapacitors are being considered as potential ESDs.

* **Corresponding author Manisha Devi:** Department of Chemistry, Gautam College Hamirpur, Himachal Pradesh, India; E-mail: manishathakur04@gmail.com

Sanjeev Verma, Shivani Verma, Saurabh Kumar & Bhawna Verma (Eds.)

In previous years, 2D materials have emerged as attractive materials for several applications depending upon their physio-chemical, optical, and electrical properties [4, 5]. Therefore, 2D materials have been widely investigated and received the Nobel Prize for the groundbreaking discovery of the first 2D material graphene [6, 7]. Due to single/few atomic layers, 2D materials exhibit properties such as nanosheet-like structures, large specific surface areas, very high aspect ratios, short ion diffusion path, and low energy barrier for electron transportation. These properties make them a potential material for the development of electrodes for electrochemical storage of energy to deal with current energy issues. 2D materials including graphene, transition metal oxides, transition metal dichalcogenides, hexagonal boron nitride, hydroxides, organics, MXenes, and black phosphorous have been extensively explored in the field of energy storage [8 - 11]. Among various 2D materials, a family of 2D materials comprising transition metal nitrides, carbides, and carbonitrides commonly recognized as MXenes, have attracted huge consideration because of their extraordinary properties including easily modulating surface chemistry, rich chemical composition possibilities, unusual combination of metallic hydrophilicity and conductivity, good thermal stability, tunable bandgap, charge carrier mobility, excellent electrical conductivity (2×10^5 Sm^{-1}), fast diffusion and easy dispersion in various solvents including water, large spacing between layers and surface area [12]. In 2011, the first MXene, $Ti_3C_2T_x$, was discovered by Yury Gogotsi *et al.* at Drexel University, USA [13]. General chemical representation $M_{n+1}X_nT_x$ corresponds to MXenes, where M = early transition metal (Sc, Ti, V, Cr, Mo, Nb, Hf, Ta, Zr, *etc.*), X = carbon and/or nitrogen atom, T_x = surface termination groups (F, O or OH) and n = 1–4. Versatile properties render MXenes as a potential candidate for energy storage in the form of batteries and supercapacitors. In this chapter, we focused on the MXene, different methods of preparation of MXenes, different properties, and utilization of MXenes and their composites in energy storage applications. Fig. (**1a**) demonstrates an overview of the full chapter.

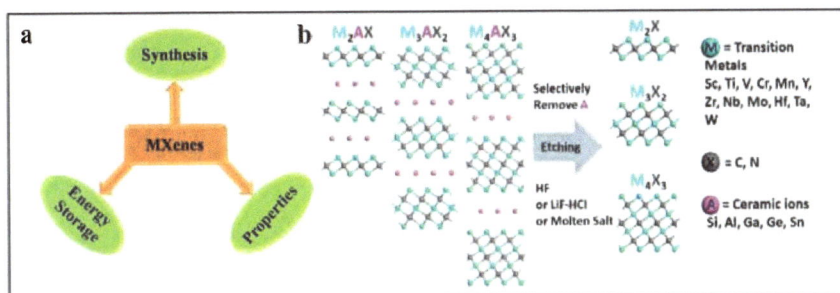

Fig. (1). (**a**) Overview of the chapter. (**b**) Schematic representation of the preparation of MXene (bottom) from MAX (top). Adapted with permission from Reference [11], Copyright (2019), American Chemical Society.

Structure of MXene and MAX Phase

MXene are collectively referred to as 2D layered transition metal nitrides, carbides, or carbonitrides [10]. MXenes (M_nX_n) were prepared by the removal of the A layer from their precursor MAX phase. The term "MAX" corresponds to the chemical composition M_nAX_n, where M corresponds to an early transition metal, A corresponds to an element from group 13 or 14 (Al, Si, Ga, *etc.*), X corresponds to carbon or nitrogen, and n = 1-4 (Figs. **1b** & **2a**) [10]. Etching is an essential step for the synthesis of MXenes. Here, the M-X bond is ionic, covalent, and metallic while the M-A bond is metallic; this leads to a higher binding energy of M-X than M-A. This suggests that the M-X bond is highly stable in comparison to the M-A. The metallic nature of both bonds inhibits mechanical exfoliation. The chemical stability of M-X bonds over M-A bonds and the difference in the binding energies of the bonds facilitate the selective etching of the A layer to prepare MXene. On removing the A layer, the corresponding MXene retains the hexagonal lattice of parent MAX. Fig. (**1b**) demonstrates the preparation of MXene (M_2X, M_3X_2, and M_4X_3) by etching of their respective MAX phases. Thin films of MXene are generally organized horizontally, which were inherited from their parent MAX phase. Multilayer MXene (m-MXene) is further converted into few-layer MXene nanosheets by delamination using intercalating agents (DMSO and LiF) that lead to increased surface area for providing redox active sites which allow the 2D MXene to be a part of energy storage applications. First MXene Ti_3C_2 was prepared by dissolving the parent MAX phase (Ti_3AlC_2) in a hydrofluoric acid (HF) solution [13]. After this, different MXenes have been prepared and explored for various applications [14, 15].

DIFFERENT APPROACHES FOR THE SYNTHESIS OF MXENES

In the synthesis of MXene etching plays a crucial role. As discussed in the previous section, due to the metallic character of M–A bonds of MAX precursor, isolation of M_nX_n layers by mechanical shearing is difficult. However, treating more chemically active M–A bonds of MAX with etching agents prompts the removal of A layers selectively. Usually, hydrofluoric acid (HF) is used to extract the "A" layer from MAX [16]. Van der Waals force of attractions between layers causes the aggregation of 2D nanosheets that restrict the high-quality 2D MXenes synthesis. Normally, two synthetic strategies, bottom-up and top-down synthesis have been employed for the preparation of 2D MXenes. Top-down synthesis is mainly the exfoliation of bulk crystalline material into single or few-layer sheets whereas bottom-up is based on the preparation of material from atoms or molecules. In general, MXene synthesis involves 3 steps; i) preparation of MAX precursor, ii) etching out the A layer, iii) intercalation and exfoliation (Fig. **2b**).

a

Etching

MAX Phase → MXene

M A X   M X T_x

H																	He
Li	Be											B	C	N	O	F	Ne
Na	Mg											Al	Si	P	S	Cl	Ar
K	Ca	Sc	Ti	V	Cr	Mn	Fe	Co	Ni	Cu	Zn	Ga	Ge	As	Se	Br	Kr
Rb	Sr	Y	Zr	Nb	Mo	Tc	Ru	Rh	Pd	Ag	Cd	In	Sn	Sb	Te	I	Xe
Cs	Ba	La	Hf	Ta	W	Re	Os	Ir	Pt	Au	Hg	Tl	Pb	Bi	Po	At	Rn

b

MAX phases are layered ternary carbides, nitrides, and carbonitrides consisting of "M", "A", and "X" layers

HF treatment

MAX phase

Sonication

Selective HF etching only of the "A" layers from the MAX phase

Physically separated 2-D MXene sheets after sonication

MXene sheets

c

Etching

Delamination

M_3AX_2

M_3X_2

$M_3X_2T_x$

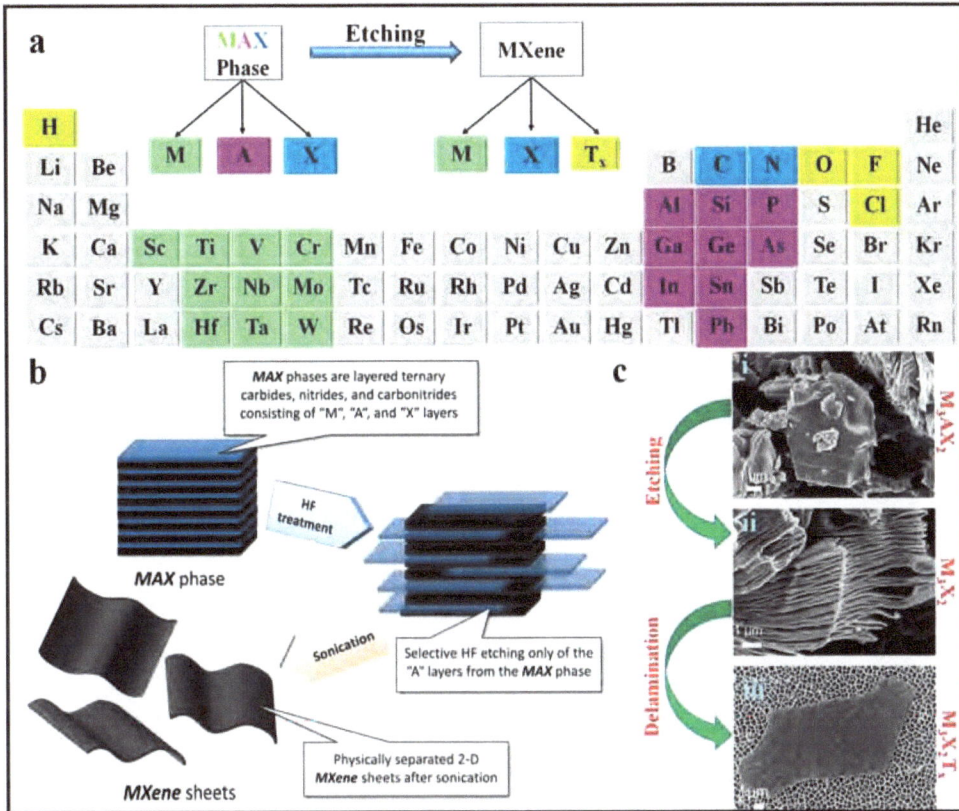

Fig. (2). (**a**) Highlighted elements of the periodic table correspond to the MAX phase and MXenes. (**b**) Schematic illustration of synthesis of MXene (M_3X_2), and MXene with surface functional groups ($M_3X_2T_x$) from MAX phase (M_3AX_2). Adapted with permission from Reference [15], Copyright (2012), American Chemical Society. (**c**) SEM images of (i) M_3AX_2 (MAX phase), (ii) non-delaminated (M_3X_2), and (iii) delaminated MXene ($M_3X_2T_x$). Adapted with permission from Reference [27], Copyright (2017), American Chemical Society.

Top-down Synthetic Approach of MXene from MAX Precursor

Here, in top-down synthesis, multi-layered MXene (m-MXene) flakes were prepared by the selective extraction/etching of specific atomic layers (A layer) from their precursors MAX. There are different methods of etching under a top-down approach as discussed below:

Wet Chemical Etching

HF Etching

The first method used for etching to synthesize the first MXene, $Ti_3C_2T_x$ is HF etching. Here, $Ti_3C_2T_x$ was synthesized by etching Al from its precursor Ti_3AlC_2

using HF solution as an etchant. Similarly, other MXenes (Nb_2CT_x, V_2CT_x, Ti_3CNT_x) were also synthesized. As shown in equation (1) MXene ($M_{n+1}X_n$) is prepared by the release of Al from the MAX precursor slowly due to weaker metallic M-A bonds retaining its inherited hexagonal lattice [17].

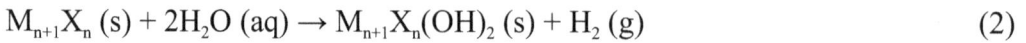

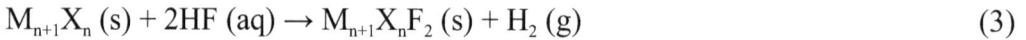

$$M_{n+1}AlX_n \text{ (s)} + 3HF \text{ (aq)} \rightarrow M_{n+1}X_n \text{ (s)} + AlF_3 \text{ (aq)} + 1.5H_2 \text{ (g)} \tag{1}$$

$$M_{n+1}X_n \text{ (s)} + 2H_2O \text{ (aq)} \rightarrow M_{n+1}X_n(OH)_2 \text{ (s)} + H_2 \text{ (g)} \tag{2}$$

$$M_{n+1}X_n \text{ (s)} + 2HF \text{ (aq)} \rightarrow M_{n+1}X_nF_2 \text{ (s)} + H_2 \text{ (g)} \tag{3}$$

The resulting m-MXene are highly reactive and unstable in H_2O or acid (HF) therefore it reacts with these and leads to surface termination by producing –F, =O, and –OH groups at the MXene surface as shown in equations 2 and 3 (Fig. **3a**). After HF etching, the solid MAX phase Fig. (**2c (i)**) is converted into accordion-like loosely packed m-MXene (Fig. **2c(ii)**). An etching process is kinetically controlled, therefore, etching conditions strongly affect the yields, particle size, defects, and the ratio of surface termination groups of MXene ($M_{n+1}X_nTx$). Delamination of surface functionalized MXene into single or few layers is of most priority because some properties are layer dependent which exist only in atomically thin 2D nanosheets [10, 17]. During HF etching, metallic M-A bonds were replaced by the weaker van der Waals and hydrogen bonds, which assist the delamination of m-MXene to a single or few layers by using sonication [10, 18]. Fig. (**2c(iii)**) shows the delaminated single transparent sheets.

Fluoride Salt Etching

Even though HF is extensively used as an etchant to generate MXene, HF is hazardous, hence to avoid its direct use, another mild liquid etchant was introduced by Gogotsi *et al.* [19] in which HF is produced *in situ* by mixing the fluoride salts with hydrochloric acid (equation 4).

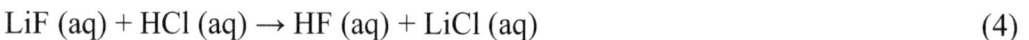

$$LiF \text{ (aq)} + HCl \text{ (aq)} \rightarrow HF \text{ (aq)} + LiCl \text{ (aq)} \tag{4}$$

The resulting MXenes from this method have fewer atomic defects leading to higher electrical conductivity. It was observed that the quality, lateral size of nanosheets, and processibility of MXene depend upon the ratio of LiF and HCl. Instead of LiF, other fluoride salts NaF, KF, CaF_2, CsF, FeF_3 *etc.* can be used in this method, with HCl concentrations ranging from 6 to 12 M [20]. The advantage of using this method is the counter ion of fluoride salt gets intercalated *in situ* in resulting MXenes that increase the interlayer space and facilitate subsequent energy storage. In recent years, various fluoride-based solutions, such as $NaHF_2$, KHF_2, and NH_4HF_2, have also been explored as mild etchants to remove A layer from MAX (Ti_3AlC_2) (equations 5 and 6) [21, 22]. After selective etching, $-NH_3$ and $-NH_4^+$ intercalate into interlayers of the obtained MXenes which results in larger interlayer space that facilitates the further delamination of MXenes. For this method, high temperature and pressure are required.

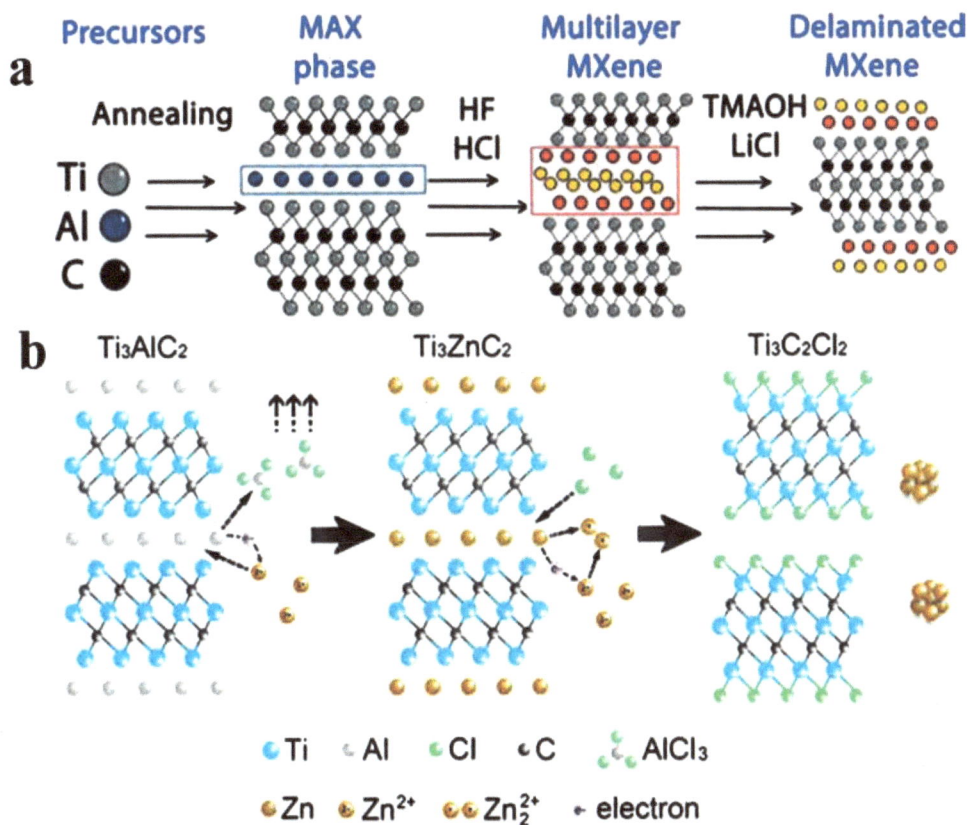

Fig. (3). Schematic illustration of the different techniques of etching. (**a**) Wet chemical. Adapted with permission from Reference [16], Copyright (2011), American Chemical Society. (**b**) Molten salts. Adapted with permission from Reference [22], Copyright (2019), American Chemical Society.

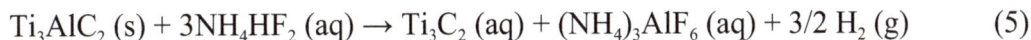

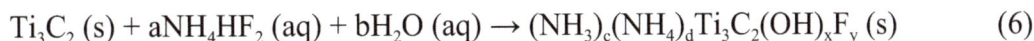

$$Ti_3AlC_2 \text{ (s)} + 3NH_4HF_2 \text{ (aq)} \rightarrow Ti_3C_2 \text{ (aq)} + (NH_4)_3AlF_6 \text{ (aq)} + 3/2\ H_2 \text{ (g)} \qquad (5)$$

$$Ti_3C_2 \text{ (s)} + aNH_4HF_2 \text{ (aq)} + bH_2O \text{ (aq)} \rightarrow (NH_3)_c(NH_4)_dTi_3C_2(OH)_xF_y \text{ (s)} \qquad (6)$$

Alkali Etching

Both the above-mentioned methods are very efficient in etching the A layer from the MAX phase, but to work with HF even at lower concentrations, it is unsafe. Therefore, the safe method to prepare good quality MXenes alkali etching of

MAX has been reported. However, alkali solution (NaOH in deaerated water) etching of the MAX phase through hydrothermal reaction becomes highly efficient [23].

$$Ti_3AlC_2 + OH^- + 5H_2O \rightarrow Ti_3C_{2(}OH)_2^- + Al(OH)_4 + 5/2H_2 \tag{7}$$

$$Ti_3AlC_2 + OH^- + 5H \rightarrow Ti_3C_2O_2 + Al(OH)_4^- + 7/2\ H_2 \tag{8}$$

Molten Salts Etching

The conditions for etching of specific layers from the MAX phase are structurally dependent. Strong etchants (HF) can dissolve MXenes by breaking both the M-A as well as M-X bonds due to their almost similar binding energies [19]. Thus, a new method of molten salt etching at very high temperatures has been reported, mainly designed for nitride MXenes. For example, Gogotsi *et al.* prepared the first multi-layered nitride MXene (m-$Ti_4N_3T_x$) by heating the MAX precursor Ti_4AlN_3 with molten fluoride salt (LiF, NaF, and KF) at 550 °C under argon flow [24]. Recently, Huang *et al.* used MAX precursors containing Zn and $ZnCl_2$ as etchants for the first time to prepare halide-terminated MXenes [22]. Using a similar strategy, a series of –Cl groups terminated MXenes ($Ti_3C_2Cl_2$ and Ti_2CCl_2) were also synthesized Fig. (**3b**). Equations 9-12 showed the related mechanism involved in the synthesis of –Cl groups terminated MXenes.

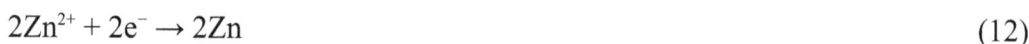

$$Ti_3ZnC_2 + ZnCl_2 \rightarrow Ti_3C_2Cl_2 + 2Zn \tag{9}$$

$$Ti_3ZnC_2 + Zn^{2+} \rightarrow Ti_3C_2 + 2Zn^{2+} \tag{10}$$

$$Ti_3C_2 + 2Cl^- \rightarrow Ti_3C_2Cl_2 + 2e^- \tag{11}$$

$$2Zn^{2+} + 2e^- \rightarrow 2Zn \tag{12}$$

Electrochemical Method of Etching

The electrochemical method can be used for the preparation of MXene nanosheets. The choice of the electrolyte is essential as it affects the completeness, yield, surface properties, and microstructure of the resulting MXene. Nowadays, a Cl^- Cl-contained electrolyte is used. Feng *et al.* [25] used a mixture of 1 M NH_4Cl and 0.2 M TMAOH as an electrolyte to etch the MAX (Ti_3AlC) for synthesizing the –OH and =O surface terminated MXene ($Ti_3C_2T_x$), as shown in equations 13-15.

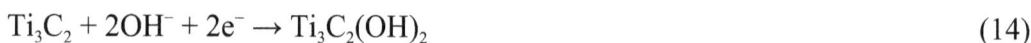

$$Ti_3AlC_2 + 3e^- + 3Cl^- \rightarrow Ti_3C_2 + AlCl_3 \tag{13}$$

$$Ti_3C_2 + 2OH^- + 2e^- \rightarrow Ti_3C_2(OH)_2 \tag{14}$$

$$Ti_3C_2 + 2H_2O \rightarrow Ti_3C_2(OH)_2 + H_2 \tag{15}$$

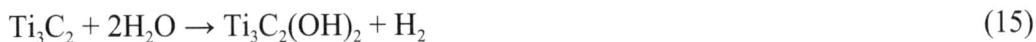

Following this, Hao *et al.* established a universal thermal-assisted electrochemical etching method [26]. They used diluted HCl without toxic intercalants as the electrolyte with small heating that resulted in the exfoliation of MXenes. This method is employed for other MXenes including V_2CT_X and Cr_2CT_X as well.

Intercalation/Delamination Method to Generate Delaminated MXenes (D-MXenes)

After completion of the etching process, multi-layered MXenes were obtained. Their specific surface area is limited which restricts the potential application of energy storage. Therefore, the energy storage capacity of MXenes can be increased by increasing the surface area of m-MXenes by intercalation or delamination to thin, few/single nanosheets. The interlayer interactions of m-MXenes are ~sixfold stronger than other 2D materials such as graphite and bulk MoS_2 [26]. Intercalating material plays a critical role in increasing the interlayer distance between 2D layers. Sonication and ion intercalation methods increased the interlayer distance that led to the delaminated MXenes.

Delamination of MXenes in Organic Solvents and Molecules

d-MXene nanosheets can be achieved by dispersing m-MXene in organic solvents mostly DMSO and large organic molecules (urea, hydrazine, isopropylamine) and organic bases (tetrabutylammonium hydroxide and choline hydroxide) [27] followed by ultrasonication and centrifugation. This is a highly efficient method to prepare delaminated MXene with microscopic structures. For example, to increase the interlayer distance between m-$Ti_3C_2T_x$ sheets, DMSO was used as intercalants. Bulky tetrabutylammonium ions exfoliate the MXene by exchanging the H^+ [28].

Delamination of MXenes with Metal Ions

During the synthesis of m-MXene, LiF and HCl are used followed by the slow addition of base up to neutralization and ultrasonication, which induce the exfoliation of multi-layered sheets into several sheets. Sodium ions also intercalate the m-MXene dispersion during the coupling reaction between phenylsulfonic acid and diazonium salt. Hence, MXene layers modified with aryl groups induce the expansion of the sheets and create a negatively charged surface. Metal cations with positively charged (Li^+, Na^+, K^+, Ca^{2+}, Mg^{2+}, $Al^{3+,}$ and Sn^{4+}) and other ions such as NH^{4+} can also be used to intercalate the negatively charged delaminated MXene layers by decreasing the Van der Waals force strength between 2D layers [29]. It is observed that sonication at higher speed results in solution enriched in single-layer MXene nanosheets with higher size, and defects.

Bottom-up Strategy

The bottom-up strategy is a controlled route synthesis of few-layer MXene epitaxial films [28, 21]. It follows two steps: the formation of MAX thin epitaxy followed by the removal of the A layer. The first time, Halim *et al.* prepared a MAX film with a bottom-up approach [21]. They first deposited MAX (Ti_3AlC_2) thin film of 15-60 nm thickness on TiC-coated sapphire substrates using DC magnetron sputtering followed by etching using HF or NH_4HF_2 etchants that lead to the formation of MXene ($Ti_3C_2T_X$) under vacuum at 780 °C. Therefore, it is evident that Ti_3C_2 thin films can be prepared by atomic layer deposition (ALD) and chemical vapor deposition (CVD) [30, 31]. Xu *et al.* synthesized high-quality, ultrathin, defect-free epitaxial thin films of a-Mo_2C, tantalum carbide (TaC), and tungsten carbide (WC) by CVD method [30]. The bottom-up approach has some limitations including complex treatment and low yield, choice of substrate and thin film transfer, *etc.* However high-quality MXenes films can be prepared by using a bottom-up synthesis route at low temperatures and improved Cu etching.

Properties of MXenes

Most of the properties of MXene including quality, structural, electronic, mechanical, optical, transport, surface, and environmental stability are affected by the synthesis process, MAX precursor, etchant intercalation methods, and sonication frequency. The surface terminated group at the MXene surface also affects the electronegativity and hydrophilicity that induce the dispersion of MXenes in polar solvents.

Theoretical Capacity

Due to pseudocapacitive sites of MXene ($Ti_3C_2T_x$) possessing a high specific capacitance. Pseudocapacitive sites and charge storage properties of MXene in acidic electrolytes arise due to changes in the valency of Ti because of the protonation of oxygen attached to Ti (equation 16) [32].

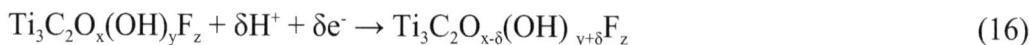

$$Ti_3C_2O_x(OH)_yF_z + \delta H^+ + \delta e^- \rightarrow Ti_3C_2O_{x-\delta}(OH)_{y+\delta}F_z \qquad (16)$$

For a voltage window of 0.55 V, theoretical capacity (~615 C g^{-1}) was observed very high in comparison to the experimental value (135 C g^{-1}). Probably active sites were not utilized completely, or incomplete redox reactions. Mostly, gold or platinum are used as current collectors [33]. Although using these metals as a current collector, the water splits due to the regular charge–discharge process and decreases the Coulombic efficiency in the potential window of interest. Gogotsi *et al.* obtained a broad potential window of 1 V by replacing the metal current collector with glassy carbon for the MXene electrode without water splitting [33]. They achieved exceptional rate performance along with 450 F g^{-1} specific capacitance in 90 nm thick electrodes. Heteroatom doping affects the theoretical capacitance of MXene. In this direction, Que *et al.* designed nitrogen-doped Ti$_3$C$_2$T$_x$ films and achieved the 2836 F cm^{-3} capacitance at 5 mV s^{-1} [34].

Electronic Band Structure

Theoretical analysis revealed the metallic character of MXenes with high electron density around the Fermi level, similar to the MAX precursor [35]. The metallic character can be modulated by forming extra Ti-X bonds. MXenes reveal a narrow band-gap semiconductor character by altering surface termination groups. Generally, MXenes have indirect band gaps except for Sc$_2$C(OH)$_2$ [7]. The electronic structure of MXene and its corresponding electronic properties can be altered by modulating surface termination groups. F and OH functionalized MXene showed similar effects on the electronic structures that accept a single electron. Among carbonitrides, nitrides, and carbide MXenes, carbonitride and nitride reveal higher metallic properties as N possesses more electrons. Topologically MXenes can either be trivial or non-trivial metallic and semiconducting electronic bands. In comparison to graphene, Ti$_3$C$_2$T$_x$ film exhibits high intrinsic electronic or ionic conductivities [34]. It is noted that Ti$_3$C$_2$Tx film showed greater experimental metallic conductivity than other 2D materials and their electrical conductivity depends on surface morphology due to more interactions between individual layers and large layers that lead to high conductivity. Ti$_3$C$_2$T$_x$ thin films with few defects showed 9880 S cm^{-1} conductivity as compared to powdered Ti$_3$C$_2$T$_x$ (1000 S cm^{-1}) [21]. Both computational and experimental studies showed the extraordinary electronic properties of MXenes that lead to their potential application as electrodes for energy storage.

Morphologies and Surface Chemistries

As discussed in previous sections, the etching method and concentration of etchant strongly affect the morphology of MXene. MXene obtained after HF etching at different HF concentrations results in accordion-like morphology (Fig. **4a**). More prominent openings of MXene lamellas were observed at higher HF concentrations as shown in Fig. (**4b-d**). while mild etching methods (LiF-HCl) result in the formation of MXenes with very less openings of lamellas (Figs. **4e,f**). However, they exhibit analogous morphology as that of the parent MAX phase. Hence, the etching of the A layer from the precursor MAX cannot be characterized by the formation of accordion-like morphology. Therefore, there are different spectroscopic techniques to characterize the MXenes. For example, d-

MXenes were characterized by XRD and EDX [28] while stacking of layers and surface terminations were characterized by TEM [35], neutron scattering [36], and NMR [37]. These techniques provide an accurate number of surface terminations groups on MXene (Ti_3C_2) sheets that help in predicting their properties. Neutron scattering measurements of $Ti_3C_2T_x$ showed that interlayer interactions between sheets are due to hydrogen bonding between surface termination groups and Van der Waals interactions [38]. Rheological properties of $Ti_3C_2T_x$ changed on intercalation of cations that caused the sliding of sheets that generated a clay-like behavior of MXenes [39].

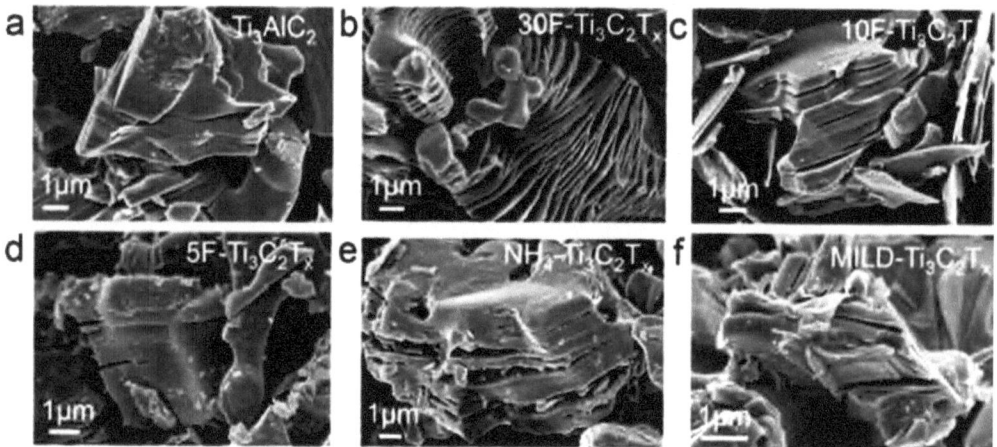

Fig. (4). SEM image of m-$Ti_3C_2T_x$ synthesized with (**a-d**) 0, 30, 10 and 5 wt.% HF; (**e**) 2M NH_4HF_2; and (**f**) mild etching LiF:HCl (10M:9M). Adapted with permission from Reference [27], Copyright (2017), American Chemical Society.

Optoelectronic Properties

MXene thin films with excellent optoelectronic properties can be prepared by solution processing followed by natural drying. MXene-based transparent films (4 nm) prepared by spin coating are highly conductive and showed an optical transmittance of ~93% [40]. It was observed on increasing the thickness of the film that the optical transmittance decreased. On decreasing 86% transmittance, the sheet resistance reduced to 330 Ω sq^{-1} [41]. Fabrication methods of thin film play a critical role in the optoelectronic performance of MXene films. For example, spin-coated transparent films of $Ti_3C_2T_x$ MXene show better optoelectronic performance in comparison to spray-coated MXene films. In the spin-coated $Ti_3C_2T_x$ film, the figure of merit (FoM) reached 15, while in the spray-coated films, it was 0.5–0.7 [40, 42]. In addition to this, spin-coated films showed lesser optical absorption, which is required for high-performance photovoltaic cells and displays [41]. For them, sheet resistance is inversely proportional to the thickness of the film similar to bulk-like films [40]. Hence, thin MXene-related transparent films with outstanding electronic conductivity can be used as supercapacitors without any other current collectors.

Mechanical Properties

d-MXene nanosheets especially monolayer exhibit extraordinary mechanical flexibility [43]. Elastic properties of $Ti_3C_2T_x$ for both monolayer and bilayer were calculated by Lipatov *et al.* [44]. For single layer $Ti_3C_2T_x$, Young's modulus was calculated as 0.33 TPa which is nearly close to the predicted value (502 GPa) of freestanding single layer Ti_3C_2 [45]. Owing to stretching and shrinking of Ti-C bonds, pure Ti_3C_2 could tolerate strains of 18% and 17% under uniaxial tension along the x and y direction, respectively, while surface functionalization of Ti_3C_2 with oxygen improved strains to 20-28%. This is attributed to the strong strength of covalent bonds between Ti and oxygen. Further, the mechanical strength of the transparent film of $Ti_3C_2T_x$ can be enhanced by the introduction of polymers like chitosan and polyethylene. Chitosan expands the dislocation of nanosheets of $Ti_3C_2T_x$ leading to an enhancement in the tensile power of films ranging from 8.2 to ~43.5 MPa [46]. Similarly, Gogotsi *et al.* incorporate polyvinylalcohol (PVA) into the $Ti_3C_2T_x$ which revealed high conductivity and exceptional mechanical properties that can tolerate 5000 times their weight [47].

Thermal Stability Properties

The thermal stability of MXene is affected by the chemical composition and it is very important for the fabrication of thin films for energy storage [45]. Thermogravimetric (TGA) analysis shows a weight loss of $Ti_3C_2T_x$ above 800 °C under Ar because it is converted into TiC [48]. In the presence of an oxygen atmosphere, at 200 °C, $Ti_3C_2T_x$ partly oxidized and converted into TiO_2 and at 1000 °C, it completely oxidized to rutile TiO_2. It was concluded that $Ti_3C_2T_x$ can be converted into TiO_2 with different structures and morphologies by controlling temperature, oxidation, and heating rate [49]. However, MXene surfaces exposed to metal atoms get oxidized as they are thermodynamically metastable [50]. The thermal conductivity of MXene depends on the lateral size [45]. For example, for Hf_2CO_2 the thermal conductivity increased from 86.25 W m^{-1} K^{-1} to 131.2 W m^{-1} K^{-1} when flake size increased from 5 μm to 100 μm [51]. MXene with N-doping also improved the thermal conductivity of single-layer Mo_2C [52].

Applications of MXenes in Energy Storage

In past years, huge expenditure on fossils for energy has raised the energy demand and caused environmental pollution. Therefore, conversion and storage of renewable energy to electricity are very important. Most of the countries have tried to develop clean energy sources. Hence, designing electrode materials for the construction of high-performance EESDs with high efficiency is essential for the electrochemical energy storage application. Owing to tremendous electrical and mechanical properties, MXene is used as an ideal electrode material. Fig. (**5**) shows the evolution of MXene as an energy storage application. In 2011, the first m-MXene came into existence [13]. In 2012, it was explored for the first time as LIBs anode, [52] and in 2013, it was investigated for the first time for its capacitive behaviour. In 2014, it was first explored for potassium/sodium storage batteries, [53] and followed by this in 2015 onward, MXene related materials have been investigated for ESDs [54].

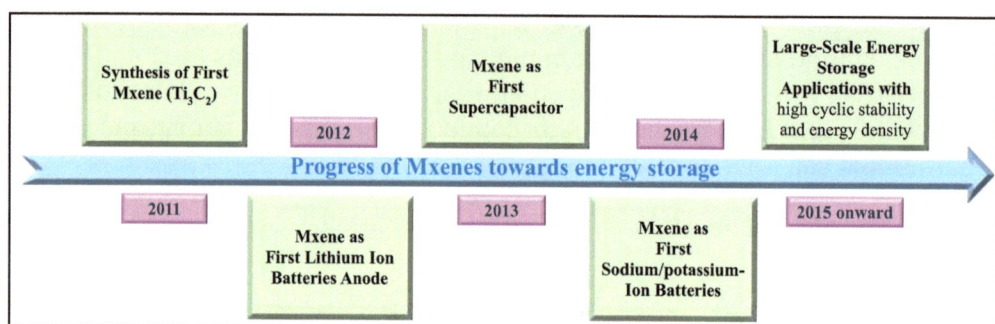

Fig. (5). History representation of MXene as energy storage applications.

MXenes for Batteries

Lithium-ion Batteries (LIBs)

LIBs have their benefits like extraordinary cycling stability and energy density, hence used in several applications including electric vehicles [55]. Previously, in rechargeable Li-ion batteries, graphite was used as an anode, but graphite anode showed low capacity which is not sufficient to fulfill today's demand for energy. In this direction, MXene becomes an attractive and effective material due to its layered structure with metallic or narrow band gap semiconducting characters. Computational studies showed that the specific capacity of MXene is affected by the formula weight and transition metals (M). M_2X has a higher specific capacity than M_3X_2 and M_4X_3 [56, 57]. This difference could be attributed to the formation of some inactive MX in M_3X_2. For example, Ti_2C has a greater specific capacity (160 mAh g^{-1}) while Ti_3C_2 has a lesser specific capacity (110 mAh g^{-1}) [53]. As per M element, Nb_2C showed greater specific capacity (180 mAh g^{-1}) in comparison to Ti_2C. This could be due to the different electronic structure of MXene [57]. Surface functionalization of MXene affects the performance of lithium storage. During HF-etching, the MXene surface is functionalized with a huge number of -OH and -F that hamper the storage and transportation of Li$^+$ and lead to decreased capacity [58]. In the high-temperature treatment of MXene, -O functional groups are dominated at the MXene surface and showed improved specific capacity [59]. The porous structure of MXene also enhances the electrochemical reaction kinetics [60]. In this direction, Gogotsi *et al.* have prepared porous Ti_3C_2 (p-Ti_3C_2) by a chemical etching. In the presence of O_2 assisted with Cu^{2+}, the layered $Ti_3C_2T_x$ was partially oxidized to TiO_2 resulting in TiO_2 nanoparticles being removed after HF treatment. As shown in Fig. (**6b**), after vacuum filtration, flexible freestanding p-$Ti_3C_2T_x$ electrode was obtained [61]. The porous structure of p-Ti_3C_2/CNT provides abundant active sites and multiple channels that decrease the pathway of ion transfer and increase the diffusion ability of Li$^+$ ions, hence they successfully achieved an improved specific capacity of 1250 mAh g^{-1} and cycle stability.

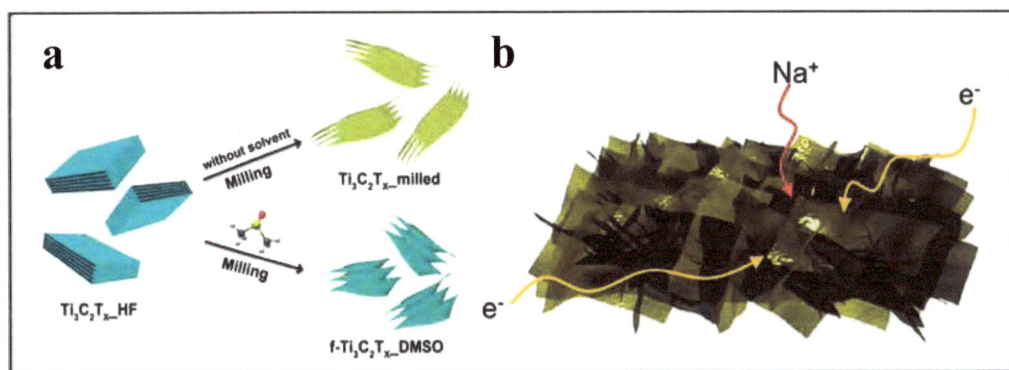

Fig. (6). (**a**) Schematic representation of f-$Ti_3C_2T_x$ synthesis. Adapted with permission from Reference [66], Copyright (2017), American Chemical Society. (**b**) Schematic illustration of the fast charge–discharge networks. Adapted with permission from Reference [67], Copyright (2018) American Chemical Society.

Modification of carbonaceous materials by heteroatoms *i.e.* doping provides a promising approach to multipurpose applications in energy fields [62]. Li *et al.* synthesized N-doped Nb_2CT_x with increased interlayer spacing and electron density through hydrothermal reaction and showed enhanced conductivity [63]. N-Nb_2CT_x anodes in LIBs revealed a higher reversible capacity of 360 mAh g^{-1} at 0.2 C than undoped Nb_2CT_x (190 mAh g^{-1}). Multiple heteroatom doping enhances the electrochemical energy storage performance of MXene. MXene-composites demonstrated improved activity of lithium storage because of the synergistic effects. Wu *et al.* synthesized MXene-SiO_2 composite by Stöber method and spray drying that showed high capacity (838 mAh g^{-1}), rate (517 mAh g^{-1}), and good cyclic stability [64]. Pure SiO$_2$ showed greater volume change and less electronic conductivity [65]. However, the incorporation of SiO$_2$ nanoparticles to the surface of 2D MXene enhanced the structural stability synergistically that buffers the changes in the volume of SiO$_2$ nanoparticles and also assists in electron/ion transfer.

Sodium Ion Battery (SIBs)

Owing to the benefits of sodium elements including low cost, suitable redox potential, and most importantly its abundance, SIBs attracted attention in energy storage on a large scale [66]. However, the larger size of Na$^+$ than Li$^+$ hinders the rate of diffusion which decreases the specific capacity. To deal with this, MXene are used as anode materials and specific capacities of SIBs can be improved by fast uptake and removal of Na$^+$. MXene has a flexible layered structure that can intercalate large-size ions and superb electrical conductivity for electron transfer. Zhang *et al.* prepared a few layers of Ti$_3$C$_2$T$_X$ consisting of greater spacing between layers that enhanced the diffusion of Na$^+$ and successfully obtained excellent specific capacity (267 mAh g^{-1}), rate performance (110 mAh g^{-1}), and cyclic stability Fig. (**6a**) [67]. In addition, Gogotsi *et al.* revealed the utilization of porous MXenes nanostructure to enhance the kinetics of Na$^+$ storage. They synthesized porous nanosheets of Ti$_3$C$_2$T$_X$*via* a sulphur loading-removal approach that revealed the enhanced electronic conductivity, increased fast ion diffusion, increased transfer of ions, and exposed active sites for increasing the rate of sodium storage Fig. (**6b**) [68]. They observed a specific capacity of Na$^+$ storage 166 mAh g^{-1} at 1 A g^{-1} and 124 mAh g^{-1} at 10 A g^{-1}.

The performance of SIB electrodes can be enhanced by heteroatom doping similar to LIBs. Pan *et al.* synthesized S-doped Ti$_3$C$_2$T$_x$ with increased interlayer spacing from 0.95 to 0.99 that showed a rate performance of 113.9 mAh g^{-1} at 4 A g^{-1} and specific capacity of 183.2 mAh g^{-1} at 0.1 A g^{-1} [69]. Yin *et al.* used *in situ* phosphorization to prepare a 3D porous Ti$_3$C$_2$/NiCoP composite in which nanoparticles of NiCoP anchored uniformly on the Ti$_3$C$_2$ nanosheets [70]. The interconnected 3D porous Ti$_3$C$_2$ avoids the aggregation of NiCoP and enhances the volume that boosts the filtration of electrolytes and decreases the pathway of Na$^+$ diffusion. Hence, the Ti$_3$C$_2$/NiCoP anode shows an extraordinary reversible specific capacity of 261.7 mAh g^{-1} and cycling stability. Nanocomposites of MXenes open new methods for the high performance of sodium storage.

Potassium Ion Battery (PIBs)

Analogous to LIBs and SIBs, PIBs worked on similar principles. With rich resources of K, low-cost, and suitable potential, PIBs attracted consideration for the development of next-generation energy storage devices on a large scale. However the larger size of K^+ than Li^+ and Na^+ results change in volume and pulverization [71]. Pure Ti_3CNT_x shows a lower storage capacity of potassium (32 mAh g^{-1} at 500 mA g^{-1}) because of the slow rate of intercalation [72]. Bao *et al.* synthesized alkalized Ti_3C_2 nanoribbons (a-Ti_3C_2) which showed a 3D porous structure and extended interlayer spacing. Therefore, it showed exceptional potassium storage with a reversible capacity of 42 mAh g^{-1} at 200 mA g^{-1} [73]. Metal-MXene attracted tremendous interest in PIBs. They showed relatively greater theoretical capacity and electronic conductivity [71, 73]. However, aggregation and an increase in the volume of electrode material lead to a decrease in capacity. Qian *et al.* developed MXene@antimony (MXene@Sb) composites by electrodeposition method as an anode for PIBs and attained a capacity of 516.8 mAh g^{-1} at 50 mA g^{-1} [73]. Zhang *et al.* synthesized a carbon-coated $MoSe_2$/MXene nanocomposite using hydrothermal reaction and annealing which showed good cyclability (Figs. **7a** & **b**) [74]. Both MXene sheets and $MoSe_2$ synergistically work for each other. MXene accommodates the increase in volume and inhibits the aggregation of $MoSe_2$, in turn, $MoSe_2$ prevents the stacking of nanosheets of 2D MXene, and in addition, carbon coating helps in structure stabilization. All these factors increased the capacity of MXene composite to 183 mAh g^{-1} at 10 A g^{-1}.

MXenes for Supercapacitors (SCs)

SC is an intermediate energy storage device between batteries and capacitors. They have more power density than batteries and more energy density than capacitors [75]. SCs have advantages in terms of safety, ease of handling, outstanding reversibility, and long cyclic stability. Based on the charge storage mechanism, SCs can be divided into two types. The first is electric double-layer capacitors, where energy storage is due to the absorption of ions electrostatically at the electrode interface that can be activated carbon, carbon nanotubes and nanofibers, porous carbon, *etc* [76]. The second is pseudocapacitors (PCs) where energy storage is due to reversible redox reactions at the surface of the electrode that can be metal oxides and conductive polymers, however, they showed a large change in volume that results in electrochemical instability [77]. To design SCs, MXene offers advantages such as smaller thickness, band gap, high packing density, large surface area, strong hydrophilicity, high electrical and thermal conductivity, tendency to intercalant the ions, and compatibility of surface functionalization. However, 2D MXene exhibits some disadvantages such as horizontal aggregation and restacking of MXene nanosheets due to strong van der Waals interactions between neighboring layers leading to the accessibility of electrolyte and ions, and the use of the entire 2D MXene surface limits these possibilities. Therefore, to overcome these disadvantages, researchers have developed an open structure of MXene nanosheets (accordion-like MXene structure, 3D porous MXene materials) by modulating the properties or morphology of MXene that decrease the ion diffusion pathway and increase the electrical conductivity for electron transfer through open channels [78]. Alshareef *et al.* synthesized 2D nanosheets of Ti_2CT_x to study the annealing effect on its structure and electrochemical performance [79]. They observed that MXene retains its layered structure even after heating in the presence of Ar, N_2, and N_2/H_2, but, it converted into TiO_2 in the presence of air. They also observed that MXene treated with N_2/H_2 showed a greater capacitance (51 F g^{-1} at 1 A g^{-1}) as a result of the stable layered structure of N_2/H_2-treated MXene where fluorine content is low and carbon content is high on the surface. Heteroatom doping also improves the SCs performance of carbon electrodes. Que *et al.* synthesized an N-Ti_3C_2 film using diethanolamine in methanol. Fig. (**8**) that showed a capacitance of 3123 F cm^{-3} at 5 mV s^{-1} [80]. Methanol facilitates the doping of N in 2D MXene sheets. It is noted here; that 100% capacity retention is achieved even after 10,000 cycles. Sun *et al.* synthesized a 1T-MoS_2/Ti_3C_2 composite by magnetic field hydrothermal method that showed 3D network structure, large space between layers, enhanced conductivity, and structural stability. It showed high capacity (386.7 F g^{-1} at 1 A g^{-1}), performance rate (207.3 F g^{-1} at 50 A g^{-1}), and 91.1% retention capacitance [81].

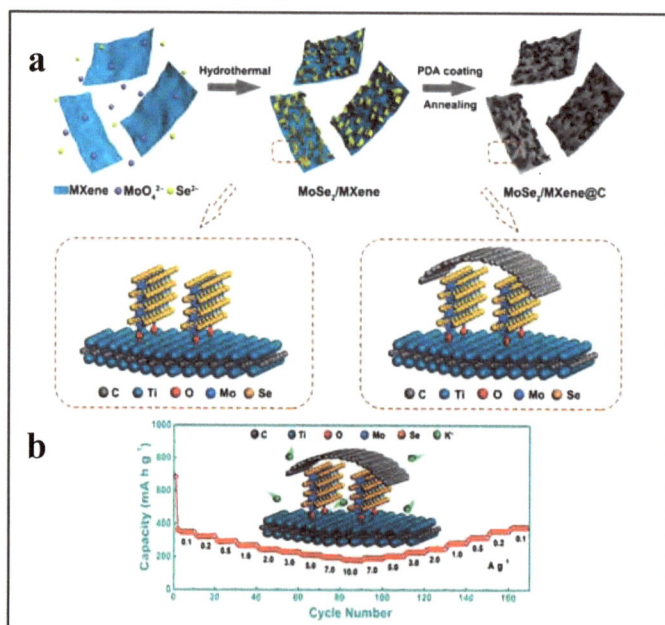

Fig. (7). (**a**) Schematic representation showed a synthesis of a-Ti_3C_2 MNRs (MXene nanoribbons) in KOH. (**b**) SEM images of a-Ti_3C_2 MNRs. Adapted with permission from Reference [74], Copyright (2019), American Chemical Society.

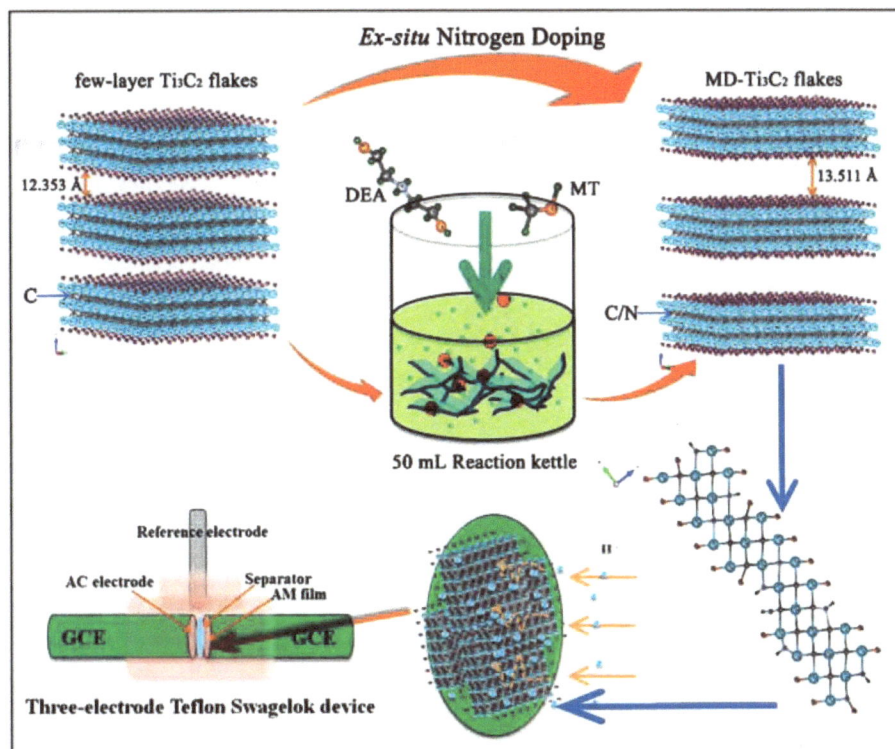

Fig. (8). Schematic illustration of the synthesis of MD-Ti$_3$C$_2$ flakes. Also shown is a schematic illustration of the intimate connection between few-layer MD-Ti$_3$C$_2$ flakes and H$^+$ facilitating charge transport in the asymmetric three-electrode Teflon Swagelok device. Adapted with permission from Reference [80], Copyright (2019), American Chemical Society.

CONCLUSION

The energy crisis demands new materials that can store energy for future needs. In this direction, owing to the excellent properties of 2D materials, especially the MXene collective term for 2D carbides, nitrides, and carbonitrides showed the high-performance energy storage application. MXenes were prepared by selective etching of A layers from the MAX precursor (M$_n$AX$_n$) where M corresponds to the early transition metal, A corresponds to the element of group 13 or 14, X corresponds to carbon or nitrogen, or n = 1-4. The preparation of MXene involves mainly 3 steps, choosing the MAX phase, etching, and delamination. There are different methods of etching including HF etching, LiF-HCl etching, molten salt etching, *etc*. Delamination of MXene is very important for energy storage applications. It should be noted that etching and delamination methods affect the surface termination groups and morphology. Surface functional groups and morphology also affect the performance of MXenes in energy storage applications. Researchers are in the process of improving the preparation methods for the enhancement of storage capacity, performance rate, and cyclic stability. They have developed MXene composites that showed the high performance of energy storage.

REFERENCES

[1]   S. Chu, and A. Majumdar, "Opportunities and challenges for a sustainable energy future", *Nature,* vol. 488, no. 7411, pp. 294-303, 2012.
[http://dx.doi.org/10.1038/nature11475] [PMID: 22895334]

[2]   B. Dunn, H. Kamath, and J.M. Tarascon, "Electrical energy storage for the grid: A battery of choices", *Science,* vol. 334, no. 6058, pp. 928-935, 2011.
[http://dx.doi.org/10.1126/science.1212741] [PMID: 22096188]

[3]   A.G. Pandolfo, and A.F. Hollenkamp, "Carbon properties and their role in supercapacitors", *J. Power Sources,* vol. 157, no. 1, pp. 11-27, 2006.
[http://dx.doi.org/10.1016/j.jpowsour.2006.02.065]

[4]   G.R. Bhimanapati, Z. Lin, V. Meunier, Y. Jung, J. Cha, S. Das, D. Xiao, Y. Son, M.S. Strano, V.R. Cooper, L. Liang, S.G. Louie, E. Ringe, W. Zhou, S.S. Kim, R.R. Naik, B.G. Sumpter, H. Terrones, F. Xia, Y. Wang, J. Zhu, D. Akinwande, N. Alem, J.A. Schuller, R.E. Schaak, M. Terrones, and J.A. Robinson, "Recent advances in two-dimensional materials beyond graphene", *ACS Nano,* vol. 9, no. 12, pp. 11509-11539, 2015.
[http://dx.doi.org/10.1021/acsnano.5b05556] [PMID: 26544756]

[5]   M. Devi, "Application of 2D nanomaterials as fluorescent biosensors", In: *ACS Symposium Series.* ACS Publications, 2020.
[http://dx.doi.org/10.1021/bk-2020-1353.ch006]

[6]   J. Pang, R.G. Mendes, A. Bachmatiuk, L. Zhao, H.Q. Ta, T. Gemming, H. Liu, Z. Liu, and M.H. Rummeli, "Applications of 2D MXenes in energy conversion and storage systems", *Chem. Soc. Rev.,* vol. 48, no. 1, pp. 72-133, 2019.
[http://dx.doi.org/10.1039/C8CS00324F] [PMID: 30387794]

[7]   X. Li, and L. Zhi, "Graphene hybridization for energy storage applications", *Chem. Soc. Rev.,* vol. 47, no. 9, pp. 3189-3216, 2018.
[http://dx.doi.org/10.1039/C7CS00871F] [PMID: 29512678]

[8]   X. Xu, Y. Zhang, H. Sun, J. Zhou, F. Yang, H. Li, H. Chen, Y. Chen, Z. Liu, Z. Qiu, D. Wang, L. Ma, J. Wang, Q. Zeng, and Z. Peng, "Progress and perspective: MXene and MXene-based nanomaterials for high-performance energy storage devices", *Adv. Electron. Mater.,* vol. 7, no. 7, p. 2000967, 2021.
[http://dx.doi.org/10.1002/aelm.202000967]

[9]   X. Li, Z. Huang, and C. Zhi, "Environmental stability of MXenes as energy storage materials", *Front. Mater.,* vol. 6, p. 312, 2019.
[http://dx.doi.org/10.3389/fmats.2019.00312]

[10]   M. Naguib, V.N. Mochalin, M.W. Barsoum, and Y. Gogotsi, "25th anniversary article: MXenes: A new family of two-dimensional materials", *Adv. Mater.,* vol. 26, no. 7, pp. 992-1005, 2014.
[http://dx.doi.org/10.1002/adma.201304138] [PMID: 24357390]

[11]   C. Zhan, W. Sun, Y. Xie, D. Jiang, and P.R.C. Kent, "Computational discovery and design of MXenes for energy applications: Status, successes, and opportunities", *ACS Appl. Mater. Interfaces,* vol. 11, no. 28, pp. 24885-24905, 2019.
[http://dx.doi.org/10.1021/acsami.9b00439] [PMID: 31082189]

[12]   M. Naguib, and Y. Gogotsi, "Synthesis of two-dimensional materials by selective extraction", *Acc. Chem. Res.,* vol. 48, no. 1, pp. 128-135, 2015.
[http://dx.doi.org/10.1021/ar500346b] [PMID: 25489991]

[13]   M. Naguib, M. Kurtoglu, V. Presser, J. Lu, J. Niu, M. Heon, L. Hultman, Y. Gogotsi, and M.W. Barsoum, "Two-dimensional nanocrystals produced by exfoliation of Ti3 AlC2", *Adv. Mater.,* vol. 23, no. 37, pp. 4248-4253, 2011.
[http://dx.doi.org/10.1002/adma.201102306] [PMID: 21861270]

[14]   P. Vattikuti, S. V., J. Shim, P. Rosaiah, A. Mauger, and C. M. Julien, "Recent Advances and Strategies in MXene-Based Electrodes for Supercapacitors: Applications, Challenges and Future Prospects", *Nanomaterials,* vol. 14, no. 1, p. 62, 2023.

[15]   M. Naguib, O. Mashtalir, J. Carle, V. Presser, J. Lu, L. Hultman, Y. Gogotsi, and M.W. Barsoum, "Two-dimensional transition metal carbides", *ACS Nano,* vol. 6, no. 2, pp. 1322-1331, 2012.
[http://dx.doi.org/10.1021/nn204153h] [PMID: 22279971]

[16]   C.E. Shuck, K. Ventura-Martinez, A. Goad, S. Uzun, M. Shekhirev, and Y. Gogotsi, "Safe synthesis of MAX and MXene: Guidelines to reduce risk during synthesis", *J. Chem. Health Saf.,* vol. 28, no. 5, pp. 326-338, 2021.
[http://dx.doi.org/10.1021/acs.chas.1c00051]

[17]   M. Naguib, and Y. Gogotsi, "Synthesis of two-dimensional materials by selective extraction", *Acc. Chem. Res.,* vol. 48, no. 1, pp. 128-135, 2015.
[http://dx.doi.org/10.1021/ar500346b] [PMID: 25489991]

[18]   M. Ghidiu, M.R. Lukatskaya, M.Q. Zhao, Y. Gogotsi, and M.W. Barsoum, "Conductive two-dimensional titanium carbide 'clay' with high volumetric capacitance", *Nature,* vol. 516, no. 7529, pp. 78-81, 2014.
[http://dx.doi.org/10.1038/nature13970] [PMID: 25470044]

[19]   X. Wang, C. Garnero, G. Rochard, D. Magne, S. Morisset, S. Hurand, P. Chartier, J. Rousseau, T. Cabioc'h, C. Coutanceau, V. Mauchamp, and S. Célérier, "A new etching environment (FeF 3 /HCl) for the synthesis of two-dimensional titanium carbide MXenes: A route towards selective reactivity vs. water", *J. Mater. Chem. A Mater. Energy Sustain.,* vol. 5, no. 41, pp. 22012-22023, 2017.
[http://dx.doi.org/10.1039/C7TA01082F]

[20]   J. Halim, M.R. Lukatskaya, K.M. Cook, J. Lu, C.R. Smith, L.Å. Näslund, S.J. May, L. Hultman, Y. Gogotsi, P. Eklund, and M.W. Barsoum, "Transparent conductive two-dimensional titanium carbide epitaxial thin films", *Chem. Mater.,* vol. 26, no. 7, pp. 2374-2381, 2014.
[http://dx.doi.org/10.1021/cm500641a] [PMID: 24741204]

[21]   A. Feng, Y. Yu, F. Jiang, Y. Wang, L. Mi, Y. Yu, and L. Song, "Fabrication and thermal stability of NH 4 HF 2 -etched Ti 3 C 2 MXene", *Ceram. Int.,* vol. 43, no. 8, pp. 6322-6328, 2017.
[http://dx.doi.org/10.1016/j.ceramint.2017.02.039]

[22]   M. Li, J. Lu, K. Luo, Y. Li, K. Chang, K. Chen, J. Zhou, J. Rosen, L. Hultman, P. Eklund, P.O.Å. Persson, S. Du, Z. Chai, Z. Huang, and Q. Huang, "Element replacement approach by reaction with lewis acidic molten salts to synthesize nanolaminated max phases and MXenes", *J. Am. Chem. Soc.,* vol. 141, no. 11, pp. 4730-4737, 2019.
[http://dx.doi.org/10.1021/jacs.9b00574] [PMID: 30821963]

[23]   L.I. Mian, L.I. You-Bing, L.U.O. Kan, L.U. Jun, E.K.L.U.N.D. Per, P.E.R.S.S.O.N. Per, R.O.S.E.N. Johanna, H.U.L.T.M.A.N. Lars, D.U. Shi-Yu, H.U.A.N.G. Zheng-Ren, and H.U.A.N.G. Qing, "Synthesis of novel MAX phase Ti3ZnC2via A-site-element-substitution approach", *J. Inorg. Mater.,* vol. 34, no. 1, pp. 60-64, 2019.
[http://dx.doi.org/10.15541/jim20180377]

[24]   P. Urbankowski, B. Anasori, T. Makaryan, D. Er, S. Kota, P.L. Walsh, M. Zhao, V.B. Shenoy, M.W. Barsoum, and Y. Gogotsi, "Synthesis of two-dimensional titanium nitride Ti 4 N 3 (MXene)", *Nanoscale,* vol. 8, no. 22, pp. 11385-11391, 2016.
[http://dx.doi.org/10.1039/C6NR02253G] [PMID: 27211286]

[25]   M. Ghidiu, M. Naguib, C. Shi, O. Mashtalir, L.M. Pan, B. Zhang, J. Yang, Y. Gogotsi, S.J.L. Billinge, and M.W. Barsoum, "Synthesis and characterization of two-dimensional Nb 4 C 3 (MXene)", *Chem.*

Commun., vol. 50, no. 67, pp. 9517-9520, 2014.
[http://dx.doi.org/10.1039/C4CC03366C] [PMID: 25010704]

[26]   O. Mashtalir, M. Naguib, V.N. Mochalin, Y. Dall'Agnese, M. Heon, M.W. Barsoum, and Y. Gogotsi, "Intercalation and delamination of layered carbides and carbonitrides", *Nat. Commun.,* vol. 4, no. 1, p. 1716, 2013.
[http://dx.doi.org/10.1038/ncomms2664] [PMID: 23591883]

[27]   M. Alhabeb, K. Maleski, B. Anasori, P. Lelyukh, L. Clark, S. Sin, and Y. Gogotsi, "Guidelines for synthesis and processing of two-dimensional titanium carbide (Ti3C2Tx MXene)", *Chem. Mater.,* vol. 29, no. 18, pp. 7633-7644, 2017.
[http://dx.doi.org/10.1021/acs.chemmater.7b02847]

[28]   H. Wang, Y. Wu, X. Yuan, G. Zeng, J. Zhou, X. Wang, and J.W. Chew, "Clay-inspired MXene-based electrochemical devices and photo-electrocatalyst: state-of-the-art progresses and challenges", *Adv. Mater.,* vol. 30, no. 12, p. 1704561, 2018.
[http://dx.doi.org/10.1002/adma.201704561] [PMID: 29356128]

[29]   J. Liu, X. Jiang, R. Zhang, Y. Zhang, L. Wu, W. Lu, J. Li, Y. Li, and H. Zhang, "MXene-enabled electrochemical microfluidic biosensor: Applications toward multicomponent continuous monitoring in whole blood", *Adv. Funct. Mater.,* vol. 29, no. 6, p. 1807326, 2019.
[http://dx.doi.org/10.1002/adfm.201807326]

[30]   L. Yu, L. Hu, B. Anasori, Y.T. Liu, Q. Zhu, P. Zhang, Y. Gogotsi, and B. Xu, "MXene-bonded activated carbon as a flexible electrode for high-performance supercapacitors", *ACS Energy Lett.,* vol. 3, no. 7, pp. 1597-1603, 2018.
[http://dx.doi.org/10.1021/acsenergylett.8b00718]

[31]   M. Hu, Z. Li, T. Hu, S. Zhu, C. Zhang, and X. Wang, "High-capacitance mechanism for Ti3C2Tx MXene by in situ electrochemical Raman spectroscopy investigation", *ACS Nano,* vol. 10, no. 12, pp. 11344-11350, 2016.
[http://dx.doi.org/10.1021/acsnano.6b06597] [PMID: 28024328]

[32]   M.R. Lukatskaya, S. Kota, Z. Lin, M.Q. Zhao, N. Shpigel, M.D. Levi, J. Halim, P.L. Taberna, M.W. Barsoum, P. Simon, and Y. Gogotsi, "Ultra-high-rate pseudocapacitive energy storage in two-dimensional transition metal carbides", *Nat. Energy,* vol. 2, no. 8, p. 17105, 2017.
[http://dx.doi.org/10.1038/nenergy.2017.105]

[33]   C. Yang, Y. Tang, Y. Tian, Y. Luo, M. Faraz Ud Din, X. Yin, and W. Que, "Flexible nitrogen-doped 2D titanium carbides (MXene) films constructed by an ex situ solvothermal method with extraordinary volumetric capacitance", *Adv. Energy Mater.,* vol. 8, no. 31, p. 1802087, 2018.
[http://dx.doi.org/10.1002/aenm.201802087]

[34]   L.H. Karlsson, J. Birch, J. Halim, M.W. Barsoum, and P.O.Å. Persson, "Atomically resolved structural and chemical investigation of single MXene sheets", *Nano Lett.,* vol. 15, no. 8, pp. 4955-4960, 2015.
[http://dx.doi.org/10.1021/acs.nanolett.5b00737] [PMID: 26177010]

[35]   N.C. Osti, M. Naguib, A. Ostadhossein, Y. Xie, P.R.C. Kent, B. Dyatkin, G. Rother, W.T. Heller, A.C.T. van Duin, Y. Gogotsi, and E. Mamontov, "Effect of metal ion intercalation on the structure of MXene and water dynamics on its internal surfaces", *ACS Appl. Mater. Interfaces,* vol. 8, no. 14, pp. 8859-8863, 2016.
[http://dx.doi.org/10.1021/acsami.6b01490] [PMID: 27010763]

[36]   K.J. Harris, "Direct measurement of surface termination groups and their connectivity in the 2D MXene V2CTx using NMR spectroscopy", *J. Phys. Chem. C,* vol. 119, no. 24, pp. 13713-13720, 2015.
[http://dx.doi.org/10.1021/acs.jpcc.5b03038]

[37]   H.W. Wang, M. Naguib, K. Page, D.J. Wesolowski, and Y. Gogotsi, "Resolving the structure of Ti3C2TX MXenes through multilevel structural modeling of the atomic pair distribution function", *Chem. Mater.,* vol. 28, no. 1, pp. 349-359, 2016.

[http://dx.doi.org/10.1021/acs.chemmater.5b04250]

[38]   N. Kurra, B. Ahmed, Y. Gogotsi, and H.N. Alshareef, "MXene-on-paper coplanar microsupercapacitors", *Adv. Energy Mater.,* vol. 6, no. 24, p. 1601372, 2016.
[http://dx.doi.org/10.1002/aenm.201601372]

[39]   C.J. Zhang, B. Anasori, A. Seral-Ascaso, S.H. Park, N. McEvoy, A. Shmeliov, G.S. Duesberg, J.N. Coleman, Y. Gogotsi, and V. Nicolosi, "Transparent, flexible, and conductive 2D titanium carbide (mxene) films with high volumetric capacitance", *Adv. Mater.,* vol. 29, no. 36, p. 1702678, 2017.
[http://dx.doi.org/10.1002/adma.201702678] [PMID: 28741695]

[40]   A.D. Dillon, M.J. Ghidiu, A.L. Krick, J. Griggs, S.J. May, Y. Gogotsi, M.W. Barsoum, and A.T. Fafarman, "Highly conductive optical quality solution processed films of 2D titanium carbide", *Adv. Funct. Mater.,* vol. 26, no. 23, pp. 4162-4168, 2016.
[http://dx.doi.org/10.1002/adfm.201600357]

[41]   C.J. Zhang, and V. Nicolosi, "Graphene and MXene-based transparent conductive electrodes and supercapacitors", *Energy Storage Mater.,* vol. 16, pp. 102-125, 2019.
[http://dx.doi.org/10.1016/j.ensm.2018.05.003]

[42]   V.N. Borysiuk, V.N. Mochalin, and Y. Gogotsi, "Molecular dynamic study of the mechanical properties of two-dimensional titanium carbides Ti n+1 C n (MXenes)", *Nanotechnology,* vol. 26, no. 26, p. 265705, 2015.
[http://dx.doi.org/10.1088/0957-4484/26/26/265705] [PMID: 26063115]

[43]   A. Lipatov, H. Lu, M. Alhabeb, B. Anasori, A. Gruverman, Y. Gogotsi, and A. Sinitskii, "Elastic properties of 2D Ti 3 C 2 T x MXene monolayers and bilayers", *Sci. Adv.,* vol. 4, no. 6, p. eaat0491, 2018.
[http://dx.doi.org/10.1126/sciadv.aat0491] [PMID: 29922719]

[44]   H. Wang, Y. Wu, X. Yuan, G. Zeng, J. Zhou, X. Wang, and J.W. Chew, "Clay-inspired MXene-based electrochemical devices and photo-electrocatalyst: State-of-the-art progresses and challenges", *Adv. Mater.,* vol. 30, no. 12, p. 1704561, 2018.
[http://dx.doi.org/10.1002/adma.201704561] [PMID: 29356128]

[45]   C. Hu, F. Shen, D. Zhu, H. Zhang, J. Xue, and X. Han, "Characteristics of Ti3C2X–chitosan films with enhanced mechanical properties", *Front. Energy Res.,* vol. 4, p. 41, 2017.
[http://dx.doi.org/10.3389/fenrg.2016.00041]

[46]   Z. Ling, C.E. Ren, M.Q. Zhao, J. Yang, J.M. Giammarco, J. Qiu, M.W. Barsoum, and Y. Gogotsi, "Flexible and conductive MXene films and nanocomposites with high capacitance", *Proc. Natl. Acad. Sci.,* vol. 111, no. 47, pp. 16676-16681, 2014.
[http://dx.doi.org/10.1073/pnas.1414215111] [PMID: 25389310]

[47]   O. Mashtalir, M.R. Lukatskaya, A.I. Kolesnikov, E. Raymundo-Piñero, M. Naguib, M.W. Barsoum, and Y. Gogotsi, "The effect of hydrazine intercalation on the structure and capacitance of 2D titanium carbide (MXene)", *Nanoscale,* vol. 8, no. 17, pp. 9128-9133, 2016.
[http://dx.doi.org/10.1039/C6NR01462C] [PMID: 27088300]

[48]   Z. Li, L. Wang, D. Sun, Y. Zhang, B. Liu, Q. Hu, and A. Zhou, "Synthesis and thermal stability of two-dimensional carbide MXene Ti3C2", *Mater. Sci. Eng. B,* vol. 191, pp. 33-40, 2015.
[http://dx.doi.org/10.1016/j.mseb.2014.10.009]

[49]   S. Lai, J. Jeon, S.K. Jang, J. Xu, Y.J. Choi, J.H. Park, E. Hwang, and S. Lee, "Surface group modification and carrier transport properties of layered transition metal carbides (Ti 2 CT x, T: –OH, –F and –O)", *Nanoscale,* vol. 7, no. 46, pp. 19390-19396, 2015.
[http://dx.doi.org/10.1039/C5NR06513E] [PMID: 26535782]

[50]   X.H. Zha, Q. Huang, J. He, H. He, J. Zhai, J.S. Francisco, and S. Du, "The thermal and electrical properties of the promising semiconductor MXene Hf2CO2", *Sci. Rep.,* vol. 6, no. 1, p. 27971, 2016.
[http://dx.doi.org/10.1038/srep27971] [PMID: 27302597]

[51] X.H. Zha, J. Yin, Y. Zhou, Q. Huang, K. Luo, J. Lang, J.S. Francisco, J. He, and S. Du, "Intrinsic structural, electrical, thermal, and mechanical properties of the promising conductor Mo2C MXene", *J. Phys. Chem. C*, vol. 120, no. 28, pp. 15082-15088, 2016.
[http://dx.doi.org/10.1021/acs.jpcc.6b04192]

[52] M. Naguib, J. Come, B. Dyatkin, V. Presser, P.L. Taberna, P. Simon, M.W. Barsoum, and Y. Gogotsi, "MXene: a promising transition metal carbide anode for lithium-ion batteries", *Electrochem. Commun.*, vol. 16, no. 1, pp. 61-64, 2012.
[http://dx.doi.org/10.1016/j.elecom.2012.01.002]

[53] D. Er, J. Li, M. Naguib, Y. Gogotsi, and V.B. Shenoy, "Ti3C2 MXene as a high capacity electrode material for metal (Li, Na, K, Ca) ion batteries", *ACS Appl. Mater. Interfaces*, vol. 6, no. 14, pp. 11173-11179, 2014.
[http://dx.doi.org/10.1021/am501144q] [PMID: 24979179]

[54] L. Wang, Z. Wu, J. Zou, P. Gao, X. Niu, H. Li, and L. Chen, "Li-free cathode materials for high energy density lithium batteries", *Joule*, vol. 3, no. 9, pp. 2086-2102, 2019.
[http://dx.doi.org/10.1016/j.joule.2019.07.011]

[55] C. Eames, and M.S. Islam, "Ion intercalation into two-dimensional transition-metal carbides: global screening for new high-capacity battery materials", *J. Am. Chem. Soc.*, vol. 136, no. 46, pp. 16270-16276, 2014.
[http://dx.doi.org/10.1021/ja508154e] [PMID: 25310601]

[56] M. Naguib, J. Halim, J. Lu, K.M. Cook, L. Hultman, Y. Gogotsi, and M.W. Barsoum, "New two-dimensional niobium and vanadium carbides as promising materials for Li-ion batteries", *J. Am. Chem. Soc.*, vol. 135, no. 43, pp. 15966-15969, 2013.
[http://dx.doi.org/10.1021/ja405735d] [PMID: 24144164]

[57] Q. Tang, Z. Zhou, and P. Shen, "Are MXenes promising anode materials for Li ion batteries? Computational studies on electronic properties and Li storage capability of Ti3C2 and Ti3C2X2 (X = F, OH) monolayer", *J. Am. Chem. Soc.*, vol. 134, no. 40, pp. 16909-16916, 2012.
[http://dx.doi.org/10.1021/ja308463r] [PMID: 22989058]

[58] Y. Xie, M. Naguib, V.N. Mochalin, M.W. Barsoum, Y. Gogotsi, X. Yu, K.W. Nam, X.Q. Yang, A.I. Kolesnikov, and P.R.C. Kent, "Role of surface structure on Li-ion energy storage capacity of two-dimensional transition-metal carbides", *J. Am. Chem. Soc.*, vol. 136, no. 17, pp. 6385-6394, 2014.
[http://dx.doi.org/10.1021/ja501520b] [PMID: 24678996]

[59] L. Chen, Y. Zhang, X. Liu, L. Long, S. Wang, W. Yang, and J. Jia, "Strongly coupled ultrasmall-Fe 7 C 3 /N-doped porous carbon hybrids for highly efficient Zn–air batteries", *Chem. Commun. (Camb.)*, vol. 55, no. 39, pp. 5651-5654, 2019.
[http://dx.doi.org/10.1039/C9CC01705D] [PMID: 31025990]

[60] C.E. Ren, M.Q. Zhao, T. Makaryan, J. Halim, M. Boota, S. Kota, B. Anasori, M.W. Barsoum, and Y. Gogotsi, "Porous two-dimensional transition metal carbide (mxene) flakes for high-performance Li-ion storage", *ChemElectroChem*, vol. 3, no. 5, pp. 689-693, 2016.
[http://dx.doi.org/10.1002/celc.201600059]

[61] X. Huang, Y. Zhang, H. Shen, W. Li, T. Shen, Z. Ali, T. Tang, S. Guo, Q. Sun, and Y. Hou, "N-doped carbon nanosheet networks with favorable active sites triggered by metal nanoparticles as bifunctional oxygen electrocatalyst", *ACS Energy Lett.*, vol. 3, no. 12, pp. 2914-2920, 2018.
[http://dx.doi.org/10.1021/acsenergylett.8b01717]

[62] R. Liu, W. Cao, D. Han, Y. Mo, H. Zeng, H. Yang, and W. Li, "Nitrogen-doped Nb2CTx MXene as anode materials for lithium ion batteries", *J. Alloys Compd.*, vol. 793, pp. 505-511, 2019.
[http://dx.doi.org/10.1016/j.jallcom.2019.03.209]

[63] G. Mu, D. Mu, B. Wu, C. Ma, J. Bi, L. Zhang, H. Yang, and F. Wu, "Microsphere-like SiO2/MXene hybrid material enabling high performance anode for lithium ion batteries", *Small*, vol. 16, no. 3, p. 1905430, 2020.

[http://dx.doi.org/10.1002/smll.201905430] [PMID: 31867880]

[64]   L. Zhang, X. Gu, C. Yan, S. Zhang, L. Li, Y. Jin, S. Zhao, H. Wang, and X. Zhao, "Titanosilicate derived SiO2/TiO2@C nanosheets with highly distributed TiO2 nanoparticles in SiO2 matrix as robust lithium ion battery anode", *ACS Appl. Mater. Interfaces,* vol. 10, no. 51, pp. 44463-44471, 2018.
[http://dx.doi.org/10.1021/acsami.8b16238] [PMID: 30516948]

[65]   N. Yabuuchi, K. Kubota, M. Dahbi, and S. Komaba, "Research development on sodium-ion batteries", *Chem. Rev.,* vol. 114, no. 23, pp. 11636-11682, 2014.
[http://dx.doi.org/10.1021/cr500192f] [PMID: 25390643]

[66]   Y. Wu, P. Nie, J. Wang, H. Dou, and X. Zhang, "Few-layer MXenes delaminated *via* high-energy mechanical milling for enhanced sodium-ion batteries performance", *ACS Appl. Mater. Interfaces,* vol. 9, no. 45, pp. 39610-39617, 2017.
[http://dx.doi.org/10.1021/acsami.7b12155] [PMID: 29039906]

[67]   X. Xie, K. Kretschmer, B. Anasori, B. Sun, G. Wang, and Y. Gogotsi, "Porous Ti3C2Tx MXene for ultrahigh-rate sodium-ion storage with long cycle life", *ACS Appl. Nano Mater.,* vol. 1, no. 2, pp. 505-511, 2018.
[http://dx.doi.org/10.1021/acsanm.8b00045]

[68]   J. Li, D. Yan, S. Hou, Y. Li, T. Lu, Y. Yao, and L. Pan, "Improved sodium-ion storage performance of Ti 3 C 2 T x MXenes by sulfur doping", *J. Mater. Chem. A Mater. Energy Sustain.,* vol. 6, no. 3, pp. 1234-1243, 2018.
[http://dx.doi.org/10.1039/C7TA08261D]

[69]   D. Zhao, R. Zhao, S. Dong, X. Miao, Z. Zhang, C. Wang, and L. Yin, "Alkali-induced 3D crinkled porous Ti 3 C 2 MXene architectures coupled with NiCoP bimetallic phosphide nanoparticles as anodes for high-performance sodium-ion batteries", *Energy Environ. Sci.,* vol. 12, no. 8, pp. 2422-2432, 2019.
[http://dx.doi.org/10.1039/C9EE00308H]

[70]   Q. Zhang, J. Mao, W.K. Pang, T. Zheng, V. Sencadas, Y. Chen, Y. Liu, and Z. Guo, "Boosting the potassium storage performance of alloy-based anode materials via electrolyte salt chemistry", *Adv. Energy Mater.,* vol. 8, no. 15, p. 1703288, 2018.
[http://dx.doi.org/10.1002/aenm.201703288]

[71]   M. Naguib, R.A. Adams, Y. Zhao, D. Zemlyanov, A. Varma, J. Nanda, and V.G. Pol, "Electrochemical performance of MXenes as K-ion battery anodes", *Chem. Commun.,* vol. 53, no. 51, pp. 6883-6886, 2017.
[http://dx.doi.org/10.1039/C7CC02026K] [PMID: 28607970]

[72]   P. Lian, Y. Dong, Z.S. Wu, S. Zheng, X. Wang, Sen Wang, C. Sun, J. Qin, X. Shi, and X. Bao, "Alkalized Ti3C2 MXene nanoribbons with expanded interlayer spacing for high-capacity sodium and potassium ion batteries", *Nano Energy,* vol. 40, pp. 1-8, 2017.
[http://dx.doi.org/10.1016/j.nanoen.2017.08.002]

[73]   Y. Tian, Y. An, S. Xiong, J. Feng, and Y. Qian, "A general method for constructing robust, flexible and freestanding MXene@metal anodes for high-performance potassium-ion batteries", *J. Mater. Chem. A Mater. Energy Sustain.,* vol. 7, no. 16, pp. 9716-9725, 2019.
[http://dx.doi.org/10.1039/C9TA02233C]

[74]   H. Huang, J. Cui, G. Liu, R. Bi, and L. Zhang, "Carbon-coated MoSe2 MXene hybrid nanosheets for superior potassium storage", *ACS Nano,* vol. 13, no. 3, pp. 3448-3456, 2019.
[http://dx.doi.org/10.1021/acsnano.8b09548] [PMID: 30817126]

[75]   J. Mao, J. Iocozzia, J. Huang, K. Meng, Y. Lai, and Z. Lin, "Graphene aerogels for efficient energy storage and conversion", *Energy Environ. Sci.,* vol. 11, no. 4, pp. 772-799, 2018.
[http://dx.doi.org/10.1039/C7EE03031B]

[76]   Y. Zhai, Y. Dou, D. Zhao, P.F. Fulvio, R.T. Mayes, and S. Dai, "Carbon materials for chemical capacitive energy storage", *Adv. Mater.,* vol. 23, no. 42, pp. 4828-4850, 2011.

[http://dx.doi.org/10.1002/adma.201100984] [PMID: 21953940]

[77] G. Wang, L. Zhang, and J. Zhang, "A review of electrode materials for electrochemical supercapacitors", *Chem. Soc. Rev.,* vol. 41, no. 2, pp. 797-828, 2012.
[http://dx.doi.org/10.1039/C1CS15060J] [PMID: 21779609]

[78] M.D. Levi, M.R. Lukatskaya, S. Sigalov, M. Beidaghi, N. Shpigel, L. Daikhin, D. Aurbach, M.W. Barsoum, and Y. Gogotsi, "Solving the capacitive paradox of 2D MXene using electrochemical quartz-crystal admittance and in situ electronic conductance measurements", *Adv. Energy Mater.,* vol. 5, no. 1, p. 1400815, 2015.
[http://dx.doi.org/10.1002/aenm.201400815]

[79] R.B. Rakhi, B. Ahmed, M.N. Hedhili, D.H. Anjum, and H.N. Alshareef, "Effect of postetch annealing gas composition on the structural and electrochemical properties of Ti2CTx MXene electrodes for supercapacitor applications", *Chem. Mater.,* vol. 27, no. 15, pp. 5314-5323, 2015.
[http://dx.doi.org/10.1021/acs.chemmater.5b01623]

[80] C. Yang, Y. Tang, Y. Tian, Y. Luo, X. Yin, and W. Que, "Methanol and diethanolamine assisted synthesis of flexible nitrogen-doped Ti3C2 (MXene) film for ultrahigh volumetric performance supercapacitor electrodes", *ACS Appl. Energy Mater.,* vol. 3, no. 1, pp. 586-596, 2020.
[http://dx.doi.org/10.1021/acsaem.9b01815]

[81] X. Wang, H. Li, H. Li, S. Lin, W. Ding, X. Zhu, Z. Sheng, H. Wang, X. Zhu, and Y. Sun, "2D/2D 1T-MoS 2 /Ti 3 C 2 MXene Heterostructure with Excellent Supercapacitor Performance", *Adv. Funct. Mater.,* vol. 30, no. 15, p. 0190302, 2020.
[http://dx.doi.org/10.1002/adfm.201910302]

CHAPTER 7

Different Supercapacitors' Characterizations

Satendra Kumar[1,2], **Hafsa Siddiqui**[2], **Netrapal Singh**[1,2], **Manoj Goswami**[1,2], **Lakshmikant Atram**[2], **S. Rajveer**[3], **N. Sathish**[1,2] and **Surender Kumar**[1,2,*]

[1] *Academy of Scientific and Innovative Research (AcSIR), Ghaziabad-201002, India*

[2] *CSIR - Advanced Materials and Processes Research Institute (AMPRI), Bhopal-462026, India*

[3] *Metallurgical and Materials Engineering, National Institute of Technology, Jamshedpur-831014, India*

Abstract: The development of new materials and technologies that can efficiently store energy while delivering power quickly has been the subject of numerous investigations. In an electrochemical supercapacitor (E-SC), the electric charge is stored in a double-layer formed at the electrode/electrolyte interface (EEI), which is based on the surface area as well as pore size availability. The high surface area provided by the micropores (pore diameter: 2 nm) is essential for charging the E-SCs and calculating the capacitance values. Mesopores (2 nm < pore diameter < 50 nm) allow good electrolyte penetration and offer a high-power density (2 nm pore diameter 50 nm). However, because a lot of non-carbonaceous materials are used to make E-SC electrodes, more *in-situ* analytical characterisation tools along with electrochemical techniques are needed. It is crucial to have at least a brief understanding of the electrochemical processes occurring at the EEI of E-SC electrodes (or devices). Variations in electrochemical, morphological and surface, and crystallographic properties will be used to categorise the data gathered by the state-of-the-art characterisation techniques. This chapter also provides a resource for researchers by outlining the methods to learn more about E-SCs and opportunities to achieve additional functionalities beyond those related to energy storage.

Keywords: Analytical tools, Electrochemical tools, Electrode/electrolyte interface, Supercapacitor, Pseudocapacitor.

INTRODUCTION OF SUPERCAPACITORS

Research into E-SCs has been a popular topic in the academic community for the better part of two decades due to their exceptional cycle stability and high power density [1]. E-SCs are intended to serve as a secondary energy storage option alongside batteries in a variety of contexts, including portable gadgets, electric

---

* **Corresponding author Surender Kumar:** Academy of Scientific and Innovative Research (AcSIR), Ghaziabad - 201002, India; CSIR - Advanced Materials and Processes Research Institute (AMPRI), Bhopal - 462026, India; E-mail: surenderjanagal@gmail.com

Sanjeev Verma, Shivani Verma, Saurabh Kumar & Bhawna Verma (Eds.)

vehicles, backup generators, and so on [2]. There have been numerous attempts to assemble high-performance E-SCs using a wide range of materials and methods (surface area/pore control, doping, heterostructure, engineering of crystal structure, and current collector modification) [3]. E-SCs have achieved enhanced levels of capacitance, rate performance, energy density, and cycling stability. Electrode materials [EMs] undergo complex structural/physical/chemical fluctuations during the charge-discharge process. Modifications to lattice parameters, ion adsorption/desorption, chemical bonds, dimensional/mass change, and other changes are all part of the electrochemical process [4].

Device performance can be estimated by using any or combination of the electrochemical techniques: cyclic voltammetry (CV), time constant (τ), galvanostatic charge-discharge (GCD), leakage current (I_{leak}), and electrochemical impedance spectroscopy (EIS) [5, 6]. This time constant defines the shortest time period over which the cell will accept a significant perturbation. It can be obtained through equation $\tau = 1/2\pi f C_{dl}$ at half of the charge-transfer resistance in the high frequency region of EIS profile. Beyond device testing, morphology and surface analysis of EMs can be analysed by scanning electron microscopy (SEM), X-ray photoelectron spectroscopy (XPS), transmission electron microscopy (TEM), Augur electron spectroscopy (AES) and so on [7]. X-ray diffraction (XRD), electrochemical quartz crystal microbalance (EQCM), nuclear magnetic resonance (NMR), scanning electrochemical microscopy (SECM), Raman spectroscopy (RS), Fourier transform infrared (FT-IR) spectroscopy, *etc.* are some of the *in-situ* characterisation methods that have been used to investigate E-SCs phase and dynamics [8, 9]. Diverse characterization methods (Fig. **1**) yield different data, but they all work together to shed light on the underlying mechanisms of EEI of E-SCs. While E-SCs are primarily used to store energy, it is important to note that new features can be created into E-SCs by modifying their configuration or integrating them. Mechanically deformable, flexible, self-healing, colour-tuneable, self-charging E-SC, *etc.*, have all been tried [10]. However, an understanding of the electrochemical phenomenon behind the various EMs is still lacking at the EEI level. Exploring new functions for E-SCs can expand their potential application domains while learning more about E-SCs can improve their understanding of their core operations at EEI level [11]. This chapter will review the advanced characterisation approaches used in E-SC investigations and provide instances of E-SCs that serve many purposes.

Double-layer Formation and Faradaic Process

As per working principles, E-SCs can be electric double-layer capacitors (EDLCs), pseudocapacitors (PCs), and hybrid capacitors (HCs) [12]. Classifica-tions of E-SCs are shown schematically in Fig. (**2**). EDLCs are capable of storing energy through the physical absorption of ions upon that electrode, while PCs use redox processes. The charged layer comprises an innermost tight layer (also called the Stern layer) and an exterior diffusion layer. The stern layer has an innermost Helmholtz plane (IHP) with solvent and adsorbed electrolyte ions and another outer Helmholtz plane (OHP) with solvated electrolyte ions [13]. On thermal disturbance, non-typically adsorbed ions may diffuse from OHP to bulk solution. Equation (1) estimates the capacitance of EDLCs:

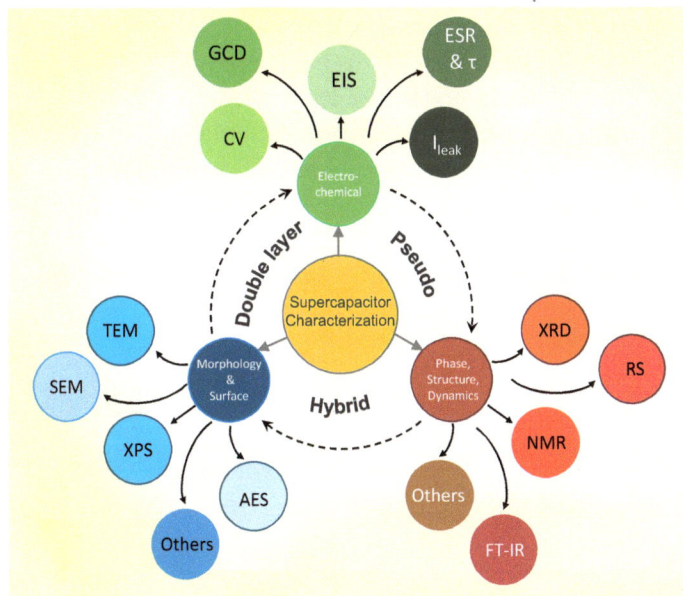

Fig. (1). A schematic representation of different characterization tools for E-SCs.

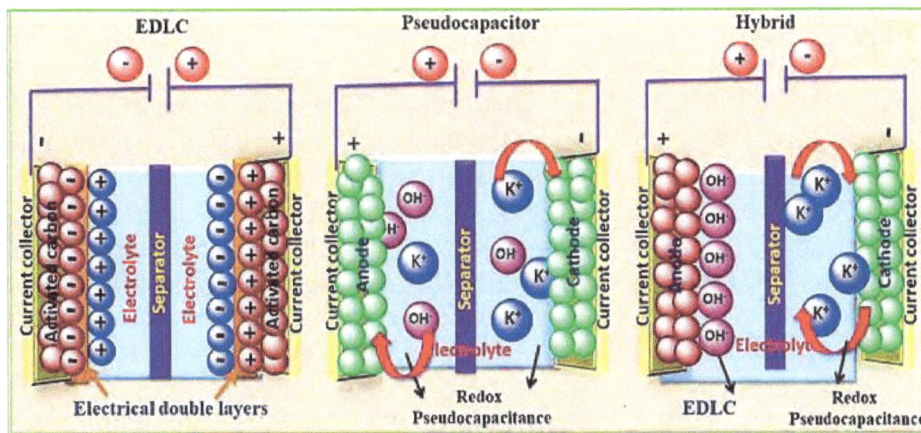

Fig. (2). Classification and charge storage mechanisms of E-SCs: EDLC, PC, and HC. *Adapted with permission from reference [14], Copyright (2022), The Royal Society of Chemistry.*

$$C = \frac{\epsilon_r \epsilon_0 A}{d} \tag{1}$$

Here, ϵ_r stands for relative permittivity, ϵ_0 stands for permittivity of free space, A stands for effective surface area, and d stands for effective charge separation in the EDLC, *i.e.*, the Debye length.

PCs store charges through quick Faradaic redox processes. Redox PCs are those in which reductive species are absorbed on oxidative electrodes (*e.g.*, RuO_2, MnO_2, and conducting polymers) in a Faradaic system [15]. Surface redox-dominated PCs behave like EDLCs. Intercalation PCs involve changeable intercalation and de-intercalation of cationic species (*e.g.*, Na^+, Li^+, *etc.*) into electrode materials. Intercalation-participated PCs (*e.g.*, Nb_2O_5) may not always undergo a phase transition. In under-potential deposition-based PCs, ions guarantee on a metal surface at potentials above their redox potential window (*e.g.*, H^+ deposit on Pt and Pd^{2+} deposit on Au) [16]. Some PCs share properties with conventional batteries, however, pseudocapacitive action happens on or near the metal-oxide electrodes.

Electrochemical Characterizations

The physical, structural, and chemical variations in EMs during the electrochemical processes can be recorded by various analytical methods. Even so, there are also electrochemical tactics to gain perceptions into the electrochemical kinds of stuff, for example using SECM to enumerate the charge transfer (heterogeneous) rate constant. This section will cover the electrochemical techniques that uncover the performance of E-SCs.

Scanning Electrochemical Microscopy

The scanning electrochemical microscopy comprises a three-electrode electrochemical cell and a bipotentiostat with a movable ultramicroelectrode. The SECM may be used for 2D mapping, kinetics, and surface alterations at specific locations. Research on E-SCs using SECM has been conducted to shed light on the kinetics at EEI. In the case of heterogeneously charge transfer, the rate constants (k_{eff}) of MnO_2 and polyaniline (PANI) based EMs were estimated by SECM in $[Fe(CN)_6]^{4-}$ (or $[Fe(CN)_6]^{3-}$) redox mediator [7]. It has been proven that the MnO_2 electrode may provide a positive feedback current. The decrease in macroscopic capacitance when using the PANI electrode (thickness > 5 μm) can be explained by correlating with the k_{eff} (with a 0.6 V electrode potential). Satpati *et al.* additionally confirmed the enhanced charge transfer when an electrochemically deposited MnO_2 electrode is biased at 0.2 V using SECM mapping [17].

Cyclic Voltammetry and Potentiometry

CV is both a standard electrochemical method and a synthesis tool. It linearly scans potential and records current as a response. Forward scan oxidises electrolyte or electrode species to create anodic current. A backward scan minimises applied potential, active components, and cathodic current. Current-voltage plots can be used to classify the kind of charge storage. Potential scanning often causes EDLCs to exhibit a current and capacitance that are both independent of the applied voltage (Figs. **3a & b**). CV provides three benefits. First, it determines the electrodeposition onset potential. Oxidation or reduction events involving charge transfer across EEI produce high current peaks [18]. The second, linear scan helps to grow homogeneous and conformal films. Third, CV can be used to synthesise transition metal oxides with various valence states. Fig. (**3c**) shows the GCD profile that when an EDLC is discharged at a constant current, the *V vs.* t plot is linear. Energy storage using pseudocapacitance is in the middle between EDLCs and batteries, which is made possible by the intercalation of charge-balancing (Li^+ or Na^+) [6]. Pseudocapacitance is shown in Figs. (**3d & f**) by the behaviour that is in between these two extremes. The peaks are caused by the oxidation and reduction of the metal centres that store charge (Figs. **3g & h**). When a battery is discharged, the *Vvs.* t plot is nonlinear and has a plateau region (Fig. **3i**). The electrochemical properties of a pseudocapacitive material will often fall into one of the following classes [19]: surface redox materials (Fig. **3b**), intercalation-type materials (Fig. **3d**), and intercalation-type materials (Fig. **3e**) exhibiting large reversible redox peaks. Materials with electrochemical responses in Figs. (**3g-i**) are similar to batteries. The capacitance calculated from the GCD profile is more accurate than the CV profile. The below equations (2-4) can be used to estimate gravimetric capacitance (C_s in F/g), energy density (E_s in Wh/kg), and power density (P_s in W/kg) from the GCD profile:

$$C_S = 4 \left(\frac{i \cdot \Delta t}{\Delta V} \right) \tag{2}$$

$$E_S = \frac{1}{7.2} C_S \cdot \Delta V^2 \tag{3}$$

$$P_S = \frac{E_S \times 3600}{\Delta t} \tag{4}$$

Here, *i* stands for discharge current density (A/g), *Δt* stands for current discharge time (s), and *ΔV* stands for actual potential window (V) [20].

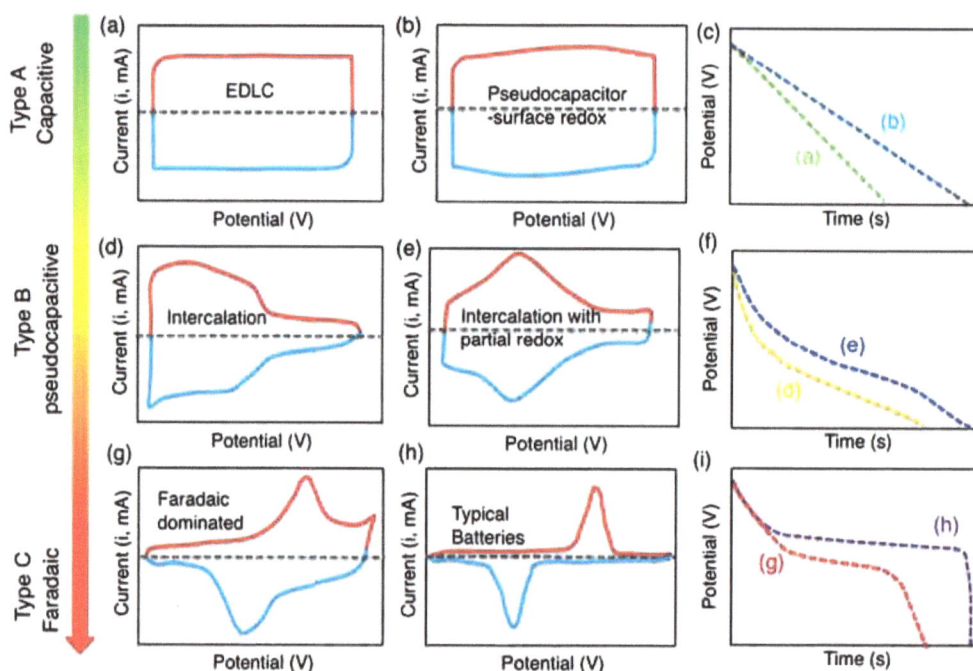

Fig. (3). Schematic CVs (a, b, d, e, g, and h) and GCD profiles (c, f, and i) for different types of energy storage media. *Adapted with permission from reference [5], Copyright (2022), American Chemical Society.*

Vanadium oxide-based nanorods produced by CV in a 100 mM $VOSO_4$ aqueous electrolyte included 50% V_5O_{12} (V^{5+} and V^{4+}) and 50% VO_2 [21]. During the anodic scan, V^{5+} was produced, and the as-deposited V_5O_{12} was largely reduced to VO_2 in the cathodic scan. Broad CV peaks in Fig. (**4a**) showed these redox reactions. Energy storage systems and materials, such as those used in E-SCs, are regularly put through GCD testing to determine how well they function. Capacitive response quality, irreversible Faradaic processes, and the derivation of various critical electrochemical figures of merit may all be assessed with the use of GCD profiles. Liu *et al.* stated that graphene-based E-SC with an elevated energy density of 85 Wk/kg at 1 A/g [22]. The GCD profile was found typically of EDLC behaviour as shown in Fig. (**4b**).

To utilize the special intercalation pseudocapacitive behaviour of $T-Nb_2O_5$ and good electrical conductivity of reduced graphene oxide (rGO), a $T-Nb_2O_5/rGO$ composite was prepared by Wu and his team [23]. Overlapped $T-Nb_2O_5/rGO$ composite outperforms pure $T-Nb_2O_5$ nanowires with a capacitance of 1492 F/g at 1.0 A/g in KOH, shown in Figs. (**4c - e**). Through hydrothermal interactions of GO with ammonia solution, Prakash and his group [24] were able to create redox-active nitrogen-grafted graphene oxynitride molecular heterostructures. For carbon-based materials, graphene oxynitride has the greatest recorded C_S of 783.5 F/g at 1.0 A/g, and it maintains about 95% even after being subjected to 10^3 charge-discharge cycles at 10 A/g (Fig. **4f**).

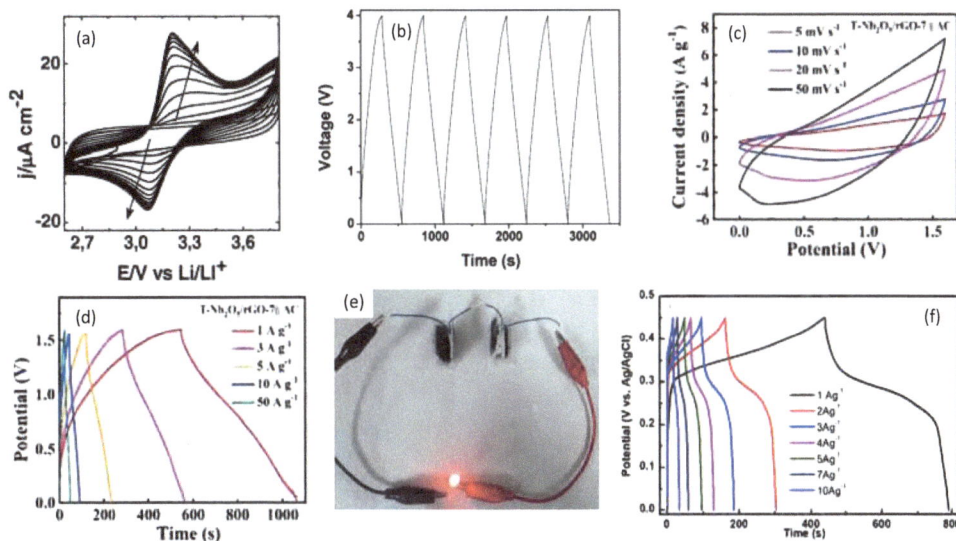

Fig. (4). (**a**) CVs of vanadium oxide-based nanorods produced in a 100 mM $VOSO_4$ electrolyte. *Adapted with permission from reference [21], Copyright (2022), Elsevier.* (**b**) GCD profile of graphene-based electrode at 1 A/g in $EMIMBF_4$ electrolyte. *Adapted with permission from reference [22], Copyright (2022), Elsevier.* (**c**) CV, (**d**) GCD profiles, and (**e**) the picture of a powered red LED. (**f**) GCD profiles of graphene oxynitride at various current densities. *Adapted with permission from reference [23], Copyright (2022), American Chemical Society.*

Electrochemical Impedance Spectroscopy and Time Constant

Electrochemical impedance spectroscopy measures electrical impedance (or resistance) across a wide-ranging frequency (10^{-4}–10^6 Hz) to determine electrode system accuracy. Rapid, non-destructive, and ready-to-go EIS assesses electrical characteristics under alternating currents. It employs a four-probe approach and Ohm's law to measure impedance (or resistance). EIS (*in-situ*) provides instantaneous data on resistance and its source throughout cathodic and anodic scans [25]. With ion adsorption/desorption on the EEI, electrolyte viscosity rises which changes the cell resistance.

Composites of activated carbon (AC) with either TiO_2 (5-20 wt.%) or manganese dioxide (5 wt.%) were analysed for their EIS spectra [42]. Due to its lower τ value, MnO_2@AC is capable of delivering electricity more efficiently. However, 5 wt.% TiO_2@AC electrodes have the lowest τ value, indicating that the active material may restore itself to electrical neutrality. Taberna *et al.* [26] examined the complicated capacitance and power of carbon-carbon E-SCs based on the results of a Nyquist analysis. Using 15 mg/cm^2 of AC, τ for a conventional 4 cm^2 cell was determined to be 10 s. Increased cell power resulted from a drop in τ from 10 to 4.8 s, as a result of less AC in the electrodes. Paula and her colleagues used EIS to reveal new details about how HCs store energy [27]. The resistance in the HCs was shown to be lowered by more than half (60%) at the redox potential of 1.25 V, demonstrating the resistance's dependency on the bias voltage for the diffusion-related component. When a redox electrolyte was used, the maximum capacitance onset frequency dropped to lower frequencies. It indicates that the kinetics processes are slower than that of fundamental capacitive facts. This also shows that the frequency at which a double layer is formed and the frequency at which a redox reaction takes place are distinct.

Leakage Current and Self-discharge

E-SCs must be well sealed to prevent electrolyte leakage and oxygen and other contaminants from entering. Impurities that go through Faradaic processes at electrodes may cause I_{leak}. Carbon compounds including impurities like transition metal ions, or dissolved oxygen deposited on AC [28]. Assembled in ambient air, the gadgets presumably include oxygen. Electrolyte breakdown due to overcharging and straight Ohmic leakage are other self-discharge methods. The I_{leak} corresponds with self-discharge. Voltage drops 13-22% per month in commercial supercapacitors. Haque *at el* [29]. has performed a detailed electrochemical comparison of E-SCs at various states of charge using an aqueous electrolyte (neutral 1 M Li$_2$SO$_4$) and along with 2% 4-n-pentyl-4'-cyanobiphenyl (5CB) that is a liquid crystal additive. Results show that compared to the behaviour of the device without liquid crystal addition 5CB, the device with 5CB displays lower self-discharge and I_{leak} short of affecting the capacitive functioning across a range of nominal voltages. Depending on the capacitance, most E-SCs in groups A, C, and E leak 10-20 µA. Better packaging (groups B, D, and F) reduced I_{leak} by 50% *i.e.*, 2-10 µA. The improvement may be attributed to lower electrolyte oxygen levels. Soavi and colleagues investigated I_{leak} and self-discharge in E-SCs based on ionic liquids (IL). The IL-based EDLCs, particularly the one with pure Pyr$_{14}$TFSI electrolyte, have lower I_{leak} and self-discharge values that are less impacted by temperature than those of the traditional organic electrolyte PC- 1 M Et$_4$NBF$_4$ cell [30].

Faradaic reactions of contaminants in E-SCs likely caused the leakage. Leakage can be avoided by vacuum-treating the electrodes before assembly to reduce adsorbed oxygen or by bubbling the electrolyte solution with inert gas to eliminate dissolved oxygen. The purity of AC is also important owing to the presence of surface functional groups and trace amounts of transition metals.

Morphology Observation and Surface Analysis

Scanning Electron Microscopy

Ex-situ characterization tools can give significant statistics, but cannot monitor kinetic characteristics, transitions in structure, or continual morphological shifts. *In-situ* approaches monitor atomic-scale changes without dismantling the testing device. SEM uses a focussed beam of 500–30000 eV to create topography and elemental composition. Liquid electrolyte cells in a vacuum are not suited for *in-situ* SEM imaging while solid-state electrolytes reduce this danger. *In-situ* SEM may be used to evaluate interfacial structural and morphological changes, fracture propagation, and volume expansion throughout the electrochemical process. In energy storage devices, *in-situ* SEM has been used to study lithiation and dendrite formation during electrochemical operations [31]. Yong *et al.* reported morphology-dependent electrochemical properties of E-SC, shown in Figs. (**5a-f**) [32]. The C_S of PANI layered floral structures was determined to be high (272 F/g) at 1.0 A/g. The effect of electrochemical cycling on a MnO_2-based aqueous E-SC was investigated by Sun *et al.* [33]. They showed the electrolyte ion intercalation during the cycling process causes significant changes in the morphologies and chemical valence state of MnO_2.

Transmission Electron Microscopy

TEM analyses surface morphology and elemental analysis at an atomic level. The accelerating voltage (80–300 keV) of TEM is substantially higher than SEM. It can magnify pictures 200 to 10^5 times with a 0.05 nm resolution. TEM offers information regarding crystal orientation, intercalation rate, and morphological changes in electrochemical devices [34]. *In-situ* TEM uses open or closed electrochemical cells. Tsai *et al.* stated the effect of MnO_2 electrodes and their CNT composites using TEM (*in-situ*) [35]. In the course of lithiation, the MnO_2 particles decreased and declined. Measurements of the CNT wall, inner spacing, and overall diameter over time are done with TEM. This enormous volume growth boosted Li-ion storage and made MnO_2 a good E-SC and battery material.

X-ray Photoelectron Spectroscopy

X-ray photoelectron spectroscopy is an unavoidable tool for understanding the underlying processes at EEI, as well as any surface organising during polarization, activation, and deactivation [36]. To gain insight into such mechanisms, *ex-situ* and *in-situ* XPS analysis of electrodes have been used. The photoelectron effect theory is the foundation of XPS, in which the core electrons of a material can be ejected by hitting it with X-rays. All potential elements existing in the electrode are qualitatively (or quantitatively) evaluated using the binding energy of electrons collected by the detector. However, XPS analysis at the EEI seems difficult due to surface sensitivity and ultra-high vacuum (UHV) requirements [37]. The XPS investigation of E-SC electrodes like PANI electrodes, alkali metal electrodes, MoS_2/PPy nanocomposites, and others has received a lot of attention. Chang *et al.* demonstrated the surface sensitivity, chemical state, and pseudocapacitive behaviour of manganese oxide-based electrodes in the IL 1-ethyl-3-methylimidazolium thiocyanate (EMI-SCN) [38].

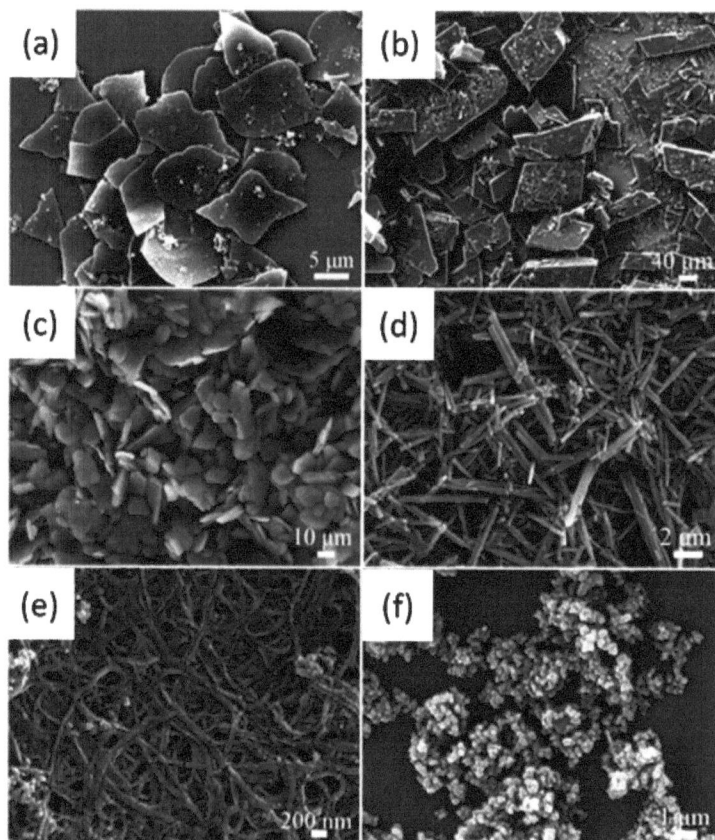

Fig. (5). PANI nanostructures: (**a**) irregular sheets (100 mg CTAB), (**b**) rhombic plates (300 mg CTAB), (**c**) irregular chunks (500 mg CTAC), and (**d**) nanorods (50 mg SDS) synthesized in 10 mM HCl. PANI nanostructures of (**e**) nanobelts (100 mg CTAC) and (**f**) nanospheres (100 mg SDS) obtained in 50 mM HCl. *Adapted with permission from reference [32], Copyright (2022), Royal Society of Chemistry.*

Ex-situ characterization is advanced, but post-analysis may result in metastable species not being analysed as they are only present during reactions, making *in-situ* analysis extremely valuable. With the advancement of XPS techniques, it is now possible to do *in-situ* electrochemical XPS measurements at pressures close to ambient using differentially pumped analysers, improved electrostatic lenses, and radiation sources based on synchrotrons [39]. Weingarth *et al.* [40] showed that utilizing an *in-situ* XPS analysis to detect the direct electrochemical shift as a function of applied potential.

Augur Electron Spectroscopy

The AES is a secondary phenomenon related to X-ray absorption spectroscopies. In absorption spectroscopy, an electron from another occupied state having higher energy eventually fills the vacant state left in the absorbing atom's core level just after the electron is stimulated by the X-ray photon. Sometimes, an excess amount of energy knocks out the electrons of the outer cores and leads to the emission of secondary excitations in the electronic system. These secondary electrons are known to be the Auger electrons and the associated spectroscopy is referred to as AES [41]. The short mean path typically in the range of 5 nm makes AES a powerful tool for the analysis of the surface of the E-SC electrodes. The in-depth compositional analysis provided by AES helps in monitoring the necessary compositions and surface thickness. Moreover, it also helps in monitoring the impurities causing the reduction of the capacitive performance. Zheng *et al.* used the AES depth sensing profiles to reconfirm their XPS results for their $Cu_{2+1}O@Cu$-MOFs hybrid clusters-based E-SCs [42]. Pinkert *et al.* used the depth profile AES along with the energy-filled TEM for the analysis of elemental distribution profiles and active component locations of metal hydroxide-based nanoparticles embedded into the mesoporous carbon [43]. The results analysis of both techniques showed that inside the carbon network, there is a highly defined spatial dispersion of metal hydroxide nanoparticles. The thorough investigation led to a significant improvement in the EMs for E-SCs application. This innovative method of studying the relationship between structure and property can be used in a variety of functional nanostructured materials and nanostructured interfaces.

Others

Apart from the above-listed tools, extended X-ray absorption fine structure (EXAFS), as well as X-ray absorption near edge structure (XANES), are used to realize the kinetics inside the electrodes. These approaches use synchrotron light to evaluate the absorption coefficient of core-level electrons stimulated to higher unoccupied states. They can provide information regarding the state of oxidation, electronic structure, material equilibrium, chemical bonding, *etc* [44]. The pseudocapacitive behaviour of MnO_2- based electrodes was investigated by Yeager *et al.* using *in-situ* XANES on $K_{0.15}MnO_2$ material [45]. Electrochemical observations with a potential regime of 0.3 - 0.6 V explained chemical kinetics by an alteration in the state of Mn. Utilizing *in-situ* XANES and EXAFS, Hsu and the team [46] described the pseudocapacitive procedure of $MnCo_2S_4$. As the electrochemical process progressed, they saw the Mn-Sulphur and Co-Sulphur bonds break apart, becoming active sites for OH⁻ ions. Research using XANES demonstrated that the state of Mn as well as Co, had changed, meanwhile, studies using EXAFS communicated details regarding Faradaic interaction between OH⁻ ions and active sites.

Phase, Structure, and Dynamics Observation

X-ray Diffraction

The XRD is crucial to comprehend the relationship between crystal structure and E-SC characteristics or performance for optimising current electrode design and developing new EMs with improved electrochemical performance. Investigate how the electrochemical reaction affects the structural characteristics of electrode materials, such as lattice parameters, phase transitions, and atomic occupancy [47]. In response to this need, various *in situ* analysis techniques, such as XRD, RS, and FT-IR, to name a few, have emerged, made possible by the fine-tuning of electrode materials. Among these, *in-situ* XRD has corroborated to be an extremely applicable technique for obtaining information on crystal structure and phase transformation of crystalline EMs during the cycle process.

Figs. (**6a-c**) depicts *in-situ* XRD configuration and corresponding XRD patterns observed for the WO_{3-x} nanowire//FTO electrodes at CV scans of 0.5 mV/s [48]. Using *in situ* XRD in an $AlCl_3$ electrolyte, Wang *et al.* investigated dual-function WO_{3-x} nanowire//FTO electrodes *via* XRD and reported the kinetic activities of intercalated Al^{3+} ion (Fig. **6d**). The WO_{3-x} nanowires 110-crystal plane maintained its position and intensity throughout charge-discharge cycling, indicating that there was no change in the interlayer distance or surface pseudocapacitance reaction (Fig. **6e**). In addition, several manganese oxide phases emerged with an enhancement in temperature, according to an *in-situ* XRD study conducted by Ghodbane *et al.* [49]. Based on *in-situ* XRD observation and subsequent thermogravimetric analyses, the calcined spinel and the ramsdellite phases of MnO_2 were developed at different temperatures. Later, they used synchrotron XRD to study the MnO_2 allotropes in two different electrolytes and noticed that the lattice spacing had changed to ~ 0.2 A. This *in-situ* method provides a satisfactory explanation for the charge of the Mn-O bond distance as it is directly related to the redox reaction.

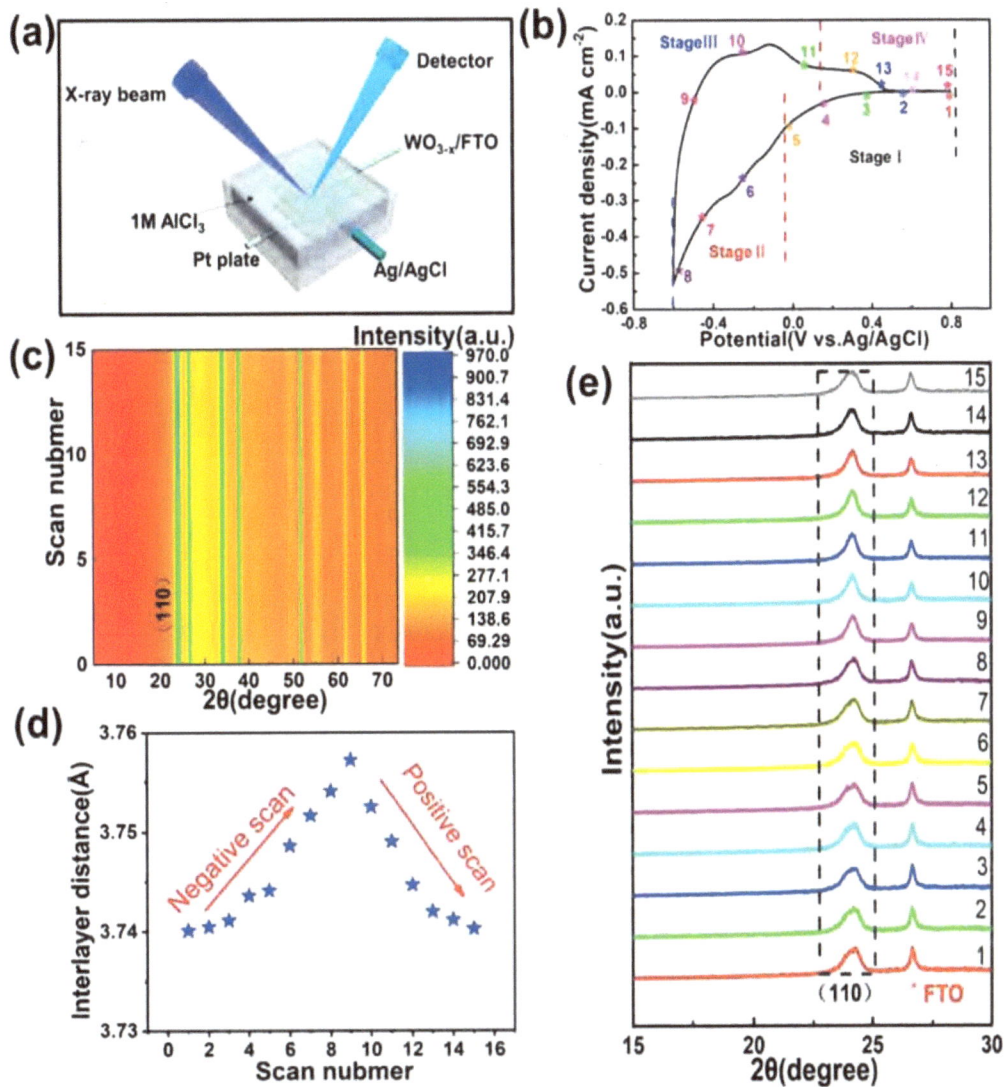

Fig. (6). (a) *in-situ* XRD set-up (b) CV profile (c) XRD plot recorded with CV cycles (d) Distance between slices *vs.* the number of scans from XRD patterns (f) *In-situ* XRD model of electrode enlarged to 2θ from 15° to 30°. *Adapted with permission from reference [48], Copyright (2022), Springer.*

According to Okubo *et al.*, MXene and electrolyte interfaces underwent partial desolvation, with a substantially lower solvation energy than hydration [50]. Following intercalation, the cation's atomic orbitals were coupled through functional groups of MXene. In nonaqueous electrolytes, MXene electrode capacitance is mostly pseudocapacitance. The change in interlayer spacing of the MXene film concerning temperature was noticed using these *in-situ* XRD patterns. They observed that the interlayer gap remained constant during the cycling test, which was sufficient for cation intercalation/deintercalation. In a different study, the electrodes were studied employing *in-situ* XRD, which showed that the MXene (Ti_3C_2) layers had changed, intercalating numerous cations, including Na^+, $Al3^+$, K^+, Mg^{2+}, and NH^{4+} [51]. As a result, countless studies using MXene as the primary material have been conducted in the field of E-SCs. Despite the effectiveness and merit of this emerging strategy, fewer major challenges still exist. Such as due to non-targeted material's ability to reflect, absorb, and diffract X-rays, incident X-rays have a moderate intensity. Therefore, the employment of X-rays may potentially result in secondary reactions.

Raman Spectroscopy

One useful non-destructive method of surface-sensitive Opto-vibrational characterisation is RS. Chemical reactions, reversible side reactions, variations in the crystal structure, lattice change, bonding/debonding, and intercalation/ deintercalation of ions are just some of the real-time measurements done by RS [52]. It details the molecular composition and short-range structure of the EMs at the EEI. A completely clear window constructed of glass, sapphire, or quartz is required for operando RS. The fluorescence effect can cause electrode deterioration and a low signal-to-noise ratio, which are the only two major drawbacks of *in-situ* RS. Gupta *et al.* availed *in-situ* RS to state the charge storage methodologies of a carbonaceous metal-oxides designed electrode (Figs. **7a & b**) [53]. Disorientation, orbital rehybridization, and a 0.2 eV change in the Fermi level were observed in hybrid electrodes made from various MnO^{2-} and CoO_3- decorated multi-layer graphene (MLG). As a result of the applied bias voltage, the EMs exhibited transitions between the D-band, G-band, and 2D-band as well as overlapping orbits of the component atomic bonds. The charge storage process may be clarified by the aforementioned observations of band placement and frequency shift and intensity variation at the EEI. In this study, *in-situ* RS was used to examine AC in NaI electrolytes first time, as shown in Figs. (**7c-f**) [54]. Minimizing their synthesis by decreasing the self-discharge state (confinement of polyiodides) as well as improving the pore-filling ability of I_2 might increase the capacitance.

Rafael *et al.* used several aqueous electrolytes to probe carbon electrodes and electrolytes under dynamic settings to understand electrochemical supercapacitor charge-discharge operations [55]. The Raman shift may be caused by solvated cations interacting with graphite under polarization. Molecular dynamics simulations were used to explain Raman spectrum variations in aqueous electrolytes comprising monovalent cationic species. In their use of RS, Yang, and team elucidated the pseudocapacitive mechanism and excellent performance of Mn_3O_4 electrodes [56]. Three key events contributed to the structural deformation of the electrode: the generation of MnO_2 from Mn_3O_4, the intercalation/deintercalation of electrolytic ions (Na^+) from the electrolyte, and the Faradaic reaction involving the insertion of Mn^{2+} into the defect sites.

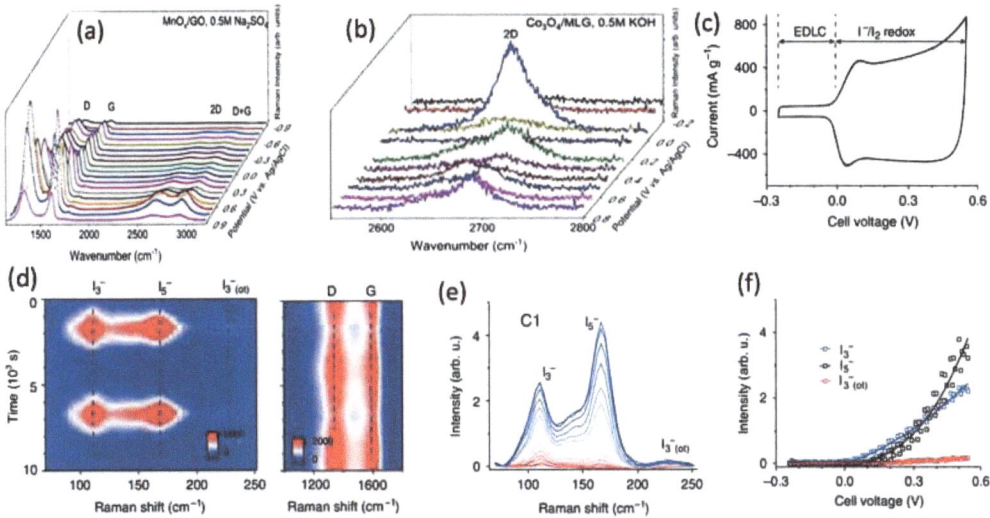

Fig. (7). Raman spectra (*in-situ*) for (**a**) MnO_x/GO and (**b**) Co_3O_4/MLG. *Adapted with permission from reference [53], Copyright (2022), American Institute of Physics.* (**c**) CV profile of *in-situ* Raman cell (0.32 mV/s). (**d**) Raman intensity *vs.* time and G- and D-bands shift to track I_3^- and I_5^- bands. (**e**) Raman spectra obtained during positive voltage sweep. (**f**) I_3^- and I_5^- intensities *vs.* cell voltage. *Adapted with permission from reference [54], Copyright (2022), American Institute of Physics.*

In-situ confocal Raman microspectrometry was used to analyse an EDLC with AC cloth electrodes and an organic electrolyte. The electrochemical studies of MXene (Ti_3C_2) electrodes in various electrolytes, as investigated by RS, revealed the electrochemical mechanism as well as the breaking/formation of O-H bonds. Yao and team used α-V_2O_5 nanometric wire, and the operando RS revealed that the charge is stored *via* a two-step phenomenon involving the intercalation/ deintercalation of electrolytic ions and the contraction and expansion of the lattice [57]. As the interlayer distance changed in response to the applied bias, the corresponding Raman band shifted. The higher power density was significantly influenced by the extensive insertion/extraction of electrolytic ions, but the significance of H^+ ions was overlooked due to the aforementioned phenomenon.

Nuclear Magnetic Resonance Spectroscopy

NMR spectroscopy can identify the structure and dynamics of EEI. NMR's success derives from its ability to discriminate intrapore ions from bulk electrolytes [8]. *In-situ* NMR is used to follow ion mobility in porous electrodes at varying voltages. Forse and team measured in-pore and ex-pore ^{19}F resonance width alteration at different temperatures to evaluate YP50F ion mobility [44]. EMITFSI's NMR spectra demonstrate temperature-dependent variations in in-pore line width. In-pore line width is caused by the dipole-dipole interface and electrolytic ion adsorption in micropores of carbon.

Compared to data recorded at 7.1 T, $Pyr_{13}TFSI$ and EMITFSI line widths increase significantly at ~9.5 T in an examined temperature regime (250–350 K), indicating distinct adsorption sites. Ion diffusion causes line shape changes. When diffusive motion surpasses the frequency band of chemical changes, a single consolidated resonance is observed. Thus, in-pore line width measures how rapidly ions flow through carbon pores; quicker diffusion, and narrower intrapore resonance. Temperature promotes ion diffusion. EMITFSI has a lower ^{19}F in-pore line width than $Pyr_{13}TFSI$ across the temperature range of interest, indicating quicker in-pore anion diffusion [58]. Comparing ^{19}F and ^1H spectra with those without solvent shows that acetonitrile substantially reduces in-pore resonance line widths.

Fourier Transform Infrared Spectroscopy

In-situ attenuation total reflection (ATR) FT-IR spectroscopy is a valuable tool for studying the electrochemical interface. ATR FT-IR uses internal and exterior reflections. Internal and external reflections are used for EEI studies [59]. Internal reflection can permeate the working electrode. In external reflection, the working electrode is situated close to the light-guiding prism. Due to a thin layer, species can absorb reflected light. This setup may be employed for an extended range of EMs and determines electrolyte and electrode species. Onion-like carbons (OLC) and carbide-derived carbons (CDC) electrodes in IL electrolytes were studied. Cation and anion's FT-IR absorbance intensity increases as voltage drops from +1.5 V to 0 V for CDC. Cyclic absorbance reveals that pure IL ions penetrate or depart from the CDC during GCD profiling. FT-IR absorption of ions remains almost unaffected for OLC at similar conditions, showing the ions are near the electrode with quick charging or discharging kinetics. CDC's larger capacitance may explain this.

In-situ FT-IR spectroscopy investigates electrolyte content. With time, the absorbance of EMIM-TFSI (in 10% propylene carbonate (Pc)) decreases, potentially due to evaporation or carbon particle trapping. The research shows that solvent-ion interactions can alter electrochemistry. Experiments with nanoporous carbon nanofibers (NCNFs) demonstrated distinct dynamics [60]. For NCNF electrodes, $EMIM^+$ and $TFSI^-$ absorbance reduces as the voltage is varied from 0 to 1.0 V, showing that ionic species concurrently enter the micro/nanopores of NCNFs owing to strong Coulombic contact between ionic species. This difference suggested that oxygen functional groups may affect ion kinetics. Positively charged nanopores expel cations but absorb anions. By functionalizing electrode surfaces, charge storage and electrochemical working of E-SC can be improved.

Others

has the capability to quantify the change in the mass of the electrode material of the E-SCs. Sauebrey's equation (5), which is the linear fashion correlation between the frequency of resonance and the change in mass, is the fundamental mathematical relation for the same [3].

$$\frac{\Delta f}{n} = -C_m \times \Delta m \tag{5}$$

Here, Δf denotes the variation in resonating frequency, n denotes the order, C_m denotes the mass sensitivity constant, and Δm denotes a change in the electrode material mass.

The following example would provide more insight into the capabilities of the EQCM process. The Δm for the N-doped rGO (N-rGO) and MnO_2 electrodes was measured using *in-situ* EQCM during the charging and discharging process [61]. The Δm was evaluated to be 8.4 µg/cm^2 for the N-rGO and 7.4µg/cm^2 for the MnO_2 electrodes. Since MnO_2 is the anode in this process, this Δm can be associated with solvent cations such as Na^+ and K^+ that are either inserted or released through the MnO_2 layers. Due to this, the MnO_2 electrode that is being charged has a higher Mn oxidation state (changes from +3.01 to +3.12 here). When fully discharged, the Δm goes back to its original value. Moreover, the inclusion of hydrodynamic spectroscopy with EQCM is helpful when the dissipation factors are associated with the resonance frequency. This process is known as EQCM with dissipation (EQCM-D). This method can be applied to the different overtones responsible for the different penetration depths. The EQCM-D technique uses very few cycles (200) compared to conventional coin cells for the screening of the electrolyte system. This corresponds to their skill to form first-rate small mass thin films; this speeds up cycle testing for extremely quick mechanical testing of interested electrodes in various electrolytes. As a result, this process is simple to deploy as a powerful *in-situ* analytical method in the domain of energy conversion and storage [62].

Moreover, *in-situ* X-ray or neutron scattering techniques are devoted to studying the ion adsorption phenomenon of the electrodes. This technique is very helpful for porous electrodes, especially in the case of pseudocapacitive systems. Prehal *et al.* [54] investigated ion adsorption in nanostructured and porous carbon electrodes under polarisation using a CsCl aqueous electrolyte. They demonstrated that ions were partially desolvated when trapped in the nanopores of carbon. The amount of desolvation and confinement was discovered to rise with the applied voltage, which consequently explained the rise in the capacitance [63]. Futamura *et al.* carried out scattering experiments to resolve the structure of IL on its confinement inside the nanoporous carbons. Their study demonstrated that when the pores are only large enough to hold one layer of ions, Coulombic ordering is reduced and finally breaks it. This, in turn, facilitates the creation of the cations and anions co-ion pairs. The creation of the non-Coulombic structure is made feasible by the image charges created in the carbon walls, which counteract the repellent electrostatic interactions between co-ions. This compensating effect results in a highly dense ionic structure of co-ions whenever the ion sizes are within the average pore size, which can explain the increase in capacitance seen in small nanopores. The study further confirmed that the density of co-ion pairs was increased in the presence of oppositely charged carbon nanopores. Thus, it provides experimental evidence of the formation of so-called superionic states in the nanopores as theoretically predicted by Kondrat [64]. Measuring the diffusivity parameters like diffusion length and jump rate is very tactical at the nanopores. Quasielastic neutron scattering (QENS) is one of the most suitable techniques to accurately measure such parameters. Moreover, the use of the QENS technique helps to examine the electrolyte movement inside the electrode material as neutrons have higher penetration power as well as show higher sensitivity towards the hydrogen-bearing species. Numerous investigations have used QENS to settle the dynamics at a microscopic level and ion immobilisation in diverse IL. The QENS method also offers an easy way to investigate how humidity impacts the immobilisation of charging on electrode walls, which impacts electrochemical performance [65].

CONCLUSION AND FUTURE OUTLOOK

The constant increase in energy use has been the impetus for research into more efficient energy storage methods. E-SCs have a lot going for them, and that includes the fact that they might help solve the energy dilemma. E-SCs are the go-to option for charge storage owing to their high power density and extended life cycle. Although their energy density is substandard, they have poor charge/discharge management and low line control quality. As we have shown, carbon-based electrodes for E-SCs, in particular, benefit from electrochemical, morphological, and analytical characterizations that shed light on the parameters impacting charging operations. Charge screening, ionic rearrangements, pore features, and confinement all have major effects on capacitance and ion dynamics. While *ex-situ* characterisation techniques do help with these problems, high-end multifunctional *in-situ* approaches are needed to cover all the bases when it comes to interpreting the charge storage phenomenon. In the case of E-SCs, *in-situ* methods are superior to *ex-situ* techniques because they can acquire data in realtime while charge-discharge cycling occurs. Perceivable physiochemical and structural changes in the electrodes of E-SCs can be observed during their circumstances of operation at the EEI. These changes include mechanisms in the redox centres, ion adsorption (or desorption), intercalation (or deintercalation), alteration of morphologies, bonding (or debonding), spreading of cracks, mechanical distortions, double-layer width, lattice parameter alterations, *etc*. To guarantee the best possible material performance in real-time, *in-situ XRD, SEM, EQCM, TEM, NMR, XPS, Raman, FT-IR, EIS, etc. have snatched the spotlight in every relevant discipline. Using findings from cutting-edge in-situ and ex-situ instruments, we detailed the factors that affect the charge storage process in E-SCs. Energy storage processes are complex and must be better understood as more and more applications, both old and new, are combined with E-SCs. This requires advances in electrode materials, knowledge of electrochemical energy conversion, in-situ methods, etc.*

REFERENCES

[1]   S. Pang, L. Gong, N. Du, H. Luo, K. Yu, J. Gao, Z. Zheng, and B. Zhou, "Formation of high-performance Cu-WOx@C tribasic composite electrode for aqueous symmetric supercapacitor", *Mater. Today Energy,* vol. 13, pp. 239-248, 2019.
[http://dx.doi.org/10.1016/j.mtener.2019.05.016]

[2]   A. Shaqsi, "Review of energy storage services, applications, limitations, and benefits", *Energy Reports,* vol. 6, pp. 288-306, 2020.

[3]   J. Chen, and P.S. Lee, "Electrochemical supercapacitors: From mechanism understanding to multifunctional applications", *Adv. Energy Mater.,* vol. 11, no. 6, pp. 2003311-2003327, 2021.
[http://dx.doi.org/10.1002/aenm.202003311]

[4]   M. Wang, and Z. Feng, "Interfacial processes in electrochemical energy systems", *Chem. Commun.,* vol. 57, no. 81, pp. 10453-10468, 2021.
[http://dx.doi.org/10.1039/D1CC01703A] [PMID: 34494049]

[5]   Y. Gogotsi, and R.M. Penner, "Energy storage in nanomaterials - capacitive, pseudocapacitive, or battery-like?", *ACS Nano,* vol. 12, no. 3, pp. 2081-2083, 2018.
[http://dx.doi.org/10.1021/acsnano.8b01914] [PMID: 29580061]

[6]   H. Banda, J.H. Dou, T. Chen, N.J. Libretto, M. Chaudhary, G.M. Bernard, J.T. Miller, V.K. Michaelis, and M. Dincă, "high-capacitance pseudocapacitors from Li + ion intercalation in nonporous, electrically conductive 2D coordination polymers", *J. Am. Chem. Soc.,* vol. 143, no. 5, pp. 2285-2292, 2021.
[http://dx.doi.org/10.1021/jacs.0c10849] [PMID: 33525869]

[7]   A. Sumboja, U.M. Tefashe, G. Wittstock, and P.S. Lee, "Monitoring electroactive ions at manganese dioxide pseudocapacitive electrodes with scanning electrochemical microscope for supercapacitor electrodes", *J. Power Sources,* vol. 207, pp. 205-211, 2012.
[http://dx.doi.org/10.1016/j.jpowsour.2012.01.153]

[8]   A.C. Forse, J.M. Griffin, C. Merlet, J. Carretero-Gonzalez, A-R.O. Raji, N.M. Trease, and C.P. Grey, "Direct observation of ion dynamics in supercapacitor electrodes using in situ diffusion NMR

spectroscopy", *Nat. Energy,* vol. 2, no. 3, pp. 16216-16223, 2017.
[http://dx.doi.org/10.1038/nenergy.2016.216]

[9]   L.J. Hardwick, P.W. Ruch, M. Hahn, W. Scheifele, R. Kötz, and P. Novák, "in situ Raman spectroscopy of insertion electrodes for lithium-ion batteries and supercapacitors: First cycle effects", *J. Phys. Chem. Solids,* vol. 69, no. 5-6, pp. 1232-1237, 2008.
[http://dx.doi.org/10.1016/j.jpcs.2007.10.017]

[10]   A. Maitra, S.K. Karan, S. Paria, A.K. Das, R. Bera, L. Halder, S.K. Si, A. Bera, and B.B. Khatua, "Fast charging self-powered wearable and flexible asymmetric supercapacitor power cell with fish swim bladder as an efficient natural bio-piezoelectric separator", *Nano Energy,* vol. 40, pp. 633-645, 2017.
[http://dx.doi.org/10.1016/j.nanoen.2017.08.057]

[11]   H. Sheng, J. Zhou, B. Li, Y. He, X. Zhang, J. Liang, J. Zhou, Q. Su, E. Xie, W. Lan, K. Wang, and C. Yu, "A thin, deformable, high-performance supercapacitor implant that can be biodegraded and bioabsorbed within an animal body", *Sci. Adv.,* vol. 7, no. 2, p. eabe3097, 2021.
[http://dx.doi.org/10.1126/sciadv.abe3097] [PMID: 33523998]

[12]   S. Najib, and E. Erdem, "Current progress achieved in novel materials for supercapacitor electrodes: mini review", *Nanoscale Adv.,* vol. 1, no. 8, pp. 2817-2827, 2019.
[http://dx.doi.org/10.1039/C9NA00345B] [PMID: 36133592]

[13]   C. Yan, H.R. Li, X. Chen, X.Q. Zhang, X.B. Cheng, R. Xu, J.Q. Huang, and Q. Zhang, "Regulating the inner helmholtz plane for stable solid electrolyte interphase on lithium metal anodes", *J. Am. Chem. Soc.,* vol. 141, no. 23, pp. 9422-9429, 2019.
[http://dx.doi.org/10.1021/jacs.9b05029] [PMID: 31117672]

[14]   B. Pal, S. Yang, S. Ramesh, V. Thangadurai, and R. Jose, "Electrolyte selection for supercapacitive devices: A critical review", *Nanoscale Adv.,* vol. 1, no. 10, pp. 3807-3835, 2019.
[http://dx.doi.org/10.1039/C9NA00374F] [PMID: 36132093]

[15]   C. Costentin, T.R. Porter, and J.M. Savéant, "How do pseudocapacitors store energy? theoretical analysis and experimental illustration", *ACS Appl. Mater. Interfaces,* vol. 9, no. 10, pp. 8649-8658, 2017.
[http://dx.doi.org/10.1021/acsami.6b14100] [PMID: 28195702]

[16]   J. Liu, J. Wang, C. Xu, H. Jiang, C. Li, L. Zhang, J. Lin, and Z.X. Shen, "Advanced energy storage devices: basic principles, analytical methods, and rational materials design", *Adv. Sci.,* vol. 5, no. 1, pp. 1700322-1700341, 2018.
[http://dx.doi.org/10.1002/advs.201700322] [PMID: 29375964]

[17]   M.K. Dey, P.K. Sahoo, and A.K. Satpati, "Electrochemically deposited layered MnO 2 films for improved supercapacitor", *J. Electroanal. Chem.,* vol. 788, pp. 175-183, 2017.
[http://dx.doi.org/10.1016/j.jelechem.2017.01.063]

[18]   E.M. Espinoza, J.A. Clark, J. Soliman, J.B. Derr, M. Morales, and V.I. Vullev, "Practical aspects of cyclic voltammetry: How to estimate reduction potentials when irreversibility prevails", *J. Electrochem. Soc.,* vol. 166, no. 5, pp. H3175-H3187, 2019.
[http://dx.doi.org/10.1149/2.0241905jes]

[19]   Y. Liu, S.P. Jiang, and Z. Shao, "Intercalation pseudocapacitance in electrochemical energy storage: recent advances in fundamental understanding and materials development", *Mater. Today Adv.,* vol. 7, pp. 100072-100104, 2020.
[http://dx.doi.org/10.1016/j.mtadv.2020.100072]

[20]   S. Kumar, M. Goswami, N. Singh, N. Sathish, M.V. Reddy, and S. Kumar, "Exploring carbon quantum dots as an aqueous electrolyte for energy storage devices", *J. Energy Storage,* vol. 55, pp. 105522-105532, 2022.
[http://dx.doi.org/10.1016/j.est.2022.105522]

[21]   W.G. Menezes, D.M. Reis, T.M. Benedetti, M.M. Oliveira, J.F. Soares, R.M. Torresi, and A.J.G.

Zarbin, "V2O5 nanoparticles obtained from a synthetic bariandite-like vanadium oxide: Synthesis, characterization and electrochemical behavior in an ionic liquid", *J. Colloid Interface Sci.,* vol. 337, no. 2, pp. 586-593, 2009.
[http://dx.doi.org/10.1016/j.jcis.2009.05.050] [PMID: 19545878]

[22]   C. Liu, Z. Yu, D. Neff, A. Zhamu, and B.Z. Jang, "Graphene-based supercapacitor with an ultrahigh energy density", *Nano Lett.,* vol. 10, no. 12, pp. 4863-4868, 2010.
[http://dx.doi.org/10.1021/nl102661q] [PMID: 21058713]

[23]   K.F. Wu, J.H. Fan, X.H. Wang, M.T. Wang, X.F. Xie, J.T. Fan, and A.Y. Chen, "Overlapped T-Nb 2 O 5 /Graphene Hybrid for a Quasi-Solid-State Asymmetric Supercapacitor with a High Rate Capacity", *Energy Fuels,* vol. 35, no. 15, pp. 12546-12555, 2021.
[http://dx.doi.org/10.1021/acs.energyfuels.1c00932]

[24]   D. Prakash, and S. Manivannan, "Unusual battery type pseudocapacitive behaviour of graphene oxynitride electrode: High energy solid-state asymmetric supercapacitor", *J. Alloys Compd.,* vol. 854, pp. 156853-156863, 2021.
[http://dx.doi.org/10.1016/j.jallcom.2020.156853]

[25]   N. Meddings, M. Heinrich, F. Overney, J-S. Lee, V. Ruiz, E. Napolitano, S. Seitz, G. Hinds, R. Raccichini, M. Gaberšček, and J. Park, "Application of electrochemical impedance spectroscopy to commercial Li-ion cells: A review", *J. Power Sources,* vol. 480, pp. 228742-228769, 2020.
[http://dx.doi.org/10.1016/j.jpowsour.2020.228742]

[26]   P.L. Taberna, P. Simon, and J.F. Fauvarque, "Electrochemical characteristics and impedance spectroscopy studies of carbon-carbon supercapacitors", *J. Electrochem. Soc.,* vol. 150, no. 3, pp. A292-A300, 2003.
[http://dx.doi.org/10.1149/1.1543948]

[27]   P. Navalpotro, M. Anderson, R. Marcilla, and J. Palma, "Insights into the energy storage mechanism of hybrid supercapacitors with redox electrolytes by Electrochemical Impedance Spectroscopy", *Electrochim. Acta,* vol. 263, pp. 110-117, 2018.
[http://dx.doi.org/10.1016/j.electacta.2017.12.167]

[28]   Y. Du, Y. Mo, and Y. Chen, "Effects of Fe impurities on self-discharge performance of carbon-based supercapacitors", *Materials,* vol. 14, no. 8, p. 1908, 2021.
[http://dx.doi.org/10.3390/ma14081908] [PMID: 33920441]

[29]   M. Haque, Q. Li, A.D. Smith, V. Kuzmenko, P. Rudquist, P. Lundgren, and P. Enoksson, "Self-discharge and leakage current mitigation of neutral aqueous-based supercapacitor by means of liquid crystal additive", *J. Power Sources,* vol. 453, p. 227897, 2020.
[http://dx.doi.org/10.1016/j.jpowsour.2020.227897]

[30]   F. Soavi, C. Arbizzani, and M. Mastragostino, "Leakage currents and self-discharge of ionic liquid-based supercapacitors", *J. Appl. Electrochem.,* vol. 44, no. 4, pp. 491-496, 2014.
[http://dx.doi.org/10.1007/s10800-013-0647-x]

[31]   M. Golozar, P. Hovington, A. Paolella, S. Bessette, M. Lagacé, P. Bouchard, H. Demers, R. Gauvin, and K. Zaghib, "In situ scanning electron microscopy detection of carbide nature of dendrites in li-polymer batteries", *Nano Lett.,* vol. 18, no. 12, pp. 7583-7589, 2018.
[http://dx.doi.org/10.1021/acs.nanolett.8b03148] [PMID: 30462516]

[32]   Y. Ma, C. Hou, H. Zhang, M. Qiao, Y. Chen, H. Zhang, Q. Zhang, and Z. Guo, "Morphology-dependent electrochemical supercapacitors in multi-dimensional polyaniline nanostructures", *J. Mater. Chem. A Mater. Energy Sustain.,* vol. 5, no. 27, pp. 14041-14052, 2017.
[http://dx.doi.org/10.1039/C7TA03279J]

[33]   Z. Sun, Y. Zhang, Y. Liu, J. Fu, S. Cheng, P. Cui, and E. Xie, "New insight on the mechanism of electrochemical cycling effects in MnO2-based aqueous supercapacitor", *J. Power Sources,* vol. 436, pp. 226795-226802, 2019.
[http://dx.doi.org/10.1016/j.jpowsour.2019.226795]

[34]   S. Kim, J. Cui, V.P. Dravid, and K. He, "Orientation-dependent intercalation channels for lithium and sodium in black phosphorus", *Adv. Mater.,* vol. 31, no. 46, p. 1904623, 2019.
[http://dx.doi.org/10.1002/adma.201904623] [PMID: 31588649]

[35]   T.C. Tsai, G.M. Huang, C.W. Huang, J.Y. Chen, C.C. Yang, T.Y. Tseng, and W.W. Wu, "*In situ* TEM investigation of the electrochemical behavior in cnts/mno $_2$ -based energy storage devices", *Anal. Chem.,* vol. 89, no. 18, pp. 9671-9675, 2017.
[http://dx.doi.org/10.1021/acs.analchem.7b00958] [PMID: 28805052]

[36]   N. Schulz, R. Hausbrand, C. Wittich, L. Dimesso, and W. Jaegermann, "XPS-Surface analysis of SEI Layers on Li-Ion Cathodes: Part II. SEI-composition and formation inside composite electrodes", *J. Electrochem. Soc.,* vol. 165, no. 5, pp. A833-A846, 2018.
[http://dx.doi.org/10.1149/2.0881803jes]

[37]   V. Augustyn, P. Simon, and B. Dunn, "Pseudocapacitive oxide materials for high-rate electrochemical energy storage", *Energy Environ. Sci.,* vol. 7, no. 5, pp. 1597-1614, 2014.
[http://dx.doi.org/10.1039/c3ee44164d]

[38]   J.K. Chang, M.T. Lee, W.T. Tsai, M.J. Deng, H.F. Cheng, and I.W. Sun, "Pseudocapacitive mechanism of manganese oxide in 1-ethyl-3-methylimidazolium thiocyanate ionic liquid electrolyte studied using X-ray photoelectron spectroscopy", *Langmuir,* vol. 25, no. 19, pp. 11955-11960, 2009.
[http://dx.doi.org/10.1021/la9012119] [PMID: 19621902]

[39]   M.A. Isaacs, J. Davies-Jones, P.R. Davies, S. Guan, R. Lee, D.J. Morgan, and R. Palgrave, "Advanced XPS characterization: XPS-based multi-technique analyses for comprehensive understanding of functional materials", *Mater. Chem. Front.,* vol. 5, no. 22, pp. 7931-7963, 2021.
[http://dx.doi.org/10.1039/D1QM00969A]

[40]   D. Weingarth, A. Foelske-Schmitz, A. Wokaun, and R. Kötz, "In situ electrochemical XPS study of the Pt/[EMIM][BF4] system", *Electrochem. Commun.,* vol. 13, no. 6, pp. 619-622, 2011.
[http://dx.doi.org/10.1016/j.elecom.2011.03.027]

[41]   B.K. Saikia, S.M. Benoy, M. Bora, J. Tamuly, M. Pandey, and D. Bhattacharya, "A brief review on supercapacitor energy storage devices and utilization of natural carbon resources as their electrode materials", *Fuel,* vol. 282, pp. 118796-118813, 2020.
[http://dx.doi.org/10.1016/j.fuel.2020.118796]

[42]   K. Zheng, H. Tan, L. Wang, J. Liu, M. Ding, and D. Jia, "Vertically Oriented Cu 2+1 O@Cu-MOFs Hybrid Clusters for High-Performance Electrochemical Capacitors", *Adv. Mater. Interfaces,* vol. 8, no. 10, p. 2002145, 2021.
[http://dx.doi.org/10.1002/admi.202002145]

[43]   K. Pinkert, L. Giebeler, M. Herklotz, S. Oswald, J. Thomas, A. Meier, L. Borchardt, S. Kaskel, H. Ehrenberg, and J. Eckert, "Functionalised porous nanocomposites: A multidisciplinary approach to investigate designed structures for supercapacitor applications", *J. Mater. Chem. A Mater. Energy Sustain.,* vol. 1, no. 15, pp. 4904-4910, 2013.
[http://dx.doi.org/10.1039/c3ta00118k]

[44]   A. Patra, N. K, J.R. Jose, S. Sahoo, B. Chakraborty, and C.S. Rout, "Understanding the charge storage mechanism of supercapacitors: in situ / operando spectroscopic approaches and theoretical investigations", *J. Mater. Chem. A Mater. Energy Sustain.,* vol. 9, no. 46, pp. 25852-25891, 2021.
[http://dx.doi.org/10.1039/D1TA07401F]

[45]   M. Yeager, W. Du, R. Si, D. Su, N. Marinković, and X. Teng, "Highly efficient K0.15MnO2 birnessite nanosheets for stable pseudocapacitive cathodes", *J. Phys. Chem. C,* vol. 116, no. 38, pp. 20173-20181, 2012.
[http://dx.doi.org/10.1021/jp304809r]

[46]   S.Y. Hsu, F-H. Hsu, J-L. Chen, Y-S. Cheng, J-M. Chen, and K-T. Lu, "The supercapacitor electrode properties and energy storage mechanism of binary transition metal sulfide MnCo 2 S 4 compared with oxide MnCo 2 O 4 studied using in situ quick X-ray absorption spectroscopy", *Mater. Chem.*

Front., vol. 5, no. 13, pp. 4937-4949, 2021.
[http://dx.doi.org/10.1039/D1QM00222H]

[47]   M. Xia, T. Liu, N. Peng, R. Zheng, X. Cheng, H. Zhu, H. Yu, M. Shui, and J. Shu, "Lab-Scale in situ X-ray diffraction technique for different battery systems: Designs, applications, and perspectives", *Small Methods,* vol. 3, no. 7, p. 1900119, 2019.
[http://dx.doi.org/10.1002/smtd.201900119]

[48]   S. Wang, H. Xu, T. Hao, P. Wang, X. Zhang, H. Zhang, J. Xue, J. Zhao, and Y. Li, "In situ XRD and operando spectra-electrochemical investigation of tetragonal WO_{3-x} nanowire networks for electrochromic supercapacitors", *NPG Asia Mater.,* vol. 13, no. 1, p. 51, 2021.
[http://dx.doi.org/10.1038/s41427-021-00319-7]

[49]   O. Ghodbane, J.L. Pascal, B. Fraisse, and F. Favier, "Structural in situ study of the thermal behavior of manganese dioxide materials: Toward selected electrode materials for supercapacitors", *ACS Appl. Mater. Interfaces,* vol. 2, no. 12, pp. 3493-3505, 2010.
[http://dx.doi.org/10.1021/am100669k] [PMID: 21114252]

[50]   M. Okubo, A. Sugahara, S. Kajiyama, and A. Yamada, "MXene as a charge storage host", *Acc. Chem. Res.,* vol. 51, no. 3, pp. 591-599, 2018.
[http://dx.doi.org/10.1021/acs.accounts.7b00481] [PMID: 29469564]

[51]   M.R. Lukatskaya, O. Mashtalir, C.E. Ren, Y. Dall'Agnese, P. Rozier, P.L. Taberna, M. Naguib, P. Simon, M.W. Barsoum, and Y. Gogotsi, "Cation intercalation and high volumetric capacitance of two-dimensional titanium carbide", *Science,* vol. 341, no. 6153, pp. 1502-1505, 2013.
[http://dx.doi.org/10.1126/science.1241488] [PMID: 24072919]

[52]   Y. Cui, H.J. Shim, Y. Gao, and S.G. Pyo, "Real-time Raman analysis of cleaning solution: Determination of Al/Alq3 residue in cleaning mixture", *J. Raman Spectrosc.,* vol. 50, no. 4, pp. 571-575, 2019.
[http://dx.doi.org/10.1002/jrs.5551]

[53]   S. Gupta, S.B. Carrizosa, J. Jasinski, and N. Dimakis, "Charge transfer dynamical processes at graphene-transition metal oxides/electrolyte interface for energy storage: Insights from in-situ Raman spectroelectrochemistry", *AIP Adv.,* vol. 8, no. 6, pp. 065225-065248, 2018.
[http://dx.doi.org/10.1063/1.5028412]

[54]   C. Prehal, H. Fitzek, G. Kothleitner, V. Presser, B. Gollas, S.A. Freunberger, and Q. Abbas, "Persistent and reversible solid iodine electrodeposition in nanoporous carbons", *Nat. Commun.,* vol. 11, no. 1, p. 4838, 2020.
[http://dx.doi.org/10.1038/s41467-020-18610-6] [PMID: 31911652]

[55]   R. Vicentini, L.M. Da Silva, D.V. Franco, W.G. Nunes, J. Fiates, G. Doubek, L.F.M. Franco, R.G. Freitas, C. Fantini, and H. Zanin, "Raman probing carbon & aqueous electrolytes interfaces and molecular dynamics simulations towards understanding electrochemical properties under polarization conditions in supercapacitors", *J. Energy Chem.,* vol. 60, pp. 279-292, 2021.
[http://dx.doi.org/10.1016/j.jechem.2021.01.003]

[56]   L. Yang, S. Cheng, X. Ji, Y. Jiang, J. Zhou, and M. Liu, "Investigations into the origin of pseudocapacitive behavior of Mn_3O_4 electrodes using in operando Raman spectroscopy", *J. Mater. Chem. A Mater. Energy Sustain.,* vol. 3, no. 14, pp. 7338-7344, 2015.
[http://dx.doi.org/10.1039/C5TA00223K]

[57]   M. Yao, P. Wu, S. Cheng, L. Yang, Y. Zhu, M. Wang, H. Luo, B. Wang, D. Ye, and M. Liu, "Investigation into the energy storage behaviour of layered α-V_2O_5 as a pseudo-capacitive electrode using operando Raman spectroscopy and a quartz crystal microbalance", *Phys. Chem. Chem. Phys.,* vol. 19, no. 36, pp. 24689-24695, 2017.
[http://dx.doi.org/10.1039/C7CP04612J] [PMID: 28861575]

[58]   H. Wang, A.C. Forse, J.M. Griffin, N.M. Trease, L. Trognko, P.L. Taberna, P. Simon, and C.P. Grey, "In situ NMR spectroscopy of supercapacitors: insight into the charge storage mechanism", *J. Am.*

Chem. Soc., vol. 135, no. 50, pp. 18968-18980, 2013.
[http://dx.doi.org/10.1021/ja410287s] [PMID: 24274637]

[59]   F.W. Richey, C. Tran, V. Kalra, and Y.A. Elabd, "Ionic liquid dynamics in nanoporous carbon nanofibers in supercapacitors measured with in operando infrared spectroelectrochemistry", *J. Phys. Chem. C,* vol. 118, no. 38, pp. 21846-21855, 2014.
[http://dx.doi.org/10.1021/jp506903m]

[60]   M.A. Gebbie, M. Valtiner, X. Banquy, E.T. Fox, W.A. Henderson, and J.N. Israelachvili, "Ionic liquids behave as dilute electrolyte solutions", *Proc. Natl. Acad. Sci.,* vol. 110, no. 24, pp. 9674-9679, 2013.
[http://dx.doi.org/10.1073/pnas.1307871110] [PMID: 23716690]

[61]   P. Iamprasertkun, A. Krittayavathananon, A. Seubsai, N. Chanlek, P. Kidkhunthod, W. Sangthong, S. Maensiri, R. Yimnirun, S. Nilmoung, P. Pannopard, S. Ittisanronnachai, K. Kongpatpanich, J. Limtrakul, and M. Sawangphruk, "Charge storage mechanisms of manganese oxide nanosheets and N-doped reduced graphene oxide aerogel for high-performance asymmetric supercapacitors", *Sci. Rep.,* vol. 6, no. 1, pp. 37560-37572, 2016.
[http://dx.doi.org/10.1038/srep37560] [PMID: 27857225]

[62]   N. Shpigel, M.D. Levi, S. Sigalov, L. Daikhin, and D. Aurbach, "In situ real-time mechanical and morphological characterization of electrodes for electrochemical energy storage and conversion by electrochemical quartz crystal microbalance with dissipation monitoring", *Acc. Chem. Res.,* vol. 51, no. 1, pp. 69-79, 2018.
[http://dx.doi.org/10.1021/acs.accounts.7b00477] [PMID: 29297669]

[63]   C. Prehal, C. Koczwara, N. Jäckel, A. Schreiber, M. Burian, H. Amenitsch, M.A. Hartmann, V. Presser, and O. Paris, "Quantification of ion confinement and desolvation in nanoporous carbon supercapacitors with modelling and in situ X-ray scattering", *Nat. Energy,* vol. 2, no. 3, pp. 16215-16223, 2017.
[http://dx.doi.org/10.1038/nenergy.2016.215]

[64]   N.C. Osti, and E. Mamontov, "Microscopic dynamics in room-temperature ionic liquids confined in materials for supercapacitor applications", *Sustain. Energy Fuels,* vol. 4, no. 4, pp. 1554-1576, 2020.
[http://dx.doi.org/10.1039/C9SE00829B]

[65]   S. Kondrat, and A. Kornyshev, "Corrigendum: Superionic state in double-layer capacitors with nanoporous electrodes", *J. Phys. Condens. Matter,* vol. 25, no. 11, p. 119501, 2013.
[http://dx.doi.org/10.1088/0953-8984/25/11/119501]

Electrolytes for Electrochemical Energy Storage Supercapacitors

Priyanka A. Jha[1,*], **Pardeep K. Jha**[1] and **Prabhakar Singh**[1]

[1] Department of Physics, Indian Institute of Technology (Banaras Hindu University), Varanasi-221005, India

Abstract: In this chapter, the types of electrolytes and the alteration in capacitance with pore size, their power density, and energy density along with the interaction of electrolytes with current collectors are discussed. The electrolytes' electrochemical stability broadly estimates the working cell voltage provided that the electrodes are stable under operating cell voltage. The electrolytes are divided into various categories such as liquid electrolyte, solid-state, and redox-active electrolyte. The liquid electrolytes are further categorized into aqueous and non-aqueous electrolytes. The critical performance parameters such as stability, lifetime, operating temperature, operating voltage, *etc.* are believed to be affected by electrolytes. Moreover, the electrolytes are believed to interact with the current collectors, additives, binders, separators, and electrode material to affect the practical performance of supercapacitors. However, the capacitance of the electrolyte depends upon the ion size and the matching between the electrode pore size and electrolyte ion size. The power density and energy density depend upon the potential window, ionic conductivity, and electrochemical stability along with concentration, respectively. Further, the ion-electrode interaction is supposed to affect the cycle life and power density as well. The thermal stability of electrolytes depends upon their boiling points, freezing points, and salt solubility and the equivalent series resistance depends upon ion conductivity, mobility, and viscosity.

Keywords: Electrolytes, Energy storage, Electrochemical supercapacitors.

INTRODUCTION

An electrolyte is an essential component (located in the center) of the electrochemical supercapacitors. It comprises solute solvated in solvent or solvent-free salt-like ionic liquids [1]. It is important because it provides (a) ionic conductivity allowing charge compensation; (b) a charge storage process *via* reversible redox process and (c) formation of double layer (electrical) in the

---

*[*] **Corresponding author Priyanka A. Jha:** Department of Physics, Indian Institute of Technology (Banaras Hindu University), Varanasi-221005, India; E-mail: priyankajha.dce@gmail.com

Sanjeev Verma, Shivani Verma, Saurabh Kumar & Bhawna Verma (Eds.)

double layer capacitors in pseudo capacitors. The electrolytes affect the power density, equivalent series resistance, lifetime, cycling stability, operational temperature range, and self-discharge rate of electrochemical supercapacitors [2 - 4]. Further, the performance of electrochemical supercapacitors depends upon the interaction of electrolytes with the other components of supercapacitors like the electrodes, current collectors, separators, binders, and additives. The factors affecting the energy density of the supercapacitors are cell voltage and capacitance as shown in Fig. (**1**). The capacitance in turn depends upon electrolyte ion size and the combination of electrode pore size and electrolyte ion size [5]. Moreover, the electrolyte and current collector interaction reduces the energy density of electrochemical supercapacitors by decreasing the operative cell voltage. Other than these, energy density is observed to depend upon the concentration of electrolytes and the electrochemical stability. It is observed that a higher concentration of the electrolyte reduces ion mobility due to less hydration (Fig. **1**) [6].

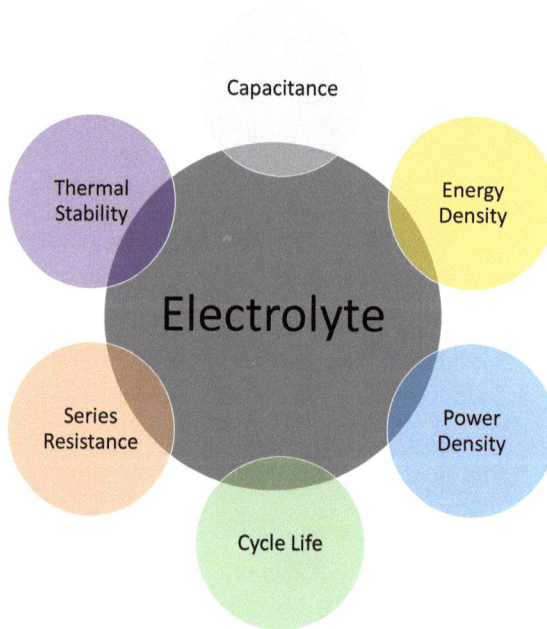

Fig. (1). The factors affecting the electrolyte and performance of electrochemical supercapacitors.

Furthermore, the number of publications based on electrolyte materials has largely increased over the last decade as shown in Fig. (**2**).

Of the well-known electrolytes, most of the supercapacitors employ organic electrolytes with acetonitrile solvents having a cell voltage of 2.5-2.8V [7, 8].

While, the electrochemical supercapacitors with graphitic carbon have shown a potential of 3.3- 3.8 V. Further, a safe, low-cost, and green electrolyte *i.e.* aqueous electrolyte possesses a ceiling voltage of 1.23V. The operational voltage depends upon the electrochemical stability of electrolytes and affects the power and energy density. Further, the ion type and the size of the electrolyte; solvent and ion concentration; interactions between solvent and ion, and the electrode-electrolyte interfacial interactions are believed to affect the pseudo capacitance and double layer capacitance. The ionic conductivity of the electrolyte affects the equivalent series resistance. The features required for a good electrolyte are (i) wide electrochemical stable window; (ii) ionic conductivity should be high; (iii) wide operating temperature range; (iv) less volatile and viscous; (v) environment-friendly; (vi) availability with high purity and (vii) less cost [9].

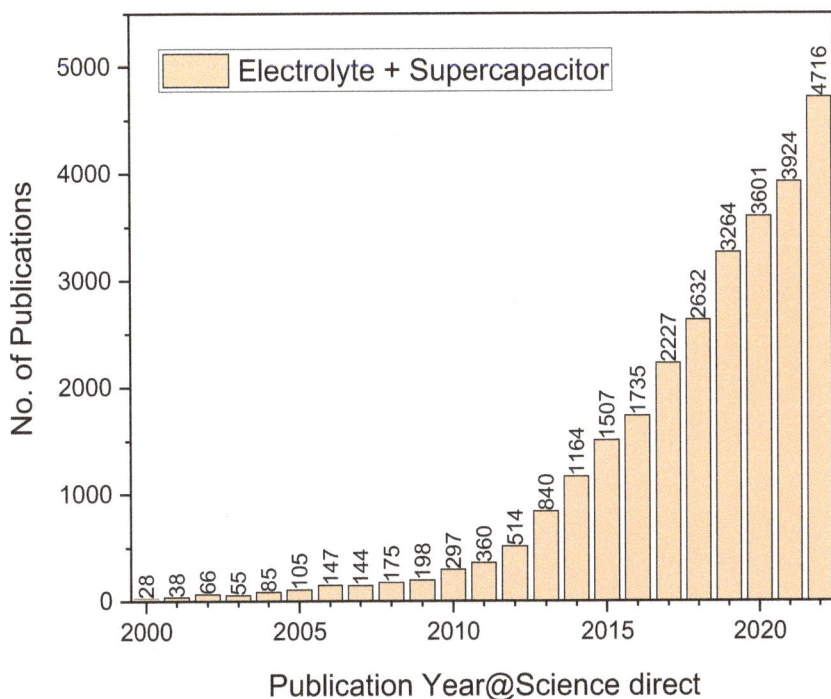

Fig. (2). Number of publications in Science Direct in the last twenty years with the keyword "electrolyte + supercapacitor".

Fig. (3) shows the classification of electrolytes into various categories and subcategories. Based on the literature survey, the electrolytes are majorly categorized into three groups solid, liquid, and redox active electrolytes (Fig. 3). Liquid electrolytes are comprised of water-in-salt, aqueous, and no-aqueous electrolytes. Quasi-solid state/solid-state comprises inorganic, dry, and gel-based polymer electrolytes. Further, Redox-active electrolytes are classified into gel polymer, ionic liquid, and aqueous electrolytes. To meet the criterion for electrolytes, no electrolyte is being reported to meet all the conditions. For instance, aqueous electrolytes have shown high conductivity and capacitance, and organic electrolytes and ionic liquids have shown higher cell voltage but have the disadvantage of lower ionic conductivity.

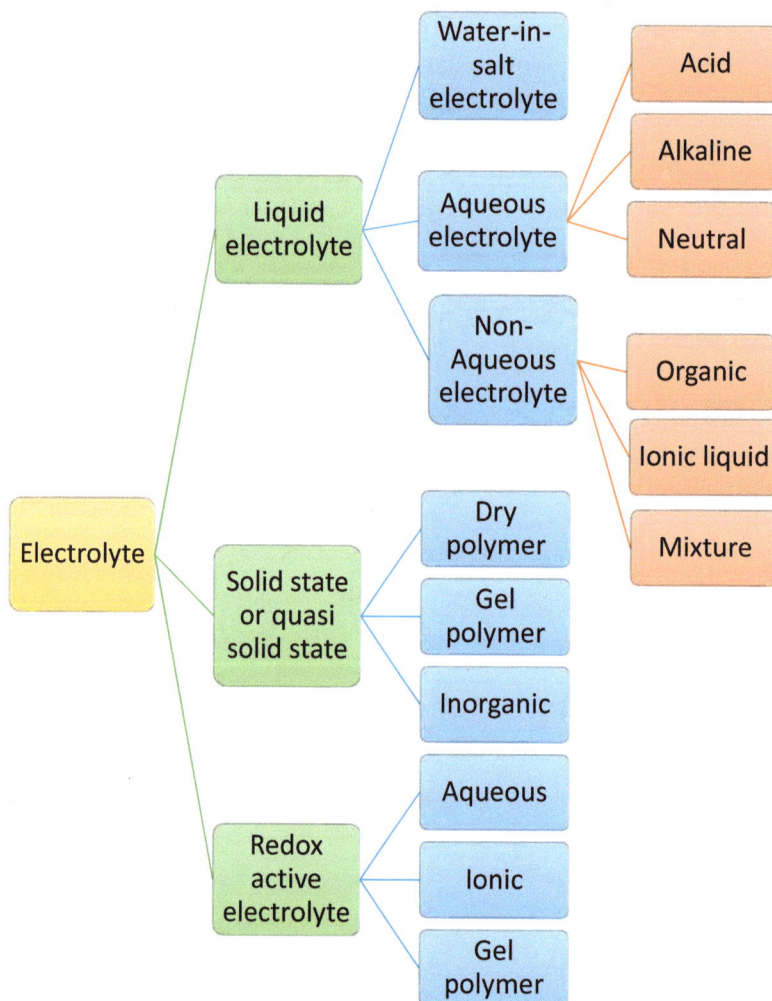

Fig. (3). Classification of electrolyte materials for supercapacitors.

The redox-active electrolytes have shown a wide operating voltage window, and increased redox moiety. Further, water-in-salt electrolyte has shown high energy storage due to a higher voltage window than aqueous and non-aqueous electrolytes. Preliminary, ionic liquids have shown low vapor pressure, non-flammable, and thermal stable nature. Thus, research focuses on the development of new electrolytes, optimizing the pore size and ion size through the processing of electrolytes, and improving interaction at the electrode-electrolyte interface. The mechanism still needs to be investigated near the interface and with varying redox components. This chapter focuses on the types of electrolytes and their interaction with inactive components such as current collectors, binders, and additives.

TYPES OF ELECTROLYTES

Liquid electrolytes

Water-in-salt Electrolyte

In water-in-salt electrolytes, anions aggregate at the interface forming a hydrophobic layer for the insulation of the electrode from water pushing more potential at the positive electrode [10, 11]. Likewise, at the negative electrode, the solid electrolyte interphase suppresses the hydrogen evolution reaction and more negative potential is produced. It is also observed that solid electrolyte interphase blocks the electronic conduction and allows the ions to pass [12]. Thus, water-in-salt electrolytes are highly attracted by researchers due to the variable potential window and great interphase stability. The factors affecting the conduction in the water-in-salt electrolyte are the concentration of electrolytes, compatible electrodes, and optimization of solid electrolyte interphase [13]. The Salt in water-electrolyte possesses a solvation sheath structure with the primary and loosely bound secondary solvation sheath. On comparing the two types of "water in salt and salt in water" electrolytes, the concentration of salt is high, and low water concentration cannot neutralize the field of cations (Fig. 4). Fig. (4) shows the difference between water-in-salt and salt-in-water electrolyte. This leads to the increase in passivation and pushing of positive potential to the more positive side and the negative to the more negative side. This further enhances the potential window and makes it a suitable candidate with a wide potential window.

Here, the solid electrolyte interphase optimizes interactions between the electrode and the electrolyte. However, the formation of solid electrolyte interphase on the negative side is controversial. It suppresses the contact of water with electrodes but allows the ions to pass, leading to the reduction in hydrogen evolution reaction [14].

Water-in-salt electrolytes are extensively investigated in lithium-ion batteries and electrochemical supercapacitors [14]. Lithium bis (trifluoromethyl sulfonyl) imide solution is a well-known "water in salt" electrolyte with an electrochemical potential window of 3V as compared to the electrochemical potential window of 1V for conventional aqueous electrolytes [15]. This concept is applied to the cations such as K^+, Na^+, Zn^{+2}, and various anions. Fig. (**5**) shows the formation of solid electrolyte interphase formation in water-in-salt electrolyte.

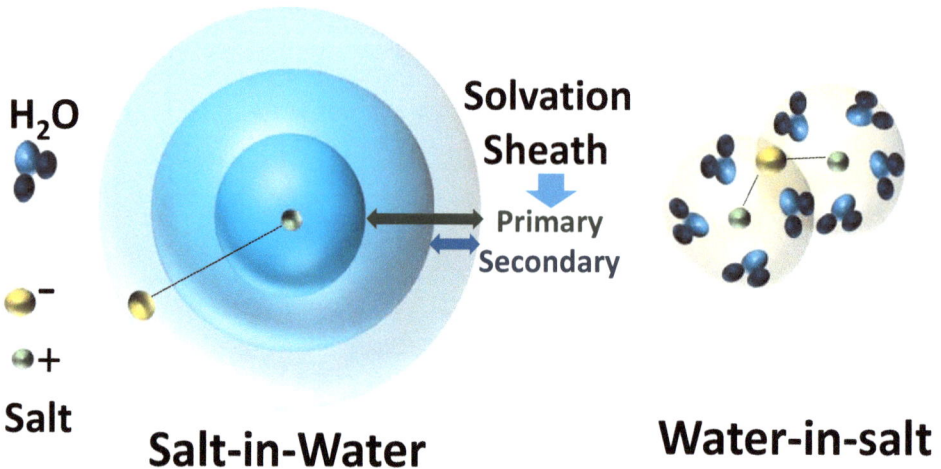

Fig. (4). Difference between water-in-salt and salt-in-water electrolytes.

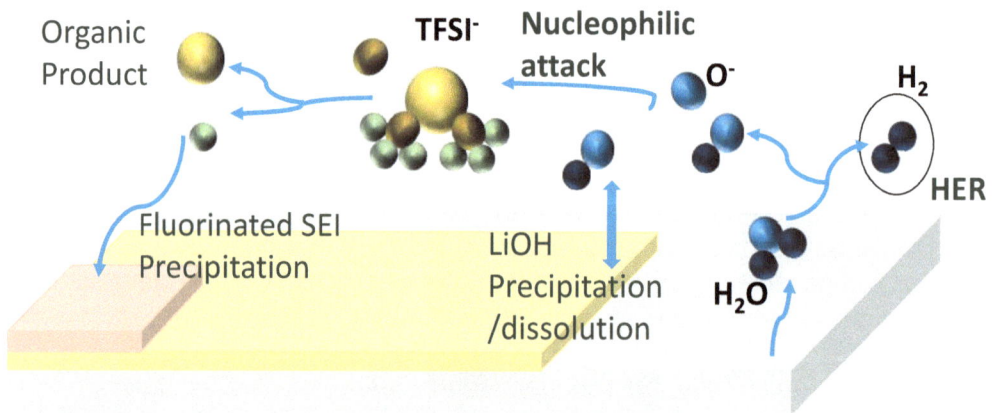

Fig. (5). Solid electrolyte interphase (SEI) formation.

Aqueous Electrolyte

Aqueous electrolytes show limited cell voltage and possess higher conductivity than Ionic liquid and organic electrolytes. They are mostly employed due to cost-effectiveness and easy handling whereas organic and ionic liquids require a moisture-free environment. Their high conductivity reduces the equivalent series resistance leading to the increase in power density. A most important feature is their high dielectric constant leading to the increase in specific capacitance [16].

This is mostly reported for carbon electrodes. The size of bare and hydrated ions, mobility of ions, ionic conductivity, electrochemical stability, and corrosive degree are the major selection criteria for electrolytes. Furthermore, the aqueous electrolytes are categorized into acid, neutral, and basic or alkaline electrolytes. The most widely practiced acid electrolyte is H_2SO_4 with an extremely high ionic conductivity of 0.8 S/cm (1M H_2SO_4 at 25 °C) [17]. It is used in supercapacitors based on pseudo-capacitive materials like RuO_2 (specific capacity ~ 1000F/g) [18]. The most commonly employed basic and neutral electrolytes are based on KOH (0.6 S/cm for 6M KOH at 25°C) and Na_2SO_4 (1M Na_2SO_4 the cell voltage increased to 1.8 V [19]). For double-layer capacitors, the specific capacitance is around 300 F/g in strong acid electrolyte *i.e.* H_2SO_4. With the change in concentration and temperature of the electrolyte, power density and specific capacitance are observed to vary. The major shortcoming of acid electrolytes is their narrow voltage window and gas leakage can rupture electrochemical supercapacitors. Further, acid electrolytes are highly corrosive for packaging materials and current collectors. Au and Ti are being used as corrosion-resistant current collectors. While Ni is used as a current collector for alkaline electrolytes [2, 20].

The developments done in the field of alkaline electrolytes are: (a) the increment in the capacitance of carbon-dependent materials *via* pseudo capacitive contribution; (b) large specific capacitance; (c) the development of the composite materials; (d) the widening of operating cell voltage window by asymmetric electrochemical supercapacitors. The temperature and concentration of alkaline electrolytes are also studied as higher electrolyte concentration increases the specific capacitance of electrodes. Further, oxygen evolution is observed to increase with temperature. Moreover, the ionic conductivity of alkaline electrolytes has been observed to decrease with an increase in the diameter of hydrated cations (K^+ (3.31Å) >Na^+ (3.58Å) >Li^+ (3.82Å)). The major concern in the case of alkaline electrolytes is cycling stability. The ion deintercalation/intercalation is linked to the diffusion of electrodes in the basic medium. On comparing the acidic, alkaline, and neutral electrolytes, neutral electrolytes have higher electrochemical stability with a large potential window. Here, again concentration of the salt plays a pivotal role, and the performance of supercapacitors increases with the increase of hydrated ions [21].

Non-Aqueous Electrolyte

Organic electrolytes have captured the supercapacitors' commercial market due to high operative voltage windows *i.e.* 2.5-2.8V. The most commercially used organic electrolyte is tetraethyl ammonium tetrafluoroborate dissolved in acetonitrile [22]. With the higher operative voltage, energy and power density increase. The salt and solvent ion size, ion and solvent interaction, viscosity, conductivity, and electrochemical stability are the factors governing the performance of supercapacitors. But the issues with the organic electrolyte are: (a) high cost, (b) low conductivity limit, (c) small specific capacitance, (d) flammability, (e) volatility and toxicity, and (f) power performance. Most importantly, they are not environment-friendly. Further, the organic solvents should have good solvation ability for conducting salt, be less viscous in the operating temperature range, highly electrochemically stable followed by non-flammability and toxicity. With acetonitrile and PC as solvents, higher conductivity with high power performance has been achieved [23]. Further, efforts are being made to improve the operative voltage. Several fluorinated carbons and cyclic carbonates have been explored as single organic solvents. Moreover, the operative voltage window has been increased by novel sulfonate solvents such as ethyl methyl sulfonates, methyl isopropyl sulfonates, butyl isobutyl sulfonate, *etc.* To overcome the low electrochemical stability possessed by ACN electrolytes, adiponitrile has been used as an organic solvent [24]. To date, no better solvent than acetonitrile has been developed. To solve the problems of single organic solvents, mixtures of solvents and additives are explored. ACN-PC-based electrolytes are being used to improve the performance of electrochemical supercapacitors. Sulfites such as diethyl sulfite and propylene sulphite are being used as additives for binary solvents. Their properties are mostly dependent on the polarizability and dielectric constant of the solvents [25 - 27].

The conducting salts in the solvent provide the cations and ions in the electrolyte. The ionic conductivity of electrolytes is remarkably affected by the concentration and mobility of electrolytes. The conducting salts alter the electrochemical and thermal stability of organic electrolytes. Further, the capacitance of supercapacitors is also altered with the conducting salts. The ion conductivity influences both the ion conductivity of electrolytes and in turn energy density of supercapacitors. The organic electrolytes possess a lower dissociation degree than the aqueous electrolytes. This dissociation degree and mobility depend on the cationic and anionic size, and symmetric or asymmetric structure in turn affecting the conductivity and viscosity of electrolytes. Some salts such as TEABF4 are sparingly soluble in PC. To increase their solubility, cyclic ammonium salts and asymmetric salts have been studied [2, 28 - 30].

Ionic liquids are room-temperature or low-temperature molten salts that are circumscribed as salts containing ions with a melting point of less than 100 °C [31, 32]. They are also termed "designer solvents" since they are customized to have better parameters such as wide operative cell voltage window, series resistance, and working temperature range. Fig. (**6**) shows the structure of types of ionic liquids as aprotic, protic, and zwitterionic liquids and the investigated ones are pyrrolidinium, ammonium imidazolium, *etc*. Imidazolium has higher ionic conductivity while pyrrolidinium has larger electrochemical stability. Further, the chain length of ions *i.e.* cations or anions, nature, and shape of ions, specific absorption, ion valence, temperature, the ratio of cation size to anion size, morphology, and electrode materials are the factors that affect the capacitance. The asymmetric CV curves are observed to be the outcome of unequal cation and anion size. The ion size is observed to affect the double-layer thickness and in turn relative permittivity. As discussed earlier, organic electrolytes are flammable, to solve this problem, solvent-free ionic liquids have been used. Aprotic ILs are suitable for lithium batteries and supercapacitors while protic ionic liquids are suitable for fuel cells. Further, zwitterionic ILs are justified for ionic liquid-based membranes [33 - 35].

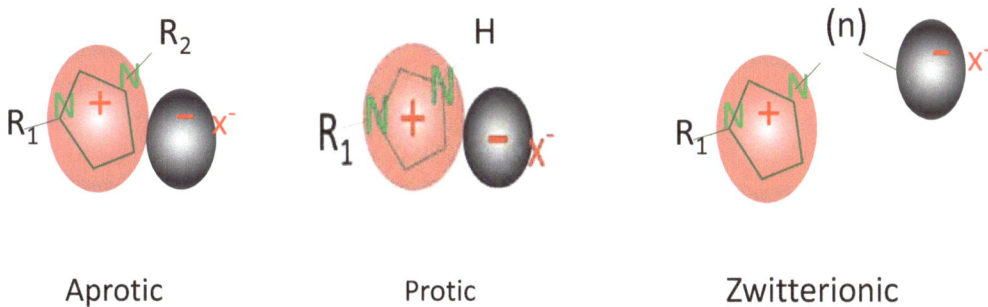

Fig. (6). Types of ionic liquids.

Solid-State

Solid-state electrolytes are useful for the development of supercapacitors in microelectronics, flexible electronic, and printable electronic devices [36, 37]. They are also used as separators and have simplified the fabrication and packaging of supercapacitors. They have reduced the leakage problem in case of liquid electrolytes. These are polymer and ceramic electrolytes which have attracted researchers. The polymer electrolytes are divided into solid/dry polymer electrolytes, gel polymer, and polyelectrolyte. Dry polymer electrolyte comprises poly ethyl oxide with LiCl and ionic conductivity is through the transfer of ions to polymer from salt. Gel polymer comprises a polymer matrix like polyvinyl alcohol with salt dissolved in a liquid solvent. The ions move in the liquid solvent than in the polymer phase in the case of gel polymer electrolyte. In polyelectrolytes, charged polymer chains constitute ionic conductivity and possess the highest ionic conductivity [38, 39].

The advantage of gel polymer is that it can be designed in a variety of structures and shapes for diverse applications. For example: flexible supercapacitors with PVA hydrogels, stretchable, origami foldable supercapacitors, *etc* [40].

Redox-Active Electrolyte

In an electric double-layer capacitor, there is a fast ion separation/aggregation at the interface while pseudo capacitors have reversible and fast redox reactions at/near the interface. Therefore, new electrolytes based on redox species and ionic liquid can be designed for wide operating voltage windows and high power. Ionic liquids have larger voltage windows with a melting point < 100 °C, high thermal stability, and ionic conductivity [41]. To store charge, the redox-active electrolytes provide the pseudocapacitive contribution to utilize faradaic reactions from electrolytes. Thus, electrolyte and electrode materials contribute to the total capacitance [42, 43].

Aqueous Electrolyte

Iodide/iodine is a typical redox example for electrolytes to be employed in carbon-based electrochemical supercapacitors. It gives a high capacitance value of 1840 F/gm in 1M KI electrolyte [44]. Here, the capacitive behavior depends upon the basic counter-ion nature of iodides. For pseudocapacitive electrodes like RuO_2, CuS, MnO_2 *etc.* redox-active electrolytes like KI are being investigated [45]. With the application of redox mediators in non-aqueous and aqueous electrolytes, cell voltage and energy density can be improved. This strategy has also been proven beneficial in solid-state electrolytes as well (ex. NaI, KI, *etc.*). Further, the ionic liquids show advantages as electrolyte and capacitance are beyond expectations [46]. Ferrocene merged with imidazolium cation has attracted researchers because it has boosted many faradaic reactions to increase capacitance [47]. This results in an increase in energy density and stability. Further, ionic liquids have been altered with redox moieties to form a high-energy supercapacitor. A new paradigm known as the "redox" mechanism is employed to study the electron transfer mechanism at the interface [48].

INFLUENCE OF PORE SIZE ON PROPERTIES OF ELECTROLYTE

Liquid Electrolyte

Aqueous Electrolyte

The micro and meso pores of $Ni_3(NO_3)_2(OH)_4$ electrode led to the deintercalation/intercalation of hydroxide (OH⁻) ions in alkaline electrolytes [49]. Here, the morphology control has improved pseudocapacitive behavior through the shortening of OH⁻ ions diffusion length [50]. Furthermore, ion intercalation/deintercalation has led to a change in the volume of the electrodes. With the change in dimensionality in $NiCo_2O_4$ nanotubes, the high specific capacitance of 1647 F/g is because its hollow nature has provided more space to OH⁻ ions and accommodated the volume change with the reduction of diffusion distance [51]. Here, the supercapacitor structure enhanced the electrochemical activity with the fast movement of ions and electrons. The cation formation in the basic/alkaline electrolytes affects the pseudocapacitance greatly.

The carbon nanomaterials acting as electrodes become a critical part of supercapacitors as they greatly affect the energy storage of the supercapacitors. They are classified based on their dimensionality as 1-D, 2D, and 3D. Of these, carbon nanotubes (CNTs) belong to 1-D materials. Graphene (GR) belongs to 2D while activated carbon (AC) and carbon nanocages (CNC) belong to 3D materials [52]. These are the widely applied electrode materials for supercapacitors. CNTs have a regular pore structure with high chemical stability and conductivity. CNTs are also known to have a large surface area [53]. GR in the 2D structure also possesses high surface area along with high conductivity with thermal and chemical stability [54]. The advantages of AC with a 3D structure are low cost due to the simple fabrication process, good conductivity, and stable electrochemical performance [55]. However, CNCs are another class of a 3D structure which shows the advantage of a uniform pore size with a multiscale pore network. They have also shown good conductivity with a large specific surface area [56]. Thus it is required to optimize their pore size to meet the performance requirements.

Furthermore, various composite methods of multilayer materials (structures) have shown their effects on the performance of supercapacitors. They are classified as binary complexes and ternary complexes. The binary complexes are formed with, (i) the composite of carbon with the metal oxides, (ii) carbon materials and metal-organic framework, (iii) carbon materials composites with polymers, and (iv) hetero-atom based carbon materials and carbon/carbon composites. Many kinds of nanomaterials are highly explored to obtain high capacitance as their low cost along with natural abundance makes them an ideal candidate. Further, their surface area, pore size, and conductivity can be tuned to improve the performance of supercapacitors.

CNTs were first synthesized in the 1900s and Hussain*et al.* prepared the vertically aligned multiwalled CNTs *via* plasma-enhanced CVD (chemical vapor deposition). In the aqueous electrolyte, the specific capacitance increases from 25 to 55 F/g upon treatment with 260 Pa and 75 W nitrogen plasma [57]. Furthermore, with the array thickness of 280 μm, the specific capacitance observed is 128 F/g [58]. 3D GR carbon nanosheets have shown a high specific capacitance of 316.8 F/g with a large specific area and good cycle stability [59]. It has also been proven that porous GR electrodes have good application prospects in microelectronics and energy storage.

ACs have a large specific area of 1000-3000 m^2/g and high capacitance. However, their pore structure needs to be optimized to have better electrochemical properties at high current densities. The proper selection and processing of carbon source materials are required as Chao used filter paper as a precursor to achieve graded carbon electrode materials with a specific capacitance of 219.6 F/g [60]. Zhu*et al* prepared KOH-activated Gingko leaf-activated carbon and showed a specific capacitance of 374F/g [61] in an alkaline medium. In Nitrogen-doped CNCs, ultra-high specific capacitance of 313 F/g at 1A/g is observed in 6M KOH solution. In these particular materials, the samples have changed from micropore to mesopore and macropore with the increase in nitrogen pressure [62]. The metal complexes such as CO-Co_3O_4 core particles encased in CNT shell have shown excellent electrochemical performance with a specific capacitance of 823.4 F/g at 1A/g [63]. Reduced graphene oxide-based layered double hydroxide nanosheets have shown a high capacitance of 1675 F/g [64]. The composites with polymers have shown a capacitance lower than the metal-organic framework (MOF) composites. Even the heteroatoms-based AC electrodes have shown lower capacitance than MOF composites. However, the ternary complexes have shown lower specific capacitance (1273 F/g) as compared to MOF-based complexes but high stability is achieved.

Non-Aqueous Electrolyte

It is very much required to match the electrode pore size with the ion size of electrolytes due to the presence of various solvated and bare ion sizes. This is very influential, especially in the case of carbon electrodes. It is also important to consider the limited disposal or deformation of the solvation shell present in ions. This greatly affects the power and energy. Further, smaller ions are observed to decrease the thickness of the double layer and increase the capacitance of the double-layer capacitors. The cations of salts such as trimethylpropylammonium BF4, and diethyl dimethyl ammonium BF4 have different hydrocarbon chain lengths and different cation sizes [51, 65]. Further, the pore size of ionic liquids has been optimized to achieve the maximum capacitance. In this study, the dissociation, adsorption, and desolvation energy are coupled with the capacitance of ionic liquids into the CNT electrode. Theoretically, various shapes and dimensions of carbon nanoparticles have been optimized to achieve better capacitance. The capacitance has also been observed to enhance when the curvature of the electrode and surface roughness length scale were comparable to electrolyte ion size. Moreover, the ratio of ion size and carbon pore width in AC electrodes has influenced their stability [66].

Redox-Active Electrolyte

In the "redox" mechanism, electrons transfer from the electrode to the molecule yielding charge stored in the electrolyte [14]. This transfer process is better described by Marcus' theory. However, this theory is applied where heterogeneous and homogeneous electron transfer occurs. In this theory, kinetics occurs *via* two processes: coupling parameter and reorganization energy. The energy required to reorganize the solvent when the solvent approaches the molecule is reorganization energy. The interaction between the electrode and molecule is the coupling parameter. Marcus' theory is applicable where the pore of the electrode is close to the size of the ion molecule. Moreover, a formalism is still required to explain the mechanism of electron transfer *via* pores. Furthermore, the capacitance in the double layer depends on the specific surface area availability as the redox/redox variants migrate to the accessible area of the electrode and not the bulk volume of the electrode. This suggests that the number of redox moieties within pores reduces thereby decreasing the redox reaction contributions. Thus, the recent research should be more focussed on the new electrolytes, electrodes, and charging mechanism at the interface so that maximum redox moieties should enter the volume of electrodes.

PERFORMANCE OF ELECTROCHEMICAL SUPERCAPACITOR DEPENDING ON ELECTROLYTE PERFORMANCE

The electrolyte's performance is important for optimization, validation, and development of electrolytes for supercapacitors. Various techniques like cyclic voltammetry (CV), electrochemical impedance spectroscopy (EIS), and galvanostatic charging/discharging have been employed to estimate capacitance, series resistance, cycle life, and power and energy density to characterize the electrolyte. These techniques are helpful in the adsorption and desorption mechanism of the movement of charges in real-time. Further, electrolyte degradation factors, their mitigation, and overcoming of the existing challenges are seen to produce durable electrolytes. The well-known problem with electrolytes is impurities as the contamination adheres to the electrode surface (electrode/electrolyte interface) and reacts during the working of the supercapacitors. These impurities can be removed through treatment processes of solvents and solutes. Of the various categories of electrolytes, non-aqueous electrolytes are difficult to purify and these impurities alter the electrochemically stable potential leading to self-discharging. The impurities effect on the electrode reactions are detected by the background currents, the smaller the currents, the better the materials. The most typical solvents include propylene carbonate, dimethyl sulfoxide, ethylene carbonate, and tetrahydrofuran. In organic and ionic liquid electrolytes, water is considered the major impurity and is being removed by high-temperature evaporation [67]. Attention is given to ionic liquids with hydrophobic anions which benefits the durability of electrolytes [2]. Another phenomenon that needs to be discussed is aging and related failure of the electrolyte-based supercapacitors. The degradation of performance of electrolyte-based supercapacitors is widely attributed to the large operation cell potential range yielding oxidation of electrodes and decomposition of electrolytes. The increase in cell voltage drastically reduces the lifetime of supercapacitors [68]. Aging also occurs due to the incompatibility of electrolyte materials with the inactive components as discussed earlier. With aqueous electrolytes, the cycle life is shorter than with organic/non-aqueous electrolytes [69]. However, the degradation of Al-based current collectors has led to the aging of organic electrolyte supercapacitors [70]. The separator materials have also caused degradation with incompatible electrolytes such as cellulose works well in organic electrolytes instead of acid electrolytes. Furthermore, electrode materials degrade with the electrolyte ion deintercalation/intercalation issues [71, 72]. A wide working voltage range and high temperature caused degradation of supercapacitor performance like high temperature decreased the life cycle of carbon due to electrode degradation in alkaline aqueous electrolytes. Further, the residue has led to the fast self-discharge and performance degradation in organic electrolyte-based supercapacitors. Thus, impurities affecting the performance of supercapacitors are mainly due to electrolytes.

It is required for the development of standard methods to measure, evaluate, and compare the performance of various electrolyte-based supercapacitors. Both experimental and theoretical models are required for the reaction processes, failure, and degradation mechanisms in a real manner.

CHALLENGES AND PERSPECTIVES OF ELECTROLYTES

The higher electrochemical stability often leads to the deterioration of other properties of electrolytes. Another mismatch is between the ionic conductivity and electrochemical stability. It is seen that the operative cell voltage of organic electrolytes is higher than aqueous electrolytes but their ionic conductivity is lower than aqueous ones. This increases the series resistance and lowers the power density of supercapacitors. One more mismatch occurs between the viscosity and electrochemical stability. As a result, higher electrochemical stability again lowers the rate and power performance. Thus, there is a challenge to make high-energy-density supercapacitors with high cyclic ability and power density.

An insight mechanism for energy storage is still required for the newly developed electrolytes. Further, double-layer capacitor charge storage mechanisms have been utilized but theoretical work is still pending with the pseudo-capacitive materials. Further, the performance of supercapacitors is evaluated without considering the quantity of electrolytes. Furthermore, some electrolytes are very costly and this high cost limits their commercialization. Aqueous electrolytes are cheap but their energy density is lowered by low electrochemical stability. Whereas, the organic and ionic liquid electrolytes increase the cell voltage and cost of supercapacitors.

The energy/power density can be improved with the evolution of innovative electrolytes and optimizing electrolytes. The charge storage capacityis increased by employing pseudo capacitance enhancement. The development of a more fundamental understanding and advanced experimental and theoretical modeling will lead to the design of new electrolytes and their optimization.

SUMMARY

In this chapter, the types of electrolytes, and their interaction with active and inactive components have been discussed. The electrolytes are classified into water-in-salt electrolytes, aqueous electrolytes, organic electrolytes, ionic liquids, redox-active, and solid-state electrolytes. The water-in-salt electrolytes are being developed to increase the energy/power density and operating voltage window. Whereas, the limited operating voltage window of traditional aqueous electrolytes hinders their utility. Further, the operating voltage window depends upon the electrochemical stability of the electrolytes. The mechanism of formation of solid-state electrolytes is yet to be explored. The work is still going on to design the road map of interphase. To enhance capacitance in redox electrolytes, an extra redox reaction is designed to store electrons in electrolytes with large voltage windows. Based on the above-mentioned concept, "redox" electrolytes are used. The matching of electrode pore size with the electrolyte ion size improves the capacitance and in turn energy/power density. Furthermore, considering the developments in the field, several challenges are there such as enhancement in the energy density with good cycling ability and power capability and standard methods' deficiency for checking the performance of electrolytes. It is more interesting to employ the pseudo-capacitive electrodes to match with ionic liquid so that the electron can enter the bulk volume instead of the surface.

ACKNOWLEDGEMENT

PS and PAJ are thankful to DRDO-NRB/4003/MAT/PG/491 and CSIR-SRA (13(9142-A)/ 2020-Pool), respectively.

REFERENCES

[1] A. Burke, and M. Miller, "The power capability of ultracapacitors and lithium batteries for electric and hybrid vehicle applications", *J. Power Sources,* vol. 196, no. 1, pp. 514-522, 2011.
[http://dx.doi.org/10.1016/j.jpowsour.2010.06.092]

[2] C. Zhong, Y. Deng, W. Hu, J. Qiao, L. Zhang, and J. Zhang, "A review of electrolyte materials and compositions for electrochemical supercapacitors", *Chem. Soc. Rev.,* vol. 44, no. 21, pp. 7484-7539, 2015.
[http://dx.doi.org/10.1039/C5CS00303B] [PMID: 26050756]

[3] B. Pal, S. Yang, S. Ramesh, V. Thangadurai, and R. Jose, "Electrolyte selection for supercapacitive devices: a critical review", *Nanoscale Adv.,* vol. 1, no. 10, pp. 3807-3835, 2019.
[http://dx.doi.org/10.1039/C9NA00374F] [PMID: 36132093]

[4] A. González, E. Goikolea, J.A. Barrena, and R. Mysyk, "Review on supercapacitors: Technologies and materials", *Renew. Sustain. Energy Rev.,* vol. 58, pp. 1189-1206, 2016.
[http://dx.doi.org/10.1016/j.rser.2015.12.249]

[5] C. Liu, X. Yan, F. Hu, G. Gao, G. Wu, and X. Yang, "Toward superior capacitive energy storage: Recent advances in pore engineering for dense electrodes", *Adv. Mater.,* vol. 30, no. 17, p. 705713, 2018.
[http://dx.doi.org/10.1002/adma.201705713]

[6] K. Fic, G. Lota, M. Meller, and E. Frackowiak, "Novel insight into neutral medium as electrolyte for high-voltage supercapacitors", *Energy Environ. Sci.,* vol. 5, no. 2, pp. 5842-5850, 2012.
[http://dx.doi.org/10.1039/C1EE02262H]

[7] Y. Shen, B. Liu, X. Liu, J. Liu, J. Ding, C. Zhong, and W. Hu, "Water-in-salt electrolyte for safe and high-energy aqueous battery", *Energy Storage Mater.,* vol. 34, pp. 461-474, 2021.
[http://dx.doi.org/10.1016/j.ensm.2020.10.011]

[8] A.P. Doherty, "Redox-active ionic liquids for energy harvesting and storage applications", *Curr. Opin. Electrochem.,* vol. 7, pp. 61-65, 2018.
[http://dx.doi.org/10.1016/j.coelec.2017.10.009]

[9] T. Xiong, T.L. Tan, L. Lu, W.S.V. Lee, and J. Xue, "Harmonizing energy and power density toward 2.7 V asymmetric aqueous supercapacitor", *Adv. Energy Mater.,* vol. 8, no. 14, p. 1702630, 2018.
[http://dx.doi.org/10.1002/aenm.201702630]

[10] L. Coustan, and D. Bélanger, "Electrochemical activity of platinum, gold and glassy carbon electrodes in water-in-salt electrolyte", *J. Electroanal. Chem.,* vol. 854, p. 113538, 2019.
[http://dx.doi.org/10.1016/j.jelechem.2019.113538]

[11] T. Quan, E. Härk, Y. Xu, I. Ahmet, C. Höhn, S. Mei, and Y. Lu, "Unveiling the formation of solid electrolyte interphase and its temperature dependence in "water-in-salt" supercapacitors", *ACS Appl. Mater. Interfaces,* vol. 13, no. 3, pp. 3979-3990, 2021.
[http://dx.doi.org/10.1021/acsami.0c19506] [PMID: 33427459]

[12] Y. Kim, M. Hong, H. Oh, Y. Kim, H. Suyama, S. Nakanishi, and H.R. Byon, "Solid electrolyte interphase revealing interfacial electrochemistry on highly oriented pyrolytic graphite in a water-i--salt electrolyte", *J. Phys. Chem. C,* vol. 124, no. 37, pp. 20135-20142, 2020.
[http://dx.doi.org/10.1021/acs.jpcc.0c05433]

[13] H.J. Xie, B. Gélinas, and D. Rochefort, "Redox-active electrolyte supercapacitors using electroactive ionic liquids", *Electrochem. Commun.,* vol. 66, pp. 42-45, 2016.
[http://dx.doi.org/10.1016/j.elecom.2016.02.019]

[14] Y. Zhu, and O. Fontaine, *Most Modern Supercapacitor Designs Advanced Electrolyte and Interface.* intechopen.
[http://dx.doi.org/10.5772/intechopen.98352]

[15]   L. Suo, "Water-in-salt' electrolyte enables high-voltage aqueous lithium-ion chemistries", *Science,* vol. 350, p. 6263, 2015.
[http://dx.doi.org/10.1126/science.aab1595]

[16]   B.E. Conway, and W.G. Pell, "Double-layer and pseudocapacitance types of electrochemical capacitors and their applications to the development of hybrid devices", *J. Solid State Electrochem.,* vol. 7, no. 9, pp. 637-644, 2003.
[http://dx.doi.org/10.1007/s10008-003-0395-7]

[17]   A. Burke, "R&D considerations for the performance and application of electrochemical capacitors", *Electrochim. Acta,* vol. 53, no. 3, pp. 1083-1091, 2007.
[http://dx.doi.org/10.1016/j.electacta.2007.01.011]

[18]   X. Liu, and P.G. Pickup, "Ru oxide/carbon fabric composites for supercapacitors", *J. Solid State Electrochem.,* vol. 14, no. 2, pp. 231-240, 2010.
[http://dx.doi.org/10.1007/s10008-009-0812-7]

[19]   L. Demarconnay, E. Raymundo-Piñero, and F. Béguin, "A symmetric carbon/carbon supercapacitor operating at 1.6V by using a neutral aqueous solution", *Electrochem. Commun.,* vol. 12, no. 10, pp. 1275-1278, 2010.
[http://dx.doi.org/10.1016/j.elecom.2010.06.036]

[20]   J. Xu, R. Zhang, P. Chen, and S. Ge, "Effects of adding ethanol to KOH electrolyte on electrochemical performance of titanium carbide-derived carbon", *J. Power Sources,* vol. 246, pp. 132-140, 2014.
[http://dx.doi.org/10.1016/j.jpowsour.2013.07.069]

[21]   R. Wang, Q. Li, L. Cheng, H. Li, B. Wang, X.S. Zhao, and P. Guo, "Electrochemical properties of manganese ferrite-based supercapacitors in aqueous electrolyte: The effect of ionic radius", *Colloids Surf. A Physicochem. Eng. Asp.,* vol. 457, pp. 94-99, 2014.
[http://dx.doi.org/10.1016/j.colsurfa.2014.05.059]

[22]   B.E. Conway, *Electrochemical Supercapacitors.* Springer US: Boston, MA, 1999.
[http://dx.doi.org/10.1007/978-1-4757-3058-6]

[23]   J.F. COETZEE, *Recommended Methods for Purification of Solvents and Tests for Impurities.* Elsevier, 1982.
[http://dx.doi.org/10.1016/C2013-0-02975-1]

[24]   M. Ue, K. Ida, and S. Mori, "Electrochemical properties of organic liquid electrolytes based on quaternary onium salts for electrical double-layer capacitors", *J. Electrochem. Soc.,* vol. 141, no. 11, pp. 2989-2996, 1994.
[http://dx.doi.org/10.1149/1.2059270]

[25]   A. Jänes, and E. Lust, "Organic carbonate–Organic ester-based non-aqueous electrolytes for electrical double layer capacitors", *Electrochem. Commun.,* vol. 7, no. 5, pp. 510-514, 2005.
[http://dx.doi.org/10.1016/j.elecom.2005.03.004]

[26]   N. Nambu, D. Kobayashi, and Y. Sasaki, "Physical and electrolytic properties of different cyclic carbonates as solvents for electric double-layer capacitors", *Electrochemistry,* vol. 81, no. 10, pp. 814-816, 2013.
[http://dx.doi.org/10.5796/electrochemistry.81.814]

[27]   S. Tian, L. Qi, M. Yoshio, and H. Wang, "Tetramethylammonium difluoro(oxalato)borate dissolved in ethylene/propylene carbonates as electrolytes for electrochemical capacitors", *J. Power Sources,* vol. 256, pp. 404-409, 2014.
[http://dx.doi.org/10.1016/j.jpowsour.2014.01.101]

[28]   M. Ue, "Chemical capacitors and quaternary ammonium salts", *Electrochemistry,* vol. 75, no. 8, pp. 565-572, 2007.
[http://dx.doi.org/10.5796/electrochemistry.75.565]

[29]   J.P. Zheng, "Theoretical energy density for electrochemical capacitors with intercalation electrodes",

J. Electrochem. Soc., vol. 152, no. 9, p. A1864, 2005.
[http://dx.doi.org/10.1149/1.1997152]

[30] J.P. Zheng, and T.R. Jow, "The effect of salt concentration in electrolytes on the maximum energy storage for double layer capacitors", *J. Electrochem. Soc.,* vol. 144, no. 7, pp. 2417-2420, 1997.
[http://dx.doi.org/10.1149/1.1837829]

[31] R.D. Rogers, and G.A. Voth, "Ionic liquids", *Acc. Chem. Res.,* vol. 40, no. 11, pp. 1077-1078, 2007.
[http://dx.doi.org/10.1021/ar700221n] [PMID: 18020399]

[32] M. Armand, F. Endres, D.R. MacFarlane, H. Ohno, and B. Scrosati, "Ionic-liquid materials for the electrochemical challenges of the future", *Nat. Mater.,* vol. 8, no. 8, pp. 621-629, 2009.
[http://dx.doi.org/10.1038/nmat2448] [PMID: 19629083]

[33] M.G. Del Pópolo, and G.A. Voth, "On the structure and dynamics of ionic liquids", *J. Phys. Chem. B,* vol. 108, no. 5, pp. 1744-1752, 2004.
[http://dx.doi.org/10.1021/jp0364699]

[34] K.N. Marsh, J.A. Boxall, and R. Lichtenthaler, "Room temperature ionic liquids and their mixtures-a review", *Fluid Phase Equilib.,* vol. 219, no. 1, pp. 93-98, 2004.
[http://dx.doi.org/10.1016/j.fluid.2004.02.003]

[35] A. Brandt, S. Pohlmann, A. Varzi, A. Balducci, and S. Passerini, "Ionic liquids in supercapacitors", *MRS Bull.,* vol. 38, no. 7, pp. 554-559, 2013.
[http://dx.doi.org/10.1557/mrs.2013.151]

[36] B.E. Francisco, C.M. Jones, S.H. Lee, and C.R. Stoldt, "Nanostructured all-solid-state supercapacitor based on Li 2 S-P 2 S 5 glass-ceramic electrolyte", *Appl. Phys. Lett.,* vol. 100, no. 10, p. 103902, 2012.
[http://dx.doi.org/10.1063/1.3693521]

[37] A.S. Ulihin, Y.G. Mateyshina, and N.F. Uvarov, "All-solid-state asymmetric supercapacitors with solid composite electrolytes", *Solid State Ion.,* vol. 251, pp. 62-65, 2013.
[http://dx.doi.org/10.1016/j.ssi.2013.03.014]

[38] H.K. Kim, S.H. Cho, Y.W. Ok, T.Y. Seong, and Y.S. Yoon, "All solid-state rechargeable thin-film microsupercapacitor fabricated with tungsten cosputtered ruthenium oxide electrodes", *J. Vac. Sci. Technol. B Microelectron. Nanometer Struct. Process. Meas. Phenom.,* vol. 21, no. 3, pp. 949-952, 2003.
[http://dx.doi.org/10.1116/1.1565348]

[39] A.A. Łatoszyńska, G.Z. Żukowska, I.A. Rutkowska, P-L. Taberna, P. Simon, P.J. Kulesza, and W. Wieczorek, "Non-aqueous gel polymer electrolyte with phosphoric acid ester and its application for quasi solid-state supercapacitors", *J. Power Sources,* vol. 274, pp. 1147-1154, 2015.
[http://dx.doi.org/10.1016/j.jpowsour.2014.10.094]

[40] C. Zhao, C. Wang, Z. Yue, K. Shu, and G.G. Wallace, "Intrinsically stretchable supercapacitors composed of polypyrrole electrodes and highly stretchable gel electrolyte", *ACS Appl. Mater. Interfaces,* vol. 5, no. 18, pp. 9008-9014, 2013.
[http://dx.doi.org/10.1021/am402130j] [PMID: 23947753]

[41] A. Balducci, R. Dugas, P.L. Taberna, P. Simon, D. Plée, M. Mastragostino, and S. Passerini, "High temperature carbon–carbon supercapacitor using ionic liquid as electrolyte", *J. Power Sources,* vol. 165, no. 2, pp. 922-927, 2007.
[http://dx.doi.org/10.1016/j.jpowsour.2006.12.048]

[42] H. Zhou, C. Liu, J-C. Wu, M. Liu, D. Zhang, H. Song, X. Zhang, H. Gao, J. Yang, and D. Chen, "Boosting the electrochemical performance through proton transfer for the Zn-ion hybrid supercapacitor with both ionic liquid and organic electrolytes", *J. Mater. Chem. A Mater. Energy Sustain.,* vol. 7, no. 16, pp. 9708-9715, 2019.
[http://dx.doi.org/10.1039/C9TA01256G]

[43] L. Miao, H. Duan, Z. Wang, Y. Lv, W. Xiong, D. Zhu, L. Gan, L. Li, and M. Liu, "Improving the

pore-ion size compatibility between poly(ionic liquid)-derived carbons and high-voltage electrolytes for high energy-power supercapacitors", *Chem. Eng. J.,* vol. 382, p. 122945, 2020.
[http://dx.doi.org/10.1016/j.cej.2019.122945]

[44]   G. Lota, and E. Frąckowiak, "Striking capacitance of carbon/iodide interface", *Electrochem. Commun.,* vol. 11, no. 1, pp. 87-90, 2009.
[http://dx.doi.org/10.1016/j.elecom.2008.10.026]

[45]   K. Lian, and C.M. Li, "Asymmetrical Electrochemical Capacitors Using Heteropoly Acid Electrolytes", *Electrochem. Solid-State Lett.,* vol. 12, no. 1, p. A10, 2009.
[http://dx.doi.org/10.1149/1.3007424]

[46]   C. Zhao, W. Zheng, X. Wang, H. Zhang, X. Cui, and H. Wang, "Ultrahigh capacitive performance from both Co(OH)2/graphene electrode and K3Fe(CN)6 electrolyte", *Sci. Rep.,* vol. 3, no. 1, p. 2986, 2013.
[http://dx.doi.org/10.1038/srep02986] [PMID: 24136136]

[47]   L.H. Su, X.G. Zhang, C.H. Mi, B. Gao, and Y. Liu, "Improvement of the capacitive performances for Co–Al layered double hydroxide by adding hexacyanoferrate into the electrolyte", *Phys. Chem. Chem. Phys.,* vol. 11, no. 13, pp. 2195-2202, 2009.
[http://dx.doi.org/10.1039/b814844a] [PMID: 19305892]

[48]   O. Fontaine, "A deeper understanding of the electron transfer is the key to the success of biredox ionic liquids", *Energy Storage Mater.,* vol. 21, pp. 240-245, 2019.
[http://dx.doi.org/10.1016/j.ensm.2019.06.023]

[49]   J. Joseph, R. Rajagopalan, S.S. Anoop, V. Amruthalakshmi, A. Ajay, S.V. Nair, and A. Balakrishnan, "Shape tailored Ni 3 (NO 3) 2 (OH) 4 nano-flakes simulating 3-D bouquet-like structures for supercapacitors: exploring the effect of electrolytes on stability and performance", *RSC Advances,* vol. 4, no. 74, pp. 39378-39385, 2014.
[http://dx.doi.org/10.1039/C4RA05054A]

[50]   D. Wang, Q. Wang, and T. Wang, "Morphology-controllable synthesis of cobalt oxalates and their conversion to mesoporous Co3O4 nanostructures for application in supercapacitors", *Inorg. Chem.,* vol. 50, no. 14, pp. 6482-6492, 2011.
[http://dx.doi.org/10.1021/ic200309t] [PMID: 21671652]

[51]   L. Li, S. Peng, Y. Cheah, P. Teh, J. Wang, G. Wee, Y. Ko, C. Wong, and M. Srinivasan, "Electrospun porous NiCo2O4 nanotubes as advanced electrodes for electrochemical capacitors", *Chemistry,* vol. 19, no. 19, pp. 5892-5898, 2013.
[http://dx.doi.org/10.1002/chem.201204153] [PMID: 23494864]

[52]   M. Zhong, M. Zhang, and X. Li, "Carbon nanomaterials and their composites for supercapacitors", *Carbon Energy,* vol. 4, no. 5, pp. 950-985, 2022.
[http://dx.doi.org/10.1002/cey2.219]

[53]   F. Liu, and J. He, "MoC nanoclusters anchored Ni@N-doped carbon nanotubes coated on carbon fiber as three-dimensional and multifunctional electrodes for flexible supercapacitor and self-heating device", *Carbon Energy,* vol. 3, no. 1, pp. 129-141, 2021.
[http://dx.doi.org/10.1002/cey2.72]

[54]   J. Zeng, C. Xu, T. Gao, X. Jiang, and X-B. Wang, "Porous monoliths of 3D graphene for electric double-layer supercapacitors", *Carbon Energy,* vol. 3, no. 2, pp. 193-224, 2021.
[http://dx.doi.org/10.1002/cey2.107]

[55]   F. Rodríguez-Reinoso, and M. Molina-Sabio, "Activated carbons from lignocellulosic materials by chemical and/or physical activation: an overview", *Carbon,* vol. 30, no. 7, pp. 1111-1118, 1992.
[http://dx.doi.org/10.1016/0008-6223(92)90143-K]

[56]   K. Xie, X. Qin, X. Wang, Y. Wang, H. Tao, Q. Wu, L. Yang, and Z. Hu, "Carbon nanocages as supercapacitor electrode materials", *Adv. Mater.,* vol. 24, no. 3, pp. 347-352, 2012.
[http://dx.doi.org/10.1002/adma.201103872] [PMID: 22139896]

[57]   S. Hussain, R. Amade, E. Jover, and E. Bertran, "Nitrogen plasma functionalization of carbon nanotubes for supercapacitor applications", *J. Mater. Sci.,* vol. 48, no. 21, pp. 7620-7628, 2013.
[http://dx.doi.org/10.1007/s10853-013-7579-z]

[58]   E.O. Fedorovskaya, L.G. Bulusheva, A.G. Kurenya, I.P. Asanov, N.A. Rudina, K.O. Funtov, I.S. Lyubutin, and A.V. Okotrub, "Supercapacitor performance of vertically aligned multiwall carbon nanotubes produced by aerosol-assisted CCVD method", *Electrochim. Acta,* vol. 139, pp. 165-172, 2014.
[http://dx.doi.org/10.1016/j.electacta.2014.06.176]

[59]   Z. Li, L. Zhang, X. Chen, B. Li, H. Wang, and Q. Li, "Three-dimensional graphene-like porous carbon nanosheets derived from molecular precursor for high-performance supercapacitor application", *Electrochim. Acta,* vol. 296, pp. 8-17, 2019.
[http://dx.doi.org/10.1016/j.electacta.2018.11.002]

[60]   Y. Chao, S. Chen, Y. Xiao, X. Hu, Y. Lu, H. Chen, S. Xin, and Y. Bai, "Ordinary filter paper-derived hierarchical pore structure carbon materials for supercapacitor", *J. Energy Storage,* vol. 35, p. 102331, 2021.
[http://dx.doi.org/10.1016/j.est.2021.102331]

[61]   X. Zhu, S. Yu, K. Xu, Y. Zhang, L. Zhang, G. Lou, Y. Wu, E. Zhu, H. Chen, Z. Shen, B. Bao, and S. Fu, "Sustainable activated carbons from dead ginkgo leaves for supercapacitor electrode active materials", *Chem. Eng. Sci.,* vol. 181, pp. 36-45, 2018.
[http://dx.doi.org/10.1016/j.ces.2018.02.004]

[62]   J. Shao, M. Song, G. Wu, Y. Zhou, J. Wan, X. Ren, and F. Ma, "3D carbon nanocage networks with multiscale pores for high-rate supercapacitors by flower-like template and in-situ coating", *Energy Storage Mater.,* vol. 13, pp. 57-65, 2018.
[http://dx.doi.org/10.1016/j.ensm.2017.12.023]

[63]   Y. Zou, "Simple synthesis of core-shell structure of Co-Co3O4@carbon-nanotube-incorporated nitrogen-doped carbon for high-performance supercapacitor", *Electrochim Acta.,* vol. Vol. 261, pp. 537-547, 2018.
[http://dx.doi.org/10.1016/j.electacta.2017.12.184]

[64]   L. Zhang, P. Cai, Z. Wei, T. Liu, J. Yu, A.A. Al-Ghamdi, and S. Wageh, "Synthesis of reduced graphene oxide supported nickel-cobalt-layered double hydroxide nanosheets for supercapacitors", *J. Colloid Interface Sci.,* vol. 588, pp. 637-645, 2021.
[http://dx.doi.org/10.1016/j.jcis.2020.11.056] [PMID: 33267956]

[65]   A.R. Koh, B. Hwang, K. Chul Roh, and K. Kim, "The effect of the ionic size of small quaternary ammonium BF4 salts on electrochemical double layer capacitors", *Phys. Chem. Chem. Phys.,* vol. 16, no. 29, pp. 15146-15151, 2014.
[http://dx.doi.org/10.1039/c4cp00949e] [PMID: 24935222]

[66]   Y. Shilina, M.D. Levi, V. Dargel, D. Aurbach, S. Zavorine, D. Nucciarone, M. Humeniuk, and I.C. Halalay, "Ion Size to Pore Width Ratio as a Factor that Determines the Electrochemical Stability Window of Activated Carbon Electrodes", *J. Electrochem. Soc.,* vol. 160, no. 4, pp. A629-A635, 2013.
[http://dx.doi.org/10.1149/2.058304jes]

[67]   S. Fletcher, V.J. Black, I. Kirkpatrick, and T.S. Varley, "Quantum design of ionic liquids for extreme chemical inertness and a new theory of the glass transition", *J. Solid State Electrochem.,* vol. 17, no. 2, pp. 327-337, 2013.
[http://dx.doi.org/10.1007/s10008-012-1974-2]

[68]   R. Kötz, M. Hahn, and R. Gallay, "Temperature behavior and impedance fundamentals of supercapacitors", *J. Power Sources,* vol. 154, no. 2, pp. 550-555, 2006.
[http://dx.doi.org/10.1016/j.jpowsour.2005.10.048]

[69]   V. Ruiz, C. Blanco, M. Granda, and R. Santamaría, "Enhanced life-cycle supercapacitors by thermal treatment of mesophase-derived activated carbons", *Electrochim. Acta,* vol. 54, no. 2, pp. 305-310,

2008.
[http://dx.doi.org/10.1016/j.electacta.2008.07.079]

[70] A.M. Bittner, M. Zhu, Y. Yang, H.F. Waibel, M. Konuma, U. Starke, and C.J. Weber, "Ageing of electrochemical double layer capacitors", *J. Power Sources,* vol. 203, pp. 262-273, 2012.
[http://dx.doi.org/10.1016/j.jpowsour.2011.10.083]

[71] L.J. Hardwick, M. Hahn, P. Ruch, M. Holzapfel, W. Scheifele, H. Buqa, F. Krumeich, P. Novák, and R. Kötz, "An in situ Raman study of the intercalation of supercapacitor-type electrolyte into microcrystalline graphite", *Electrochim. Acta,* vol. 52, no. 2, pp. 675-680, 2006.
[http://dx.doi.org/10.1016/j.electacta.2006.05.053]

[72] S. Ishimoto, Y. Asakawa, M. Shinya, and K. Naoi, "Degradation Responses of Activated-Carbo--Based EDLCs for Higher Voltage Operation and Their Factors", *J. Electrochem. Soc.,* vol. 156, no. 7, p. A563, 2009.
[http://dx.doi.org/10.1149/1.3126423]

Graphene-Based Fiber Shape Supercapacitors for Flexible Energy Storage Applications

Ankit Tyagi[1,*], **Bhuvaneshwari Balasubramaniam**[2] and **Raju Kumar Gupta**[3,4,5]

[1] *Department of Chemical Engineering, Indian Institute of Technology Jammu, Jammu, 181221, J & K, India*

[2] *Department of Material Science Programme, Indian Institute of Technology Kanpur, Kanpur 208016, UP, India*

[3] *Department of Chemical Engineering, Indian Institute of Technology Kanpur, Kanpur-208016, UP, India*

[4] *Center for Environmental Science and Engineering, Indian Institute of Technology Kanpur, Kanpur-208016, UP, India*

[5] *Department of Sustainable Energy Engineering, Indian Institute of Technology Kanpur, Kanpur-208016, UP, India*

Abstract: Energy storage devices are essential because of ever-worsening fossil fuel depletion, increasing energy demand, and increasing environmental pollution. Maxwell Technologies, NessCap, Ashai Glass, and Panasonic commercialize carbon-based conventional supercapacitor devices. Carbon materials like graphene, carbon nanotubes, and activated carbon, are considered favourable materials for bendable and wearable electronic devices. The graphene, because of its high conductivity (thermal $\sim 5 \times 10^3$ W m^{-1} K^{-1}, electrical $\sim 10^2$ to 10^8 S m^{-1}), extraordinary surface area (theoretically ~ 2630 m^2 g^{-1}), outstanding electrochemical performance (100 to 200 F g^{-1}), less weight compared to transition metal oxides (because of less molecular weight of the carbon), outperform every other carbon material [1, 2]. The fiber-shaped supercapacitors are considered a potential future candidate for electrochemical energy storage systems and have gained considerable attention from the energy storage research community. This chapter discusses the importance of fiber-shaped supercapacitors, their evolution, various forms of their device structures, and electrolytes used for fiber-shaped supercapacitors. Further, the wet-spinning technique for synthesizing graphene fibers and their composites with pseudo-capacitive materials are also discussed.

Keywords: Graphene, Supercapacitor, Carbon-based material, Energy storage, Flexible devices.

---

* **Corresponding author Ankit Tyagi:** Department of Chemical Engineering, Indian Institute of Technology Jammu, Jammu, 181221, J & K, India; E-mail: ankit.tyagi@iitjammu.ac.in

Sanjeev Verma, Shivani Verma, Saurabh Kumar & Bhawna Verma (Eds.)

INTRODUCTION

Continuous technological improvements pave the way for developing flexible, wearable, and lightweight electronic devices that can be rolled, folded, bent, and stretched [3]. Flexible electronic textiles and fiber-shaped storage devices have been developed for smart cloth, electronic paper, environmental monitoring devices, implantable medical devices, entertainment, health monitoring sensors, and fashion technologies (Fig. **1**) [4]. The rapid development of wearable displays, memory devices, sensors, and transistors on different textiles has recently gained enormous attention [5]. The latest global market survey report projected that the market for wearable electronic devices will grow at a rate of ~15% annually from 2022 to 2030, which was evaluated at USD 52.14 billion in 2021. After COVID-19, the demand for health monitoring sensors built into wristwear, eyewear, and headwear will be expected to increase [6]. The main challenge is developing a large energy density light weight energy storage system with large areal capacitance, without sacrificing wearable properties.

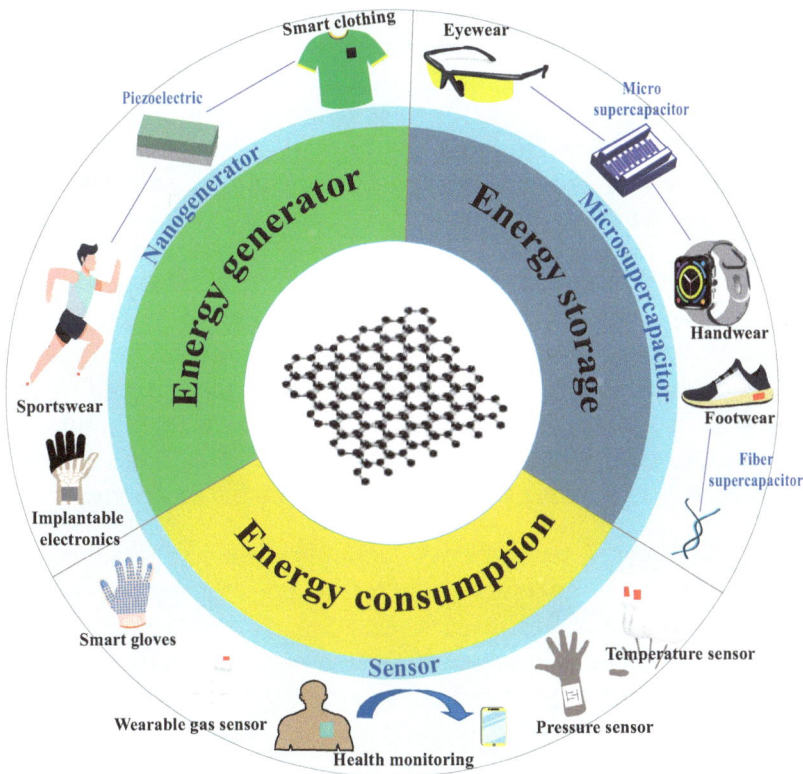

Fig. (1). Graphene-based electronics devices for future applications.

The traditional supercapacitors (SCs) contain two active material-coated metal current collectors, with a separator, and an electrolyte (Fig. **2a**). The conventional SCs based on liquid electrolytes are rigid, too heavy, and have large-size device structures to ensure safety, making them unsuitable for portable and flexible electronics [7, 8]. Alternative to conventional SCs, thin film Li-ion batteries, flexible micro-supercapacitors, and micro batteries with various shapes and sizes have been developed extensively to power flexible electronic devices, which are two-dimensional (2D) planner structures [9]. Micro batteries significantly improve charge storage performance because of their smaller size in the micro-meter range *via* reducing transportation length and enhancing active material surface area exposed to the electrolyte solution [10]. However, micro-supercapacitors (MSCs) are considered the utmost favorable micro-energy storing unit for a range of miniaturized and portable devices because of their extraordinary power density, outstanding cycling (greater than 100,000 cycles), fast charging-discharging (within seconds or several minutes), flexibility, ease of fabrication, and environment-friendly nature compared to micro batteries [11, 12].

Although in-plane interdigital MSCs are flexible and portable, there is an extensive need to further improve these MSCs for wearable and stretchable electronic devices [3]. When in-plane interdigital MSCs are woven into textiles or cloth, they require more space than 1D fiber shape SCs, limiting their potential in wearable electronics. Further, in-plane interdigital MSCs devices can also restrict the natural airflow or convection necessary for a person's health [13].

To address these issues, the focus of scientists has considerably shifted from developing 2D planer structures to fabricating 1D fiber shape devices having lightweight, increased flexibility, and foldability [14, 15]. 1D fiber shape SC devices are better for flexible, stretchable, and wearable electrical devices than traditional 2D planner devices because of their shape diversity, outstanding foldability, twisting-ability, lightweight, and ease of weaving into textiles to develop smart fabrics for wearable electronics [14]. They have omnidirectional flexibility, making them suitable for portable and flexible electronics [16].

Hydrothermal techniques, laser-reduction techniques, and wet spinning techniques are used to synthesize graphene fibers. However, the wet-spinning method outclassed other synthesis techniques due to its considerably lower cost, lightweight and mechanically robust fibers, control over properties of fibers (electrical and mechanical), and the capability of large industrial production [17].

This chapter highlights the topics related to synthesizing graphene oxide (GO) fibers following wet-spinning techniques and the reduction and modification of GO fibers to reduced GO (rGO) fibers for developing rGO-based fiber shape SCs.

We also discussed the evolution of fiber-shaped SCs and the different configurations used to make fiber-shaped SCs. Further, the performance evaluation techniques and electrolytes used in fiber-shaped SCs are discussed. We have also elaborated on the performance of some recently reported wet-spinning fiber-shape SCs.

Evolution of Fiber Shape SCs

Flexible thin-film SCs and conventional SCs are similar in terms of their device structure and possess several structural similarities in terms of their large sizes (Fig. 2). In thin-film SCs, flexible substrates were used along with the reduced overall thickness of the device compared to conventional SCs.

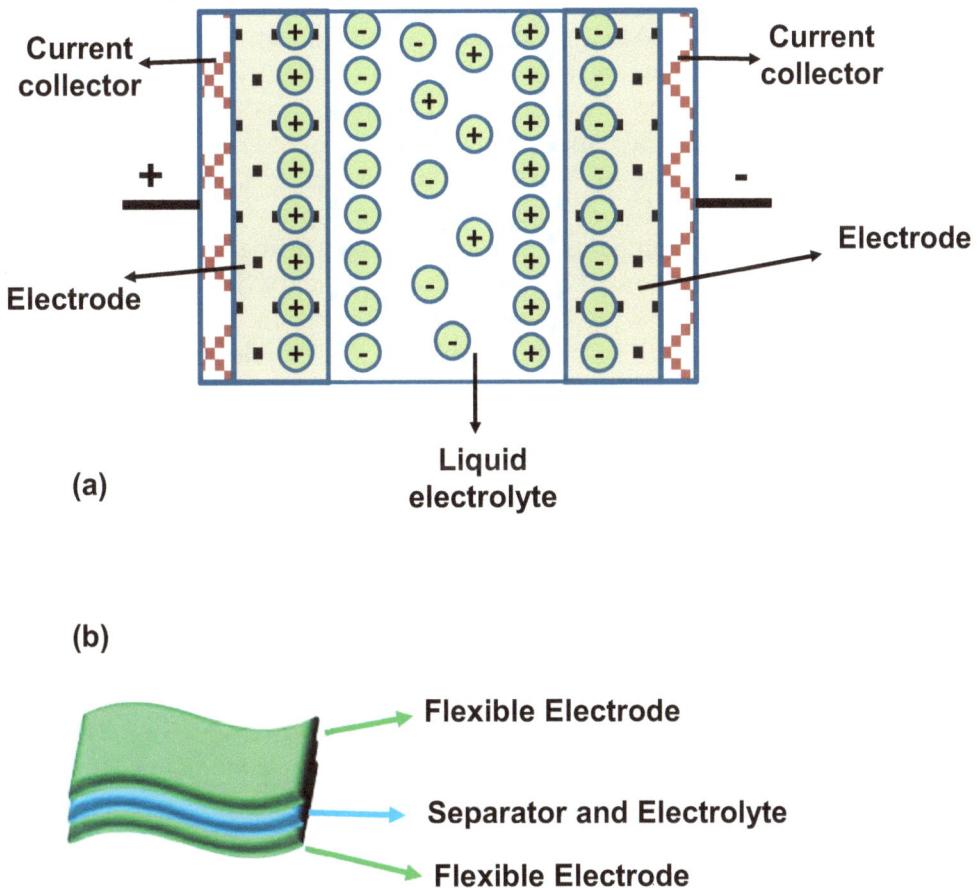

Fig. (2). Schematic for (**a**) conventional supercapacitors, (**b**) flexible supercapacitors.

These flexible SCs can show stable specific capacitance and cycling performance at various bend and twist angles and exhibit promising performance for various flexible electronic applications [18]. Laser-scribed graphene (LSG) was synthesized by Maher *et al.* by using laser irradiation to GO. The LSG sheets have an electrical conductivity of ~1738 S m^{-1} caused by the exfoliation of LSG sheets. The binder-free LSG electrode sheets were directly used with poly(vinyl alcohol) PVA-H$_3$PO$_4$ gel type electrolyte for synthesizing thin film all-solid-state symmetric supercapacitor (SSC) devices having thickness <100 μm and 1 V operating voltage window. The device showed ~5% degradation in performance after 1000 bending cycles at 150°. They also demonstrated the SSC devices with larger voltage windows with organic electrolyte (3 V, tetraethylammonium tetrafluoroborate TEA-BF4 in acetonitrile ACN) and ionic liquid (4 V, 1-ethyl-3-methylimidazolium tetrafluoroborate EMIMBF4) and run the red LED for ~24 min continuously [19]. Transparent and stretchable thin film SSC based on polydimethylsiloxane (PDMS) supported chemical vapor deposition (CVD) grown wrinkled graphene showed ~60% transparency at 550 nm and no degradation in performance when stretched up to 40% strain. The SSC device exhibited 5.8 μF cm^{-2} of areal capacitance at 8 μA current [20]. The thin film SSC devices based on flexible rGO/PVA/H$_2$SO$_4$ paper exhibit 183.5 F g^{-1} of gravimetric capacitance at 50 mV s^{-1} and a bending angle of 180°. The device also demonstrated almost no change in electrochemical performance when loaded with weights up to 200 g and showed 7.18 mW h cm-3 energy density at 2.92 W cm^{-3} of power density [21]. Gao *et al.* prepared electrospun polyamide fiber film support and modified it with carbon nanotubes (CNT) to electro-deposited MnO$_2$ nanowires and FeOOH over it. The thin film ASC was assembled using positive electrode as MnO$_2$/CNT, negative electrode as FeOOH/CNT, and LiCl/PVA gel electrolyte (Fig. **3**). The all-solid-state device was 90 μm thick. The device showed no degradation in specific capacitance (70 F g^{-1} before bending) upto 5000 bending cycles at an angle of 90° (68 F g^{-1} upto 5000 cycles) at 0.5 A g^{-1}, confirm excellent flexibility of CNT/polyamide thin films. The energy density of ASC device was 1.1 W h L^{-1} at 22.2 mA cm^{-3}, along with a steady performance at several bend angles (0°, 45°, 90°) [22].

Zhang *et al.* synthesized cellulose nanofibers/rGO/polypyrrole (CNFs/rGO/PPy) aerogel using a 3D porous structure (electrical conductivity ~2.5 S cm^{-1}). The thin film SSC device exhibited 720 mF cm^{-2} at 0.25 mA cm^{-2} using PVA/H$_2$SO$_4$ electrolyte. The energy density was 60.4 μW h cm^{-3} at 0.1 mW cm^{-2} power density, with holding of 95% of capacitance up to 2000 cycles. The device also showed stable performance at different bending (0°, 45°, 90°, and 180°) at 50 mV s^{-1} [23]. Recently, Fan *et al.* synthesized flexible thin film ASC using positive electrodes as Mn$_3$O$_4$/graphene aerogels and negative electrodes as carbon nanohorns/graphene aerogels, with the gel type NaOH/PVA electrolyte. The operating potential of the device was 1.4 V, with 17.4 μW h cm^{-2} energy density at 14.1 mW cm^{-2} power density, and holding of 87.8% capacitance up to 5000 cycles. The capacitance reported was 11.6 mF cm^{-2} at 1.1 mA cm^{-2} current density and stayed stable at various angles (0°, 45°, 90°, 180°, and 360°) [24].

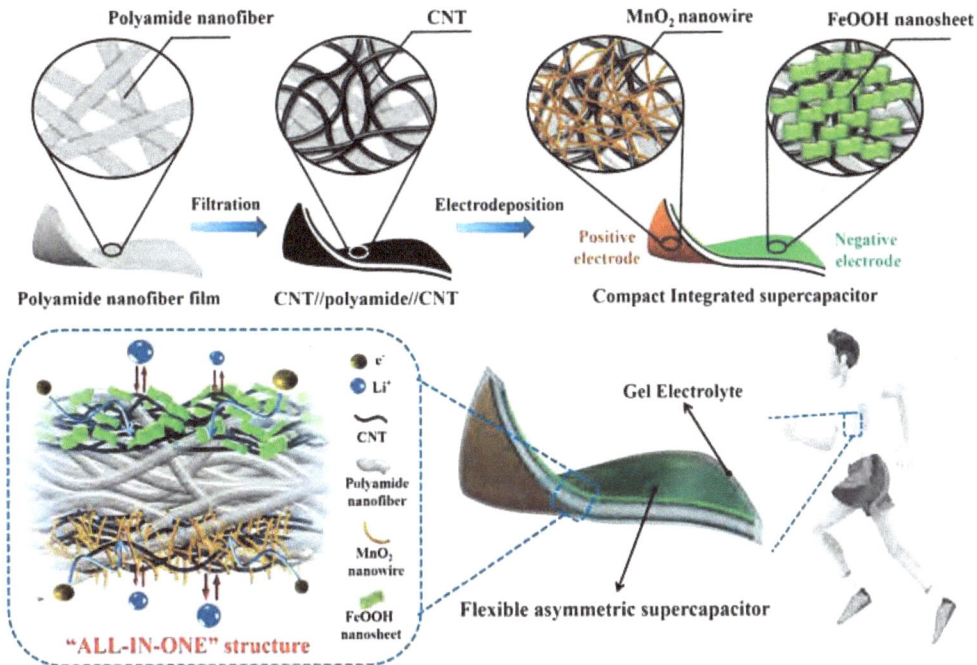

Fig. (3). Schematic showing the flexible all-solid-state ASC based on CNT/polyamide thin film. Adapted with permission from Reference [22], Copyright (2018), American Chemical Society.

Recent portable and wearable electronic device development requires an energy storage system with a small overall area, lightweighted, and easily integrated with devices. To meet this demand, fiber shape SCs are highly suitable because of their more flexible device structure, lightweight, and large specific volume power densities compared to two-dimensional planer device structures [25].

The first fiber shape SCs were presented by Bae *et al.* in 2011. They fabricated fiber-shaped SCs using hydrothermally grown ZnO nanowires over flexible plastic and Kevlar fibers as substrates [26]. The gel polymer electrolyte also works as a separator between the two electrodes so they will not come in contact, which avoids the short-circuit problem. However, in this type of device structure, two electrodes directly come in contact during the bending and twisting of devices, which results in a short circuit of the device. Fu *et al.* introduced a tri-helix device structure to overcome short-circuiting during bending, and they used two fibers as working electrodes and an extra spacer wire. The specific capacitance reported was low (11.9 mF cm^{-2} at 16.7 mA cm^{-2}) because of the small conductivity of plastic compared to metal wire as current collectors [14].

Metallic wires and Polymeric fibers are used as conventional fibers in SCs [27, 28]. Metal wires were used to increase the fiber's conductivity, which increases the device's overall weight and limits the flexibility of fiber-shape SCs. Carbon fibers such as carbon nanotubes (CNTs), graphene, multi-walled CNTs (MCNTs), and their composites with transition metal ions were extensively explored as lightweight and flexible SCs electrodes and as conductive support for flexible fiber shape SCs. Carbon fiber-based SCs exhibited outstanding gravimetric capacitance and stable cycling. However, graphene is an eye-catching material for fiber-shape energy storage devices owing to its exceptional mechanical, electrical, and thermal properties [29]. The graphene fibers have a sufficiently large surface area, chemically stable nature, outstanding electrical conductivity, comparably lightweight, and good mechanical characteristics compared to conventional fibers shape MSCs. So, graphene fibers are now being explored extensively as backbones of the electrode for MSCs [30].

Fibers shape SCs withdifferent types of metal wires, polymer fibers, and carbon fibers were demonstrated more frequently [27, 28]. However, because of outstanding mechanical, electrical, chemical, and thermal properties, enormous surface area (theoretically ~2630 m^2 g^{-1}), and large operating windows, graphene is an excellent candidate for fiber-shape SCs [29]. The graphene fibers have a sufficiently large surface area, chemically stable nature, outstanding electrical conductivity, comparably less weight, and mechanical characteristics compared to conventional fibers shape SCs [30].

Electrolyte used for Fiber Shape SCs

The electrodes, electrolytes, and separators are the three main constituents of SCs. However, the performance and device architecture of SCs are affected majorly by the choice of electrolyte. The gel polymer electrolytes, because of their excellent mechanical strength under bending, stretching, and 10^{-4} to 10^{-3} S cm^{-1} compared to 10^{-8} to 10^{-7} S cm^{-1} ionic conductivity for solid electrolytes, emerge as the most attractive electrolyte for fiber shape SCs. Poly(ethylene oxide) (PEO), polyacrylonitrile (PAN), poly(vinylidene fluoride) (PVDF), polyacrylate (PAA), poly(methyl methacrylate) (PMMA), and poly(vinyl alcohol) (PVA) are frequently used polymer for gel electrolytes. Out of these electrolytes, PVA forms stable gel electrolytes with acidic (H$_2$SO$_4$ and H$_3$PO$_4$), basic (KOH and LiOH), and neutral salts (LiCl and KCl). Further, the low cost, good mechanical properties, excellent electrochemical stability, and environment-friendly nature of PVA make it an attractive polymer for gel polymer electrolyte [31]. The

PVA/H$_2$SO$_4$, PVA/H$_3$PO$_4$, PVA/KOH, PVA/LiOH, PVA/KCl, and PVA/LiCl have mainly used PVA water-based gel electrolytes for fiber shape SCs. The operating potential of SCs depends on electrolyte decomposition potential. Organic electrolytes contain quaternary salts in organic solvents having a decomposition voltage greater than aqueous electrolytes and can achieve more than 2.7 V. Electrolyte conductivity has a noticeable influence on equivalent series resistance (ESR), which determines the power output. Aqueous electrolytes have higher conductivity than other types of electrolytes and have low viscosity, due to which they possess a fast rate of charging and discharging. To maximize the power density and to reduce ESR high concentration of electrolytes is used. However, concentrated acids and bases pose limitations to the cell construction material. Also, higher electrolyte concentrations increased the rate of self-discharge of SCs [32, 33].

To increase the SCs operating potential, many researchers have used organic liquid and ionic liquid electrolytes for fiber shape SCs, but their use is limited because of their high cost, low conductivity, and difficulties in packing. On the other hand, aqueous electrolytes can be acidic (H$_2$SO$_4$), basic (KOH), or neutral salts (LiCl, KCl, or Na$_2$SO$_4$), having their decomposition potential of 1.23 V, at which water decomposed. Ionic liquids are recently used for fiber shape SCs to attain higher voltage windows up to a maximum of 4.2 V [34 - 36].

Type of Fiber Shape SCs Device Structures

The device structure of fiber shape SCs plays a vital role in the device's performance. For transforming graphene fibers into SCs, the porous microstructure of graphene, the capacitive nature of material added with fibers, and the type of electrolytes are the main factors that need to be well-thought-out along with assembling techniques of fiber shape SCs [37, 38]. The primary device structures used for fibers-shape SCs are as follows:

1. **Parallel assembly:** Two fiber electrodes are placed parallel to each other and separated *via* a separator or gel electrolyte. As a result of the gap between two fibers, parallel assembly is safe, and there is no chance of short-circuiting. Still, this gap increases the charge transportation resistance and reduces the device's overall performance (Fig. **4a**).
2. **Coaxial assembly:** The core and shell-type fiber structure were used to fabricate fibers shape SCs. Separator or gel polymer electrolytes separate core and shell fibers (Fig. **4b**).
3. **Twisted assembly:** Fiber electrodes are twisted over each other, with gel polymer electrolytes used between them. This assembly increases electrode-electrolyte interaction and, thus, the overall performance of SC (Fig. **4c**).

4. **Coaxial-helix assembly:** One fiber electrode wrapped in coil form around another fiber electrode separated by gel polymer electrolyte (Fig. **4d**).
5. **All-in-one type assembly:** Two fibers as positive and negative electrodes with separator are combined into one fiber without binding agent (Fig. **4e**).
6. **Coaxial stretchable assembly:** This type of assembly is similar to the coaxial assembly of fiber-shape SCs but uses elastic fiber, which gives stretchability to the device (Fig. **4f**).

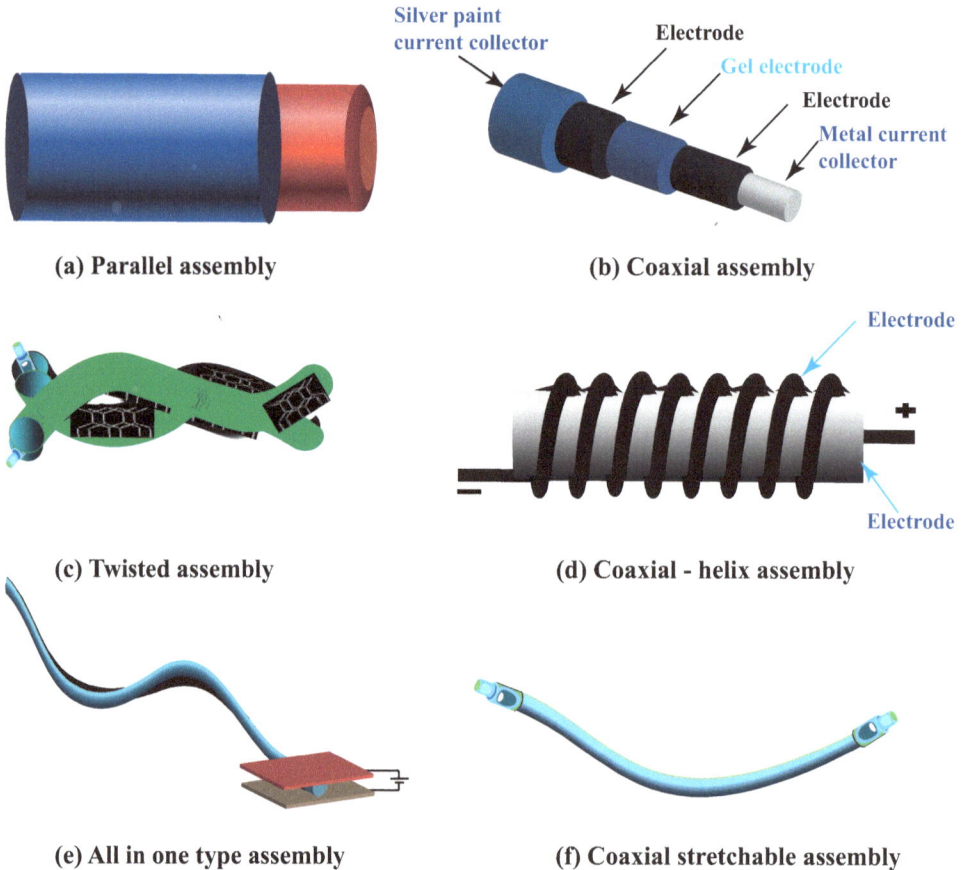

(a) Parallel assembly   (b) Coaxial assembly

(c) Twisted assembly   (d) Coaxial - helix assembly

(e) All in one type assembly   (f) Coaxial stretchable assembly

Fig. (4). Device structures for fiber shape supercapacitors. (**a**) parallel assembly, (**b**) coaxial assembly, (**c**) twisted assembly, (**d**) coaxial-helix assembly, (**e**) all-in-one type assembly, (**f**) coaxial stretchable assembly.

Performance Evaluation of Fiber Shape SCs

The performance of SCs is generally analyzed using a 2-electrode setup along with a 3-electrode setup. The 3-electrode setup includes the active material coated working electrode, whereas Ag/AgCl/saturated KCl or Calomel electrode as reference and Pt rod as counter electrodes. In a two-electrode setup, there are no reference and counter electrodes. Two electrodes can be similar or different depending on symmetric or asymmetric device configuration and are placed one after another, parted by a separator and electrolyte, as shown in Fig. (**2**). For the cells in which the electrodes are identical, the configuration is termed symmetric SCs (SSCs), while for those cells in which the electrodes are different, the cell configuration is termed asymmetric SCs (ASCs). A membrane, called a separator, separates these two electrodes. Separators inhibit the conduction of electrons, but ions can flow through them, thus preventing short-circuiting. The electrolyte filled in the space between two electrodes.

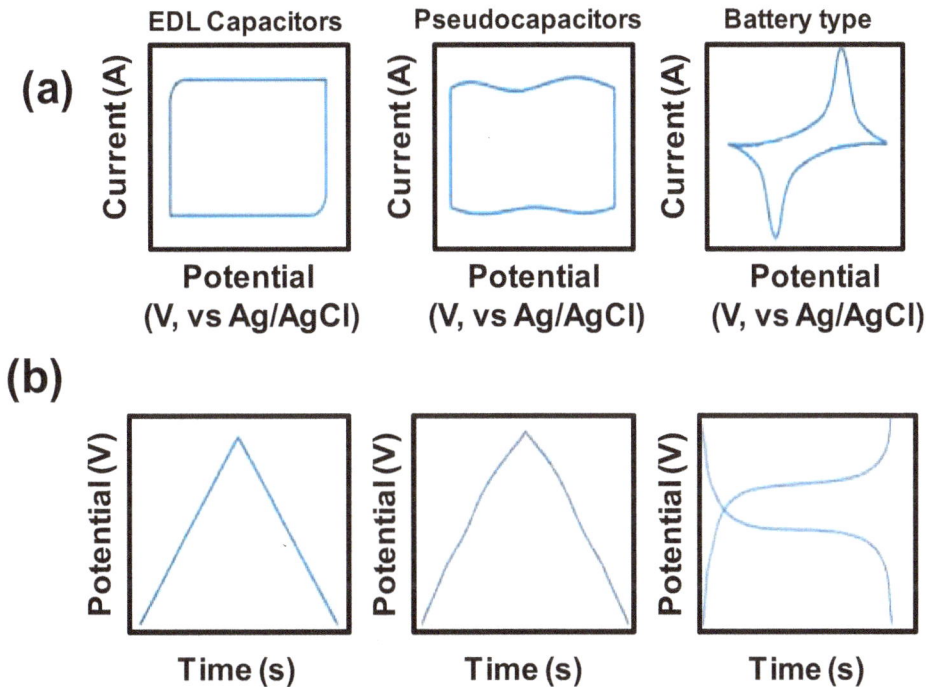

Fig. (5). (**a**) CV, (**b**) GCD plots for EDLCs, pseudocapacitors, and battery-type materials.

The nature of the cyclic voltammetry (CV) curve and the Galvanostatic charge-discharge (GCD) curve decides that the material will show EDLC or pseudocapacitive or battery-type behavior. Fig. (**5**) shows the representative shape of CV and GCD plots for EDLC, pseudocapacitance, and battery-type behavior of materials. Fig. (**5a**) showed that the CV plot is rectangular and does not have a peak for the material showing EDLCs, whereas clear peaks are present for a battery-type material. Pseudocapacitive materials show rectangular CV curves with small peaks. Fig. (**5b**) reveals that the GCD profile is triangular for EDLCs with a constant slope. Whereas, in the case of pseudocapacitors, the GCD profile is triangular with non-linearity. Battery-type materials show a GCD profile having a constant potential region and a sudden drop in potential. Therefore, one can find out the storage mechanism by looking at the nature of the CV and GCD curves of the material [39].

The fiber shape SC device's total capacitance (CT)s is found from GCD curves *via* equation 1 [40].

$$C_T = \frac{I \times \Delta t}{\Delta V} \tag{1}$$

Here I is discharging current (A), Δt- discharge time (s), and ΔV- operating potential window (V). Few researchers calculate C_T using CV curves as per equation 2.

$$C_T = \frac{Q/V}{2} = \frac{1}{2 \times V \times \mathcal{V}} \int_{V_1}^{V_2} i\,(V)\,dV \tag{2}$$

Where Q - total charge stored in both electrodes. V - scan rate during CV, V_1, V_2 - cut-off voltages during CV (*i.e.*, $V = V_1$-V_2), and i (V) - sweeping current during the CV. The fiber shape SCs is made of two active electrodes, and total device capacitance was thought because of two capacitors in series *i.e.*,

$$\frac{1}{C_T} = \frac{1}{C_1} + \frac{1}{C_2} \tag{3}$$

C1 (positive electrode) and C2 (negative electrode) are the capacitance of fiber shape SCs device. For SSCs configuration of fiber shape SC devices, C_1 and C_2 are the same, *i.e.*, $C_1 = C_2 = C$ and $C = 2C_T$. For a more practical understanding, the performance of fiber shape SCs is evaluated in terms of specific capacitance (Csp).

$$C_{sp} = \frac{C_T}{X} \tag{4}$$

Where X can be mass, area, volume, or length for gravimetric (F g^{-1}), areal (F cm^{-2}), volumetric (F cm^{-3}), or length capacitance (F cm^{-1}), respectively, of fiber shape SCs. The fiber shape SCs can be considered a cylinder for calculating their area or volume. The area of cylinder = $2\pi rL$ and volume of cylinder = $\pi r^2 L$ will be used to estimate the areal and volumetric capacitance. Here r - radius, and L -

length of the fiber. Length L can be used for calculating its length capacitance. The energy density (E, W h kg^{-1}, W h cm^{-2}, W h cm^{-3}, or W h cm^{-1}) and power density (P, W kg^{-1}, W cm^{-2}, W cm^{-3}, or W cm^{-1}) of fiber shape SC devices are calculated using following equations [40]:

$$E = \frac{1}{2} \times C_T \times \Delta V^2 \tag{5}$$

$$P = \frac{E}{\Delta t} \tag{6}$$

In the case of wearable electronics, where we have limited area or volume to fix the fiber shape SCs, areal or volumetric capacitance, energy, and power density are more appropriate for measuring the SC's performance. Electrochemical impedance spectroscopy (EIS) is used to know the resistances offered by systems, affecting the device's performance and cycling stability. So, it is essential to know the stability of SCs for prolonged use in real applications. Since fiber shape SCs are designed for flexible electronics applications, it is logical to test the electrochemical performance of these devices for various bending angles and twisted stages. The same set of equations will be used for the testing of flexible SCs [41].

Wet-spinning of GO Fibers

One of the most common wet-spinning setups includes a syringe pump that maintains GO gel flow from a known diameter needle into the rotating coagulation bath, as presented in Fig. (**6**) [42]. In wet-spinning techniques, formation of a thermodynamically stable GO liquid crystal phase is a crucial step. One can draw the long and robust GO fibers only when a stable liquid crystal phase is formed. The liquid crystal phase represents the state of material showing properties between anisotropic liquid and isotropic solid phases [43]. The formation of liquid crystals also depends on GO solution concentration, GO nanosheets aspect ratio and ionic concentration of GO solution.

As shown in Fig. (**7**), GO sheets with large sizes tend to align in an ordered fashion (nematic liquid crystal state) from the isotropic state as the concentration of GO sheets in solution exceeds the critical value. This behavior is observed because as the concentration of GO sheets in isotropic solution exceeds the critical value, there is a constraint in the freedom of particles due to the excluded volume effect. In order to compensate for this change in entropy, GO sheets align themselves in an ordered way. However, if the size of GO sheets is smaller than a specific value, it is tough to achieve the LC state or nematic liquid crystal state by increasing the concentration of the system [44]. According to Onsager, the aspect ratio of GO sheets (width/thickness = $\sim 2.3 \times 10^4$) determined the concentration of the GO gel for the stable liquid crystal phase [45]. GO nanosheets with lateral sizes greater than ~37 µm can form a liquid crystal phase in polar solutions. Also, critical concentration is not a constant value. It depends upon the lateral size of individual sheets inside the GO solution. GO dispersion with a high average lateral size of individual sheets can exhibit LC at low concentrations. GO sheets are aligned in a particular order in the liquid crystal phase. Due to this pre-alignment, the process energy required to draw these fibers decreases, resulting in the formation of less deformed fibers [43]. In the wet-spinning technique, these GO liquid crystal solutions can be spun into continuous meter-long macroscopic fibers [46].

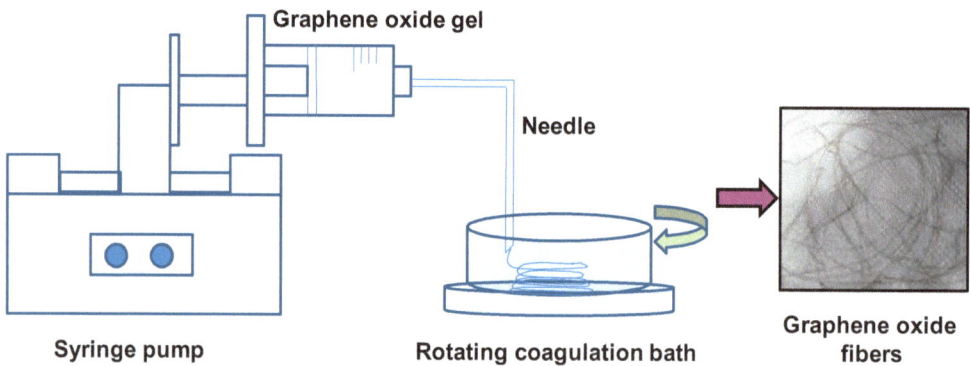

Fig. (6). Schematic illustration for production of GO fibers *via* wet-spinning technique.

Wet-spun graphene fibers have the following advantages:

1. Can utilize the specific surface area of graphene sheets effectively because well-aligned graphene sheets give an inner fiber structure that will not destroy the electrically conductive paths in the fibers [47].
2. The flexibility of adding porous carbon materials or growing porous structures onto the graphene fiber surface increases the accessibility of surface area for electrolyte ions [48].
3. Conductive spacers like CNT and heteroatoms like N, P, and S, *etc.*, can be added to increase the electrolyte wettability of fibers [49, 50].

GO Fibers with Doping and Composite with Carbon and Polymer Materials

Many research reports are about the performance of doped GO fibers along with carbon composites and polymers for the creation of flexible SCs. Yang *et al.* have combined a three-ply core-shell structure using the direct spinning method. They used a three-channel spinneret with a coaxial arrangement. As prepared, rGO/sodium carboxymethyl cellulose (CMC)/rGO SCs showed 249 mF cm^{-2} along with 96% retention in performance up to 10000 cycles [51]. Chang *et al.* synthesized N-doped graphene fibers through heat treatment of wet-spun GO fibers in an ethanol coagulation bath containing hydroxylamine. They reported that N-doped graphene fibers have larger interior pores with a small entrance. The fiber met was heated to increase electric conductivity and used as an electrode for synthesizing SSCs. They found 188 F g^{-1} gravimetric capacitance at 5 mV s^{-1} using a KOH electrolyte. As prepared, SSCs possess 48.7 kW kg^{-1} of power density at 300 A g^{-1} of current density and showed 2.24 W h kg^{-1} of energy density [52]. Zhu and co-workers mixed various ratios of porous carbon black in a GO solution and synthesized carbon black/rGO hybrid fibers. They reported that flexible SCs assembled with these fibers displayed 97.5 F cm^{-3} gravimetric capacitance at 2 mV s^{-1} of scan rate and 2.8 mW h cm^{-3} energy density at 80 mW cm^{-3} [53]. The same group also mixed a few walled carbon nanotubes (FWCNT) in GO gel during the wet-spinning of fibers and found that FWCNT helps in increasing the electrical conductivity and mechanical properties [54]. Recently, Zhu's lab mixed porous carbon nanotube in GO gel and followed a similar wet-spinning procedure to synthesize CNT/rGO fibers for all-solid-state SCs. They reported 54.9 F cm^{-3} of volumetric capacitance at 2 mV s^{-1} of scan rate and 4.9 mW h cm-3 of energy density at 15.5 W cm^{-3} of power density. CNT acts as a spacer for graphene sheets and helps in increasing the conductivity [55]. Lu *et al.* mixed CNT directly with graphene sheets dispersed in chlorosulfonic acid prepared a spinnable solution to wet-spun highly elastic CNT/graphene fibers, and found a capacitance of 138 F g^{-1} at 1 A g^{-1} [56]. Graphene sheets have very strong π-π interactions along with van der Waals force among two layers due to which stacking of layers takes place, which reduces the space between the layers. Therefore, it is challenging for electrolytes to penetrate between graphene layers, which causes the poor performance of graphene sheets in SC devices. Recently, Sheng *et al.* synthesized the polypyrrole-coated 2-2-6-6-tetramethyl piperidine- 1-oxyl (TEMPO)-oxidized bacterial cellulose (TOBC)/rGO fibers through wet-spinning for all-solid-state SCs.

TOBC acts as a spacer in the middle of graphene layers to ensure no stacking and easy infiltration of electrolytes. Polypyrrole acts as a pseudocapacitive material and increases the electrode's electrical conductivity. The specific capacitance was found to be 259 F g^{-1} (258 F cm^{-3}) at 0.2 A g^{-1} (0.2 A cm^{-3}) of current density along with 8.8 mW h cm^{-3} of energy density at 49.2 mW cm^{-3} of power density which remained 4.1 mW h cm^{-3} at 429.3 mW cm^{-3} of power density [57]. Poly(3,4-ethylene dioxythiophene): poly(styrene sulfonate) (PEDOT: PSS) based fibers showed excellent flexibility as SCs electrodes but suffered from low electrical conductivity. Li *et al.* synthesized PEDOT: PSS/rGO fibers with wrinkled and porous surface structures. They reported that SSCs fabricated from these fibers showed areal capacitance of 131 mF cm^{-2} at 20 μA cm^{-2} of current density. The device exhibited 4.55 μW h cm^{-2} of energy density [58].

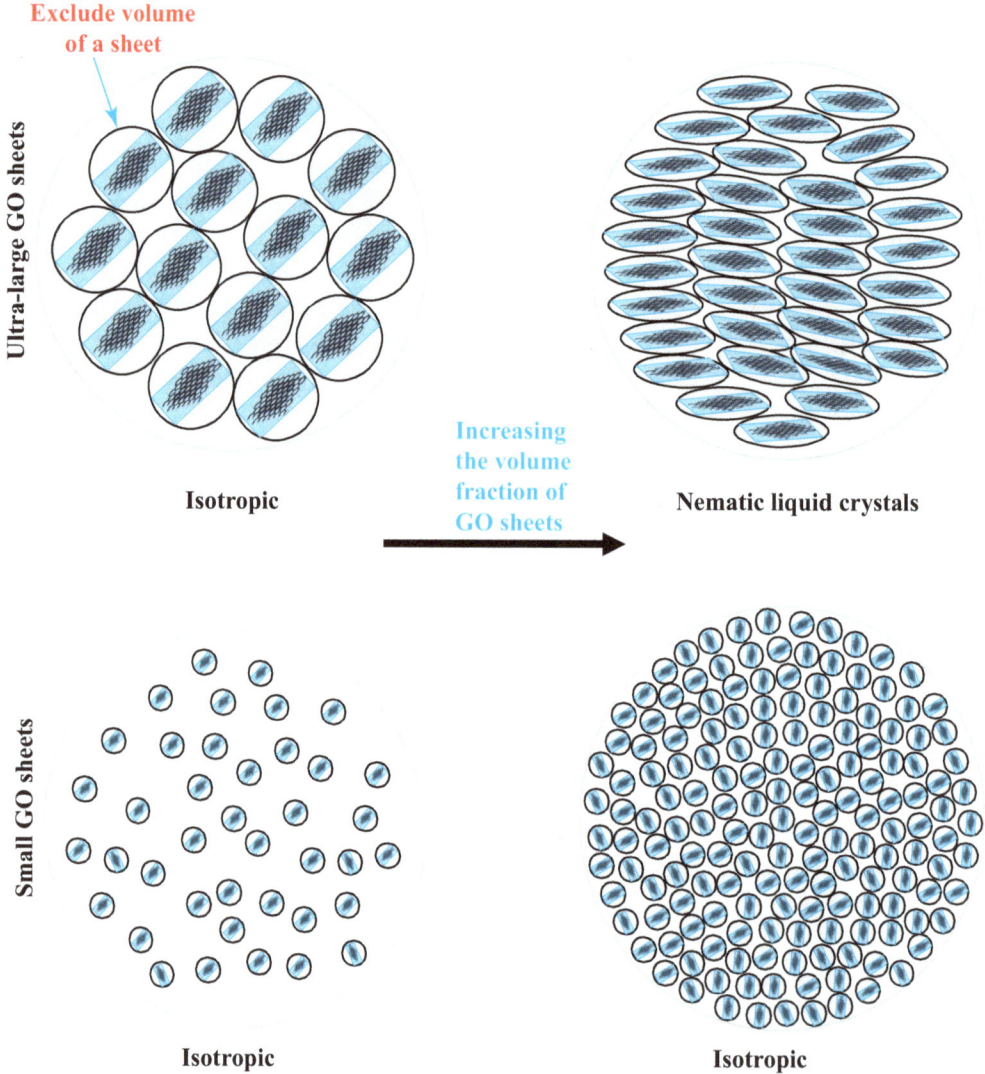

Fig. (7). Phase transition of GO solution with different GO sheet sizes as the concentration of solution increases.

GO Fibers Composite with Metal Oxides, Metal Sulfide, and MXene

Activated carbon, CNTs, carbon black, and other carbon-based ingredients were mixed with graphene oxide during wet-spinning to form symmetric fiber shape

SCs. Still, they suffer low operating potential and, therefore, a low energy density of devices [59]. Many measures have been taken to overcome the drawback of synthesizing asymmetric fiber pseudocapacitors. Ma *et al.* used MnO_2/rGO positive and MoO_3/rGO fibers as negative electrodes for asymmetric fiber SCs with 1.6 V operating potential window and exhibited 53.5 F cm^{-3} of volumetric capacitance at 100 mA cm^{-3} which remained 26.3 F cm^{-3} at increased volumetric current density of 4000 mA cm^{-3}. The authors reported an energy density of 18.2 mWh cm-3 at corresponding 76.4 mW cm^{-3} of power density [60]. All-solid-state ASCs fabricated using V_2O_5-single walled CNT (SWCNT) fiber and rGO-SWCNT fiber as positive and negative electrodes, respectively, showed 1.95 mWh cm^{-3} of energy density at 7.5 mW cm^{-3} of power density [61]. MnO_2-graphene fiber and graphene-CNT fiber-based ASC showed 11.9 mWh cm^{-3} of energy density [62]. Yu *et al.* reported ASC with a positive electrode as MnO_2 decorated rGO-SWCNT fiber and a negative electrode as N-doped rGO-SWCNT fibers, having the working potential of 1.8 V [63]. Zhu and co-workers mixed MnO_2 nanoparticles in GO gel and wet-spun to synthesize MnO_2/rG-O fibers having various ratios of MnO_2. They reported that gel formation for fiber spinning takes place with a maximum of 40% MnO_2. It was found that increasing the amount of MnO_2 increases the performance of MnO_2/rGO fiber owing to more pseudocapacitance property due to more amount of MnO_2. The specific capacitance was 66.1 F cm^{-3} at 60 mA cm^{-3} of current density and remained 96% of the initital value after 10000 cycles. The device exhibited 5.8 mW h cm^{-3} of energy density at 0.51 W cm^{-3} of power density [64]. A similar method of mixing nanoparticles in GO gel was implemented by Chen *et al.* for the production of Mn_3O_4/rGO fiber. They reported the volumetric capacitance, 45.5 F cm^{-3} at 50 mA cm^{-3} for an SSC fabricated with such fibers. The energy density was 4.05 mW h cm^{-3} at 10 mW cm^{-3} of power density [50]. Li *et al.* also mixed MoS_2 nanoparticles with GO gel to synthesize MoS_2/rGO fibers. At a current of 50 mA capacitance was 221.9 F cm^{-3}. The flexibility test showed no degradation in the volumetric capacitance after 30 and 60-degree bending [65]. MXene nanosheets are comparatively newer materials of the two-dimensional materials family than graphene. They possess excellent electric conductivity of ~9880 S cm^{-1} and high theoretical volumetric capacitance of ~1500 F cm^{-3}, which makes them eye-catching materials for pseudocapacitive energy storage devices. Researchers can't make MXene fibers for fiber-shape SCs because of their smaller lateral sheet size (less than 1 μm). Due to the small sheet size, they have a weak or insufficient interaction between the sheets for making spinnable gel for fiber synthesis. Razal's group makes the spinnable gel with 88% of the MXene mixed with GO gel and formed MXene/rGO fibers to make SSCs. They found the volumetric capacitance of 256 F cm^{-3} at 0.1 A cm^{-3} of volumetric current density along with 100% capacitance retention up to 20000 charge-discharge and no decay in columbic efficiency at 5 A cm^{-3}. They reported 5.1 mW h cm^{-3} of energy density at 20 mW cm^{-3} of power density [66]. In the same group, a decrease in the MXene contained in the MXene/rGO fibers increases the electrochemical performance of SSCs. They reported the capacitance of 361.4 F cm^{-3} at 2 mV s^{-1}. Energy density increased to 7.13 mW h cm^{-3} at the power density of 142.2 mW cm^{-3} in comparison to 88% MXene contained in MXene/rGO fibers [67]. MXene/rGO fibers' electrical conductivity increases from 72.3 S cm^{-1} to 290 S cm^{-1} when MXene decreases from 88% to 70% in the Mxene/rGO fibers, leading to better electrochemical performance of 70% MXene/rGO fibers.

CONCLUSION

Carbon fiber materials, particularly graphene fiber, are extensively used in wearable electronic devices as flexible electrode material. GO gel arrived from GO can be converted into fiber morphology by the wet-spinning method in a coagulation bath. The GO fibers were synthesized by mixing carbon materials like carbon black, carbon nanotubes, and some dopants like N, and S, also used to increase the conductivity and electrochemical performance of rGO fibers. Conducting polymers (polypyrrole, PEDOT:PSS), transition metal oxides (MnO_2, V_2O_5, MoO_3), transition metal sulfides (MoS_2), and MXene were used as pseudocapacitive materials to make hybrid fibers for better electrochemical performance. Although textile and wearable energy storage have excellent potential for future electronic devices, there are still so many challenges in terms of safety to humans because these devices come into direct contact with the human body. Further, disposal of such devices after use is also a critical issue for environmental safety. Hence, the challenges lie in the judicious selection of electrode materials and economically *via*ble synthesis processes for developing future energy storage devices.

REFERENCES

[1] H.J. Choi, S.M. Jung, J.M. Seo, D.W. Chang, L. Dai, and J.B. Baek, "Graphene for energy conversion and storage in fuel cells and supercapacitors", *Nano Energy,* vol. 1, no. 4, pp. 534-551, 2012.
[http://dx.doi.org/10.1016/j.nanoen.2012.05.001]

[2] B. Fang, D. Chang, Z. Xu, and C. Gao, "A review on graphene fibers: Expectations, advances, and prospects", *Adv. Mater.,* vol. 32, no. 5, p. e1902664, 2020.
[http://dx.doi.org/10.1002/adma.201902664] [PMID: 31402522]

[3] Y. Shao, M.F. El-Kady, L.J. Wang, Q. Zhang, Y. Li, H. Wang, M.F. Mousavi, and R.B. Kaner, "Graphene-based materials for flexible supercapacitors", *Chem. Soc. Rev.,* vol. 44, no. 11, pp. 3639-3665, 2015.
[http://dx.doi.org/10.1039/C4CS00316K] [PMID: 25898904]

[4] X. Pu, L. Li, M. Liu, C. Jiang, C. Du, Z. Zhao, W. Hu, and Z.L. Wang, "Wearable self-charging power textile based on flexible yarn supercapacitors and fabric nanogenerators", *Adv. Mater.,* vol. 28, no. 1, pp. 98-105, 2016.
[http://dx.doi.org/10.1002/adma.201504403] [PMID: 26540288]

[5]   P. Sundriyal, and S. Bhattacharya, "Textile-based supercapacitors for flexible and wearable electronic applications", *Sci. Rep.,* vol. 10, no. 1, p. 13259, 2020.
[http://dx.doi.org/10.1038/s41598-020-70182-z] [PMID: 32764660]

[6]   Wearable Technology Market Share & Trends Report Available from: https://www.grandvie wresearch.com/industry-analysis/wearable-technology-market (Accessed Jan. 14, 2023).

[7]   D. Pech, M. Brunet, H. Durou, P. Huang, V. Mochalin, Y. Gogotsi, P-L. Taberna, and P. Simon, "Ultrahigh-power micrometre-sized supercapacitors based on onion-like carbon", *Nat. Nanotechnol.,* vol. 5, no. 9, pp. 651-654, 2010.
[http://dx.doi.org/10.1038/nnano.2010.162] [PMID: 20711179]

[8]   A. Tyagi, and R.K. Gupta, "Carbon nanostructures from biomass waste for supercapacitor applications", In: *Nanomaterials* CRC Press, 2017, pp. 261-282.

[9]   C. Zhang, "Liquid exfoliation of interlayer spacing-tunable 2D vanadium oxide nanosheets: High capacity and rate handling Li-ion battery cathodes", *Nano Energy,* vol. 39, pp. 151-161, 2017.
[http://dx.doi.org/10.1016/j.nanoen.2017.06.044]

[10]   N.A. Kyeremateng, T. Brousse, and D. Pech, "Microsupercapacitors as miniaturized energy-storage components for on-chip electronics", *Nat. Nanotechnol.,* vol. 12, no. 1, pp. 7-15, 2017.
[http://dx.doi.org/10.1038/nnano.2016.196] [PMID: 27819693]

[11]   Y.Y. Peng, B. Akuzum, N. Kurra, M-Q. Zhao, M. Alhabeb, B. Anasori, E.C. Kumbur, H.N. Alshareef, M-D. Ger, and Y. Gogotsi, "All-MXene (2D titanium carbide) solid-state microsupercapacitors for on-chip energy storage", *Energy Environ. Sci.,* vol. 9, no. 9, pp. 2847-2854, 2016.
[http://dx.doi.org/10.1039/C6EE01717G]

[12]   A. Tyagi, Y. Myung, K.M. Tripathi, T. Kim, and R.K. Gupta, "High-performance hybrid microsupercapacitors based on Co–Mn layered double hydroxide nanosheets", *Electrochim. Acta,* vol. 334, p. 135590, 2020.
[http://dx.doi.org/10.1016/j.electacta.2019.135590]

[13]   K. Wang, Q. Meng, Y. Zhang, Z. Wei, and M. Miao, "High-performance two-ply yarn supercapacitors based on carbon nanotubes and polyaniline nanowire arrays", *Adv. Mater.,* vol. 25, no. 10, pp. 1494-1498, 2013.
[http://dx.doi.org/10.1002/adma.201204598] [PMID: 23300025]

[14]   Y. Fu, X. Cai, H. Wu, Z. Lv, S. Hou, M. Peng, X. Yu, and D. Zou, "Fiber supercapacitors utilizing pen ink for flexible/wearable energy storage", *Adv. Mater.,* vol. 24, no. 42, pp. 5713-5718, 2012.
[http://dx.doi.org/10.1002/adma.201202930] [PMID: 22936617]

[15]   Y. Li, Y. Zhao, J. Zhang, N. Li, H. Zhang, G. Huang, and H. Wang, "Hierarchical porous carbon fiber for fiber-shaped supercapacitor", *Funct. Mater. Lett.,* vol. 14, no. 4, p. 2150016, 2021.
[http://dx.doi.org/10.1142/S1793604721500168]

[16]   F. Meng, W. Lu, Q. Li, J.H. Byun, Y. Oh, and T.W. Chou, "Graphene-based fibers: A review", *Adv. Mater.,* vol. 27, no. 35, pp. 5113-5131, 2015.
[http://dx.doi.org/10.1002/adma.201501126] [PMID: 26248041]

[17]   L. Chen, Y. Liu, Y. Zhao, N. Chen, and L. Qu, "Graphene-based fibers for supercapacitor applications", *Nanotechnology,* vol. 27, no. 3, p. 032001, 2016.
[http://dx.doi.org/10.1088/0957-4484/27/3/032001] [PMID: 26655379]

[18]   X. Wang, K. Jiang, and G. Shen, "Flexible fiber energy storage and integrated devices: recent progress and perspectives", *Mater. Today,* vol. 18, no. 5, pp. 265-272, 2015.
[http://dx.doi.org/10.1016/j.mattod.2015.01.002]

[19]   M. F. El-Kady, V. Strong, S. Dubin, and R. B. Kaner, "Laser scribing of high-performance and flexible graphene-based electrochemical capacitors", *Science,* vol. 355, no. 6074, pp. 1326-1330, 2012.
[http://dx.doi.org/10.1126/science.1216744]

[20] T. Chen, Y. Xue, A.K. Roy, and L. Dai, "Transparent and stretchable high-performance supercapacitors based on wrinkled graphene electrodes", *ACS Nano,* vol. 8, no. 1, pp. 1039-1046, 2014.
[http://dx.doi.org/10.1021/nn405939w] [PMID: 24350978]

[21] J. Cao, C. Chen, K. Chen, Q. Lu, Q. Wang, P. Zhou, D. Liu, L. Song, Z. Niu, and J. Chen, "High-strength graphene composite films by molecular level couplings for flexible supercapacitors with high volumetric capacitance", *J. Mater. Chem. A Mater. Energy Sustain.,* vol. 5, no. 29, pp. 15008-15016, 2017.
[http://dx.doi.org/10.1039/C7TA04920J]

[22] T. Gao, Z. Zhou, J. Yu, D. Cao, G. Wang, B. Ding, and Y. Li, "All-in-one compact architecture toward wearable all-solid-state, high-volumetric-energy-density supercapacitors", *ACS Appl. Mater. Interfaces,* vol. 10, no. 28, pp. 23834-23841, 2018.
[http://dx.doi.org/10.1021/acsami.8b06143] [PMID: 29956918]

[23] Y. Zhang, Z. Shang, M. Shen, S.P. Chowdhury, A. Ignaszak, S. Sun, and Y. Ni, "Cellulose nanofibers/reduced graphene oxide/polypyrrole aerogel electrodes for high-capacitance flexible all-solid-state supercapacitors", *ACS Sustain. Chem. Eng.,* vol. 7, no. 13, pp. 11175-11185, 2019.
[http://dx.doi.org/10.1021/acssuschemeng.9b00321]

[24] L. Fan, Y. Zhang, Z. Guo, B. Sun, D. Tian, Y. Feng, N. Zhang, and K. Sun, "Hierarchical Mn3 O4 Anchored on 3D Graphene Aerogels via C-O-Mn linkage with superior electrochemical performance for flexible asymmetric supercapacitor", *Chemistry,* vol. 26, no. 42, pp. 9314-9318, 2020.
[http://dx.doi.org/10.1002/chem.201903947] [PMID: 31523882]

[25] S. Qin, S. Seyedin, J. Zhang, Z. Wang, F. Yang, Y. Liu, J. Chen, and J.M. Razal, "Elastic fiber supercapacitors for wearable energy storage", *Macromol. Rapid Commun.,* vol. 39, no. 13, p. e1800103, 2018.
[http://dx.doi.org/10.1002/marc.201800103] [PMID: 29774612]

[26] J. Bae, M.K. Song, Y.J. Park, J.M. Kim, M. Liu, and Z.L. Wang, "Fiber supercapacitors made of nanowire-fiber hybrid structures for wearable/flexible energy storage", *Angew. Chem. Int. Ed. Engl.,* vol. 50, no. 7, pp. 1683-1687, 2011.
[http://dx.doi.org/10.1002/anie.201006062] [PMID: 21308932]

[27] Q. Wang, X. Wang, J. Xu, X. Ouyang, X. Hou, D. Chen, R. Wang, and G. Shen, "Flexible coaxial-type fiber supercapacitor based on NiCo2O4 nanosheets electrodes", *Nano Energy,* vol. 8, pp. 44-51, 2014.
[http://dx.doi.org/10.1016/j.nanoen.2014.05.014]

[28] G.A. Snook, P. Kao, and A.S. Best, "Conducting-polymer-based supercapacitor devices and electrodes", *J. Power Sources,* vol. 196, no. 1, pp. 1-12, 2011.
[http://dx.doi.org/10.1016/j.jpowsour.2010.06.084]

[29] C. Lee, X. Wei, J. W. Kysar, and J. Hone, "Measurement of the elastic properties and intrinsic strength of monolayer graphene", *Science,* vol. 321, no. 5887, pp. 385-388, 2008.
[http://dx.doi.org/10.1126/science.1157996]

[30] L. Huang, D. Santiago, P. Loyselle, and L. Dai, "Graphene-based nanomaterials for flexible and wearable supercapacitors", *Small,* vol. 14, no. 43, p. e1800879, 2018.
[http://dx.doi.org/10.1002/smll.201800879] [PMID: 30009468]

[31] J. Sun, Y. Huang, Y.N. Sze Sea, Q. Xue, Z. Wang, M. Zhu, H. Li, X. Tao, C. Zhi, and H. Hu, "Recent progress of fiber-shaped asymmetric supercapacitors", *Mater. Today Energy,* vol. 5, pp. 1-14, 2017.
[http://dx.doi.org/10.1016/j.mtener.2017.04.007]

[32] P.J. Hall, M. Mirzaeian, S.I. Fletcher, F.B. Sillars, A.J.R. Rennie, G.O. Shitta-Bey, G. Wilson, A. Cruden, and R. Carter, "Energy storage in electrochemical capacitors: designing functional materials to improve performance", *Energy Environ. Sci.,* vol. 3, no. 9, pp. 1238-1251, 2010.
[http://dx.doi.org/10.1039/c0ee00004c]

[33]   R. Kötz, and M. Carlen, "Principles and applications of electrochemical capacitors", *Electrochim. Acta,* vol. 45, no. 15-16, pp. 2483-2498, 2000.
[http://dx.doi.org/10.1016/S0013-4686(00)00354-6]

[34]   A. Schneuwly, and R. Gallay, "Properties and applications of supercapacitors from the state-of-the-art to future trends", *Proceeding PCIM,* 2000.

[35]   M. Shi, H. Zhu, C. Yang, J. Xu, and C. Yan, "Chemical reduction-induced fabrication of graphene hybrid fibers for energy-dense wire-shaped supercapacitors", *Chin. J. Chem. Eng.,* vol. 47, pp. 1-10, 2022.
[http://dx.doi.org/10.1016/j.cjche.2021.05.045]

[36]   B. Bhuvaneshwari, A. Tyagi, and R.K. Gupta, "Ionic liquid-based electrolytes for supercapacitor applications", In: *Ceramic and Specialty Electrolytes for Energy Storage Devices.* CRC Press, 2021, pp. 285-305.

[37]   S.T. Senthilkumar, Y. Wang, and H. Huang, "Advances and prospects of fiber supercapacitors", *J. Mater. Chem. A Mater. Energy Sustain.,* vol. 3, no. 42, pp. 20863-20879, 2015.
[http://dx.doi.org/10.1039/C5TA04731E]

[38]   F. Wang, X. Wu, X. Yuan, Z. Liu, Y. Zhang, L. Fu, Y. Zhu, Q. Zhou, Y. Wu, and W. Huang, "Latest advances in supercapacitors: From new electrode materials to novel device designs", *Chem. Soc. Rev.,* vol. 46, no. 22, pp. 6816-6854, 2017.
[http://dx.doi.org/10.1039/C7CS00205J] [PMID: 28868557]

[39]   Y. Gogotsi, and R.M. Penner, "Energy storage in nanomaterials-capacitive, pseudocapacitive, or battery-like?", *ACS Nano,* vol. 12, no. 3, pp. 2081-2083, 2018.
[http://dx.doi.org/10.1021/acsnano.8b01914]

[40]   T. Chen, and L. Dai, "Carbon nanomaterials for high-performance supercapacitors", *Mater. Today,* vol. 16, no. 7-8, pp. 272-280, 2013.
[http://dx.doi.org/10.1016/j.mattod.2013.07.002]

[41]   D. Yu, Q. Qian, L. Wei, W. Jiang, K. Goh, J. Wei, J. Zhang, and Y. Chen, "Emergence of fiber supercapacitors", *Chem. Soc. Rev.,* vol. 44, no. 3, pp. 647-662, 2015.
[http://dx.doi.org/10.1039/C4CS00286E] [PMID: 25420877]

[42]   K. Wang, Y. Chao, Z. Chen, S. Sayyar, C. Wang, and G. Wallace, "Wet spinning of hollow graphene fibers with high capacitance", *Chem. Eng. J.,* vol. 453, p. 139920, 2023.
[http://dx.doi.org/10.1016/j.cej.2022.139920]

[43]   R. Narayan, J.E. Kim, J.Y. Kim, K.E. Lee, and S.O. Kim, "Graphene oxide liquid crystals: Discovery, evolution and applications", *Adv. Mater.,* vol. 28, no. 16, pp. 3045-3068, 2016.
[http://dx.doi.org/10.1002/adma.201505122] [PMID: 26928388]

[44]   R. Jalili, S.H. Aboutalebi, D. Esrafilzadeh, K. Konstantinov, J.M. Razal, S.E. Moulton, and G.G. Wallace, "Formation and processability of liquid crystalline dispersions of graphene oxide", *Mater. Horiz.,* vol. 1, no. 1, pp. 87-91, 2014.
[http://dx.doi.org/10.1039/C3MH00050H]

[45]   L. Onsager, "Anisotropic solutions of colloids", *Phys. Rev.,* vol. 62, no. 11–12, pp. 558-559, 1942.

[46]   Z. Xu, and C. Gao, "Graphene chiral liquid crystals and macroscopic assembled fibres", *Nat. Commun.,* vol. 2, no. 1, p. 571, 2011.
[http://dx.doi.org/10.1038/ncomms1583] [PMID: 22146390]

[47]   S.H. Aboutalebi, R. Jalili, D. Esrafilzadeh, M. Salari, Z. Gholamvand, S. Aminorroaya Yamini, K. Konstantinov, R.L. Shepherd, J. Chen, S.E. Moulton, P.C. Innis, A.I. Minett, J.M. Razal, and G.G. Wallace, "High-performance multifunctional graphene yarns: Toward wearable all-carbon energy storage textiles", *ACS Nano,* vol. 8, no. 3, pp. 2456-2466, 2014.
[http://dx.doi.org/10.1021/nn406026z] [PMID: 24517282]

[48]   Y. Meng, Y. Zhao, C. Hu, H. Cheng, Y. Hu, Z. Zhang, G. Shi, and L. Qu, "All-graphene core-sheath microfibers for all-solid-state, stretchable fibriform supercapacitors and wearable electronic textiles", *Adv. Mater.*, vol. 25, no. 16, pp. 2326-2331, 2013.
        [http://dx.doi.org/10.1002/adma.201300132] [PMID: 23463634]

[49]   H. Sun, X. You, J. Deng, X. Chen, Z. Yang, J. Ren, and H. Peng, "Novel graphene/carbon nanotube composite fibers for efficient wire-shaped miniature energy devices", *Adv. Mater.*, vol. 26, no. 18, pp. 2868-2873, 2014.
        [http://dx.doi.org/10.1002/adma.201305188] [PMID: 24464762]

[50]   G. Wu, P. Tan, X. Wu, L. Peng, H. Cheng, C-F. Wang, W. Chen, Z. Yu, and S. Chen, "High-performance wearable micro-supercapacitors based on microfluidic-directed Nitrogen-doped graphene fiber electrodes", *Adv. Funct. Mater.*, vol. 27, no. 36, p. 1702493, 2017.
        [http://dx.doi.org/10.1002/adfm.201702493]

[51]   Z. Yang, W. Zhao, Y. Niu, Y. Zhang, L. Wang, W. Zhang, X. Xiang, and Q. Li, "Direct spinning of high-performance graphene fiber supercapacitor with a three-ply core-sheath structure", *Carbon*, vol. 132, pp. 241-248, 2018.
        [http://dx.doi.org/10.1016/j.carbon.2018.02.041]

[52]   Y. Chang, G. Han, D. Fu, F. Liu, M. Li, and Y. Li, "Larger-scale fabrication of N-doped graphene-fiber mats used in high-performance energy storage", *J. Power Sources*, vol. 252, pp. 113-121, 2014.
        [http://dx.doi.org/10.1016/j.jpowsour.2013.11.115]

[53]   W. Ma, S. Chen, S. Yang, and M. Zhu, "Hierarchically porous carbon black/graphene hybrid fibers for high performance flexible supercapacitors", *RSC Advances*, vol. 6, no. 55, pp. 50112-50118, 2016.
        [http://dx.doi.org/10.1039/C6RA08799J]

[54]   A. Tyagi, K.M. Tripathi, and R.K. Gupta, "Recent progress in micro-scale energy storage devices and future aspects", *J. Mater. Chem. A Mater. Energy Sustain.*, vol. 3, no. 45, pp. 22507-22541, 2015.
        [http://dx.doi.org/10.1039/C5TA05666G]

[55]   W. Ma, M. Li, X. Zhou, J. Li, Y. Dong, and M. Zhu, "Three-dimensional porous carbon nanotubes/reduced graphene oxide fiber from rapid phase separation for a high-rate all-solid-state supercapacitor", *ACS Appl. Mater. Interfaces*, vol. 11, no. 9, pp. 9283-9290, 2019.
        [http://dx.doi.org/10.1021/acsami.8b19359] [PMID: 30762337]

[56]   Z. Lu, J. Foroughi, C. Wang, H. Long, and G.G. Wallace, "Superelastic hybrid CNT/graphene fibers for wearable energy storage", *Adv. Energy Mater.*, vol. 8, no. 8, p. 1702047, 2018.
        [http://dx.doi.org/10.1002/aenm.201702047]

[57]   N. Sheng, S. Chen, J. Yao, F. Guan, M. Zhang, B. Wang, Z. Wu, P. Ji, and H. Wang, "Polypyrrole@TEMPO-oxidized bacterial cellulose/reduced graphene oxide macrofibers for flexible all-solid-state supercapacitors", *Chem. Eng. J.*, vol. 368, pp. 1022-1032, 2019.
        [http://dx.doi.org/10.1016/j.cej.2019.02.173]

[58]   B. Li, J. Cheng, Z. Wang, Y. Li, W. Ni, and B. Wang, "Highly-wrinkled reduced graphene oxide-conductive polymer fibers for flexible fiber-shaped and interdigital-designed supercapacitors", *J. Power Sources*, vol. 376, pp. 117-124, 2018.
        [http://dx.doi.org/10.1016/j.jpowsour.2017.11.076]

[59]   X. Zhao, J. Zhang, K. Lv, N. Kong, Y. Shao, and J. Tao, "Carbon nanotubes boosts the toughness and conductivity of wet-spun MXene fibers for fiber-shaped super capacitors", *Carbon*, vol. 200, pp. 38-46, 2022.
        [http://dx.doi.org/10.1016/j.carbon.2022.08.045]

[60]   W. Ma, S. Chen, S. Yang, W. Chen, W. Weng, Y. Cheng, and M. Zhu, "Flexible all-solid-state asymmetric supercapacitor based on transition metal oxide nanorods/reduced graphene oxide hybrid fibers with high energy density", *Carbon*, vol. 113, pp. 151-158, 2017.
        [http://dx.doi.org/10.1016/j.carbon.2016.11.051]

[61]   H. Li, J. He, X. Cao, L. Kang, X. He, H. Xu, F. Shi, R. Jiang, Z. Lei, and Z-H. Liu, "All solid-state V2O5-based flexible hybrid fiber supercapacitors", *J. Power Sources,* vol. 371, pp. 18-25, 2017. [http://dx.doi.org/10.1016/j.jpowsour.2017.10.031]

[62]   B. Zheng, T. Huang, L. Kou, X. Zhao, K. Gopalsamy, and C. Gao, "Graphene fiber-based asymmetric micro-supercapacitors", *J. Mater. Chem. A Mater. Energy Sustain.,* vol. 2, no. 25, pp. 9736-9743, 2014.
[http://dx.doi.org/10.1039/C4TA01868K]

[63]   D. Yu, K. Goh, Q. Zhang, L. Wei, H. Wang, W. Jiang, and Y. Chen, "Controlled functionalization of carbonaceous fibers for asymmetric solid-state micro-supercapacitors with high volumetric energy density", *Adv. Mater.,* vol. 26, no. 39, pp. 6790-6797, 2014.
[http://dx.doi.org/10.1002/adma.201403061] [PMID: 25182340]

[64]   W. Ma, S. Chen, S. Yang, W. Chen, Y. Cheng, Y. Guo, S. Peng, S. Ramakrishna, and M. Zhu, "Hierarchical MnO2 nanowire/graphene hybrid fibers with excellent electrochemical performance for flexible solid-state supercapacitors", *J. Power Sources,* vol. 306, pp. 481-488, 2016.
[http://dx.doi.org/10.1016/j.jpowsour.2015.12.063]

[65]   J. Li, Y. Shao, P. Jiang, Q. Zhang, C. Hou, Y. Li, and H. Wang, "1T-Molybdenum disulfide/reduced graphene oxide hybrid fibers as high strength fibrous electrodes for wearable energy storage", *J. Mater. Chem. A Mater. Energy Sustain.,* vol. 7, no. 7, pp. 3143-3149, 2019.
[http://dx.doi.org/10.1039/C8TA09328H]

[66]   S. Seyedin, E.R.S. Yanza, and J.M. Razal, "Knittable energy storing fiber with high volumetric performance made from predominantly MXene nanosheets", *J. Mater. Chem. A Mater. Energy Sustain.,* vol. 5, no. 46, pp. 24076-24082, 2017.
[http://dx.doi.org/10.1039/C7TA08355F]

[67]   J. Zhang, S. Seyedin, S. Qin, Z. Wang, S. Moradi, F. Yang, P.A. Lynch, W. Yang, J. Liu, X. Wang, and J.M. Razal, "Highly conductive Ti3C2Tx MXene hybrid fibers for flexible and elastic fiber☐shaped supercapacitors", *Small,* vol. 15, no. 8, p. 1804732, 2019.
[http://dx.doi.org/10.1002/smll.201804732]

CHAPTER 10

Quantum Dots-Based Nanostructures for Supercapacitors

Himadri Tanaya Das[1,*], Swapnamoy Dutta[2], T. Elango Balaji[3] and **Nigamananda Das[1,3]**

[1] *Centre of Excellence for Advance Materials and Applications, Utkal University, Bhubaneswar 751004, Odisha, India*

[2] *University of Tennessee, Bredesen Center for Interdisciplinary Research and Graduate Education, Knoxville, TN, 37996, USA*

[3] *Department of Chemical Engineering, National Taiwan University of Science and Technology, Taipei, 10607, Taiwan*

Abstract: Recently, Quantum dot nanomaterials have been explored to a great extent for their exciting properties. Their application as electrodes to produce clean energy and hazardous chemical-free components of supercapacitors is the most interesting. The quantum dot electrodes are found to offer high charge storage capacity as well as stability that ultimately boosts their demand in advanced hybrid supercapacitors. This chapter discusses the synthesis, physiochemical features, properties, and electrochemical performance of various quantum dots. Additionally, insights into their electrochemical properties in different supercapacitors are illustrated. The best operational parameters are highlighted to provide readers with the future scope of this research area.

Keywords: Electrochemical, Energy storage devices, Nanostructured, Supercapacitors, Quantum dots.

INTRODUCTION

In the present and future global progress, the development of efficient and sustainable energy sources is and will be playing a great role. Both the environment and economy are presently getting impacted due to excessive dependency on fossil fuels, and this problem needs to be deciphered as soon as possible to ensure stable situations. High-performance and eco-friendly renewable energy sources and storage devices have rapidly developed which gave signs of

---

* **Corresponding author Himadri Tanaya Das:** Centre of Excellence for Advance Materials and Applications, Utkal University, Bhubaneswar 751004, Odisha, India, E-mail: himadridas@utkaluniversity.ac.in

hope to remove the dependency on conventional fuels. Electrochemical energy is a vital segment that plays a revolutionary role in the advancement of energy scenarios. Most advanced and efficient energy devices like supercapacitors (SCs), batteries, and fuel cells are the major fractions that are constructed based on the concepts of electrochemical energy storage and conversion. Owing to excellent characteristics and benefits such as high specific capacitance (C_s), durability, safety, rapid charge-discharge, very negligible maintenance cost, almost no memory effect, and the bridge between power-energy difference which prevails in the capacitor and fuel cells/batteries, *etc.*, SCs have grabbed the attention of several researchers [1 - 4]. Structures of SCs can be altered depending on their target usage; for instance, compact, lightweight, and flexible structures can be used for devices like smartphones, tablets, laptops, smartwatches, and digital cameras; on the other hand, they can also be used in the form of heavier structures to use as the power supplier for the applications which require immediate release of energy for the shorter duration like in solar arrays, wind turbines, automotive industries, *etc.* However, the major issue with supercapacitors is their lesser energy density (ED). Since the ED of capacitors is strongly correlated with Cs and voltage, if both increase then ED will also increase

SCs consist of two electrodes which are separated by a separator to prevent the connection between the electrodes and also provides good ionic transportation in the electrolyte. SCs accumulate charge in the electrode-electrolyte interface, typically the charge storage process highly relies on the structure and characteristics of the electrode material. For instance, a non-faradaic double-layer charge storage process was proposed for the carbon-based electrodes, while a faradaic rapid, reversible redox reaction mechanism was proposed for metal oxide-based electrodes. So, modifying electrode materials and electrolytes can lead to achieving better outcomes like fast charging, high power capabilities, high C_s, a proper life cycle with high-capacity retention, *etc.* Electrode materials became a largely concentrated topic in SCs which are broadly investigated either by chemical composition alteration or *via* changing their physical factors like surface area, pore size, *etc.* Due to adjustable electronic and optical properties, high cycling, and excellent biocompatibility, QDs are considered one of the prominent electrode materials. For example, carbon quantum dots (CQDs) are quasi-spherical particles that have shown good conductivity and efficient supercapacitor performances. Similarly, they exhibit strong quantum confinement and great capacitance outcome. For both graphene quantum dots (GQDs) and CQDs, performance can be elevated by using various surface functional groups [5]. Apart from these 0D, 1D, and 2D, QDs like g-C_3N_4, h-BN, MoS_2, silicene, antimonide, TMDCs, MXenes-QDs, *etc.* have also shown great results [6 - 8].

Synthesis of Quantum Dots for Electrodes in Supercapacitors

H. Devendrappa and the group prepared GQDs composited with a conjugated polymer by employing a combination of hydrothermal synthesis and in situ chemical polymerization techniques. Firstly, a hydrothermal approach was utilized to prepare GQDs where distilled water was used to dissolve 12 wt% of glucose precursor, heated at 180 °C for a period of 3 h with the pressure maintained at 110 psi inside an autoclave. Finally, they were centrifuged to derive GQDs. Afterward, the chemical-oxidative polymerization technique was implemented to prepare polypyrrole, $FeCl_3$, and HCl. GQDs were mixed, stirred and heated, and cooled by placing them in an ice bath. PPY–GQDs composite was obtained after the oxidizing agent $FeCl_3$ was poured and stirred, followed bypolymerization by keeping the mixture at rest for 24 h [7, 9]. In another work, Peihui Luo and the group utilized hydrothermal treatment of GQDs and graphene oxide to incorporate GQDs into a 3D graphene (3-DG) framework to form GQDs/3DG hydrogels [10]. Firstly, 3-DG hydrogels were prepared by using aqueous GO as a precursor under hydrothermal conditions of 180 °C for 4 h. Afterwards different concentrations of GQDs (prepared by oxidizing carbon fibers) were added to the same GO solution and autoclaved at 180 °C for 4 h to get the GQDs/3DG hydrogels which were used as a supercapacitor electrode. In the study, several composites were developed by compositing GQDs/GO with various weight ratios such as 10%, 20%, 40%, and 80%. Hydrothermal synthesis was also used by M. Ashourdan *et al.*, to synthesize $CuMnO_2$/graphene QD composite [11]. The author used citric acid, and sodium hydroxide to synthesize GQDs, and $Cu(NO_3)_2$. $Mn(NO_3)_2$ precursors were used to prepare $CuMnO_2$. Afterward, both the GQDs and $CuMnO_2$ were mixed and allowed for a hydrothermal reaction for 24 h at 80 °C. Finally, the product was purified using water and ethanol and then dried in an oven at 55 °C for 8 h to obtain the composite. Not only the GQDs, nickel-cobalt oxide ($NiCo_2O_4$) QDs were also prepared using the hydrothermal method. Hydrothermal method has also been used to develop graphitic carbon nitride (g-C_3N_4) QDs/graphene hydrogel where melamine is used to produce g-C_3N_4 QDs [12]. Poonam Siwatch *et al.* [13], used DI water and ethanol to dissolve the following precursors Ni/Co-acetate tetra-hydrate in a urea solution and kept for stirring. Afterward, the mixture was hydrothermally heated at 125 °C for 6 h, then washed with DI water and ethanol and dried in a vacuum oven for 10 h. Finally, it was calcined for 4 h at various temperatures ranging from 300 °C to 500 °C, however the author found that the sample calcined at 300 °C exhibited superior capacitive behavior. Although hydrothermal is the frequent method employed to prepare QDs for supercapacitor applications. Few other synthesis approaches are also employed. Abu Jahid Akhtar and the group used the wet-precipitation method to prepare different Co-doped ZnO QD samples [$Zn_{1-x}Co_xO$ (x = 0, 0.02, 0.04)] [14]. In this method, firstly, zinc acetate dihydrate was dispersed in an alcoholic medium having a solution of methanol and NaOH constantly stirred at 333 K for 2 h to prepare ZnO QD. The QDs are precipitated using *n*-hexane. Afterwards, cobalt chloride dissolved in methanol was added to the ZnO QD solution and heated at 343 K for 1 h. Finally, the obtained precipitate was freeze-dried followed by heating at 353 K for 24 h to attain Co-doped ZnO QD. Other than that, methods like the soft chemical method, typical one-pot chemical synthesis, electrochemical approach, hot injection method, and ultra-sonication methods have also been employed to prepare QDs to use as supercapacitor electrodes [15 - 20]. Yuan *et al.* [20], adopted a temporally and spatially shaped femtosecond strategy to prepare MXene quantum dot(MQD)/laser-reduced graphene(LRGO). This approach is a simple one-step-in situ process, making the synthesis process less time-consuming. When the MXene target was focused using the temporally and spatially shaped (Bessel) laser (TSBL), plasma emission and specific pulse delay (0–15 ps) were observed. Repeatedly ablating the laser pulse sequence to the freshly generated particles resulted in a decrease in the particle size, and hence a homogenous pattern was observed. Hei *et al.* [21], prepared V_2O_3 quantum dots (~6.64 nm) which were utilized for supercapacitor application. The author first refluxed crushed corn straws in dilute H_2SO_4 (1 h at 95 °C), washed them with DI and dried them; afterward, they were impregnated in $VOSO_4$ solution and kept under ultrasound. They were then treated under vacuum. Finally, they were filtered and heat treated under Ar atmosphere at 850 °C.

Performances or Reported Articles with Electrolyte

Yuan *et al.* [20], achieved outstanding electrochemical activity along with good transparency when preparing a transparent composite electrode using the MQD/LRGO for flexible and transparent supercapacitors. The fabricated supercapacitor using the prepared MQD/LRGO electrode exhibited high energy storage capabilities with ultrahigh ED of 2.04×10^{-3} mWh cm^{-2}, stable cycle-life with 97.6% (12 000 cycles), and superb C_s (10.42 mF cm^{-2}) with high transparency (transmittance over 90%) and good flexibility and durability. MQDs were uniformly distributed onto few-layered LRGO that was useful in obtaining better electrochemical performance compared to individual compound activities as this unique composite structure provided plentiful active edge sites and thus led to efficient carrier mobility. A transparent flexible supercapacitor was prepared where sandwiching gel electrolyte (PVA/H_2SO_4) was placed with MQD/LRGO coating. The authors inferred that the reason for the low-frequency region is due to the charge transfer process at the electrode-electrolyte interface and the electrolyte ion diffusion. Hei *et al.* [21], investigated a unique supercapacitor electrode (0D/2D nanostructure) prepared with V_2O_3 QDs and mesoporous carbon (MC) nanosheets. The well-distributed V_2O_3 QDs on MC nanosheets exhibited C_s

of 270 F g^{-1} at 1 A g^{-1} and for almost 5000 cycles, it showed reasonable durability at 10 A g^{-1}. The nano-sized V$_2$O$_3$-QDs ensure a fast ionic/electronic transport which leads to high-rate capabilities and long-term cycling stability. Na$_2$SO$_4$ aqueous electrolyte was in a SC prepared with V$_2$O$_3$ QDs /MC nanosheets electrode [21]. The author specified that this electrolyte along with the electrode (uniformly distributed ultra small V$_2$O$_3$-QDs on the carbon substrates) provided adequate accessible sites for electrochemical activity, which is why reversible capacity was enhanced.

H. Devendrappa and the group prepared a PPY–GQDs composite which exhibited high C$_s$ and cyclic life of SCs (hybrid supercapacitor device) [9]. To examine the electrochemical performance of individual PPY and GQDs, as well as the composite PPY–GQDs, a 3-electrode system containing Pt as a counter electrode, and Ag/AgCl as a reference electrode. To construct the working electrode GQDs, PPY and PGC-modified ITO electrodes were utilized where the electrolyte solution was NaCl (1 M). Furthermore, PGC$_3$ was identified as the optimized sample, and the supercapacitor of individual PPY (pure) and optimized sample (PGC$_3$) was assembled *via* employing the 2-electrode system where PVA-LiCl$_2$ gel was used as an electrolyte and for separator, a small piece of filter paper was placed. The enhanced current density was observed from the cyclic voltammetry (CV) outcomes and the area of the CV loop seemed to be improved with the expanding scan rate and the GQDs concentration. Using these electrodes, separately assembled supercapacitor device, the electrodes PGC$_1$ and PGC$_3$ exhibited a high ED of 67.8 W h kg^{-1} and 93 W h kg^{-1} at a power density (PD) of 1210 W kg^{-1} and 1430 W kg^{-1}. PGC$_1$ and PGC$_3$ exhibited the highest C$_s$ values of 467.32 and 647.54 F g^{-1}. The PPY-GQDs composites attained improved cycle life (2000 cycles). Similar to GQDs, CQDs are also used to form CQD/polypyrrole composite electrodes which exhibited areal capacitance of 315 mF cm^{-2} at a current density of 0.2 mA cm^{-2} when the electrodes were assembled in all-solid-state supercapacitor device with good stability and 85.7% capacitance retention over 2000 cycles [18].

Peihui Luo and the group prepared GQDs/3DG composites by simply altering the feeding ratio of GQDs/GO by different weight percentages (10%, 20%, 40%, and 80%), out of which, the feeding ratio of 40% as electrode demonstrated a C$_s$ of 242 F g^{-1} [10]. The outcome was quite higher (almost 22%) than the individual 3DG electrodes (198 F g^{-1}). Such enhanced activity mostly resulted because the prepared composite seemed to have superior electrical conductivity as well as larger surface area (SA) while using a moderate quantity of GQDs. The fabricated GQDs/3DG composites exhibited good electrochemical stability. In the study, the working electrode is GQDs/3DG hydrogel coated on Platinum foil and 1 M KOH aqueous solution as an electrolyte which exhibited capacitance retention of 93%

after 10,000 charge-discharge cycles. Ashourdan *et al.*, reported CuMnO2/GQD which showed significant improvement in the reversibility and discharge time (Figs. **1a** & **b**). The as-fabricated asymmetric supercapacitor using $CuMnO_2$ nanocrystals and $CuMnO_2$/GQD composite electrodes with 3 M KOH was used as electrolyte and polyethylene (PE) paper was used as separator [11]. When comparing the $CuMnO_2$ nanocrystals and $CuMnO_2$/GQD composite, that better C_s and improved efficiency of the fabricated supercapacitors were observed using these electrodes separately. The $CuMnO_2$/GQD electrode has a high capacity of 520.2 C g^{-1} at a current density of 1 A g^{-1}. The asymmetric supercapacitor of $CuMnO_2$/GQD//AC has an excellent specific energy of 47.9 W h Kg^{-1} at a specific power of 1108.1 W kg^{-1}. After 5000 charge/discharge cycles, the capacitance reduced by only less than 13.3%. Siwatch *et al.* [13], prepared a highly porous and hollow $NiCo_2O_4$-QDs which exhibited a C_s of 362 F g^{-1}. The as-synthesized structure has especially exhibited a highly porous arrangement which leads to excellent electrochemical activity and also provides a good electroactive surface that can ensure favorable faradaic redox reactions. Also, an asymmetric supercapacitor with NCO-QDs as positive and rGO as negative electrodes, and Na_2SO_4 aqueous solution as an electrolyte has been fabricated which exhibits a high discharge time (Fig. **1c**) and good charge storage capacity along with a stable lifetime showing capacity retention of 86% (1000 cycles) (Fig. **1d**). The as-fabricated device has demonstrated adequate ED of 69.5 W h kg^{-1} with a PD of 2.22 kW kg^{-1}. Owing to thermal and chemical stability, ZnO QDs are also considered a great choice for exploring energy applications. In addition, they have exhibited good higher surface-to-volume ratio (SVR), satisfactory ED, and swift charge transportation activities, but their inadequate ionic and electronic conductivity limits their energy applications, especially for supercapacitor devices. Abu Jahid Akhtar and the group Co-doped ZnO QD which exhibited superior capacitive behavior compared to individual ZnO QDs [14]. After doping, there is an increased active site, high specific SA, and increased diffusion of electrons and ions which are due to the charge transfer difference between the doping metal element and the ZnO. As a result, the x = 0.04 electrode exhibited in 1 M KOH excellent supercapacitor behavior, with a specific capacitance of 697 F g^{-1}, excellent PD and ED (1026 W kg^{-1} and 24 W h kg^{-1}), and better stability (97% after 2000 cycles).

Transition metal-based QDs have also exhibited promising outcomes in supercapacitor applications. In a study, Wenyong Chen *et al.*, synthesized monodisperse Ni_3S_4 QDs to use as electrodes along with KOH aqueous electrolyte [15]. The hybrid supercapacitor was assembled as shown in Fig. (**2a**) which demonstrated good reversibility with high discharge time (Fig. **2b** & **c**) and the device exhibited ED and PD of 49.3 W h kg^{-1} and 21718 W kg^{-1} (Fig. **2d**) with good stability of only 8.3% loss of initial capacitance after 8000 cycles. In addition, the material was tested as the anode material for lithium energy storage, the results portrayed excellent electrochemical property of the material with a high reversible specific capacity of 647.5 mAh·g^{-1} and only 2% deterioration of reversible capacity after 500 cycles. Midya and group prepared WS$_2$ QDs and counterpart nanosheets, which showed higher areal C$_s$ of 28 mF cm^{-2} and an ED higher than the WS$_2$ nanosheets [16]. The group used PVA–H$_3$PO$_4$ as electrolyte and electrodes comprising WS$_2$ NSs and QDs as active materials and acetylene black with a ratio of 85:15, respectively. Another group has approached surface treatment of WS$_2$ QDs with 1,2-ethanedithiol, which resulted in high active edge atoms [19]. The as-synthesized modified WS$_2$ QDs demonstrated C$_s$ of 457 F/g where sodium sulfate was used as an electrolyte with good stability and efficiency. In a different study, Das and colleagues synthesized SnO$_2$ QDs with an average size of 2.4 nm delivering 1.4 times higher C$_s$ and high stability when compared with 25 nm large-particle SnO$_2$ materials [17]. The C$_s$ of the QDs is 10 F g^{-1} and 91% retention was obtained when the scan rate was increased to 500 mV s^{-1} at 0.5 M KOH electrolyte. SnO$_2$ QDs exhibited 98% stability even after 1000 cycles.

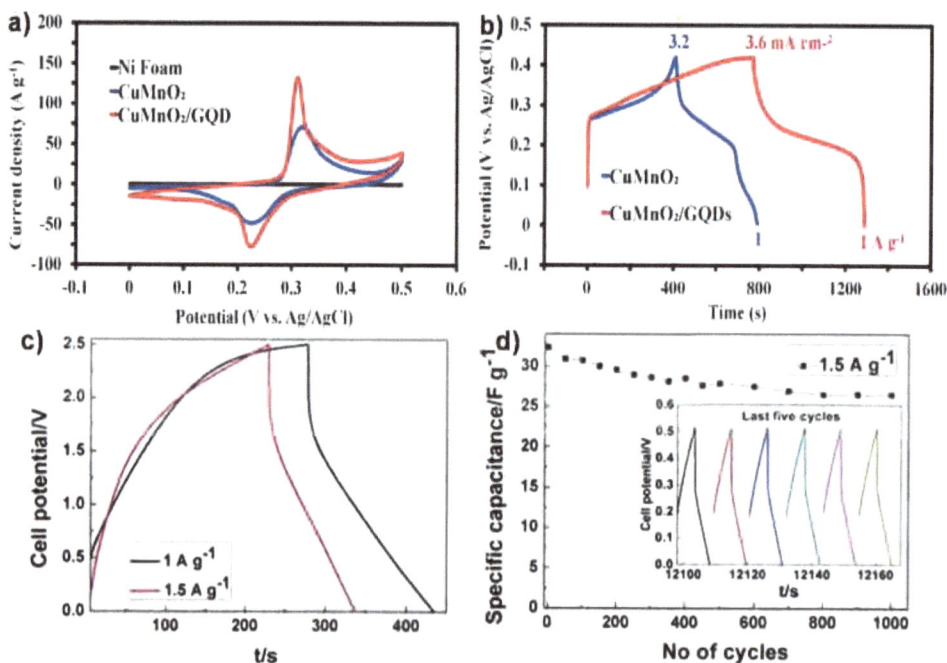

Fig. (1). a) CVs of different electrodes at scan rate 40 mV s^{-1}, b) GCD curves of CuMnO$_2$/GQD and CuMnO$_2$ electrodes at current density as 1 A g^{-1}, c) GCD curves of ASC at different current densities after 1000 cycles of charge-discharge, d) Cycling performance at current density of 1.5 A g^{-1} of ASC based on NCO-QDs//rGO. Copyrights (Elsevier, 2021) (Elsevier, 2019). Reproduced from Ref [11, 13].

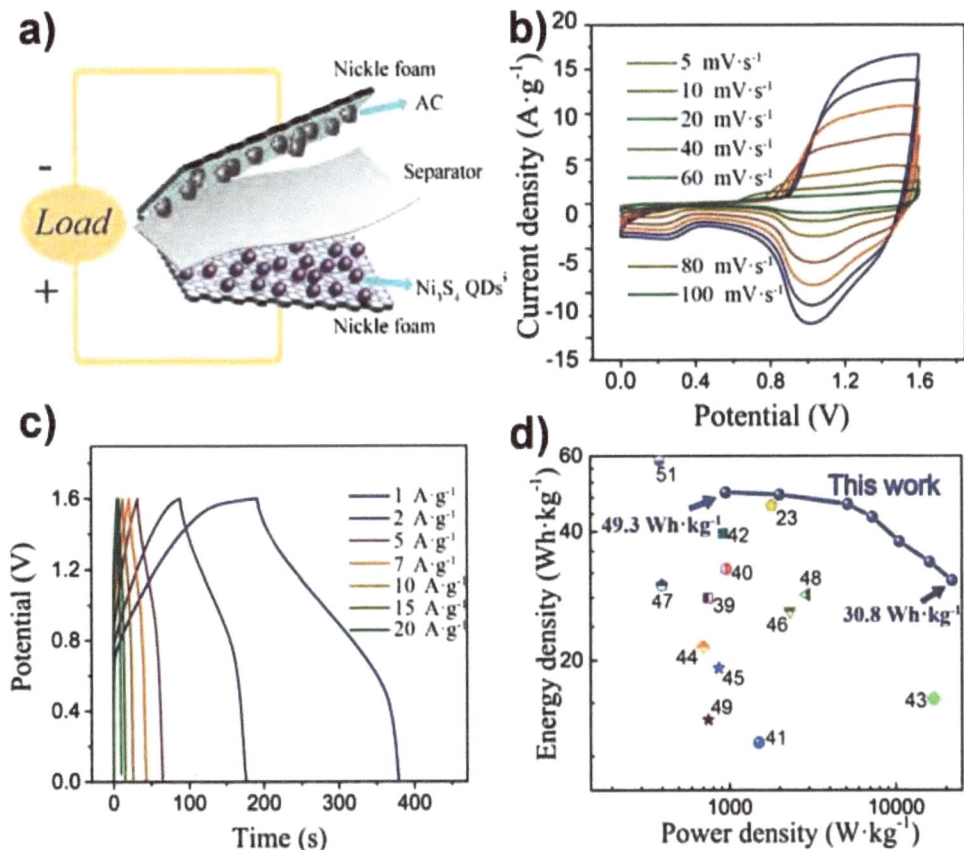

Fig. (2). a) Model schematic of Ni_3S_4 QDs/NF//AC/NF hybrid supercapacitors, **b)** CV curves scanned in different rates, **c)** GCD curves under various current densities of the hybrid supercapacitor, **d)** Ragone plot for the hybrid supercapacitors. Copyrights (Elsevier, 2019). Reproduced from Ref [15].

Sim *et al*. proposed functionalizing of graphene quantum dots with fluorine and nitrogen by a simple two-step method. The as-prepared material exhibited a specific capacitance of 244 F g^{-1} at a current density of 3 mA cm^{-2}. Also it showed good catalytic activity for water splitting [22]. Kharangarh *et al*. prepared spinel $NiCo_2O_4$ which was decorated by graphene quantum dots which increased the electrical conductivity. The as-fabricated electrode exhibited a specific capacitance of 481.4 F g^{-1} under alkaline conditions, which is significantly higher than the graphene quantum dots having a specific capacitance of 45.6 F g^{-1} [23]. Ashourdan *et al*. reported a nanocomposite of $CuMnO_2$ and graphene quantum dots. The nanocomposite exhibits a high specific capacity of 520.2 C g^{-1} with the device showing a significantly high energy density of 47.9 Wh kg^{-1} [24]. Zhang *et al*. reported the incorporation of graphene quantum dots into $MnCo_2O_{4.5}$ while the quantum dots function as a conductive material and metal oxide is a strong structural unit. The resulting composite has a synergy of stable structure and high conductivity with the highest capacitance of 1625 F g^{-1} attained due to the novel composite. Also, the device exhibits an excellent energy density of 46 Wh kg^{-1} [25]. Qiu *et al.* reported novel flower ball-like graphene quantum dots incorporated into the layers of the NiCo-LDH through a simple microwave synthesis. The electrode exhibited a high specific capacitance of 1526 F g^{-1} not only because of the LDH but also due to the functionalization of graphene quantum dots using histidine. The assembled device showed an improved energy density of 48.89 Wh kg^{-1} [26]. Tian *et al.* incorporated graphene quantum dots into a porous carbon to increase the conductivity further than the commercially available activated carbon. The as-prepared material exhibited a high specific capacitance of 170 F g^{-1} even at a higher current density and loading of 100 A g^{-1} 20 mg cm^{-2}, which makes it suitable for commercialization [24].

Discussing the Pros and Cons of Supercapacitor

QDs have a wide series of applications viz. energy storage devices, optical devices, photocatalysts, sensors, biomedicine, *etc.* In this chapter, we have discussed several QDs like CQDs, GQDs, TMD QDs, *etc.* Their tunable characteristics, size and structure variations, and quantum confinement effect have made them attractive for supercapacitor applications. Importantly, quantum confinement is linked to the substantial rise in SVR by decreasing the QD diameter which is beneficial for the application purpose [27]. However, a few aspects are still required to be investigated using both theoretical and experimental approaches. Firstly, the complexity is high for understanding the resistance between QDs and the corresponding current collector, as based on this aspect charge storage and the delivery of charge features would be comprehended. Secondly, optimizing cost-effective and simple synthesis routes and efficient purification procedures to develop QDs will be beneficial to take the lab-scale production to commercialize production. In addition, the surface functionalization of QDs needs to be explored as it can enhance electric conductivity which in turn will be impacting characteristics like Cs, PD, ED, and long-term cycling durability. There are several drawbacks of specific QDs, which need to be addressed in upcoming research as ZnO QDs exhibit poor water stability and easy agglomeration whereas controlling size and particles in GQDs is difficult. Also, FeOOH QDs exhibited smaller SA and poor electronic conductivity. These drawbacks often lead to unsatisfactory outcomes. These problems have already been deciphered in different ways, but efficient and economical solutions can make them more impactful for SC applications. Also, degradability is an issue in QDs which also requires adequate attention.

Future Scope for Supercapacitor

As the supercapacitor technology progresses, flexible and conformal supercapacitors are getting more attraction to make the upcoming technology more compact and efficient. Studies on GQDs/Ni(OH)$_2$, an electrode for a flexible supercapacitor, have shown outstanding electrochemical activity where nearly no major difference in the performance and CV curve was recorded when the supercapacitor was bent up to 180° [28 - 30]. In another study, MoS2 QD/PANI hydrogel has shown great performance when fabricated in an all-solid-state flexible supercapacitor where the electrode has exhibited excellent flexibility which indicates bending and winding and have a negligible impact on the capacitive activity of the material [31]. Similarly, ZnO QD/N-doped porous carbon/carbon nanofiber (CNF) composite has demonstrated nearly no change even after bending at an angle of 180°, which indicates flexible and good electrochemical stability, and proved that the assembled asymmetric flexible supercapacitors are compatible for use in flexible electronic equipment [32]. However, still, studies need to focus precisely on understanding the robustness of the bending process between 0° to 180°, and their correlation with the charge storage of the assembled supercapacitor devices. Proper investigation related to stability for SCs though varying the bending angle is an essential verification of wearability of QD SCs. The usage of transparent electrodes prepared with QDs can also offer added value to wearable devices. In this case, CQDs can be used as they are highly transparent materials. Apart from these, more optimization is required for utilizing dopants, and functional groups to make QD composite economically and efficiently.

CONCLUSION

Next-generation technologies are growing in a very swift manner and in this progress, energy conversion and storage devices have immense contributions. In the upcoming days, new edge electronics will be more demanding to support battery and SC technologies. Due to characteristics like outstanding power density, longer cycle life, and cost-effectiveness, SCs are being considered for several real-time applications. In recent times, we have already witnessed several applications such as solar systems, faster trains, electric vehicles, and wind power equipment, involving the utilization of SCs to get stable and efficient results. Several active materials are employed as electrode materials. However, materials are also observed to have dimensional aspects, as every morphology has its advantages. QDs are generally 0-D materials that are employed in several applications because of their beneficial properties such as greater SVR, abundant specific CA with active sites, quantum confinement effect, simple doping conditions, *etc.* In this chapter, we have illustrated the role of QDs in SC applications. Several approaches to the synthesis along with the fabricated SCs with the help of the prepared QD electrodes were demonstrated with their respective performances (PD, cycle life, *etc*). Different types of individual QDs and composites formed with the QDs were demonstrated along with their morphologies in this chapter. In addition, we tried to identify studies regarding numerous SC configurations that involve the use of QDs. This approach comes hand in hand with batteries which potentially resolve many energy-related crises and still be able to act as a substantial alternative to conventional fossil fuels.

REFERENCES

[1] H.T. Das, and S. Dutta, "Biodegradable electrode materials for sustainable supercapacitors as future energy storage devices", In: *Handbook of Biodegradable Materials.*, G.A.M. Ali, A.S.H. Makhlouf, Eds., Springer: Cham, 2022, pp. 1-25.
[http://dx.doi.org/10.1007/978-3-030-83783-9_41-1]

[2] H.T. Das, "Recent trends in carbon nanotube electrodes for flexible supercapacitors: A review of smart energy storage device assembly and performance", *Chemosensors,* vol. 10, no. 6, p. 223, 2022.
[http://dx.doi.org/10.3390/chemosensors10060223]

[3] H. T. Das, "Recent trend of CeO2-based nanocomposites electrode in supercapacitor: A review on energy storage applications", *J. Energy Storage,* vol. 50, p. 104643, 2022.
[http://dx.doi.org/10.1016/j.est.2022.104643]

[4] H.T. Das, T.E. Balaji, S. Dutta, N. Das, and T. Maiyalagan, "Recent advances in MXene as electrocatalysts for sustainable energy generation: A review on surface engineering and compositing of MXene", *Int. J. Energy Res.,* vol. 46, no. 7, pp. 8625-8656, 2022.
[http://dx.doi.org/10.1002/er.7847]

[5] F.A. Permatasari, M.A. Irham, S.Z. Bisri, and F. Iskandar, "Carbon-based quantum dots for supercapacitors: Recent advances and future challenges", *Nanomaterials,* vol. 11, no. 1, p. 91, 2021.
[http://dx.doi.org/10.3390/nano11010091]

[6] H. Abdelsalam, and Q. F. Zhang, "Properties and applications of quantum dots derived from two-dimensional materials", *Adv. Phys.: X,* vol. 7, no. 1, p. 2048966, 2022.
[http://dx.doi.org/10.1080/23746149.2022.2048966]

[7] T. Maiyalagan, "Developing potential aqueous Na-ion capacitors of Al2O3 with carbon composites as electrode material: Recycling medical waste to sustainable energy", *J. Alloys Compd.,* vol. 931, p. 167501, 2023.
[http://dx.doi.org/10.1016/j.jallcom.2022.167501]

[8] Q. Abbas, A. Mateen, A.J. Khan, G.E. Eldesoky, A. Idrees, A. Ahmad, E.T. Eldin, H.T. Das, M. Sajjad, and M.S. Javed, "Binder-free zinc–iron oxide as a high-performance negative electrode material for pseudocapacitors", *Nanomaterials,* vol. 12, no. 18, p. 3154, 2022.
[http://dx.doi.org/10.3390/nano12183154] [PMID: 36144942]

[9] M. Vandana, H. Vijeth, S. P. Ashokkumar, and H. Devendrappa, "Hydrothermal synthesis of quantum dots dispersed on conjugated polymer as an efficient electrodes for highly stable hybrid supercapacitors", *Inorg. Chem. Commun.,* vol. 117, p. 107941, 2020.
[http://dx.doi.org/10.1016/j.inoche.2020.107941]

[10] P. Luo, X. Guan, Y. Yu, X. Li, and F. Yan, "Hydrothermal synthesis of graphene quantum dots supported on three-dimensional graphene for supercapacitors", *Nanomaterials,* vol. 9, no. 2, p. 201, 2019.
[http://dx.doi.org/10.3390/nano9020201]

[11] M. Ashourdan, A. Semnani, F. Hasanpour, and S. E. Moosavifard, "Synthesis of CuMnO2/graphene

quantum dot nanocomposites as novel electrode materials for high performance supercapacitors", *J. Energy Storage,* vol. 36, p. 102449, 2021.
[http://dx.doi.org/10.1016/j.est.2021.102449]

[12]   D. Liu, T. Van Tam, and W.M. Choi, "Facile synthesis of g-C3N4 quantum dots/graphene hydrogel nanocomposites for high-performance supercapacitor", *RSC Adv.,* vol. 12, no. 6, pp. 3561-3568, 2022.
[http://dx.doi.org/10.1039/D1RA08962E]

[13]   P. Siwatch, K. Sharma, and S. K. Tripathi, "Facile synthesis of NiCo2O4 quantum dots for asymmetric supercapacitor", *Electrochimica. Acta.,* vol. 329, p. 135084, 2020.
[http://dx.doi.org/10.1016/j.electacta.2019.135084]

[14]   A. Dutta, K. Chatterjee, S. Mishra, S. K. Saha, and A. J. Akhtar, "An insight into the electrochemical performance of cobalt-doped ZnO quantum dot for supercapacitor applications", *J. Mater. Res.,* vol. 37, no. 22, pp. 3955-3964, 2022.
[http://dx.doi.org/10.1557/s43578-022-00654-7]

[15]   W. Chen, "One-pot scalable synthesis of pure phase Ni3S4 quantum dots as a versatile electrode for high performance hybrid supercapacitors and lithium ion batteries", *J. Power Sources,* vol. 438, p. 227004, 2019.
[http://dx.doi.org/10.1016/j.jpowsour.2019.227004]

[16]   A. Ghorai, A. Midya, and S. K. Ray, "Superior charge storage performance of WS2 quantum dots in a flexible solid state supercapacitor", *New J. Chem.,* vol. 42, no. 5, pp. 3609-3613, 2018.
[http://dx.doi.org/10.1039/C7NJ03869K]

[17]   V. Bonu, B. Gupta, S. Chandra, A. Das, S. Dhara, and A. K. Tyagi, "Electrochemical supercapacitor performance of SnO2 quantum dots", *Electrochimica Acta.,* vol. 203, pp. 230-237, 2016.
[http://dx.doi.org/10.1016/j.electacta.2016.03.153]

[18]   X. Jian, H.-m. Yang, J.-g. Li, E.-h. Zhang, L.-l. Cao, and Z.-h. Liang, "Flexible all-solid-state high-performance supercapacitor based on electrochemically synthesized carbon quantum dots/polypyrrole composite electrode", *Electrochimica Acta.,* vol. 228, pp. 483-493, 2017.
[http://dx.doi.org/10.1016/j.electacta.2017.01.082]

[19]   W. Yin, D. He, X. Bai, and W. W. Yu, "Synthesis of tungsten disulfide quantum dots for high-performance supercapacitor electrodes", *J. Alloys Compd.,* vol. 786, pp. 764-769, 2019.
[http://dx.doi.org/10.1016/j.jallcom.2019.02.030]

[20]   Y. Yuan, L. Jiang, X. Li, P. Zuo, X. Zhang, Y. Lian, Y. Ma, M. Liang, Y. Zhao, and L. Qu, "Ultrafast shaped laser induced synthesis of mxene quantum dots/graphene for transparent supercapacitors", *Adv. Mater.,* vol. 34, no. 12, p. 2110013, 2022.
[http://dx.doi.org/10.1002/adma.202110013] [PMID: 35072957]

[21]   J. Hei, "Uniformly confined V2O3 quantum dots embedded in biomass derived mesoporous carbon toward fast and stable energy storage", *Ceram. Int.,* p. 197, 2023.
[http://dx.doi.org/10.1016/j.ceramint.2023.01.197]

[22]   Y. Sim, "The synergistic effect of nitrogen and fluorine co-doping in graphene quantum dot catalysts for full water splitting and supercapacitor", *Appl. Surf. Sci.,* vol. 507, p. 145157, 2020.
[http://dx.doi.org/10.1016/j.apsusc.2019.145157]

[23]   P. R. Kharangarh, N. M. Ravindra, R. Rawal, A. Singh, and V. Gupta, "Graphene quantum dots decorated on spinel nickel cobaltite nanocomposites for boosting supercapacitor electrode material performance", *J. Alloys Compd.,* vol. 876, p. 159990, 2021.
[http://dx.doi.org/10.1016/j.jallcom.2021.159990]

[24]   W. Tian, "Micelle-induced assembly of graphene quantum dots into conductive porous carbon for high rate supercapacitor electrodes at high mass loadings", *Carbon,* vol. 161, pp. 89-96, 2020.
[http://dx.doi.org/10.1016/j.carbon.2020.01.044]

[25]   M. Zhang, W. Liu, R. Liang, R. Tjandra, and A. Yu, "Graphene quantum dot induced tunable growth

of nanostructured MnCo2O4.5 composites for high-performance supercapacitors", *Sustain. Energy Fuels,* vol. 3, no. 9, pp. 2499-2508, 2019.
[http://dx.doi.org/10.1039/C9SE00341J]

[26] H. Qiu, "Microwave synthesis of histidine-functionalized graphene quantum dots/Ni-Co LDH with flower ball structure for supercapacitor", *J. Coll. Interface Sci.,* vol. 567, pp. 264-273, 2020.
[http://dx.doi.org/10.1016/j.jcis.2020.02.018]

[27] H.T. Das, "Polymer composites with quantum dots as potential electrode materials for supercapacitors application: A review", *Polymers,* vol. 14, no. 5, p. 1053, 2022.
[http://dx.doi.org/10.3390/polym14051053]

[28] G. Wei, "Flexible asymmetric supercapacitors made of 3D porous hierarchical CuCo2O4@CQDs and Fe2O3@CQDs with enhanced performance", *Electrochimica Acta.,* vol. 283, pp. 248-259, 2018.
[http://dx.doi.org/10.1016/j.electacta.2018.06.153]

[29] Y. Huang, "Graphene quantum dots-induced morphological changes in CuCo2S4 nanocomposites for supercapacitor electrodes with enhanced performance", *Appl. Surf. Sci.,* vol. 463, pp. 498-503, 2019.
[http://dx.doi.org/10.1016/j.apsusc.2018.08.247]

[30] E. Duraisamy, H.T. Das, A. Selva Sharma, and P. Elumalai, "Supercapacitor and photocatalytic performances of hydrothermally-derived Co 3 O 4 /CoO@carbon nanocomposite", *New J. Chem.,* vol. 42, no. 8, pp. 6114-6124, 2018.
[http://dx.doi.org/10.1039/C7NJ04638C]

[31] S. Das, R. Ghosh, D. Mandal, and A. K. Nandi, "Self-assembled nanostructured MoS2 quantum dot polyaniline hybrid gels for high performance solid state flexible supercapacitors", *ACS Appl. Energy Mater.,* vol. 2, no. 9, pp. 6642-6654, 2019.
[http://dx.doi.org/10.1021/acsaem.9b01171]

[32] Z. Li, "Electrospun carbon nanofibers embedded with MOF-derived N-doped porous carbon and ZnO quantum dots for asymmetric flexible supercapacitors", *New J. Chem.,* vol. 45, no. 24, pp. 10672-10682, 2021.
[http://dx.doi.org/10.1039/D1NJ01369F]

CHAPTER 11

Metal-Organic Frameworks (MOFs) Based Nanomaterials for Supercapacitor Applications

Pardeep K. Jha[1,*], Priyanka A. Jha[1] and Prabhakar Singh[1]

[1] *Department of Physics, Indian Institute of Technology (Banaras Hindu University), Varanasi-221005, India*

Abstract: In the last two decades, nanomaterials with enhanced active sites and better surface kinetics as compared to their bulk counterpart, have been significantly studied for supercapacitor electrode materials. Contemporarily, Metal-organic frameworks (MOFs) by virtue of versatile structure, charge conduction, high porosity, and redox-active functionality have also emerged as the most potential materials for next-generation energy storage technologies. Despite these excellent features, the bulk phase inorganic-MOFs have some chemical and physical limitations that hinder cell performance and thus novel materials are required. Recently, MOFs-based nanomaterials(nMOF) got due attention leading to the discovery of a variety of properties not observed or relevant in bulk systems, such as well-defined 3D structures, permanent porosity, and accelerated adsorption/desorption kinetics. That's why nMOFs are considered an emerging class of modular nanomaterials. However, understanding of nMOFs is still in its infancy, film uniformity along with the unstable structure in a highly corrosive electrolyte is still a bottleneck problem. In this chapter, the recent developments of pristine MOF and MOF-derived porous nanocomposites for the next-generation supercapacitor applications will be discussed.

Keywords: Metal-organic frameworks (MOFs), Nanomaterials, Supercapacitors.

INTRODUCTION

Supercapacitors are an electrochemical device that has the best energy storage performance features of batteries and capacitors. It has higher power density, charge/discharge efficiency and life cycle than batteries, and higher energy density than capacitors. The first electrochemical cell was patented in 1957 (by General Electric's H.I. Becker) and commercially viable by 1975. Since then, supercapacitors have been serving society with their applications in the automotive industry, hybrid transportation systems, traction, and grid stabilization. Supercapacitors are combined with various other energy sources to

---

* **Corresponding author Pardeep K. Jha:** Department of Physics, Indian Institute of Technology (Banaras Hindu University), Varanasi-221005, India; E-mail: pardeepjha.jiit@gmail.com

Sanjeev Verma, Shivani Verma, Saurabh Kumar & Bhawna Verma (Eds.)

improve performance. Their combination with fuel cells helps in rapid charging. Its essentiality will soon be reflected at charging stations with the expansion of electrical vehicles in the global market. However, for next-generation supercapacitors, we still require better performance than recent supercapacitors in use at a lower cost.

The supercapacitors consist of an electrolyte in which two porous electrodes (along with its separator in between and current collectors on both ends) are immersed/placed. For better capacitive performance, porosity and surface area of electrodes play pivotal roles. On the basis of charge storage mechanism capacitive or Faradaic, electrode materials for supercapacitors can be broadly categorized into three groups: capacitive (electrochemical double layer capacitor, EDLC), battery-like materials, and pseudo-capacitive (pseudo-capacitor) [1, 2]. The carbon materials-based electrode (CE) is generally capacitive in nature and therefore used in EDLC. This is one of the basic features of CE, by which, we, in general, distinguish EDLC from PC [1 - 3]. The capacitive features of EDLCs are a function of electrical conductivity and surface area. The active electrode materials employed in EDLCs are porous carbons like carbon nanotubes, activated carbon, and cross-linked or holey graphene [4, 5]. However, the fostering of carbon materials synthesis technique is not only a way to improve EDLCs but requires out of the box thinking, *i.e.*, a conception of entirely different class materials with optimized surface area and electrical conductivity. Considering this, the nanomaterials (NMs) were used for SCs fabrication in the year ~2002 as shown in the timeline in Fig. (**1**). Contemporarily metal–organic frameworks [6 - 8] (MOFs, with surface areas exceeding 7,000 m^2 g^{-1} [7]) and another innovative class of porous crystalline materials have been developing since 1995 [9, 10]. MOFs with their versatile structure with open metal sites, high porosity, electrical conductivity, and redox-active functionality turned up as promising materials for energy storage technologies. That is why soon after NMs, MOFs-based SCs were also introduced in ~2006 (see timeline in Fig. (**1**)). MOFs are the structural combination of metals and organic linkers. This structural modularity permits not only creative design but also gives plenty of scope to engineer the porous structure, morphology, stability, and functionality for desired applications [11, 12]. The MOFs are directly employed as supercapacitor electrodes or as a template for fabricating different MOF-composites for supercapacitor electrodes. Some MOFs (pristine) are listed in Table **1**. In order to increase the functionality of these MOFs, understanding the structure-property relation in these MOFs plays a pivotal role. For instance, MOF-5, being one of the earliest MOFs, possess a ZnO_4- tetrahedral-type entity originating from Zn ions and 1,4-benzene-dicarboxylate (BDC). It is also termed IRMOF-1 (here IRMOFs means isoreticular MOFs) as its frameworks are formed with the same structural topology, by organic linkers with different lengths and functionalities [13]. For instance, pore size in MOF-5 is altered *via* using BDC and its derivative linkers and hence functionality.

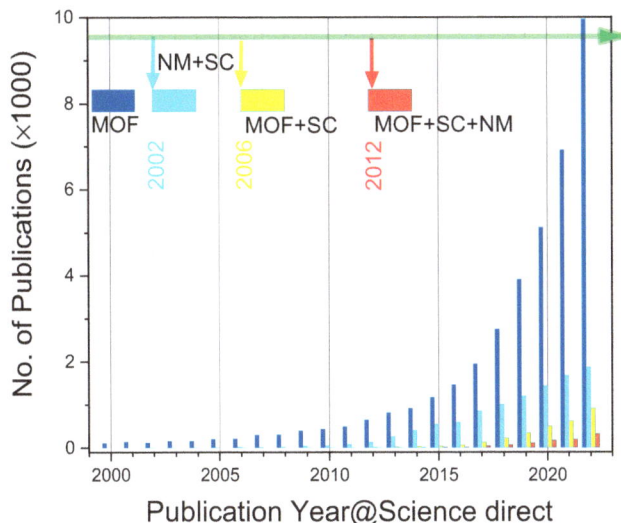

Fig. (1). Journey of MOF in supercapacitors along with nanomaterials.

In spite of excellent features, the bulk phase MOFs have some physiochemical limitations that hinder cell performance. Very soon in ~2012 as shown in Fig. (**1**) it was realized that MOF-based nanomaterials (nMOF) could enhance the storage capacity after the introduction of MOFs to superconductors and therefore, got due attention. Consequently, many novel observations were found which were absent in the bulk, and it was understood that the decreased particle size of nMOF is the prime requisite for the creation of large surface area electrodes along with small diffusion paths [11]. For instance, UiO-66-based supercapacitors exhibited significantly higher charge storage for smaller particles [12]. That is why nMOFs are considered an emerging class of modular nanomaterials.

In this chapter, we will first cover some basics of nanomaterials synthesis for MOFs. It will be followed by morphological classification of MOFs: dimensional morphology (0-3D nanomaterials) and compositional morphology (pristine MOF, carbon and non-carbon based nanocomposites). In the last, we will discuss and summarize the recent developments for the next-generation supercapacitor applications.

SYNTHESIS OF METAL-ORGANIC FRAMEWORKS BASED NANOMATERIALS

The synthesis of nanomaterials plays a pivotal role in the properties of the system. The size of the nanoparticle and monodispersity are the two major prerequisites to achieving controlled features in the materials. The crystallisation process, in general, is three-stage process – induction, nucleation, and crystal growth [14]. The kinetics of phase formation is not very different from Avrami model [15 - 17]. The temperature–time relation plays a critical role in the synthesis of the particles. As per Avrami eqn., the fraction of crystalline phase formation is defined as:

$$f(t,T) = 1 - e^{-k_0\left(\frac{-E}{kT}\right).t^n} \tag{1}$$

Here, t, T, k_0, E, and n are time, temperature, pre-factor constant, activation energy, and Avrami exponent, respectively. The exponent n and activation energy vary from system to system; therefore, crystallization behaviour is quite different for two systems *e.g.*, MOF-5 and HKUST-1. The studies [18, 19] find that the nucleation process of HKUST-1 is relatively slow but continuous; whereas growth process is fast and induction period is absent too. In most cases, for MOF formation, the induction period is either almost instantaneous or much smaller as compared to the nucleation and growth process. That is why MOF crystallization is explained as per the LaMer model, which separates crystal nucleation and growth. Also, it discloses that the high content of precursor can activate thermodynamically derived forces [20, 21]. Therefore, understanding the nucleation process is very crucial for morphology control. However, nucleation in case of MOF formation is a complex process due to the formation of metastable phases and their agglomeration [20, 22], which is beyond our present understanding. Along with nucleation, crystal growth is also an important process that depends on the crystal symmetry of nuclei, crystal surface structure growth, and supersaturation of crystal habit. Apart from this, irrespective of the synthesis technique, some common parameters that affect the crystal growth are the nature of solvent and metal precursors, the concentration and the ratio of the reactants, stirring condition, and temperature.

Synthesis Techniques

The crystal growth kinetics is affected by the diffusion of nuclei and jumping frequency at the interface (say liquid-solid interface)- thus two mechanisms – diffusion-controlled growth and interface-controlled growth are used to explain the formation of monodispersed nanoparticles. Though for the formation of nanoparticles < 10 nm, synthesis methods based on the diffusion control growth have an edge over the interface control growth, which is in accordance with LaMer's model. Therefore, synthesis methods for nMOFs with good monodispersity are- solvothermal and microemulsion [5]. For solvothermal synthesis, the high-throughput (HT) method is a very important tool for optimization. Other alternative synthesis routes include "sonochemical synthesis, electrochemical synthesis, microwave-assisted synthesis and mechanochemical synthesis" [23].

Recently some new strategies such as fast precipitation induced by initiation solvent, morphological and textural control by using surfactants, morphology, and porosity controlled by hard template have been adopted and termed as modulation methods. In the synthesis for nMOF, the modulators and stabilizing agents control the morphology and functionality and therefore, the desired shape of nanocrystals. Fig. (**2**) shows the role of modulators and stabilizers on the morphology control of nanocrystals. Depending on control over the nucleation rate and nucleation process, there are two types of modulators-coordination and deprotonation modulators [24].

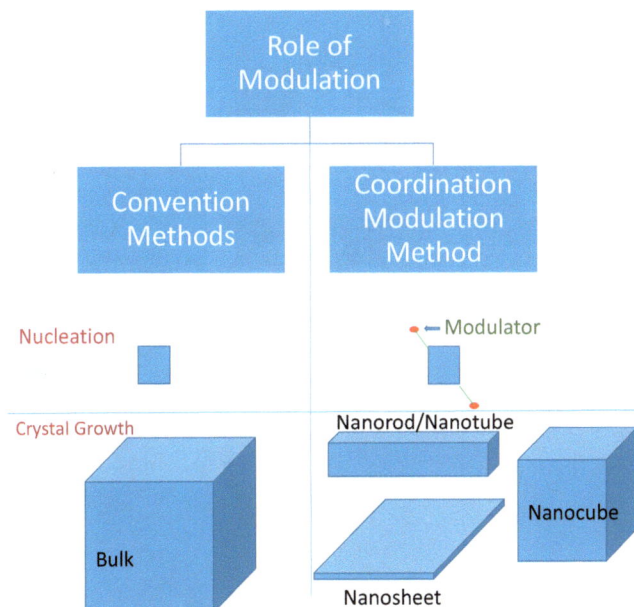

Fig. (2). Role of Modulation for nMOF fabrication.

CLASSIFICATION

The main criterion to classify any subject generally depends upon the structure, morphology, and physical behavior. MOFs are not separated from this fact. For MOFs, the eruption development is observed in the last two decades, which raised the number of MOFs structures to huge ~10^6 (as per CCDC) which is difficult to classify or to find the proper basis for classification [25]. Still, at present, morphology is only a reasonable and arguable basis for the classification of MOFs. On the basis of morphology, MOFs can be broadly classified into two categories (i) dimensional morphology and ii) compositional morphology.

Dimensional Morphology

The parameters required for the designing of high-performance supercapacitors-electrodes are higher specific capacitance along with long capacitance retention and primarily cyclic stability. Further, the above-mentioned parameters are further related to surface area, surface-to-volume ratio (S/V), conductivity, and chemical stability [26]. The novel nanostructures can provide high S/V and therefore increase the accessible active sites along with structural robustness, and mass-charge carrier transfer [27]. The nanostructures on the basis of spatial dimensionality can be classified into four categories 0D, 1D, 2D, and 3D as shown in Fig. (**3**).

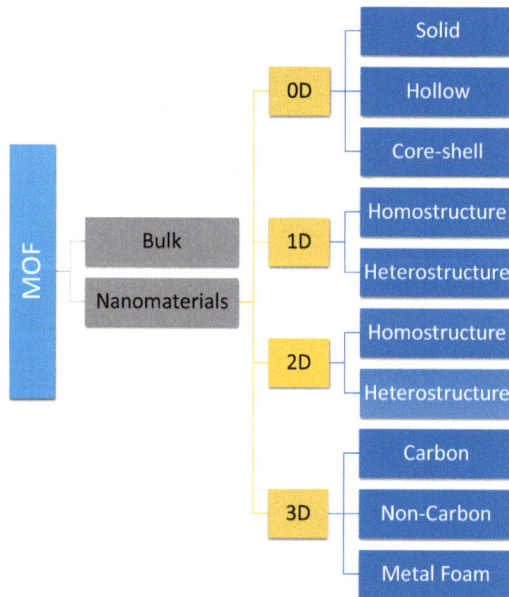

Fig. (3). Morphological classification of MOF on the basis of dimensionality.

0D

The three-dimensional nanostructures with dimension < 100 nm (it is ideal, in practical < 1 μm [26]) are considered as 0D nanostructures. The 0D nanostructures can be classified into three categories: "solid, hollow, and core-shell nanostructures" [27].

Solid 0D nanostructures are the nanospheres of carbon and transition metal compounds that have been intensified and studied as a material for and supercapacitor electrodes. Depending on the pore size distribution, it is also termed macroporous (> 50 nm), mesoporous systems (2-50 nm), and microporous systems (< 2nm). Further, in microporous systems, even for moderate specific areas, better performance can be achieved due to increased accessible surface for the ions.

Hollow 0D nanostructures are promising candidates for electrode design as it has low density, high S/V, and reduced path for mass transportation. On the basis of the template, the hollow 0D nanostructures synthesis method is of three kinds: hard, soft, and template -free methods. Thanks to its ability to control the morphology of the product, hard template methods are commonly used. For example, hollow carbon nano-spheres (HCNS) synthesis is performed by carbonizing its hollow nanospheres precursor, which is extracted from silica sphere-assisted hard template method [26].

Core-shell 0D nanostructures is in general defined as a nanoparticle (hollow or solid) coated with a thin shell. The core-shell nanostructure can combine faradaic and non-faradaic materials and thus possess considerable advantages such as robust mechano-chemical stability, higher conductivity, and less agglomeration [26, 27]. For example, optimized HCS–PAni core–shell structures specific capacitance of 525 F g^{-1}, as compared to 268 F g^{-1} for pristine HCS [28]. Core-shell is further extended to multi-shell 0D nanostructures.

1D

1D homostructures are nanostructures with only a singular structure [26]. The parameter termed as aspect ratio (*i.e.*, length/diameter), classifies 1D nanostructures into three morphological terms: nanorods/nanopillars with AR < 10, nanowires if AR > 10, and nanotubes if their interior is hollow. The 1D homostructures synthesis methods are divided on the basis usage of templates either template-assisted and template-free. To get initial microstructure/ morphology, the sacrificial templates are employed in template-assisted synthesis methods and later templates are removed with an acidic or alkaline/basic solution. While in template-free methods, other synthetic techniques such as hydrothermal synthesis, electrospinning, chemical vapour deposition, *etc.*, are used [26, 29].

1D Heterostructures is a multi-component structure, which gives a synergetic effect on the MOF's performance parameters. 1D Heterostructures further can be classified as –1D-1D composite core-shell (*e.g.*, CNT@PPy–MnO_2) and 0D-1D composites (*e.g.*, MnO_2/CNT arrays). 1D-1D composite can further be divided into the core and shell materials as current collector core, active core—active shell, and pseudocapacitive core.

2D

2D structures, due to surficial (thin film) structure always remain the subject of interest and therefore unique among nanostructures. They exhibit unique mechanical and chemical stability, flexibility, and transparency. Anisotropicity [30] is another unique feature that can increase the functionality of electrode materials.

Like 1D nanostructures, 2D nanostructures are also categorized as homo- and hetero-nanostructures. On the basis of application, 2D homo-structures are broadly classified into: graphene-based (EDLC) electrodes, metal oxide/ hydroxide (Pseudo capacitors) electrodes and recently developed transition metal chalcogenides, carbides, and nitrides-based electrodes.

2D Heterostructures are basically a selective combination of EDLC and pseudocapacitance materials to get robust electrode that possesses high power density along with high energy. Graphene metal oxide/hydroxides and graphene-dichalcogenides are the combinations used for the purpose. Along with this, some recent hybrid 2D hetero-structures are specially designed for the advanced supercapacitors *e.g.*, micro-supercapacitors.

3D

The major limitations of low dimensional materials are aggregation issues and restacking issues, which might result in low permeability of the electrolyte along with the accessibility of interior active sites. Therefore, for practical electrochemical storage (ECS) applications, it is more desirable to synthesize the nMOF with hierarchically porous 3D skeletons. The proper assembling of the low dimension (*i.e.*, 0D, 1D, and 2D)- MOFs dependent building blocks over multiple length scales leads to hierarchically porous 3D nMOFs. The resultant, hierarchically porous 3D nMOFs are propitious for the next generation of ECS devices [27]. Further, in 3D nanostructures, the pore size always remains critical and can be varied by altering the functional organic group *e.g.*, octahedral Zn-O-C clusters with benzene links from MOF-5. It is shown that various organic groups such as $-Br$, $-NH_2$, $-OC_3H_7$, $-OC_5H_{11}$, $-C_2H_4$, and $-C_4H_4$ can functionalize MOF-5. Further with the incorporation of long molecular struts biphenyl, tetrahydropyrene, pyrene, and terphenyl, the pore size of MOF-5 can be expanded [13]. The studies suggest the efficiency of porous carbon as supercapacitor electrodes would increase the coexistence of macropores and mesopores [26] and confirmed [30] in the case of ordered carbon structure with meso/macro/ micropores with a high specific capacitance of ~ 350 F g^{-1}.

The 3D nanostructures can be further classified on the basis of composition into three categories- carbon-based EDLC materials, non-carbon (metal oxides, chalcogenides, phosphate) based pseudocapacitive materials, and classical current collector-based metal foam electrode materials. Generally, it can form hierarchical structures, superstructures, and arrays. The hybrid nanostructures can play a critical role in enhancing the storage capability of MOFs *e.g.*, significantly exposed active sites on the 2D surface, improved conductivity and permeability with a cell voltage of 1.55 V at a current density of 10 $mAcm^{-2}$ is obtained for an ultrathin NiFe-MOF nanosheet array, vertically grown on 3D Ni foam [31].

From the above discussion, it is clear that every dimensional morphology has its own merit and demerits, *viz.*, tuning of the pore size < 10 nm is easier in the case of 0D carbon materials but in the presence of an insulating polymer binder, these materials show low conductivity. This issue can be resolved by using active materials decorated 1D materials, But, here the limit is the precise control of accessible space between high AR structures. A feasible future direction is the hybrid heterostructure which can provide high conductivity, high specific capacitance due to surface area, and mechano-chemical stability. It is believed that such hybrid-hetero nanostructures can be a new generation of supercapacitor electrodes with high energy along with high power densities in the near future.

Compositional Classification

The basic limitation of MOFs for direct use in supercapacitors, as mentioned above is low conductivity and which can be enhanced in two ways: (i) designing MOFs itself by modifying either organic ligands or metal centre or both; (ii) construction of MOFs/X composites *via* the introduction of conductive functional components into pristine MOFs, where X represent the conducting polymers, metal, metal oxide/hydroxide/sulfide and other porous carbon materials (graphene, CNT, *etc.*) [32].

By definition, a composite is a multiple-component material with various phases possessing at least one continuous phase [33]. No clear definition regarding the classification on the basis of composition is provided in the literature. Thus we define pristine MOF as the MOFs synthesized by the first approach (*e.g.*, MOF-5, MOF 74, UIO-66, ZIF- 8, ZIF-67 *etc.*,) and MOF composites to those MOFs synthesized by the second approach (*e.g.*, ZIF@PANI).

Pristine-MOF

Pristine MOF, as the name suggests it is the MOF prepared without any additive materials for better conduction. Many transition metal ions like Fe, Co, Ni, Mn, Cu, Zn, Zr, along with Al, Ce and organic linkers such as imidazoles, and carboxylic acids, are applied in the synthesis of pure MOFs [34]. However, due to its low conductivity and structural degradation in conventional aqueous electrolytes, pristine MOFs are rarely used as electrode materials for supercapacitor applications [35]. The pores, structures, and morphology of MOFs (and their derivatives) severely affect the ECS performance. Therefore, the conductivity of MOFs could be improved by the construction of 3D network structures and conjugate systems [36]. The first MOF, which is directly used as an active material in supercapacitors is, perhaps $Ni_3(HITP)_2$ [37] and delivered an areal capacitance compatible to the majority of carbon-based materials [8, 38]. Otherwise, it is MOF composites which are used as electrode materials for superconductors. A list of some pristine MOFs is provided in Table **1**.

MOF Composites

Fig. (**4**) shows the classification of MOFs as pristine and composites. In an effort to enhance conductivity and chemical stability, pristine MOFs are synthesized with another suitable material, and the product is termed as MOF composite. Depending upon the second constituent material, the MOF-composites can be divided into three categories: conducting polymer (CP-MOF), carbon (C- MOF), and non-carbon (NC-MOF). C-MOF and NC MOF are basically MOF composites but in the literature, they are also classified independently as MOF-derivatives. In MOF composites, MOFs can play a dual role -as substrate (template) or as active materials.

MOF-CP

The conducting polymers like polyaniline (PAni) [39], polypyrrole (PPy) [40] [41], polythiophene (PTh) [42], and PEDOT [43] can provide non-covalent interaction (between the host and the guest polymer) which results in the formation of the conductive pathways [44]. Such conductive pathways can significantly increase the MOF's conductivity [45].

MOF- Derivatives

C-MOF

In EDLCs, the microporous structure allows ions centre to proceed towards the surface of carbon electrodes leading to enhanced specific capacitance. While mesoporous structures provide a passage for the ions into the narrow pores of the double-layer capacitor carbons without affecting the solvent shells. Owing to its large surface area, high conductivity, and porous structure, the MOFs-derived porous carbon nanotube (CNT) could be used as the electrode of solid-state supercapacitors. That is why these electrodes have not only been intensively employed in the EDLCs but also gained popularity for their employability in hybrid as well as asymmetric supercapacitors.

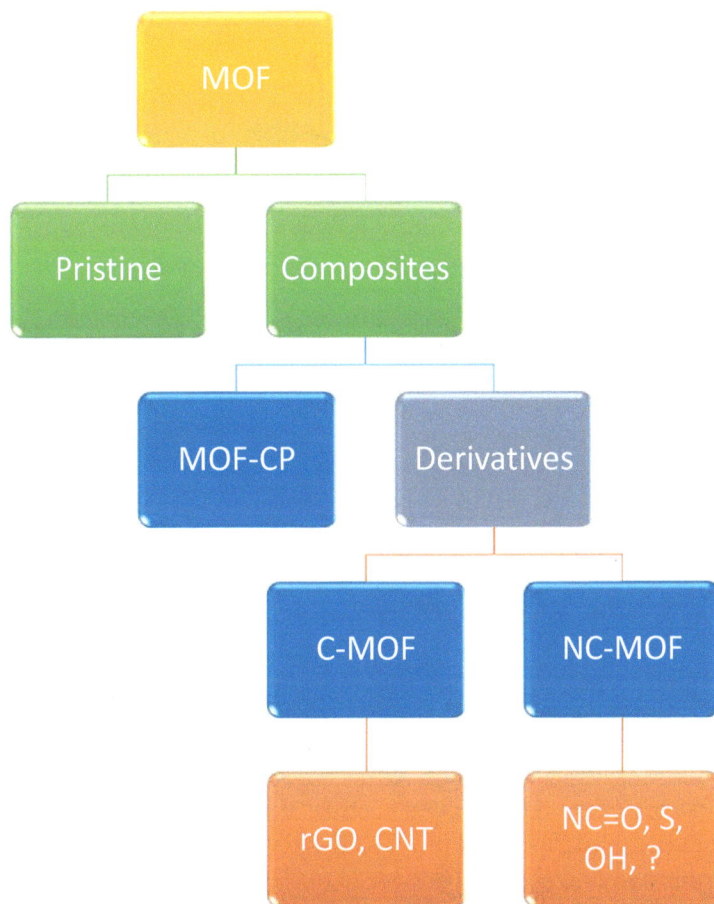

Fig. (4). Morphological classification of MOF on the basis of composition.

The electrical conductivity can be increased if carbon (in the form of graphene or reduced graphene oxide (rGO)) is introduced into the MOFs-derived metal oxide electrodes. In addition to it, MOFs-derived carbon nanowall arrays (CNWAs) can be used as substrates for electrodepositing metal compounds for the composite electrode's preparation [46].

NC-MOF

The MOF-derived metal compounds work as pseudocapacitors. Therefore, these composites possess better electron capacity and energy density as compared to porous carbon materials. Many metal compound-based MOFs possess almost all the optimized features- high electrical conductivity, better porosity, huge surface area and robust thermo-chemical stability for ECS devices.

Metal Oxides Composites

Owing to promising pseudo-capacitive properties, metal oxide composites are the most studied NC-MOF. However, the low specific surface area of metal oxides is still a major limitation for ion diffusion through electrodes [34]. Recently along with binary oxides, the use of ternary oxides was also observed. It is observed that porous Co_3O_4 fabricated from ZIF-67 is highly stable with high specific capacitance [47]. In pseudocapacitors, nanostructured and nanosized transition metal oxides (RuO_2 and IrO_2) [48 - 51] have shown high specific capacitance at high current density (for example $NiCo_2O_4$ nanosheets with 1450 Fg^{-1} at a high current density of 20 Ag^{-1}) [52].

Metal Hydroxides Composites

Apart from oxides, hydroxide of metals also gains research interest *e.g.*, the MOFs-derived $Co(OH)_2$ with unique porous structure provides the swift ion transportation, rich redox reaction, and robust structure stability of supercapacitor electrodes. This MOF-hydroxide has shown a high specific capacitance of 604.5 F g^{-1} at 0.1 A g^{-1} along with excellent cycling stability [53]. However, the bottleneck problem is severe agglomeration and relatively low conductivity which affects the output of energy and power in supercapacitors. Similarly, the MOF-derived $Ni(OH)_2$ electrode (used in asymmetric supercapacitor), with an energy density of 36.2 W h kg^{-1} at a power density of 436.1 W kg^{-1}, as well as a capacitance of 98.0 F g^{-1} at a current density of 0.5 A g^{-1} can be considered as a promising candidate for future electrode materials. But, MOF-derived NiO and NiS electrodes show better performance [54].

Metal Sulfides Composites

In metal sulfides, due to larger anion size of di-valent sulphur ion S^{2-} (as compare to O^{2-}) [55], the better access of inner void sites of porous materials in the liquid electrolyte, is available. Further, the lower electronegativity of sulphur, nickel and sulfide shows better structural stability and faster electron transfer as compared to nickel oxides [56]. Therefore, metal sulfide-composites can prove a promising alternative to oxides.

Other NC MOF Derivatives

Beyond the oxide, hydroxide, and sulphide, there are a few metal NC MOFs derivatives that have been studied. Recently, phosphide, nitrides, and carbides-based MOFs have been studied for various applications but it is too early to extract much information [56, 57]. For instance, in case of metal phosphide, though high theoretical capacitance is reported for CoP but poor cyclic stability and limited rate capability are major limitations. However, it is suggested that transitional metal substitution can work for CoP [58].

Hybrid MOF Derivatives

In the continuation effort to increase the electrical conductance of MOF, MOF-composites are further extended by the use of multi constituents such as conducting polymers and carbon nanotubes. With improved overall features, composites can be used in many diverse applications *viz.*, energy storage, catalysis and sensors. Such composites due to the integration of large pseudo-capacitance of the polymers and the mechanical and structural properties of the nanotubes are very promising candidates for novel supercapacitor electrodes with ultra-high capacitance and power density [59]. These kinds of composites are termed hybrid MOF-composites. For example, high-density lignin-derived carbon nanofiber supercapacitors show enhanced volumetric energy density (3 Wh L^{-1} at 10 kW L^{-1}) [60].

Porous carbon materials based MOF showed excellent supercapacitance due to unique morphology and porous structure. Their large surface area leads to adequate contact at the electrode/electrolyte interface. The migration and diffusion of electrolyte ions depend on pore morphology. Furthermore, nitrogen-based carbon MOFs have improved surface properties and pseudo capacitance. This also enhances the wettability of microspores in carbon electrodes. Furthermore, activated carbon, and carbon nanofibers modified MOFs have shown low density, excellent conductivity, high strength, electrochemical stability, and fast charging/discharging process [61].

MOF-derived metal hydroxide/oxide/sulfides' porous and hollow structures have enhanced reaction kinetics due to shortened pathways. The enhanced surface area and optimized micro/nanostructure with optimized pore size increase the stability of capacitance at a high current density. With the coupling of carbon with MOFs, the pseudo-capacitance and cycling performance can be enhanced. Moreover, with the tuning of precursors, conductivity and redox capability can be increased. The nanostructured MOF with the conductive substrate provides more efficient materials for pseudo-capacitance.

CHALLENGES

There are several challenges that are creating hindrances in the commercialization of MOFs. (a) The unstable structure along with the uniformity of films in the highly corrosive electrolytes, (b) opportunities for coordination modulation, (c) high cost of catalysts and short lifetime of electro-catalysts, (d) severe shuttle effect and sluggish oxygen evolution reaction, (e) excessive internal resistance of MOFs, (f) microstructural effect on capacitance behaviour, (g) integrated theory and experimental characterization, and (h) lack of optimization of synthesis methods and strategies.

In order to overcome the above-mentioned shortcomings, the strategies to be followed are; (a) theoretical studies on microstructure effects and chemical pressure on the energy storage of supercapacitors, (b) synergistic effects of MOFs with other energy storage materials for novel supercapacitors, (c) novel composite materials with MOF derivatives as electrodes, and (d) structure-property correlation for electrode characteristics.

Other than these, some factors that need to be understood are durability, slow synthesis process, cost and reliability of electrode materials. The novel synthesis routes are being undertaken to scale the development of MOF materials

CONCLUDING REMARKS

MOFs due to tunable structure, designer composition, controllable porosity, bifunctionality, and high S/V ratio can play a pivotal role for supercapacitor electrode materials. The formation of films and composites with large aspect ratios increases the performance and energy storage capacity. However, the composites have limitations regarding the insulating behavior of MOF and chemical instability. Efforts are being made to design conductive MOFs with significantly high porosity and chemical stability. From above, it is clear that every dimensional morphology has its own merit and demerits, *viz.*, tuning of the pore size < 10 nm is easier in case of 0D carbon materials but with low conductivity in the presence of an insulating polymer binder, which can be improved by using active decorated 1D materials. But, the limit is the precise control on accessible space between high AR-structures. A feasible future direction is a hybrid heterostructure based on 2D MOF materials that have promoted cycling ability and conductivity. However, their electrode mechanism is still unclear. Furthermore, the nanocomposites with 0D, 1D, 2D, and 3D materials along with heterostructures using conductive substrates lead to an increase in active sites and in turn redox activity. Thus, better designing and fabrication techniques for such hybrid-hetero nanostructures will lead to a new generation of supercapacitor electrodes with high energy and power densities in the near future.

ACKNOWLEDGEMENT

PS and PAJ are thankful to DRDO-NRB/4003/MAT/PG/491 and CSIR-SRA (13(9142-A)/ 2020-Pool), respectively.

REFERENCES

[1]   Y. Gogotsi, and R.M. Penner, "Energy storage in nanomaterials – capacitive, pseudocapacitive, or battery-like?", *ACS Nano,* vol. 12, no. 3, pp. 2081-2083, 2018.
        [http://dx.doi.org/10.1021/acsnano.8b01914] [PMID: 29580061]

[2]   Y. Jiang, and J. Liu, "Definitions of pseudocapacitive materials: A brief review", *Energy Environ. Mater.,* vol. 2, no. 1, pp. 30-37, 2019.
        [http://dx.doi.org/10.1002/eem2.12028]

[3]   S. Roldán, D. Barreda, M. Granda, R. Menéndez, R. Santamaría, and C. Blanco, "An approach to classification and capacitance expressions in electrochemical capacitors technology", *Phys. Chem. Chem. Phys.,* vol. 17, no. 2, pp. 1084-1092, 2015.
        [http://dx.doi.org/10.1039/C4CP05124F] [PMID: 25412831]

[4]   Y. Xu, Z. Lin, X. Zhong, X. Huang, N.O. Weiss, Y. Huang, and X. Duan, "Holey graphene frameworks for highly efficient capacitive energy storage", *Nat. Commun.,* vol. 5, no. 1, p. 4554, 2014.
        [http://dx.doi.org/10.1038/ncomms5554] [PMID: 25105994]

[5]   X. Cai, Z. Xie, D. Li, M. Kassymova, S.Q. Zang, and H.L. Jiang, "Nano-sized metal-organic frameworks: Synthesis and applications", *Coord. Chem. Rev.,* vol. 417, p. 213366, 2020.
        [http://dx.doi.org/10.1016/j.ccr.2020.213366]

[6]   H. Furukawa, K.E. Cordova, M. O'Keeffe, and O.M. Yaghi, "The chemistry and applications of metal-organic frameworks", *Science,* vol. 341, p. 6149, 2013.
        [http://dx.doi.org/10.1126/science.1230444]

[7]   O.K. Farha, I. Eryazici, N.C. Jeong, B.G. Hauser, C.E. Wilmer, A.A. Sarjeant, R.Q. Snurr, S.T. Nguyen, A.Ö. Yazaydın, and J.T. Hupp, "Metal-organic framework materials with ultrahigh surface areas: Is the sky the limit?", *J. Am. Chem. Soc.,* vol. 134, no. 36, pp. 15016-15021, 2012.
        [http://dx.doi.org/10.1021/ja3055639] [PMID: 22906112]

[8]   D. Sheberla, J.C. Bachman, J.S. Elias, C.J. Sun, Y. Shao-Horn, and M. Dincă, "Conductive MOF electrodes for stable supercapacitors with high areal capacitance", *Nat. Mater.,* vol. 16, no. 2, pp. 220-224, 2017.
        [http://dx.doi.org/10.1038/nmat4766] [PMID: 27723738]

[9]   O.M. Yaghi, and H. Li, "Hydrothermal synthesis of a metal-organic framework containing large rectangular channels", *J. Am. Chem. Soc.,* vol. 117, no. 41, pp. 10401-10402, 1995.
        [http://dx.doi.org/10.1021/ja00146a033]

[10]   O.M. Yaghi, G. Li, and H. Li, "Selective binding and removal of guests in a microporous metal-organic framework", *Nature,* vol. 378, no. 6558, pp. 703-706, 1995.
        [http://dx.doi.org/10.1038/378703a0]

[11]   K.M. Choi, H.M. Jeong, J.H. Park, Y.B. Zhang, J.K. Kang, and O.M. Yaghi, "Supercapacitors of nanocrystalline metal-organic frameworks", *ACS Nano,* vol. 8, no. 7, pp. 7451-7457, 2014.
        [http://dx.doi.org/10.1021/nn5027092] [PMID: 24999543]

[12]   A.E. Baumann, D.A. Burns, B. Liu, and V.S. Thoi, "Metal-organic framework functionalization and design strategies for advanced electrochemical energy storage devices", *Commun. Chem.,* vol. 2, no. 1, p. 86, 2019.
        [http://dx.doi.org/10.1038/s42004-019-0184-6]

[13]   M. Eddaoudi, "Systematic design of pore size and functionality in isoreticular MOFs and their application in methane storage", *Science,* vol. 295, p. 5554, 2002.

[http://dx.doi.org/10.1126/science.1067208]

[14] T.T.T. Nguyen, L.H.T. Nguyen, N.X.D. Mai, H.K.T. Ta, T.L.T. Nguyen, U-C.N. Le, B.T. Phan, N.N. Doan, and T.L.H. Doan, "Mild and large-scale synthesis of nanoscale metal-organic framework used as a potential adenine-based drug nanocarrier", *J. Drug Deliv. Sci. Technol.,* vol. 61, p. 102135, 2021. [http://dx.doi.org/10.1016/j.jddst.2020.102135]

[15] M. Avrami, "Kinetics of Phase Change. I General Theory", *J. Chem. Phys.,* vol. 7, no. 12, pp. 1103-1112, 1939. [http://dx.doi.org/10.1063/1.1750380]

[16] M. Avrami, "Kinetics of phase change. II transformation-time relations for random distribution of nuclei", *J. Chem. Phys.,* vol. 8, no. 2, pp. 212-224, 1940. [http://dx.doi.org/10.1063/1.1750631]

[17] M. Avrami, "Granulation, phase change, and microstructure kinetics of phase change. III", *J. Chem. Phys.,* vol. 9, no. 2, pp. 177-184, 1941. [http://dx.doi.org/10.1063/1.1750872]

[18] D. Zacher, J. Liu, K. Huber, and R.A. Fischer, "Nanocrystals of [Cu3(btc)2] (HKUST-1): A combined time-resolved light scattering and scanning electron microscopy study", *Chem. Commun.,* no. 9, pp. 1031-1033, 2009. [http://dx.doi.org/10.1039/b819580c] [PMID: 19225626]

[19] F. Millange, M.I. Medina, N. Guillou, G. Férey, K.M. Golden, and R.I. Walton, "Time-resolved in situ diffraction study of the solvothermal crystallization of some prototypical metal-organic frameworks", *Angew. Chem. Int. Ed.,* vol. 49, no. 4, pp. 763-766, 2010. [http://dx.doi.org/10.1002/anie.200905627] [PMID: 20017176]

[20] V.K. LaMer, and R.H. Dinegar, "Theory, production and mechanism of formation of monodispersed hydrosols", *J. Am. Chem. Soc.,* vol. 72, no. 11, pp. 4847-4854, 1950. [http://dx.doi.org/10.1021/ja01167a001]

[21] C.B. Whitehead, S. Özkar, and R.G. Finke, "LaMer's 1950 model for particle formation of instantaneous nucleation and diffusion-controlled growth: A historical look at the model's origins, assumptions, equations, and underlying sulfur sol formation kinetics data", *Chem. Mater.,* vol. 31, no. 18, pp. 7116-7132, 2019. [http://dx.doi.org/10.1021/acs.chemmater.9b01273]

[22] M.G. Goesten, F. Kapteijn, and J. Gascon, "Fascinating chemistry or frustrating unpredictability: Observations in crystal engineering of metal–organic frameworks", *CrystEngComm,* vol. 15, no. 45, pp. 9249-9257, 2013. [http://dx.doi.org/10.1039/c3ce41241e]

[23] N. Stock, and S. Biswas, "Synthesis of metal-organic frameworks (MOFs): Routes to various MOF topologies, morphologies, and composites", *Chem. Rev.,* vol. 112, no. 2, pp. 933-969, 2012. [http://dx.doi.org/10.1021/cr200304e] [PMID: 22098087]

[24] D. Jiang, C. Huang, J. Zhu, P. Wang, Z. Liu, and D. Fang, "Classification and role of modulators on crystal engineering of metal organic frameworks (MOFs)", *Coord. Chem. Rev.,* vol. 444, p. 214064, 2021. [http://dx.doi.org/10.1016/j.ccr.2021.214064]

[25] P.Z. Moghadam, A. Li, X.W. Liu, R. Bueno-Perez, S.D. Wang, S.B. Wiggin, P.A. Wood, and D. Fairen-Jimenez, "Targeted classification of metal–organic frameworks in the Cambridge structural database (CSD)", *Chem. Sci.,* vol. 11, no. 32, pp. 8373-8387, 2020. [http://dx.doi.org/10.1039/D0SC01297A] [PMID: 33384860]

[26] Z. Yu, L. Tetard, L. Zhai, and J. Thomas, "Supercapacitor electrode materials: nanostructures from 0 to 3 dimensions", *Energy Environ. Sci.,* vol. 8, no. 3, pp. 702-730, 2015. [http://dx.doi.org/10.1039/C4EE03229B]

[27] T. Qiu, Z. Liang, W. Guo, H. Tabassum, S. Gao, and R. Zou, "Metal-organic framework-based materials for energy conversion and storage", *ACS Energy Lett.,* vol. 5, no. 2, pp. 520-532, 2020.
[http://dx.doi.org/10.1021/acsenergylett.9b02625]

[28] Z. Lei, Z. Chen, and X.S. Zhao, "Growth of polyaniline on hollow carbon spheres for enhancing electrocapacitance", *J. Phys. Chem. C,* vol. 114, no. 46, pp. 19867-19874, 2010.
[http://dx.doi.org/10.1021/jp1084026]

[29] A. Saad, S. Biswas, E. Gkaniatsou, C. Sicard, E. Dumas, N. Menguy, and N. Steunou, "Metal–organic framework based 1d nanostructures and their superstructures: synthesis, microstructure, and properties", *Chem. Mater.,* vol. 33, no. 15, pp. 5825-5849, 2021.
[http://dx.doi.org/10.1021/acs.chemmater.1c01034]

[30] M. Azami-Ghadkolai, M. Yousefi, S. Allu, S. Creager, and R. Bordia, "Effect of isotropic and anisotropic porous microstructure on electrochemical performance of Li ion battery cathodes: An experimental and computational study", *J. Power Sources,* vol. 474, p. 228490, 2020.
[http://dx.doi.org/10.1016/j.jpowsour.2020.228490]

[31] J. Duan, S. Chen, and C. Zhao, "Ultrathin metal-organic framework array for efficient electrocatalytic water splitting", *Nat. Commun.,* vol. 8, no. 1, p. 15341, 2017.
[http://dx.doi.org/10.1038/ncomms15341] [PMID: 28580963]

[32] S. Huang, X.R. Shi, C. Sun, Z. Duan, P. Ma, and S. Xu, "The application of metal–organic frameworks and their derivatives for supercapacitors", *Nanomaterials (Basel),* vol. 10, no. 11, p. 2268, 2020.
[http://dx.doi.org/10.3390/nano10112268] [PMID: 33207732]

[33] W.J. Work, K. Horie, M. Hess, and R.F.T. Stepto, *Secretary from 1998); K. Hatada (Japan, to 1997, Associate Member to 1999),* 1985.

[34] X. Liu, and J. Song, "Metal-organic framework materials for supercapacitors", *J. Phys. Conf. Ser.,* vol. 2021, no. 1, p. 012008, 2021.
[http://dx.doi.org/10.1088/1742-6596/2021/1/012008]

[35] Z. Liang, C. Qu, W. Guo, R. Zou, and Q. Xu, "Pristine metal-organic frameworks and their composites for energy storage and conversion", *Adv. Mater.,* vol. 30, no. 37, p. 201702891, 2018.
[http://dx.doi.org/10.1002/adma.201702891]

[36] H. Gao, H. Shen, H. Wu, H. Jing, Y. Sun, B. Liu, Z. Chen, J. Song, L. Lu, Z. Wu, and Q. Hao, "Review of pristine metal-organic frameworks for supercapacitors: Recent progress and perspectives", *Energy Fuels,* vol. 35, no. 16, pp. 12884-12901, 2021.
[http://dx.doi.org/10.1021/acs.energyfuels.1c01722]

[37] D. Sheberla, L. Sun, M.A. Blood-Forsythe, S. Er, C.R. Wade, C.K. Brozek, A. Aspuru-Guzik, and M. Dincă, "High electrical conductivity in Ni3(2,3,6,7,10,11-hexaiminotriphenylene)2, a semiconducting metal-organic graphene analogue", *J. Am. Chem. Soc.,* vol. 136, no. 25, pp. 8859-8862, 2014.
[http://dx.doi.org/10.1021/ja502765n] [PMID: 24750124]

[38] D. Tian, C. Wang, and X. Lu, "Metal-organic frameworks and their derived functional materials for supercapacitor electrode application", *Adv. Energy Sustain. Res.,* vol. 2, no. 7, p. 2100024, 2021.
[http://dx.doi.org/10.1002/aesr.202100024]

[39] S. Xiong, F. Yang, H. Jiang, J. Ma, and X. Lu, "Covalently bonded polyaniline/fullerene hybrids with coral-like morphology for high-performance supercapacitor", *Electrochim. Acta,* vol. 85, pp. 235-242, 2012.
[http://dx.doi.org/10.1016/j.electacta.2012.08.056]

[40] V. Gupta, and N. Miura, "Electrochemically deposited polyaniline nanowire's network", *Electrochem. Solid-State Lett.,* vol. 8, no. 12, p. A630, 2005.
[http://dx.doi.org/10.1149/1.2087207]

[41] J. Zang, S-J. Bao, C.M. Li, H. Bian, X. Cui, Q. Bao, C.Q. Sun, J. Guo, and K. Lian, "Well-aligned

cone-shaped nanostructure of polypyrrole/ruo 2 and its electrochemical supercapacitor", *J. Phys. Chem. C,* vol. 112, no. 38, pp. 14843-14847, 2008.
[http://dx.doi.org/10.1021/jp8049558]

[42]   N. Marshall, W. James, J. Fulmer, S. Crittenden, A.B. Thompson, P.A. Ward, and G.T. Rowe, "Polythiophene doping of the cu-based metal-organic framework (MOF) HKUST-1 using innate MOF-initiated oxidative polymerization", *Inorg. Chem.,* vol. 58, no. 9, pp. 5561-5575, 2019.
[http://dx.doi.org/10.1021/acs.inorgchem.8b03465] [PMID: 30950603]

[43]   J. Bobacka, A. Lewenstam, and A. Ivaska, "Electrochemical impedance spectroscopy of oxidized poly(3,4-ethylenedioxythiophene) film electrodes in aqueous solutions", *J. Electroanal. Chem.,* vol. 489, no. 1-2, pp. 17-27, 2000.
[http://dx.doi.org/10.1016/S0022-0728(00)00206-0]

[44]   I. Dědek, V. Kupka, P. Jakubec, V. Šedajová, K. Jayaramulu, and M. Otyepka, "Metal-organic framework/conductive polymer hybrid materials for supercapacitors", *Appl. Mater. Today,* vol. 26, p. 101387, 2022.
[http://dx.doi.org/10.1016/j.apmt.2022.101387]

[45]   B. Dhara, S.S. Nagarkar, J. Kumar, V. Kumar, P.K. Jha, S.K. Ghosh, S. Nair, and N. Ballav, "Increase in electrical conductivity of MOF to billion-fold upon filling the nanochannels with conducting polymer", *J. Phys. Chem. Lett.,* vol. 7, no. 15, pp. 2945-2950, 2016.
[http://dx.doi.org/10.1021/acs.jpclett.6b01236] [PMID: 27404432]

[46]   Y. Wang, Y. Liu, H. Wang, W. Liu, Y. Li, J. Zhang, H. Hou, and J. Yang, "Ultrathin NiCo-MOF nanosheets for high-performance supercapacitor electrodes", *ACS Appl. Energy Mater.,* vol. 2, no. 3, pp. 2063-2071, 2019.
[http://dx.doi.org/10.1021/acsaem.8b02128]

[47]   Y.Z. Zhang, Y. Wang, Y.L. Xie, T. Cheng, W.Y. Lai, H. Pang, and W. Huang, "Porous hollow Co 3 O 4 with rhombic dodecahedral structures for high-performance supercapacitors", *Nanoscale,* vol. 6, no. 23, pp. 14354-14359, 2014.
[http://dx.doi.org/10.1039/C4NR04782F] [PMID: 25329598]

[48]   T. Audichon, T.W. Napporn, C. Canaff, C. Morais, C. Comminges, and K.B. Kokoh, "IrO 2 Coated on RuO 2 as efficient and stable electroactive nanocatalysts for electrochemical water splitting", *J. Phys. Chem. C,* vol. 120, no. 5, pp. 2562-2573, 2016.
[http://dx.doi.org/10.1021/acs.jpcc.5b11868]

[49]   L.Å. Näslund, Á.S. Ingason, S. Holmin, and J. Rosen, "Formation of RuO(OH) 2 on RuO 2 -based electrodes for hydrogen production", *J. Phys. Chem. C,* vol. 118, no. 28, pp. 15315-15323, 2014.
[http://dx.doi.org/10.1021/jp503960q]

[50]   A. Gomez Vidales, J. Kim, and S. Omanovic, "Ni0.6-xMo0.4-xIrx-oxide as an electrode material for supercapacitors: investigation of the influence of iridium content on the charge storage/delivery", *J. Solid State Electrochem.,* vol. 23, no. 7, pp. 2129-2139, 2019.
[http://dx.doi.org/10.1007/s10008-019-04311-8]

[51]   C. Yuan, J. Li, L. Hou, X. Zhang, L. Shen, and X.W.D. Lou, "Ultrathin mesoporous NiCo 2O 4 nanosheets supported on Ni foam as advanced electrodes for supercapacitors", *Adv. Funct. Mater.,* vol. 22, no. 21, pp. 4592-4597, 2012.
[http://dx.doi.org/10.1002/adfm.201200994]

[52]   H.B. Wu, and X.W.D. Lou, "Metal-organic frameworks and their derived materials for electrochemical energy storage and conversion: Promises and challenges", *Sci. Adv.,* vol. 3, no. 12, p. eaap9252, 2017.
[http://dx.doi.org/10.1126/sciadv.aap9252]

[53]   Z. Wang, Y. Liu, C. Gao, H. Jiang, and J. Zhang, "A porous Co(OH) 2 material derived from a MOF template and its superior energy storage performance for supercapacitors", *J. Mater. Chem. A Mater. Energy Sustain.,* vol. 3, no. 41, pp. 20658-20663, 2015.

[http://dx.doi.org/10.1039/C5TA04663G]

[54]   S. Zhang, Z. Yang, K. Gong, B. Xu, H. Mei, H. Zhang, J. Zhang, Z. Kang, Y. Yan, and D. Sun, "Temperature controlled diffusion of hydroxide ions in 1D channels of Ni-MOF-74 for its complete conformal hydrolysis to hierarchical Ni(OH) 2 supercapacitor electrodes", *Nanoscale,* vol. 11, no. 19, pp. 9598-9607, 2019.
       [http://dx.doi.org/10.1039/C9NR02555C] [PMID: 31063163]

[55]   J. Yu, X. Gao, Z. Cui, Y. Jiao, Q. Zhang, H. Dong, L. Yu, and L. Dong, "Facile synthesis of binary transition metal sulfide tubes derived from NiCo-MOF-74 for high-performance supercapacitors", *Energy Technol.,* vol. 7, no. 6, p. 1900018, 2019.
       [http://dx.doi.org/10.1002/ente.201900018]

[56]   H. Liu, H. Guo, L. Yue, N. Wu, Q. Li, W. Yao, R. Xue, M. Wang, and W. Yang, "Metal-Organic Frameworks-Derived NiS 2 /CoS 2 /N-Doped Carbon Composites as Electrode Materials for Asymmetric Supercapacitor", *ChemElectroChem,* vol. 6, no. 14, pp. 3764-3773, 2019.
       [http://dx.doi.org/10.1002/celc.201900746]

[57]   B. Ren, Q. Yi, F. Yang, Y. Cheng, H. Yu, P. Han, Y. Yang, G. Chen, I. Jeerapan, Z. Li, and J.Z. Ou, "MOF-Derived Zero-Dimensional Cu 3 P Nanoparticles Embedded in Carbon Matrices for Electrochemical Hydrogen Evolution", *Energy Fuels,* vol. 36, no. 15, pp. 8381-8390, 2022.
       [http://dx.doi.org/10.1021/acs.energyfuels.2c01708]

[58]   J. Gu, L. Sun, Y. Zhang, Q. Zhang, X. Li, H. Si, Y. Shi, C. Sun, Y. Gong, and Y. Zhang, "MOF-derived Ni-doped CoP@C grown on CNTs for high-performance supercapacitors", *Chem. Eng. J.,* vol. 385, p. 123454, 2020.
       [http://dx.doi.org/10.1016/j.cej.2019.123454]

[59]   C. Peng, S. Zhang, D. Jewell, and G.Z. Chen, "Carbon nanotube and conducting polymer composites for supercapacitors", *Prog. Nat. Sci.,* vol. 18, no. 7, pp. 777-788, 2008.
       [http://dx.doi.org/10.1016/j.pnsc.2008.03.002]

[60]   S. Hérou, J.J. Bailey, M. Kok, P. Schlee, R. Jervis, D.J.L. Brett, P.R. Shearing, M.C. Ribadeneyra, and M. Titirici, "High-density lignin-derived carbon nanofiber supercapacitors with enhanced volumetric energy density", *Adv. Sci.,* vol. 8, no. 17, p. 2100016, 2021.
       [http://dx.doi.org/10.1002/advs.202100016] [PMID: 34014597]

[61]   Y. Liu, X. Xu, Z. Shao, and S.P. Jiang, "Metal-organic frameworks derived porous carbon, metal oxides and metal sulfides-based compounds for supercapacitors application", *Energy Storage Mater.,* vol. 26, pp. 1-22, 2020.
       [http://dx.doi.org/10.1016/j.ensm.2019.12.019]

CHAPTER 12

MXene-based Nanomaterials for High-performance Supercapacitor Applications

Zaheer Ud Din Babar[1,2], Ayesha Zaheer[2], Jahan Zeb Hassan[3], Ali Raza[2] and Asif Mahmood[4,5,*]

[1] *Scuola Superiore Meridionale (SSM), University of Naples Federico II, Largo S. Marcellino, 10, 80138, Italy*

[2] *Department of Physics "Ettore Pancini", University of Naples Federico II, Piazzale Tecchio, 80, 80125 Naples, Italy*

[3] *Department of Physics, Riphah Institute of Computing and Applied Sciences (RICAS), Riphah International University, 14 Ali Road, Lahore, Pakistan*

[4] *School of Chemical and Biomolecular Engineering, The University of Sydney, Sydney, Australia*

[5] *Center for Clean Energy Technology, School of Mathematical and Physical Sciences, Faculty of Science, University of Technology Sydney, Sydney, Australia*

Abstract: Technological advances in recent decades have augmented the demand for durable and inexpensive energy storage devices with higher charge capacity. Owing to their unique charge storage and surface capability, a recent class of two-dimensional (2D) materials known as MXenes has been widely used in energy storage devices. MXenes are the layered transition metal carbides, nitrides, and/or carbonitrides produced *via* selective etching of interleaved "A" layers from parent MAX phases. Unlike other 2D materials, MXenes earned great attention because of their intrinsic surface functional groups, hydrophilicity, unique electrochemical nature, high conductivity, and superior charge storage capacity. Such features render MXenes as the ultimate material from the 2D family, thus inspiring researchers to delve further into experimental and theoretical realms. Numerous attempts have been made to elucidate synthesis strategies to produce MXene and its fundamental characteristics. The current chapter emphasizes the recent advancements in MXene-based electrochemical energy storage applications using supercapacitors which are recognized as a dominant source. The effect of MXene's morphology and electrode growth on the charge-storage mechanism has also been highlighted in subsequent sections. In addition, this chapter outlines the current state-of-the-art on the supercapacitors compromised of the MXene-based composites. A discussion of relevant challenges associated with such materials for energy storage applications is also presented, and future perspectives provide additional insight into their practical aspects.

---

* **Corresponding author Asif Mahmood:** School of Chemical and Biomolecular Engineering, The University of Sydney, Sydney, Australia, Center for Clean Energy Technology, School of Mathematical and Physical Sciences, Faculty of Science, University of Technology Sydney, Sydney, Australia,: E-mail: asif.mahmood@uts.edu.au

Keywords: Composite, Energy storage, MXene, Supercapacitor, 2D material.

INTRODUCTION

MXenes are 2D layered nitrides, carbides, or carbonitrides of primary transition metals produced *via* selective etching of interleaved "A" layers from their parent MAX phases. They were named "MXene" due to their structural and morphological similarities with graphene [1]. MAX phases serve as the precursor to MXene. They are denoted as Mn+1AXn (where n = 1 to 4), where "M" is the transition metal that belongs to the elements from groups III-VI of the periodic table (*i.e.*, Nb, Mo, V, Mn, Ta, Cr, Ti, Zr, *etc.*), "A" represents group IIIA/IVA elements (*i.e.*, In, Ga, Ge, Al, As, Pb, Sn, and Si, *etc.*). At the same time, "X" indicates carbide (C), nitride (N) and/or carbonitride (CN). The layers of "M" atoms are interweaved with the "A" layers, whereas the "X" atoms reside in octahedral sites in MAX phases with D_6h_4-$P6_3$/mmc close-packed symmetry [2]. As the layers in transition metal dichalcogenide and Graphene have van der Waals (vdWs) bonding, the MX layers in MXenes have mixed ionic, metallic, and covalent nature. In contrast, MA layers have only metallic bonds. Due to the reactive nature and faint bonding among M-A layers for IIIA/IVA elements with fluoride-containing acids, it is easy to remove the A layers selectively *via* a wet-chemical etching route to yield MXenes [3]. MXenes are formed after this discriminating chemical engraving of A layers (covalently associated) using MAX phases and represented as $M_{n+1}X_nT_x$. After etching, the MXenes surface is terminated with a distinct surface group, *e.g.*, -Cl, -F, -OH, or -O as represented by "T" whereas the "n" value specifies the number of layers [4]. These surface terminal groups impart hydrophilic behaviour to MXene's sheets [5]. Furthermore, ion transport and other electronic features (*e.g.*, bandgap) of MXene are essentially governed by surface termination groups [6]. Also, the environmental stability of MXenes (*e.g.*, oxidation) depends on the "n" values. Higher values of "n" result in the yield of more stable MXene [7].

In 2011, a research group at Drexel University led by Professors M. W. Barsoum and Yuri Gogotsi described the experimental synthesis of $Ti_3C_2T_x$ [8] *via* selective removal of "Al" by immersing MAX phase precursor (Ti_3AlC_2) in hydrofluoric acid as an etchant [9]. Ti_2CT_x, $Zr_3C_2T_x$, Nb_2CT_x, $Nb_4C_3T_x$, V_2CT_x, Ti_3CNT_x, Mo_2CT_x, $Ti_4N_3T_x$, $Mo_4VC_4T_x$, $Mo_2ScC_2T_x$, $(Ti_{0.5}, Nb_{0.5})_2CT_x$, $(Nb_{0.8}, Ti_{0.2})_4C_3T_x$, and $(Nb_{0.8}, Zr_{0.2})_4C_3T_x$ are examples of the MXenes that produced experimentally till date. Due to its rare blend of ceramic (C/N atoms) and metallic characteristics, a wide range of applications have been studied using MXenes. More than 100 MAX phases have been designed using different recipes of MAX phase components [10]. Nevertheless, about 30 MXenes are synthesized experimentally, with their characteristics thoroughly evaluated. They created a pathway for research and the

MXene community to realize new MXene structures with distinct features. Due to better surface chemistry, MXene has large metallic conductivity ranges from ~ 6000-8000 S.cm^{-1} [11], better thermal conductivity [12], exceptional mechanical stability [13], promising optical properties [14], excellent electric and magnetic properties [15], intrinsic hydrophilicity [16], abundant surface terminations [17]. Moreover, MXenes can sufficiently intercalate external species among its layers [18, 19]. These distinctive qualities and easy post-processing have made MXenes ideal candidates for numerous applications, including nuclear waste treatment [20], antibacterial activities and water purification [21], sensors [22], photocatalysis [23], solid lubricants [24], electromagnetic interference (EMI) shielding [25], energy conversion devices and transparent conductors in electronics [26], among others. Owing to the micro-scale lateral dimension, higher hydrophilicity, and atomic layer thickness of MXenes, it is possible to recognize thin layers of MXene *via* vacuum-aided filtration or simply by spray coating/printing. The hydrophilicity of Ti_3C_2T was studied by Ghidiu *et al.* by measuring the contact angle of water on MXene film [11]. As compared to semiconductor-type metal oxides, MXene has a greater power density due to strong conductivity, thus facilitating the devices to charge quickly [27]. In addition to these features, the strong electronegativity and rich surface chemistry of MXenes have made it possible to fabricate composite electrodes [28, 29]. In addition, MXenes are better substrate materials for catalysis applications because of the fast electron transport between their layers [30]. Also, MXenes exhibit various characteristics depending on their compositions, elemental stoich-iometries, synthesis routes, interlayer spacing, and lateral flake sizes. For example, according to Seh *et al.*, Mo_2CT_x MXenes have higher catalytic activity than Ti_2CT_x [31]. Nitride-assisted MXenes show better electrical conductivity than carbide-assisted MXenes; conversely, carbide-assisted MXenes have superior structural stability [32]. Moreover, the configuration of the exterior terminal groups affects the characteristics of MXene. Amid several surface terminations on Ti_3C_2 MXene, O-terminated MXene ($Ti_3C_2O_2$) is more stable than the others, while H-terminated ($Ti_3C_2H_2$) is the least stable. Moreover, the order of stability is $Ti_3C_2F_2 > Ti_3C_2(OH)_2 > Ti_3C_2H_2 > Ti_3C_2$ [33]. Various devices have been proposed through the advancement of technology, from microchips to massive electronic devices that require energy to function.

To overcome this, devices with adequate energy storage must be durable and have minimal production and processing costs. It includes a wide-ranging use of batteries for energy storage applications, including sodium-sulfur [34], lithium-ion [35], sodium-ion batteries [36], and supercapacitors [37]. Batteries have an enhanced energy density but low power density though the case of capacitors conflicts with batteries. Supercapacitors perform far better as energy storage devices than batteries owing to their increasing power density, significant energy storage ability, and longer life span [38]. However, these devices are often non-flexible, so limiting their applications. For instance, wearable electronics (*i.e.*, health monitors, smart textiles, displays, and other essentials) are incredibly flexible and have been developed effectively over the past several decades [39]. Researchers have created flexible, robust, and more significant energy storage ability devices to meet such demand [40]. Concerning this, MXene, with an excellent layered structure, greater electrical conductivities, hydrophilicity, several surface terminations, and a larger specific surface area, has offered considerable potential for flexible electrochemical energy storage applications [41]. The present chapter highlights the latest progress in the MXene-based electrochemical energy storage systems *via* supercapacitors as a power basis. Different strategies, as well as their characteristics, are considered meticulously. Additionally, giving a precise focus on charge-storing pathways of MXene-assisted supercapacitors, the impact of electrode fabrication and morphological influence on the electrochemical progress has been underlined. This chapter offers existing state-of-the-art development on MXenes-composite based supercapacitors. Moreover, relevant experiments related to this material and future viewpoints have also been presented that provide a further understanding of its practical application [41].

Charge Storage Pathways in the MXene-based Supercapacitors

Aqueous Media

The reversible electrosorption of ions on high-surface-area electrode materials causes electric double-layer capacitance (EDLC) in energy storage devices. Simultaneously, the near-surface redox activities and reversible surface drive the pseudocapacitance. Based on their electrochemical capacitive properties, pseudocapacitance is classified into three forms [42]:

1. Under-potential deposition,

2. Redox pseudocapacitance

3. Intercalation pseudocapacitance.

Researchers have also been curious about the phenomenon occurring with 2D layered MXenes during their electrochemical process. MXenes, particularly the first ever discovered Ti_3C_2 MXene, have high conductivity and have been investigated thoroughly to ensure the fast transfer of electrons in supercapacitor applications [43]. It is known that metallic ions can be intercalated into MXenes either electrochemically or through the spontaneous reduction of precursor salts (Fig. **1a**), which signifies their capacity to store charge. For instance, several cations, including K^+, Na^+, Li^+, TEA^+, Ca^{2+}, Mg^{2+}, Cs^+, $NH^{4+,}$ and Al^{3+}, can readily intercalate into MXenes *via* reversible electrochemical route. Electrochemically active sites on the surface of MXene host such cations, which contribute to energy storage phenomenon [44]. Typically, ion adsorption sites can cover trivial and deep adsorption sites with MXenes (Fig. **1b**). Usually, the adsorption occurs at external adsorption sites closer to particle edges in addition to deep adsorption sites with a considerable activation energy [45]. As a result, the ion may quickly bind to shallow adsorption sites to ensure a substantial degree of charge storing that enables MXene to display good progress even at high scan rates. Along with this intercalation of cation, confined water molecules in between the layers also showed a significant shift. Li^+, Mg^{2+}, and Al^{3+} are common cosmotropic cations that intercalate into MXene interlayers in their partly hydrated form; however, the introduction of chaotropic cations (*i.e.*, TEA^+ and Cs^+) substantially dehydrates material [46].

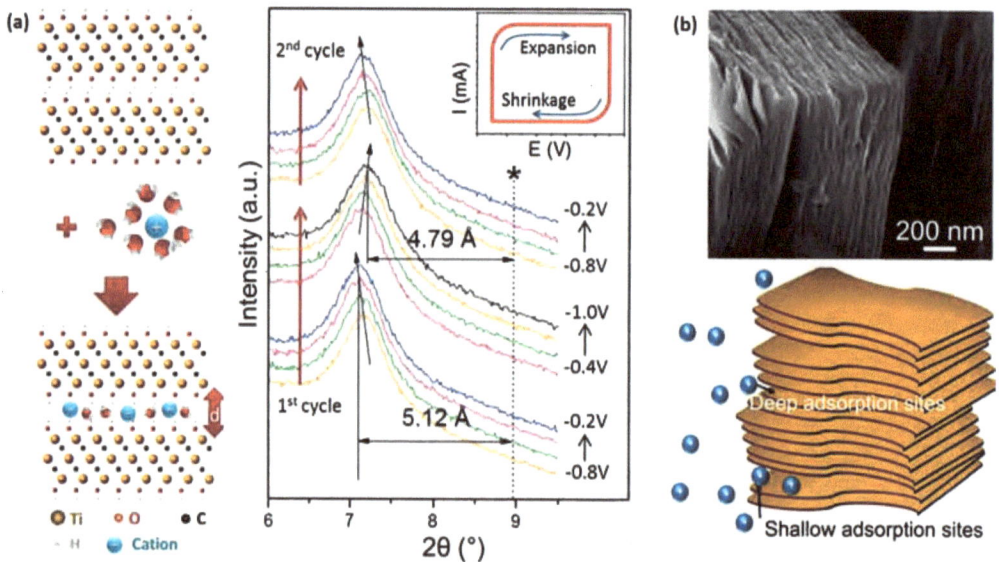

Fig. (1. (a) Representation for electrochemical *in-situ* XRD analysis and cation intercalation of multilayered exfoliated $Ti_3C_2T_x$ MXene in KOH. Adapted with permission from Reference [44] Copyright 2013 AAAS. **(b)** The SEM micrograph of $Ti_3C_2T_x$ multilayered MXene, whereas the bottom photograph displays shallow and deep adsorption sites for MXene. Adapted with permission from Reference [45] Copyright 2015 John Wiley and Sons, Ltd.

Similar to interlayer variations in carbon materials, MXenes exhibit various deformation behaviours when intercalating with different ions [45]. Generally, cations with more charge and small ionic radii reduce the void of $Ti_3C_2T_x$ MXene. Contrarily, cations with lesser charge increase the d-spacing, indicating the expansion due to cation intercalation. Furthermore, A negative charge on the

$Ti_3C_2T_x$ surface attracts the positive charge cations by electrostatic attraction. As a result, the expansion caused by cation intercalation could be related to an electrostatic interface among MXene (negatively charged) and cations (positively charged). Furthermore, these variations in MXene's structure are also influenced by the magnitude of intercalated cations. So, many intercalated H^+ can effectively reverse the contraction of the interlayer gaps caused by specific numbers of intercalated H^+ ions. As investigated through many techniques, including *in-situ* AFM [47] and *in-situ* electrochemical quartz-crystal admittance [45], such variation (both the contraction and expansion) in MXene structure is reversible to a large extent in the intercalation and deintercalation processes [48]. The cycling performance of the electrodes depends on their degree of mechanical stability and cyclic reversibility. However, despite being highly reversible, mechanical deformations during discharging/charging series tend to change the structural framework of MXenes, possibly affecting their lifetime. Even though MXene has excellent cycle stability; still, it is possible to significantly increase its cycle life and reduce energy loss by developing an electrolyte with almost no volume change upon charging and discharging events.

One must be curious enough to see the MXene interspacing after ion insertion. MXene cyclic voltammetry (CV) graphs reveal prominent redox peaks [11]. Due to its layered structure, MXenes have significant ion-accessible electrochemical surface even if their specific surface areas are low (20-100 m^2 g^{-1}) as estimated *via* the Brunauer-Emmett-Teller method (BET) [49]. As a result, it is challenging to determine merely from CV profiles or particular surface area analysis whether the energy-storing actions are either pseudocapacitive or electrosorption-capacitive. As visualized in Fig. (**2a**), *in-situ* electrochemical X-ray absorption spectroscopy (XAS) was performed in case of Ti_3C_2T MXene (in 1M H_2SO_4, commonly used electrolyte) to determine the reasons that lead to variations in oxidation states of transition metal during electrochemical measurements [50]. Combining this with experimental findings, the electrochemical behaviour $Ti_3C_2T_x$ in an electrolyte such as H_2SO_4 occurs in a mode that alters titanium oxidation state through oxygen functional groups protonation and electron transfer as explained by the following chemical process (Figs. **2b** & **c**) [51].

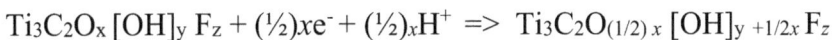

$$Ti_3C_2O_x[OH]_y F_z + (½)xe^- + (½)_xH^+ => Ti_3C_2O_{(1/2)\,x}[OH]_{y\,+1/2x} F_z$$

After discharging, an 1190 Fg^{-1} ultrahigh capacitance might be ideally attained for O-terminated $Ti_3C_2T_x$ MXene (where x = 2) as predicted by theoretical studies (DFT) [49]. The change in atomic Bader charge was observed on a Ti atom at varying H exposures and electrode potentials, as displayed in Fig. (**2d**) [52]. Here, it should be noted that the MXenes function extremely well electrochemically when exposed to negative potentials. However, an irreversible oxidation strategy was acquired when exposed to positive potentials (greater than 0.2 to 0.4 V exceeding for open circuit), thus increasing the resistance and declining the capacitance.

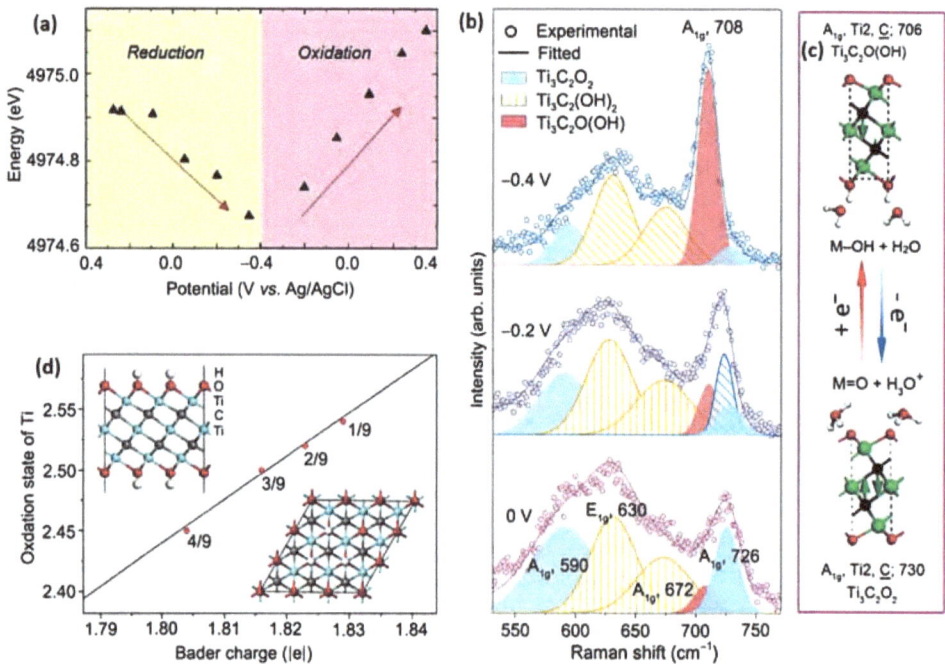

Fig. (2). (a) Data from an electrochemical *in-situ* XAS. Ti edge energy variation concerning potential (at half the height of the normalized X-ray absorption near the edge structure spectra) during a full potential sweep in the range of -0.35V to 0.35 V. Adapted with permission from Reference [50] Copyright John Wiley & Sons, Ltd. **(b)** Raman spectrum, Adapted with permission from Reference [49] Copyright 2018 American Chemical Society. **(c).** Subsequent Lorentzian fits of the $Ti_3C_2T_x$ electrode in H_2SO_4 electrolyte during charging. Adapted with permission from Reference [51] Copyright 2016 American Chemical Society. **(d)** The Bader charge on the oxidation state of Ti fluctuates with various levels of H coverages. $Ti_3C_2O_2$ that has been partially protonated is depicted from the top and side in the insets. Adapted with permission from Reference [52] Copyright 2018 American Chemical Society.

It can be concluded from the previous explanation that $Ti_3C_2T_x$ is an intercalation pseudocapacitive negative electrode material that reacts in H_2SO_4 electrolyte. It contains the following prospects:

1. Fast intercalation of ions and their transport through open channels are made possible by the layered structure.

2. Quick electron transport is made possible by higher conductivities of the MX layer.

3. Transition metal (conductive core) offers better charge transfer due to variable valency.

4. Redox-active sites were accessible *via* -O terminated transition metal surface, which is reliable through the property of intercalation pseudocapacitive materials [53].

Apart from this, the process of electrochemical storage is not constrained *via* diffusion fit for minimum scan rates (*i.e.*, 20 mV/s) [11]. The contrary is noticed in different electrolytes, *e.g.*, $MgSO_4$ and $(NH_4)_2SO_4$ combinations, where the entire Raman bands associated with $Ti_3C_2T_x$ MXene remain constant throughout the cycling process [49]. As a result, the hydrated ions are intercalated into MXene interlayers without being dehydrated. In comparison, the hydration shell serves as dielectric media between the electrode and the intercalated ions with -ve dielectric constant (K). The capacitance is often more significant in the case of a hydration shell with small radii. The EDLC capacitive energy storage process shown here explains why these electrolytes have lower capacitance than H_2SO_4 and why H_2SO_4 media is commonly utilized. When associating the electrochemical behaviour of H_2SO_4 with other media, a question needed to be answered: why does the pseudocapacitance only happen in H_2SO_4 electrolytes? Is it because the protons are the slightest cations and have better contact with electrochemically active sites? Clarification will indeed require further research. According to a recent investigation, methanesulfonic acid and seawater [54] both exhibit good electrochemical performance when used with MXene [55]. It is worth noting that the above studies on the charge-storing pathways nearly concentrate on $Ti_3C_2T_x$ MXene because of its easy synthesis.

Non-aqueous Media

Non-aqueous systems, for instance, ionic-liquid gel as well as organic electrolytes, were correspondingly investigated in addition to aqueous systems because of their more expansive voltage windows that can result in larger energy densities as given by the formula $E = 1/2 \, CV^2$. In this formula, E represents the energy density; V corresponds to the potential window, whereas C indicates capacitance. Compared to the aqueous solutions, these non-aqueous electrolytes exhibit unique electrochemical behaviour. In 1-ethyl-3-methylimidazolium bis-(trifluoromethylsulfonyl)-imide in acetonitrile electrolyte, $Ti_3C_2T_x$ MXene display CV trends as a pair of extensive peaks (~ 0.2V and 0.6V) vs. Ag [56]. Since -F or -O terminal groups are not intended as electrochemically energetic in aprotic ionic liquids, unlike MXenes behaviour in the aqueous electrolyte. Hence these properties cannot arise from pseudocapacitive assistances of surface functional groups [57].

Furthermore, intercalation of bis[(trifluoromethyl)sulfonyl]imide (TFSI⁻) anions as well as EMI⁺ cations in the respective –ve and +ve potential ranges have been confirmed by *in-situ* XRD analysis [57]. These peaks can be ascribed to the insertion of TFSI/EMI⁺ among $Ti_3C_2T_x$ layers. After the intercalation process, charges were deposited by means of a redox process supported due to hybridized orbital of surface terminal groups and cations [58]. Intriguingly, *ex-situ* XRD configurations showed a significant increase in the *d*-lattice parameter (8.7 to 10.5 Å) throughout the primary charging period in a 1M $LiPF_6$/ethylene carbonate (EC)-dimethyl carbonate (DMC) electrolyte (Fig. **3a**). A group of researchers [58] reported that partly solvated Li⁺ ions are intercalated in the initial stage. Due to this additional charging, a solid columbic attraction induces among MXene sheets. Li⁺ ions that result in solvation shell of intercalated Li⁺ decrease to provide an insignificant d-value; therefore, it causes a charge transferal strategy for MXene sheets (Fig. **3b**) [58]. The potential uses of MXenes in Li-ion battery applications are severely constrained by the requirement of a wide operating potential window which produces an extraordinary operating potential for -ve electrodes. However, for MXene etched by lewis acidic etching route, the charge storage at a current (constant) vs. potential is comparable to what was realized in pseudocapacitive specie [59]. Intuitively, before intercalation, in this case, lithium ions are also disolvated. However, it should be noted that not all non-aqueous electrolytes desolvate lithium ions. For instance, when using propylene carbonate as the solvent in a lithium bis(trifluoromethylsulfonyl) amine (LiTFSI) organic electrolyte; moreover, before intercalation Li ions were completely desolvated that yields rapid charge migration, hence causes to enhanced electrochemical progress [60]. The non-aqueous electrolytes are crucial to MXenes' ability to store charges. Hence, forthcoming studies should wage more consideration on the fundamental roles of the solvents. Owing to their excellent power and energy densities, organic electrolytes are used in most commercial supercapacitor devices. Therefore, it is crucial to research MXene-based organic devices. However, minor literature is available that reveals MXenes consideration in non-aqueous media, and perhaps more research is urgently required.

MXenes as Supercapacitor Electrodes

Surface Chemistry

MXenes concerning surface chemistry can be drastically affected by synthesis conditions and subsequent post-processing techniques. As a result, the electrochemical performance varies following the variations in chemical composition on the surface. The surface of transition metals in MXenes is terminated with specific functional groups that function as redox-active sites. When using the HF etching technique, MXene etched with low HF concentrations has additional -O terminations involved in the oxidation-reduction process through the charge-storing route [49]. Further Nuclear magnetic resonance (NMR) findings suggest more significant content of -O functional groups for advanced ability compared to -F and -OH terminations when etched with HCl/LiF (Fig. **4a**) [61]. Improved electrochemical progress of HCl/LiF etched MXene is partly due to substantial aggregate of -O functional groups, larger d-value is because of Li^+ intercalating with water, and adsorption of Cl on MXenes edges. Using the alkali etching technique, MXenes lacking -F functional groups can be produced [62]. The $Ti_3C_2T_x$ MXene produced *via* alkali-assisted hydrothermal technique in NaOH (aqueous media) showed a gravimetric capacitance of 314 F/g at 2 mVs^{-1} is more than multilayer $Ti_3C_2T_x$ obtained by HF treatment in 1M H_2SO_4 [63].

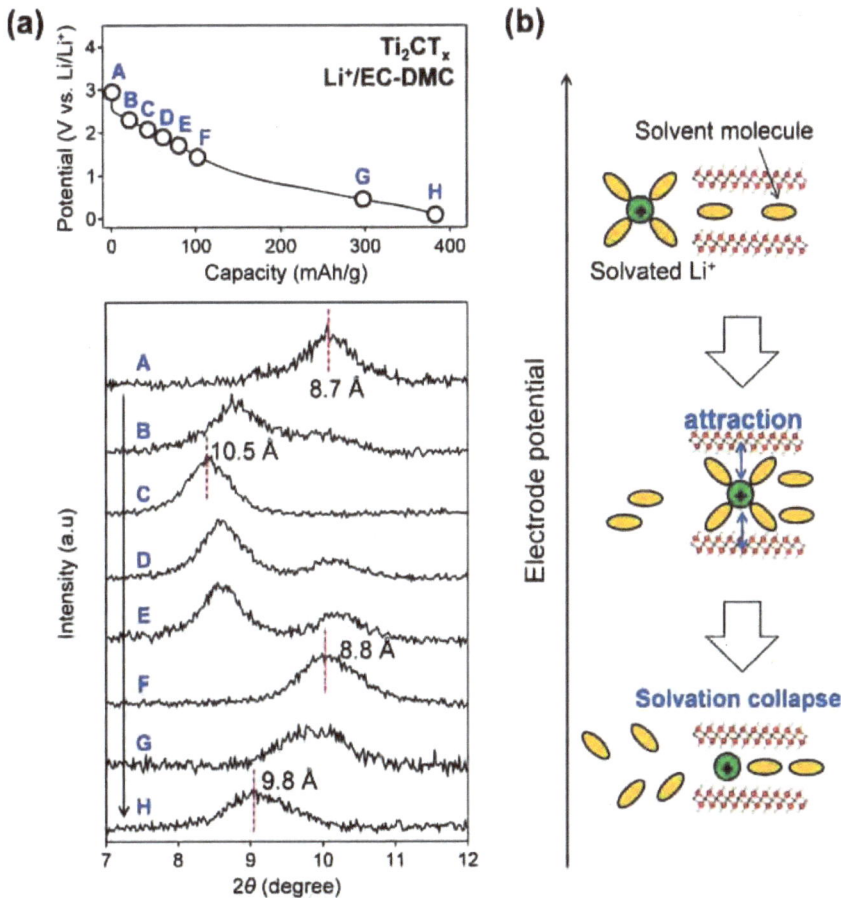

Fig. (3). **(a)** First charge curve for Ti_2CT_x (at 20 mA/g and 1 M LiPF$_6$/EC-DMC media) together with the respective X-ray diffraction peaks (*ex-situ*). **(b)** An illustrative representation of electrode reaction for MXene in non-aqueous media at the beginning of the charging process. Adapted with permission from Reference [58] Copyright 2018 American Chemical Society.

The Utilization of organic bases such as tetra butyl ammonium hydroxide and/or tetramethyl ammonium hydroxide (TMAOH) is used to produce MXene; it leads to the production of aluminum-oxoanion-functionalized MXene sheet. This material does not serve as a good electrode material but exhibits broad and more robust optical absorption behavior in near-infrared [62]. Fluoride-free MXene (without -F functional groups) with surface terminations, for instance -Cl, -OH, and –O groups, can also be attained by adopting electrochemical etching. More –Cl and -O groups without -OH groups are present when etched *via* the Lewis acid etching route, resulting in an enhanced Li^+ storage capacity [59]. As discussed above, the type of etchant (methods) and its concentration greatly influence MXene's terminations and quality, thus effecting its electrochemical progress. Therefore, the etching procedure and the environment must be carefully considered to make MXenes with even better electrochemical performance. The gravimetric performance of Lewis acidic melt or alkali etched MXene is superior to that of the HF etched MXene. Since they cannot delaminate MXene and as a result, their volumetric performance is lower than that of the HCl/LiF etched MXene. Further study is required since the use of Lewis acidic melts or alkalies for etching has not been adopted significantly by the MXene community. Additionally, post-etching techniques like intercalation, heat treatment, and others can be employed to tune their surface chemistry. For instance, the O-containing functional groups preferred in redox processes can be considered as a substitute for the -F groups through the CH_3COOK or KOH intercalation route, resulting in an increasing capacitance. The surface chemistry of MXenes is similarly altered by hydrazine intercalation, with a reduction in -OH surface groups (Fig. **4b**). The O/F ratio can vary by dimethyl sulfoxide (DMSO) delamination as well [64].

The Surface terminations can be removed by K^+ ion intercalation, and subsequent calcination allows the MXene to experience a notable increase in capacitance (over 500 Fg^{-1} at 1 Ag^{-1}) [65]. Due to the low content of -F group after annealing, Ti_2CT_x MXene exhibits improved capacitance when heated in Ar, N_2, and N_2/H_2 atmospheres [66]. Small contents of -F terminations can significantly enhance the robust bonding interaction among MXene surface and ions. For example, $Ti_3C_2T_x$ MXene with a decreased amount of F contents (~ 0.65 atom%) than MXenes with excess F contents (~ 8.09 atom%) has a lower self-discharge rate (reduced by ~ 20%) [37]. The capacitance of $Ti_3C_2T_x$ MXene was significantly increased by replacing or doping functional groups by N, as obtained by urea-assisted solvothermal processing (Fig. **4c**) [67] or ammonia annealing [68], or NH_4F-assisted heating [69]; as a result, MXene's capacitance was further improved [70]. Following this, we further hypothesize that "N", due to lowermost electronegativity, has superior charge mobility and better e⁻ donor properties that tend to share e⁻ with intercalated cations, which gives incremented pseudo capacitance. The theoretical measurements (first principle) showed that -Si and -P functionalized MXenes revealed minor ion diffusion obstructions and energy than one with -OH and-F functionalization, increasing the metal ion capacity [71]. Further research is needed to understand how MXenes are functionalized using heteroatoms completely. Presumably, the MXene materials have an impact on electrochemical progress. In addition, Ti_2CT_x has a substantially larger gravimetric capacitance than $Ti_3C_2T_x$ due to its lower relative molecular weight. However, reports suggest Ti_2CT_x is less stable than $Ti_3C_2T_x$ because it has more carbon vacancies. In addition, 2D $Mo_{4/3}C$ sheets (for well-ordered metal di-vacancies) possess ~ 65% higher volumetric capacitance compared to corresponding Mo_2C sheets with no vacancies [72].

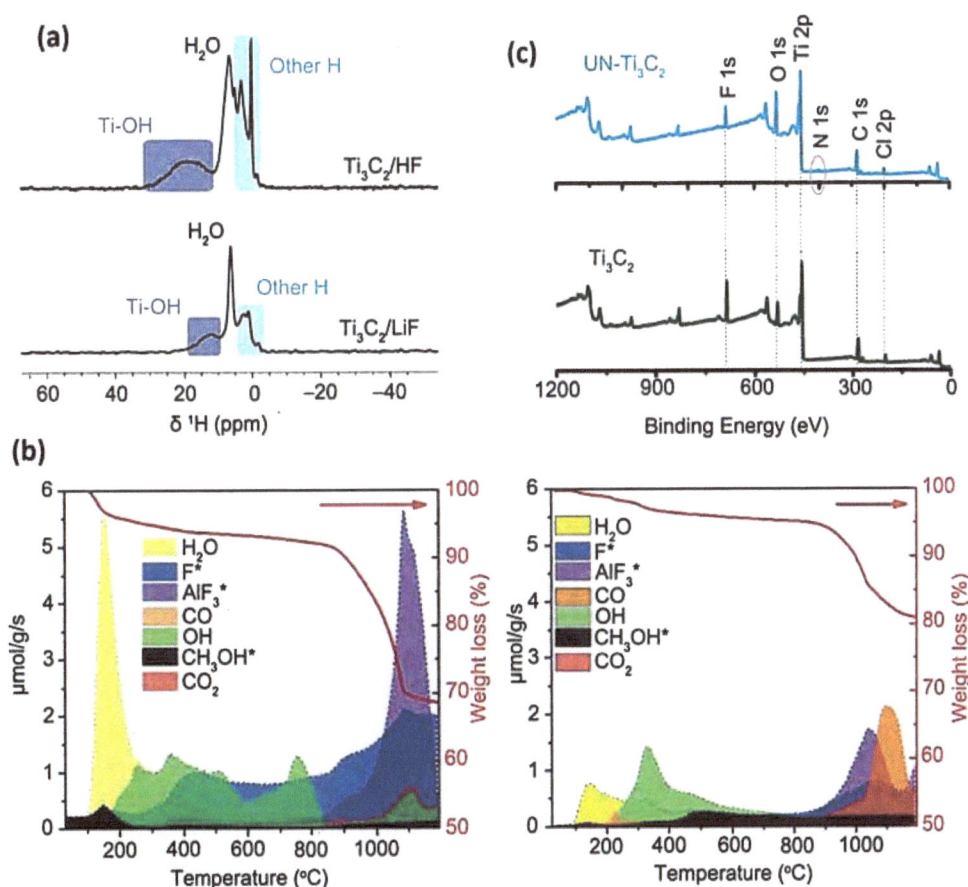

Fig. (4). (a)^{1}H NMR spectrum of MXenes synthesized in HF and LiF/HCl. Adapted with permission from Reference [61] Copyright 2016 The Royal Society of Chemistry. **(b)** Data from temperature-programmed desorption mass spectrometry measurement on MXene following hydrazine treatment. Adapted with permission from Reference [73] Copyright 2016 The Royal Society of Chemistry. **(c)** XPS spectra for bare $Ti_3C_2T_x$ MXene and $Ti_3C_2T_x$ that have been doped with urea-assisted nitrogen. Adapted with permission from Reference [67] Copyright 2018 John Wiley & Sons, Ltd.

Fabrication and Design

Due to fast ion intercalation and bulk redox reactions in the H_2SO_4 solution, MXene, a typical layered electrode material, displays intercalation pseudo-capacitance. High-density charge storage might be possible if there was a large interlayer spacing since it might speed up ion intercalation and increase the amount of active surface exposed to the electrolytes. Therefore, interlayer expansion techniques were utilized to enhance the charge-storing potential for MXenes. Their structure is always influenced by the synthesis techniques, which ultimately results in variations in electrochemical performance. Since higher mobility water molecules are present among MXene interspaces in low-concentration HF-etched MXene, the interlayer spacing is more accessible, and the capacitance is increased [49]. Strikingly, the MXene electrochemical behavior appears unaffected by the etching duration or temperature [74]. There was only a slight increase in capacitance from 100 to 120 Fg^{-1} upon increasing the etching duration more than 100 times. Compared to HF-etched MXenes, an ultrahigh capacitance (900 Fcm^{-3}) was attained in the H_2SO_4 electrolyte when etched in HCl/HF. This may be owing to water intercalation molecules and lithium ions through manufacturing procedures that prevent the restacking of sheets. Moreover, -Cl terminations that are deposited on MXenes corners might increase the d-value because of their greater ionic radii, which indeed encourages the quick diffusion of ions [75]. The restacking and lower conductivity of MXene sheets can be solved simultaneously by partially removing Aluminum (A-layer). Partially etched Al provides a channel for ion transport; however, the remaining A acts as "electron bridges", thus ensuring excellent interlayer conductivity and improving capacitance. Post-processing techniques can also adjust the interlayer configuration, such as intercalation, heat treatment, and many others.

The interlayer becomes wider with multilayered MXene delamination by sonication or larger-size molecule intercalation, thereby making it easier for the ions to diffuse and increase capacitance. The interlayer spaces can be increased by cation intercalation after KOH or CH_3COOK treatment, which increases the capacitance [64]. Compared to pristine MXene, K^+ ions intercalation produces a homogeneous layered structure of MXene and creates a rapid ion diffusion channel. Furthermore, the carbon intercalated N-doped $Ti_3C_2T_x$ MXene showed significant electrochemical properties for 3 Vs^{-1} [76]. A pillar between interlayers was formed when intercalated with hydrazine [73] or polymer (polyvinylalcohol (PVA) and poly diallyldimethylammonium chloride) [77], which facilitated the charge transport and led to a significantly higher capacitance. While several other large organic compounds, *e.g.*, trimethyl alkylammonium salts, could be intercalated in multilayered MXenes, due to higher molecular radius, they seem to hinder the mobility of the ions [78]. The selection of the intercalation species must therefore be carefully considered. In addition to ion intercalation, the ammonia annealing behavior has the potential to expand interlayer spacing. Surprisingly, better electrochemical performance can be achieved using an organized anodic oxidation approach to boost MXene d-spacing and create pores without compromising electrochemically active sites [79]. Along with interlayer structure, MXenes exhibit electrochemical characteristics that are size-dependent. MXenes of smaller sizes improve the performance of electrochemical processes because they lessen ion diffusion routes and enable electrolytes' easier access at the active regions. In addition to the lateral size of the MAX phase precursor, the power and sonication time can be used to modify the MXene flake size [80]. Ti_3AlC_2 particles that have been subjected to ball milling can significantly improve the capacitance of the resulting $Ti_3C_2T_x$ MXene. Optimization of the electrode architecture is also crucial in increasing capacity. In case of Hydrogel $Ti_3C_2T_x$, the volumetric capacitance of MXene electrodes was greater, reaching up to ~ 1500 F cm^{-3} [81]. An open-structure macroporous MXene electrode configuration created using a PMMA microsphere [81] or MgO nanoparticle [82] template exhibits outstanding rate performance [83]. Moreover, $Ti_3C_2T_x$ film density could be efficiently enhanced when external pressures were applied, resulting in great volumetric performances [84].

Factors Affecting the Electrochemical Performance of Supercapacitors

Synthesis

MXenes exhibit pseudo-capacitance due to a bulk redox process with rapid ion intercalation. To increase ion intercalation and improve accessibility to the electrolyte, interlayer spacing must be increased. By delaminating the MXene sheets with sonication, intercalation with cations, or both, the interlayer spaces and, consequently, the capacitance are increased [64]. Replacing the HF etchant with HCl/LiF in H_2SO_4 solution results in Li$^+$ ions intercalation and water molecules that prevent MXene sheets from restacking and increase the capacitance considerably from 300 to 900 Fcm^{-3} [11]. When the Ti_3AlC_2 MAX phase was etched with a FeF_3/HCl, the resulting Ti_3C_2 MXene had small fluorine terminations, more Fe^{3+} cations, and water intercalation among layers, causing better-quality electrochemical progress of Ti_3C_2 electrode [85]. Gogotsi *et al.* demonstrated a higher volumetric capacitance of $Ti_3C_2T_x$ MXene electrodes by intercalating various cations, including K$^+$, Al^{3+}, Na$^+$, and Mg^{2+} [86]. They explained that cation intercalation, which raises the c-lattice parameter of MXenes, is the cause of increased capacitance. Lu *et al.* [87] found that high valence cation Al^{3+} pillared MXene performed better electrochemical than low valence cations like Mg^{2+} and Na$^+$ pillared MXene. According to the DFT calculations, M_2C MXene intercalated with Li$^+$ or Mg^{2+} exhibit enhanced gravimetric capacitances because of their large radii compared to the Na^+ or K^+ intercalated M_2C MXenes [88].

The interlayer separation is further increased by pre-pillaring MXene with cationic surfactants like cetyltrimethylammonium bromide (CTAB), which have prolonged hydrophobic chains . The ion exchange in Li^+ intercalated $Ti_3C_2T_x$ MXene with TMAOH cations induces a significant upsurge in MXene's resistivity as the large size of TMAOH makes it difficult for the ions to diffuse and was examined in electrochemical capacitors as an interlayer framework [78]. Hantanasirisakul *et al.* [89] suggested a simple technique to make transparent and conductive nanosheet flakes by spray coating the delaminated $Ti_3C_2T_x$ MXene. As seen in Fig. (**5a**), $Ti_3C_2T_x$ thin films were created by spray-coating the d-$Ti_3C_2T_x$ suspension onto a transparent and flexible substrate. Fig. (**5b**) depicts a digital photograph of the $Ti_3C_2T_x$-coated flexible substrate. Uniformly thin, transparent, and conductor MXene films based on Ti_3C_2 were produced on flexible substrates. The sheet resistance for Ti_3C_2 MXene films were reported to be 0.5 and 8 $k\Omega$ sq^{-1} at a transmittance of 40% and 90%, respectively. It suggests that flexible devices can be created using synthesized $Ti_3C_2T_x$ thin films. Recently, Zhang *et al.* created MXene-based membranes employing vacuum-assisted filtration [90]. Aramid nanofibers (ANFs) have received interest as an intriguing nanomaterial in varity of applications owing to their exceptional chemical and thermal stability and have been combined with other materials [91]. As shown in Fig. (**5c**), Ti_3C_2-based MXene was initially made by selectively etching Al from Ti_3AlC_2 through HF etchant. Kevlar yarns were used to produce charged ANF yarns, as shown in Fig. 5d. As-prepared MXenes and ANFs were suspended in DMSO to form stable suspensions with good dispersibility to produce MXene/ANF hybrid ink. The filtration technique created the flexible and conductive MXene/ANF composite film shown in Fig. (**5e**). According to Mashtalir *et al.* [73], hydrazine intercalation causes the pillaring effects in $Ti_3C_2T_x$ MXene and dramatically reduces the amount of water and the fluorine terminations and increasing the porosity of the layered surface as well, thus facilitating better access to the electrolytes.

The H_2SO_4 electrolyte exhibits a capacitance increase of 250 F/g and a longer cycle life with 75 mm thick electrodes [73]. Mashtalir *et al.* [18] reported that $Ti_3C_2T_x$ MXene delamination *via* DMSO intercalation enhances the specific surface area from 23 m^2/g to 100 m^2/g. The volumetric capacitance of this delaminated MXene increased by up to 520 Fcm^{-3} and had improved specific surface area with decreased thickness. $Ti_3C_2T_x$ MXene intercalated with ethylenediamine showed a specific surface area of 59.2 m^2/g, further enhancing these electrodes' electrochemical efficiency [92]. In addition to organic bases, various 0-D nanoparticles (*e.g.*, quantum dots, MnO_x, RuO_2, and Ag), one-dimensional materials (*e.g.*, nanotubes, nanowires, and 2D nanotube materials) (*e.g.*, MoS_2, reduced graphene oxide, nanosheets of phosphorus); and several heterospecies (*e.g.*, double layered hydroxides (LDH), polymers, *etc.*) caused the pilling effect and showed better electrochemical performance than the MXene-composite based supercapacitors [93]. Besides the various intercalation approaches, advanced synthesis techniques have been demonstrated in the literature to improve interlayer separation and protect the MXene sheets from restacking, which also contributes to better electrochemical performance. Contrarily, recent research has revealed that the volumetric capacitance of MXene electrodes decreases upon the introduction of nanomaterials and other heterostructures. A facile-controlled H_2SO_4 oxidation technique was used to produce porous and small flake-size Ti_3C_2Tx MXene films with enhanced interlayer spacing to overcome these issues. The produced electrodes demonstrated improved electrochemical efficiency without forfeiting the volumetric capacitance. In a different investigation [94], using a low-cost, scalable laser writing technique, interlayer water molecules and partial functional groups were photothermally gasified to create pores in the restacked sheets.

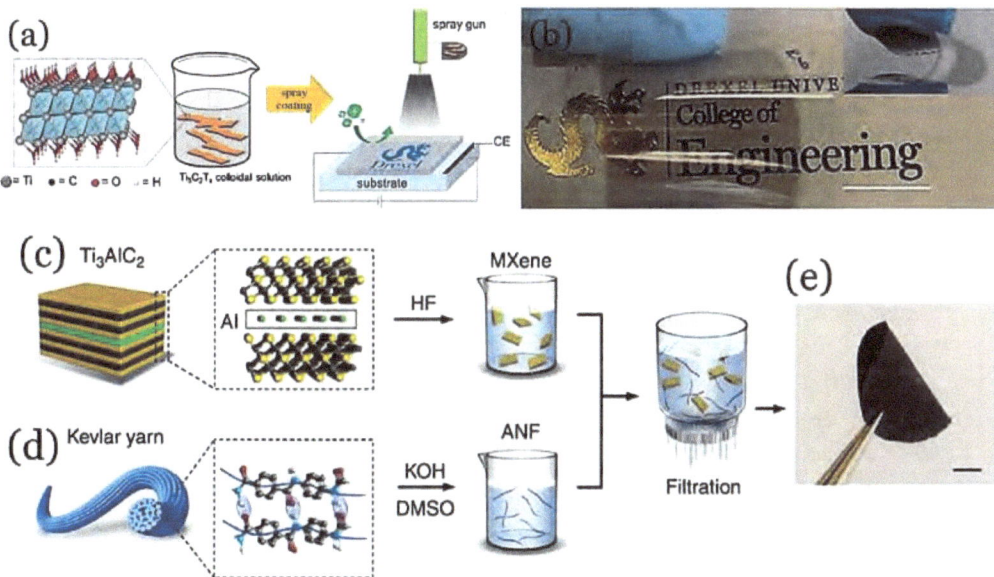

Fig. (5). (a) Preparation of conductive and transparent $Ti_3C_2T_x$ electrodes *via* spray coating technique. **(b)** Images of the MXene films over the flexible polyester substrate (Inset shows a high degree of the $Ti_3C_2T_x$ MXene film bending over that substrate). Adapted with permission from Reference [89] Copyright 2016 John Wiley & Sons, Ltd. **(c)** Ti_3C_2 MXene was synthesized *via* HF etching from Ti_3AlC_2 MAX phase precursor. **(d)** The ANFs were produced by using KOH/DMSO exfoliation method from commercial Kevlar yarns. **(e)** Image of the free-standing, flexible, paper-like MXene/ANF composite nanosheets synthesized *via* vacuum-assisted filtration. Adapted with permission from Reference [90] Copyright 2019 Nature Publishing Group.

The symmetric MXene-based micro-supercapacitor showed improved capacitance and rate capabilities due to the reduction in MXene layer restacking. Upon removal of surface terminations, a carbon intercalated sandwich structure was created by the *in-situ* carbonization of long-chain fatty amines in the spaces between the layers of $Ti_3C_2T_x$ MXene. Consequently, the interlayer spacing increased because of this carbon intercalation, and the MXene electrodes showed more excellent electronic conductivity and improved specific capacitance than pristine MXene [95]. Partial elimination of Al layers can resolve the issue of restacking in MXene sheets and weak conductivity. As a result of the leftover layers, there is a bridge of electrons that facilitates interlayer conduction and increases capacitance significantly.

According to Hu *et al.* [96], the capacitance is increased by increasing the interlayer spacing and integrating heteroatoms with reduced electronegativity between the interlayers. Interlayer separation provides a more significant ion diffusion channel, allowing faster ion transfer and a greater capacitance due to the abundance of redox-active sites. Thus, ammonia-annealed Ti_2CT_x MXene etched with HCl/LiF revealed an exceptional specific capacitance of 570 F/g. Due to more excellent electron donor capability, the less electronegative N atoms share their electrons with intercalated cations, and the enhanced charge mobility further enhances pseudocapacitance [96]. It has been reported that a $Ti_3C_2T_x$ hybrid MXene intercalated with N-doped carbon displays better electrochemical performance at scan rates up to 3 V/s and retains its specific capacitance of 82.8 F/g for 1.0 A/g even after 5000 cycles [76]. It was underlined that N-$Ti_3C_2T_x$ at the temperature of 300 °C displayed a considerably low slope in the short frequency range and much superior corresponding series resistance and R_{ct}; this clearly expresses the present's substandard conductivity electrode compared to N-$Ti_3C_2T_x$ at 200 °C [68]. They are in good agreement with the electrochemical progress enhancement, which was less marked in N-$Ti_3C_2T_x$ at 300 °C than N-$Ti_3C_2T_x$ at 200 °C as visualized in Fig. (6). Furthermore, the electrochemical progress of $Ti_3C_2T_x$ and N-$Ti_3C_2T_x$ electrodes were verified using the three-electrode arrangement in which cells were occupied with H_2SO_4 and $MgSO_4$ (1 M for both). In Fig. (6a), observed CV results show a perfect rectangular shape, which is usually realized for pure EDLC on carbon electrodes. Unexpectedly, N-$Ti_3C_2T_x$ at 200 °C electrode preserved exceptional capacitance progress at higher voltage scan rates (200 mV s^{-1}) with Cg = 128 F g^{-1} as displayed in Fig. (6b). The $Ti_3C_2T_x$ electrodes exhibit a rectangular-shaped CV at 1 mV s^{-1} using $MgSO_4$ (1 M) as visualized in Fig. (6c), and rectangular form maintained at ~ 50 mV s^{-1}. Finally, $Ti_3C_2T_x$ electrodes had a definite gravimetric capacitance (52 F g^{-1}) at 1 mV s^{-1} using $MgSO_4$ (1 M) that weakened with growing voltage scan rates as demonstrated in Fig. (6d).

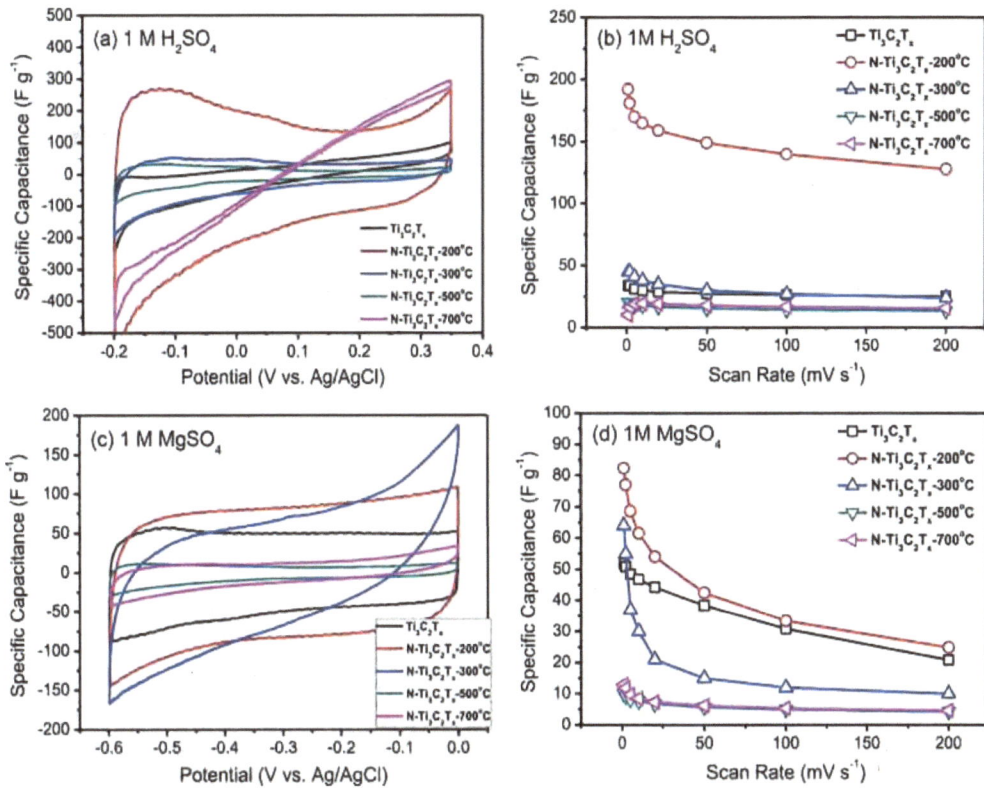

Fig. (6). (a) CV images at 1 mV s^{-1} and **(b)** Ti$_3$C$_2$T$_x$ and N-Ti$_3$C$_2$T$_x$ specific capacitances in 1 M H$_2$SO$_4$ electrolyte at several scan rates. **(c)** CV models at 1 mV s^{-1}, and **(d)** measurements of Ti$_3$C$_2$T$_x$ and N-Ti$_3$C$_2$T$_x$ specific capacitances at various scan rates in 1 M MgSO$_4$. Adapted with permission from Reference [68] Copyright 2017 Elsevier B.V.

When comparing *ex-situ* N solvothermal doping to *in-situ* "N" solid-solution doping (TN-Ti$_3$C$_2$), *ex-situ* N solvothermal doping allows a drastic control over nitrogen content [67]. The N-doped MXenes resulting from this process are more reliable in obtaining higher capacitance. In 3M H$_2$SO$_4$, N-doped Ti$_2$CT$_x$ MXenes synthesized using the urea (UN-Ti$_3$C$_2$) and monoethanolamine (MN-Ti$_3$C$_2$) supported solvothermal technique exhibited outstanding volumetric capacitances of 2836 F/cm^3 and 2643 F/cm^3, respectively. In UN-Ti$_3$C$_2$ films, the capacitance retention was 81.7% after 20,000 cycles, while in MN-Ti$_3$C$_2$ films, the capacitance retention was 100% after 20,000 cycles [67]. UN-Ti$_3$C$_2$ films-based symmetric supercapacitors attained remarkable energy densities of 76 Wh L^{-1} and proved a highly effective method for engineering supercapacitors [67]. An analysis of the electrochemical performance was presented for the production of MXene electrodes in 1 M H$_2$SO$_4$ synthesized from aqueous suspensions by vacuum filtering, electrophoretic precipitation (EPD) from an aqueous suspension (EPD

H_2O), and propylene carbonate-based suspension (EPD PC), respectively. In 1M H_2SO_4, the electrochemical performance of MXene electrodes produced from aqueous suspensions by vacuum filtering, electrophoretic precipitation (EPD) of an aqueous suspension (EPD H_2O), and propylene carbonate-based suspension (EPD PC) was examined [97]. The maximum capacitance of the EPD PC film was observed at 2 mV/s, which rapidly decreased as the scan rate increased. Even though water-based films such as EPD H_2O and H_2O had significantly lower capacitances, they exhibited almost identical scan rate behavior [97].

Structure and Size of MXene

In addition to the interlayer structure and the corresponding MXene precursor, the flake size also affects their electrochemical behavior. The electrochemical behavior can be improved by using MXenes with a smaller lateral size because they reduce the path for ion diffusion, thereby making the ion electrolytes more readily available on the electrode surface. Small flakes, conversely, have a lower electrical conductivity because of enhanced interflake/interfacial contact resistance and sonication-induced defects, which adversely affects the capacitance. An improved gravimetric (435 F/g) and volumetric capacitance (1513 F/cm^3) were reported by combining the different sizes of MXene flakes (both small and large) at a relatively fixed ratio of 1:1 M at a slow scan rate [80]. According to thorough research, larger MXene flakes may offer lower interflake/interfacial contact resistance and exhibit rapid electrical conductance.

The capacitance of submicron-size MXene particles produced by molten salt etching was nearly double that of the large flakes produced through conventional high-temperature etching techniques [98]. Extensive structural, spectroscopic, and electrochemical analyses revealed that the decrease in the lateral size of MXene particles is responsible for the rise in capacitance. In contrast, the smaller flakes arrange a more direct route for ion diffusion, increasing ionic conductivity and making the combination suitable for producing electrodes. In MXenes, the transition metal has surface functional terminations that provide active sites for typical redox reactions followed by ion intercalation (as described earlier). As a result, the surface morphology significantly influences the electrochemical properties of MXene-based supercapacitors. Gravimetric capacitance is larger for MXenes with -O surface functional groups than those with -F, -H, or -O.H. surface terminations [88].

MXenes with light transition atom and -O surface functional groups were found to have excellent capacitance. Additionally, the researchers reported 28.2% and 214% capacitances greater than those of HF [64] and LiF/HCl etched $Ti_3C_2T_x$ clay [11], respectively. Negatively charged $Ti_3C_2T_x$ MXene with significantly high uptake of Li^+ ion were generated by their MAX phase precursor using redox-controlled etching with lewis acidic melt, which exhibits a solely pseudocapacitive performance in 1 M $LiPF_6$ carbonate-based electrolyte [59]. Although MXenes produced through alkali and Lewis acid etchants have better specific capacitance than HF-etched MXenes; however, their volumetric capacitance is still lower than that of HCl/LiF etched MXenes due to their inability to delaminate [38]. Also, the fluoride-free $Ti_3C_2T_x$ MXene electrodes produced subsequently after electrochemically etching route was utilized to construct the symmetric solid-state supercapacitor that showed improved cycle stability. Additionally, they showed a high areal capacitance of 20 mF/cm^2 along with a volumetric capacitance 439 F/cm^3 at 10 mV/s, respectively. They found that low fluorine concentration and large oxygen surface groups of MXenes are responsible for their high capacitance [99].

Additionally, various post-processing approaches like intercalation, heat treatment, and other similar techniques can change the composition of surface functional groups. Such as, at 2 mV/s, the volumetric capacitance increases to 520 F/cm^3 by replacing the -F terminal groups with -O *via* CH_3COOK or KOH intercalation [64]. Comparatively, the potassium hydroxide and potassium acetate intercalated $Ti_3C_2T_x$ MXene, which possesses O and/or OH surface groups that subsidize the surface redox reactions and produce pseudocapacitance, the -F terminated $Ti_3C_2T_x$ MXene exhibits a lower capacitance. The O/F ratio of LiF/HCl etched $Ti_3C_2T_x$ exhibits an O/F ratio of 3.6 after DMSO intercalation, which is significantly greater than that of the non-delaminated LiF/HCl etched $Ti_3C_2T_x$. Compared to HF-etched delaminated $Ti_3C_2T_x$ MXene (333 F/cm^3), a considerably higher volumetric capacitance of 508 F/cm^3 was achieved with -O surface terminations [100]. Researchers found that delaminated $Ti_3C_2T_x$ MXene electrodes etched with LiF/HCl and HF retained 81% and 61% of their capacitance, respectively. The $Ti_3C_2T_x$ MXene synthesized through K+ ion intercalation and subsequent calcination demonstrated an acceptable specific capacitance of 500 F/g at 1 mV/s in 1M H_2SO_4 with retention of 99% up to 10,000 cycles [65]. Symmetrical supercapacitors fabricated by as-synthesized MXene electrodes depict a greater energy density of 27.4 W h/kg in a combination of two electrolytes with 1M H_2SO_4 and 1M Li_2SO_4. The intercalation of K^+ ions increased the gap between the layers; the calcination technique significantly eliminated the surface terminations to improve the exposure of the active sites to the electrolytes and leads to higher capacitance. Due to the increased carbon content and fewer fluorine terminations, HF-etched Ti_2CT_x MXene performed better electro-chemically after heat treatment under N_2, N_2/H_2, or Ar environment. Fluorine reduces the tight-binding force between the surface of MXene electrode and electrolyte ions, resulting in a 20% decrease in self-discharge (0.65 atom%) over MXenes with higher F levels (8.09 atom%) [37].

Architecture of Electrodes

The proposed electrode design significantly impacts the electrochemical behavior of the MXene electrode. According to Zhu *et al.* [82], MgO nanoparticles can extend MXene sheets. It demonstrated a capacitance of 180 F/g at 1 A/g and showed capacitance retention of 99% over 5,000 charge-discharge cycles at 5 A/g. In addition, they found that heating the MXene-Urea composite to 550 °C produced porous MXene foam, which exhibits a specific capacitance of 203 F/g at 5 A/g with 99% retention capacity after 5000 cycles. In contrast, pure MXene displayed a capacitance of 82 F/g under identical conditions. There is a significant improvement in rate performance and stability of symmetric and asymmetric supercapacitors made with MXene-foam/MXene and MXene-foam/MnO$_2$ electrodes. Furthermore, the asymmetric electrode outperformed the symmetric electrode and showed an energy density of 16.5 Wh/kg and a power density of 160 W/kg. MXene hydrogel supercapacitor was designed by freezing the MXene slurry unidirectionally and then introducing the suggested thawing technique in the H$_2$SO$_4$ electrolyte [101]. Due to its open 3D morphology, the hydrogel electrode is highly permeable to electrolyte and intercalated protons, facilitating electrolyte transport and ion absorption. This approach can help significantly reduce MXene sheet restacking, a common issue for the poor electrocatalytic activity of most MXene-based supercapacitors (Fig. 7).

When mechanically sheared and self-assembled into a discotic lamellar liquid crystal phase, the 2D MXene exhibits electrochemical performance independent of film thickness up to 200 mm [83]. Electrochemical kinetics are improved, and ion transit times are decreased substantially under vertically arranged MXene arrays. Moreover, increasing the surface area of MXene electrodes can increase their electrochemical capacitance, and also pseudo capacitance is enhanced because electrolytes have greater access to electrodes. Furthermore, the specific surface area of electrodes can be altered by changing the synthesis conditions. Tang *et al.* [102] studied Ti$_3$C$_2$ MXene by subjecting it to prolonged etching times and observed that MXene sheets start delaminating upon continued exposure to C atoms. For the 216-hour etching period, there were no appreciable improvements in the capacitance value at a scan rate of 5 mV/s. After 216 hours, the capacitance increased dramatically, producing a specific capacitance of 118 F/g. The scientists attribute the increased capacitance to the higher C-content, which improves overall conductivity, accelerates electron transport, increases the effective surface area, and ensures a better electrolyte availability to the electrodes. The ball milling method was chosen because of the higher carbon content and the creation of porous surfaces. It increases the capacitance of Ti$_3$C$_2$ MXene by improving the electrochemical efficiency of MXene electrodes [74]. It is observed that macroporous and hydrogel Ti$_3$C$_2$T$_x$ MXene electrode designs exhibit increased volumetric and gravimetric capacitances of up to 1500 F/cm³, respectively [81]. The interlayer spacing increases and defects are introduced into the $Ti_3C_2T_x$ MXene electrode under controlled anodic oxidation in an acidic electrolyte, thereby improving their electrochemical performance [79]. The resulting MXene electrode increased the performance rate by 30% and showed 2 times increased capacitance retention [79].

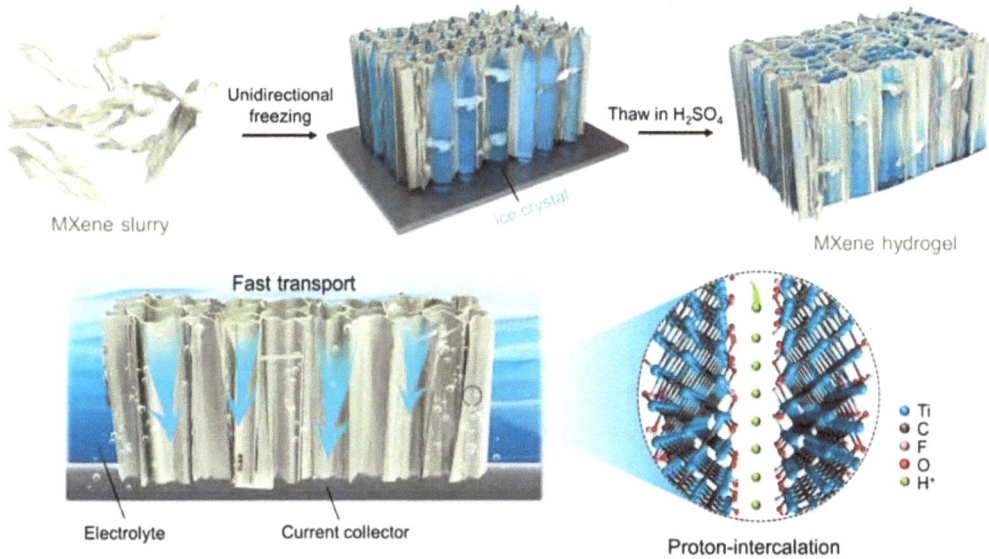

Fig. (7). **(a)** Schematic representation of the multi-scale structural engineering strategy that combines unidirectional freezing and thawing in H_2SO_4 for designing the MXene hydrogel supercapacitor electrode. Adapted with permission from Reference [101] Copyright 2021 John Wiley & Sons, Ltd.

Electrolyte

The electrochemical performance of MXene-based supercapacitors may be altered by changing the electrolyte. The electrolyte can percolate the space between the MXene layers through cation intercalation. The electrolyte also influences the voltage window, conductivity, pseudocapacitive functioning, and efficiency of the supercapacitor. A variety of electrolytes, such as potassium sulphate (K_2SO_4) and aluminum sulphate ($Al_2(SO_4)_3$) are used along with the aluminum nitrate ($Al(NO_3)_3$). Compared with Al^{3+} ions, K^+ ions produced a more significant rectangular CV curve, illustrating how cation valencies affect electrochemical efficiency of supercapacitor [86]. Furthermore, experiments showed that using the more conductive electrolyte, $Al(NO_3)_3$, resulted in better CV performance than using the less conductive electrolyte, $Al_2(SO_4)_3$. The H_2SO_4 electrolyte has been found to limit the electrode potential of asymmetric supercapacitors to 1.3 volts and for symmetric supercapacitors to 0.9 volts [103]. It is possible to overcome this narrow voltage window by employing a neutral LiCl salt electrolyte, which broadens the potential window from 1.2 V – 0.3 V. The asymmetric supercap-acitor was developed using $Mo_{1.33}CMn_xO$ electrode in a 5M LiCl and exhibited a remarkable volumetric energy density of 58 mWh/cm^3 [103]. Comparatively speaking about acidic and salt electrolyte solutions, basic electrolytes like NaOH and KOH performed more efficiently in the electrochemical analysis [44]. A three-electrode asymmetric supercapacitor with Ag/AgCl reference electrode and LiOH, KOH, and NaOH electrolytes demonstrated a volumetric capacitance of up to 350 F/cm^3 [44]. Both $Ti_3C_2T_x$ MXene with KOH intercalation and delaminated $Ti_3C_2T_x$ MXene exhibited stable capacitances of 215 F/cm^3 and 415 F/cm^3, respectively, during galvanostatic cycling at 5 A/g [64].

Upon partial oxidation of MXene with H_2O_2 and flocculation in acid, it formed a porous material with a higher surface area and exhibited a gravimetric capacitance of 307 F/g at a 20 mV/s in 1M H_2SO_4. It has been proved from the literature that oxidation and structural crumpling led to the creation of mesopores and macropores, respectively. These pores are reliable for boosting the diffusion of ions and preventing the restacking of MXene layers, raising the electrochemical performance of electrodes [104]. MXene films become denser as the pressure rises, which increases the volumetric capacitance [84]. At a pressure of 40 MPa, the volumetric energy densities of the delaminated $Ti_3C_2T_x$ MXene supercapacitor electrode in 1M Li_2SO_4 and 1M 1-ethyl -3- methyl imidazolium tetra fluoroborate/ acetonitrile organic electrolytes were reported to 22 WhL^{-1} and 41 WhL^{-1}, respectively. It was also observed that the electrode's capacitance is incredibly reliant on its thickness and drops progressively as thickness increases [11]. This could be due to the electrode's geometry, which shows that thinner flakes have a better-layered structure while thicker flakes have a poorly arranged layered structure with defects. Through the breakdown of 3D hydrogel into microgels and rebuilding on $Ti_3C_2T_x$ nanosheets in variable mass proportions, the $Ti_3C_2T_x$ MXene film developed a very dense and porous 3D matrix. It resulted in an enhanced energy density of up to 40 Wh/L when used as an electrode for a supercapacitor [105]. Lin *et al.* [106] investigated the electrochemical performance of working electrodes with varying mass loadings of Ti_3C_2 MXene ranging from 1.8 to 7.6 mg. An electrode loaded with 1.8 mg of Ti_3C_2 MXene produces 117 F/g of specific capacitance at a scan rate of 2 mV/s with 579 mF/cm^2 was the highest areal capacitance of 7.6 mg loaded Ti_3C_2 MXene electrode. It is surprising that both electrodes retain 97% and 98% of their capacitance after 10,000 cycles.

Current Advances in the MXene-based High-performance Supercapacitors

MXene-based two-dimensional materials seem promising in supercapacitors due to their layered nature, exceptional electrical and mechanical properties, and extended surface areas. Spontaneous intercalation in the $Ti_3C_2T_x$ MXene sheets was first studied by Lukatskaya *et al.* [44] by intercalating the cations (Na^+, NH^{4+} Al^{3+}, and Mg^{2+}) from their aqueous salt solution. These findings pave new pathways toward the fabrication and application of MXene-based supercapacitors. Among various MXenes discovered so far, $Ti_3C_2T_x$ has been the most explored material for supercapacitors and usually exhibits a pseudocapacitive response in H_2SO_4 [50]. Hu *et al.* [51] studied the capacitive behavior of Ti_3C_2 MXene using *in-situ* Raman spectroscopy in aqueous electrolytes such as H_2SO_4, $MgSO_4$, and $(NH_4)_2SO_4$ containing three kinds of sulfate ions. Pseudocapacitance is produced in acidic electrolytes by the hydronium in the H_2SO_4 electrolyte, participating in bonding/debonding with the terminated O atoms on $Ti_3C_2T_x$ during discharging/charging. Lukatskaya *et al.* [50] synthesized self-assembled $Ti_3C_2T_x$ films *via* a dropping-mild baking technique and $Ti_3C_2T_x$ MXene flake suspension. In acidic solutions, an excellent gravimetric capacitance of 499 F/g was observed with great cyclability. Xu *et al.* [107] recently created a modified electrophoretic deposition process to create binder-free $Ti_3C_2T_x$ electrodes for supercapacitors that exhibit outstanding electrochemical characteristics. Because the Al atom layers in the original Ti_3AlC_2 phase were deliberately removed during the etching process, the obtained etched $Ti_3C_2T_x$ bricks exhibit a separated layered structure, as shown in Fig. (**8a**).

The SEM micrograph of $Ti_3C_2T_x$ at higher magnification (Fig. **8b**) further supports the fact that layers of Al atom layers can be preferentially removed by etching, leaving only the periodic 2D interlayer space. Due to interlayer coupling in MXenes, the cross-section of $Ti_3C_2T_x$ bricks (Fig. **8c**) between these individual MXene $Ti_3C_2T_x$ nanosheets still shows the irreversible stacking phenomenon. Additionally, the cross-sectional TEM image (Fig. **8d**) explicitly indicates the phenomenon of irreversible stacking [107]. Two pieces of $Ti_3C_2T_x$ film on conductive fabric substrates were used to assemble a flexible symmetric all-solid-state supercapacitor, which can be used to create a watchband-shaped supercapacitor to power an LED (Figs. **8e-f**). As shown in Fig. (**8i**), the electrochemical effectiveness of the device holds almost constant throughout the bending process, demonstrating that bending the devices has little impact on the functionality of flexible solid-state supercapacitors. These supercapacitors exhibit exceptional electrochemical behaviour, as shown by CV (Fig. **8g**) and galvanostatic charge-discharges (Fig. **8h**), with conventional capacitive effectiveness and highly reversible electrochemical redox processes when the flexible solid-state supercapacitors were charged and discharged. The cyclic efficiency of flexible supercapacitor is also assessed as a crucial factor for supercapacitors in application areas (Fig. **8j**).

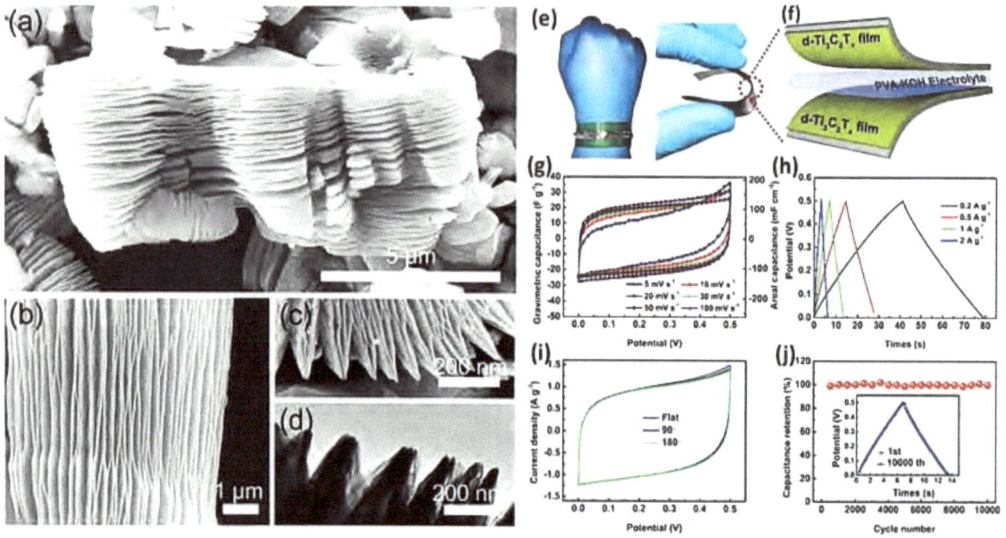

Fig. (8). **(a-c)** Characteristic SEM and **(d)** TEM photographs of $Ti_3C_2T_x$ nanoflakes at various magnifications. **(e)** The digital image depicts three flexible solid-state supercapacitors connected in series to drive an LED and a $Ti_3C_2T_x$ film electrode on the wearable conductive fabric substrate. **(f)** Schematic diagram of the assembled flexible supercapacitors. **(g)** Cyclic voltammogram of the assembled flexible supercapacitor at scan rate ranging from 5-100 mV s^{-1}**(h)** galvanostatic charge-discharge curves at various current densities. **(i)** Capacitive efficiency of flexible supercapacitors for several bending angles at a scan rate of 100 mVs^{-1}. **(j)** Cycling performance of the assembled flexible solid-state supercapacitors. Inset showing the 1st and 10,000th charge-discharge curves at 1 A g^{-1}. Adapted with permission from Reference [107] Copyright 2017 Elsevier B.V.

With its distinctive layered porous structure, strong electrical conductivity, no binders' requirements, and/or conductive additives, all solid-state MXene-based symmetrical micro-supercapacitors display outstanding electrochemical progress. Zhang and colleagues [108] reported that the fabrication of solid-state supercapacitors assisted in greatly conductive and transparent $Ti_3C_2T_x$ MXene, showing significant volumetric capacitance (~ 676 F/cm^3). According to Lukatskaya *et al.* [81], macroporous $Ti_3C_2T_x$ MXene has a high capacitance (~ 100 F/g) with 40 V/s scan rate and 210 F/g at 10 V/s. Moreover, $Ti_3C_2T_x$ MXene hydrogels also attained a significant volumetric capacitance of ~1500 F/cm^3. Dall'Agnese *et al.* studied the effect of surface properties on the electrochemical behaviour of $Ti_3C_2T_x$ MXene supercapacitors in H_2SO_4 electrolyte [64]. F-terminations must be replaced as they do not contribute to any pseudocapacitive activities. The $Ti_3C_2T_x$ layers can be delaminated using DMSO, thus allowing the water to interact and oxidize the $Ti_3C_2T_x$ layers. Owing to the O-containing groups, a high volumetric capacitance of 415 F/cm^3 was observed at 5 Ag^{-1} without any evident degradation after the 10,000 cycles. The findings show that appropriate surface modification of MXene can tune their electrochemical performance. As illustrated in Fig. **(9a)**, Li *et al.* recently produced intercalated

$Ti_3C_2T_x$ MXene with modified surface terminations (400-KOH-Ti_3C_2) by K^+ intercalation and subsequent calcination [65]. Unlike pristine Ti_3C_2 electrodes, 400-KOH Ti_3C_2 electrodes in 1M H_2SO_4 showed outstanding gravimetric capacitance of 517 F/g at 1 A/g (almost 211% from the origin) with enhanced cycling stability of around 99% after 10,000 cycles (Figs. **9b-d**). This considerable improvement can be accredited to the large interlayer spacing and the reduced concentration of surface terminal groups.

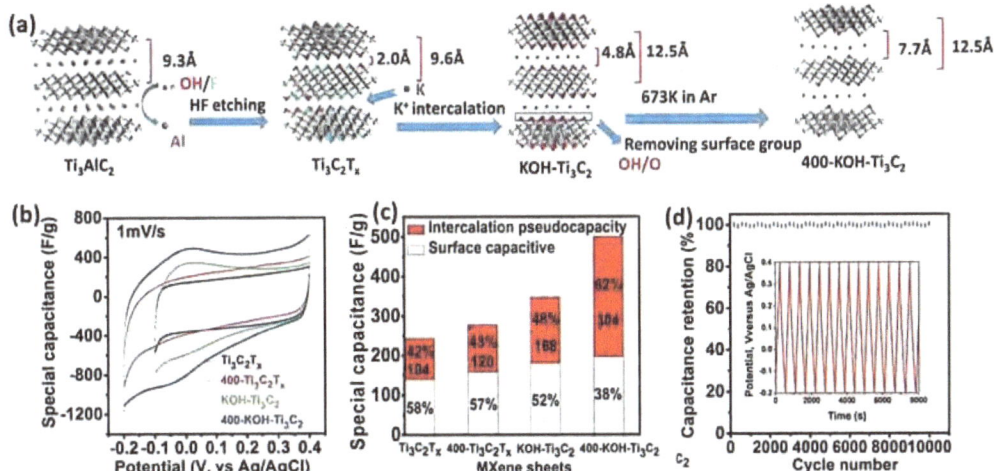

Fig. (9). **(a)** Diagrammatic representation of the modified MXene synthesis technique **(b)** Cycle Voltammogram at 1 mV s^{-1} in 1 M H_2SO_4; **(c)** capacitive comparison; **(d)** Retention tests of 400-KOH-Ti_3C_2 electrode in 1 M H_2SO_4 electrolyte. Inset represents the galvanostatic charge/discharge cycling data at 1 A g^{-1}. Adapted with permission from Reference [65] Copyright 2017 John Wiley & Sons, Ltd.

Further studies showed that the intercalation of hydrazine into the $Ti_3C_2T_x$ MXene sheets could alter the surface chemistry by reducing the -F and -OH groups and the intercalated water. Additionally, the hydrazine intercalation improves accessibility by creating pillars between the MXene layers. Therefore, the MXene electrode treated with hydrazine exhibited enhanced capacitance of 215 F/g at 5 A/g in 1 M H_2SO_4, having about zero degradation following 10,000 cycles. Vacuum-assisted filtration and immersion of $Ti_3C_2T_x$ in the LiCl solution allows the preparation of Li^+ intercalated $Ti_3C_2T_x$, which has an excessive volumetric capacitance of 892 F/cm^3 in addition to reliable cyclic stability [109]. By ion exchange with trimethyl alkyl ammonium cations and Li^+ intercalation, MXene's interlayer spacing is also increased. According to Levi *et al.* [45], the cationic insertion appears rapidly in MXene and is followed by significant deformation of MXene particles. Those adsorbed cations are then electrochemically incorporated between swelled layers. So, the argument leads to the conclusion that the $Ti_3C_2T_x$ interlayer has shallow and deep cationic adsorption sites (as discussed previously). As a result, a high capacitance is produced by the high rate of intercalation of ions into the $Ti_3C_2T_x$ MXene.

Additionally, MXene-based composites have been investigated as supercapacitor electrodes. Ling *et al.* studied the electrochemical performances of $Ti_3C_2T_x$/polymer (poly diallyl dimethyl ammonium chloride or polyvinyl alcohol) composite films, and the composite electrodes displayed better volumetric capacitance in KOH electrolyte [110]. In another study, Boota *et al.* [111] utilized $Ti_3C_2T_x$ MXene with an electrochemically active polypyrrole (PPy) polymer that can intercalate between the MXene sheets and enlarge the interlayer spacing. The PPy chains can offer a good arrangement and strong Pseudocapacitance. Such PPy/$Ti_3C_2T_x$ composite demonstrated a high volumetric capacitance of 1000 F/cm^3 when used as the electrode for supercapacitors with 92% retention after 25,000 cycles.

Zhu *et al.* synthesized conductive and freestanding hybrid films of PPy/$Ti_3C_2T_x$ with good electrochemical efficiency for supercapacitor electrodes. Research on the interaction between the polyfluorene derivatives and $Ti_3C_2T_x$ has recently been performed [112]. It shows that novel organic materials can intercalate among MXene layers and can be used to create hybrid structures for high-performance energy storage devices. Besides the carbon-based substances and polymer composites to enable their efficient surface utilization and prevent MXenes restacking, other materials like carbon nanotubes (CNTs) and graphene have also been added to increase the interlayer spacing. Another approach was reported to create sandwich-like MXene/CNT composite paper electrodes [77]. Additionally, the technique was utilized to develop sandwich-like MXene/OLC and MXene/rGO papers using zero-dimensional onion-like carbon (OLC) and 2D rGO. The as-synthesized sandwich-like paper electrodes showed exceptional electrochemical behaviour in the 1M $MgSO_4$ aqueous electrolyte solution. After that, Yan *et al.* [113] developed a technique for producing flexible MXene ($Ti_3C_2T_x$)/rGO films by electrostatic self-assembly between the positively charged rGO and negatively charged MXene nanosheets. The resulting self-supporting MXene/rGO-5% film displayed exceptional volumetric capacitance (1040 F/cm^3) and exceptional cycle stability [57]. The CNT/$Ti_3C_2T_x$ electrodes showed decent electrochemical progress in alkaline and organic media. Wang *et al.* created $Ti_3C_2T_x$ MXene/nickel-aluminum composites by growing LDH *in situ* on an MXene substrate [114]. A prepared 3D network provides firm transport of ions, offers copious active sites for oxidation-reduction reactions, and shows a higher specific capacitance ($\sim$ 1061 F/g). The electrochemical progress of MXene was improved by adopting transition metal oxides such as MnO_x and TiO_2 as composites.

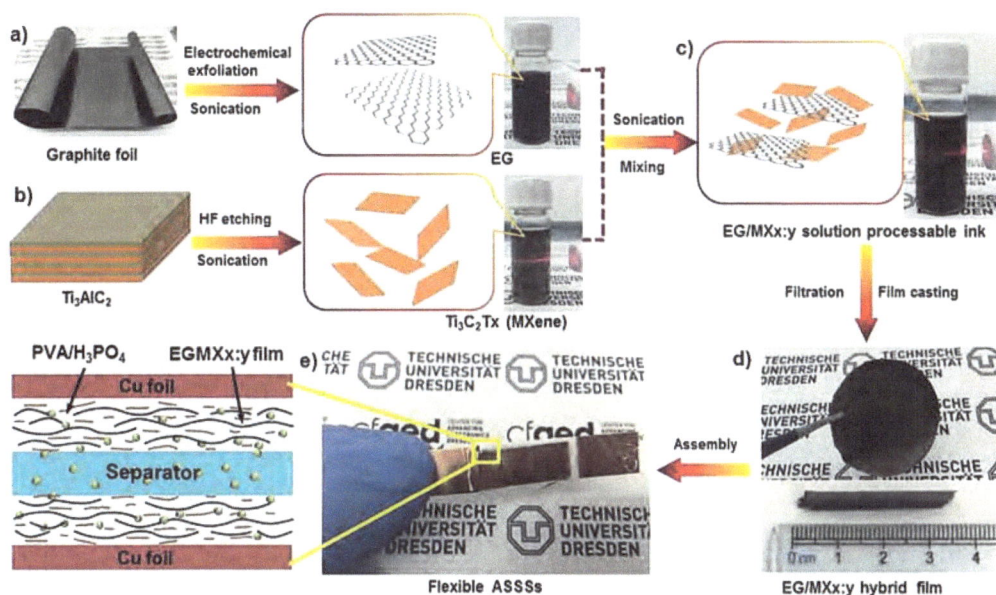

Fig. (10). (a) Schematic illustration of graphene and stable graphene- isopropanol dispersion using electrochemical exfoliation and sonication technique. **(b)** A pictorial description of $Ti_3C_2T_x$ MXene nanosheets and corresponding MXene-water dispersion *via* etching and sonication. **(c)** Schematics showing a fabrication method of EGMX1:3 nanohybrid dispersion. **(d, e)** Digital images of free-standing, flexible, EGMX1:3 film and associated ASSS assembly. Adapted with permission from Reference [118] Copyright 2017 John Wiley & Sons, Ltd.

MXene-on-paper coplanar micro-supercapacitors fabricated by Kurra *et al.* [115] utilizing the Meyer rod coating process and direct laser machining demonstrated remarkable electrochemical activity and suggested potential uses for adaptable on-paper energy storage devices. $Ti_3C_2T_x$ MXene as supercapacitor electrodes in ionic liquid electrolytes (1-ethyl-3-methylimidazolium bis (trifluoromethyl sulfonyl) imide (EMI-TFSI)) with a vast potential window of 3 V was first suggested by Lin *et al.* [116]. The electrochemical properties of $Ti_3C_2T_x$ electrodes were further examined using *in-situ* XRD in pure EMI-TFSI electrolytes. Furthermore, MXene-based solid-state supercapacitors were also investigated. Peng *et al.* [117] created all-$Ti_3C_2T_x$ on-chip inter-digital solid-state micro-supercapacitors using two layers of $Ti_3C_2T_x$ with various flake sizes, and the prepared micro supercapacitor exhibited high volumetric (357 F/cm^3) and areal capacitances (27 mF/cm^2) at 20 mV/s. Moreover, the proposed device shows 100% capacitance retention after 10,000 cycles at 50 mV/s. Li *et al.* [118] produced on-chip micro-supercapacitors using electrochemically exfoliated graphene (EG) and MXene nanosheets (Figs. **10a-c**). The flexible micro supercapacitor delivered high areal (2.26 mF/cm^2) and volumetric capacitances (33 F/cm^3) at 5 mV/s. PVA/H$_3$PO$_4$ saturated glass fiber membrane was placed between the two film electrodes to create an all-solid-state supercapacitor (Fig. **10 e**). Fig. (**10d**) showed that vacuum-assisted filtration was also used to assemble EGMX x:y thin film electrodes for all-solid-state supercapacitors, where x:y denotes the weight ratio of EG to MXene [118].

Besides $Ti_3C_2T_x$, Ti_2CT_x MXene was also used as supercapacitors electrode material. Rakhi *et al.* [66] studied the effects of post-etch forging environment (Ar and N_2/H_2 *etc.*) on the operational and electrochemical performance of Ti_2CT_x MXenes. MXene annealed in N_2/H_2 was reported to be the most effective when utilized in symmetric two-electrode arrangements and exhibited a specific capacitance of 51 F/g at 1 A/g. This development can be recognized by the high C content and decreased F-terminations over the surface of MXene [75]. Recently, Ti_2CT_x-based wire-type supercapacitors showed good cycle stability, higher definite capacitance (3.09 mF/cm), and an elevated energy density ($\sim$ 210 nW h cm^{-1}) [119]. The effects of MXene's surface termination groups on electrochemical performance were examined using first-principles calculations [120]. The pillared Ti_3C_2 MXene (CTAB-Sn(IV)@Ti_3C_2) was created *via* liquid-phase CTAB pre-pillaring and subsequent Sn^{4+} pillaring. The supercapacitors achieved a high-power density of 10.8 kW/kg while delivering an energy density of 45.31 Wh/kg using industrial activated carbon as the cathode and CTAB-Sn(IV)@Ti_3C_2 as the anode. Dall'Agnese *et al.* [121] reported the electrochemical performance of V_2CT_x MXene when used as a cathode for Na-ion capacitors.

CONCLUSION AND OUTLOOK

This chapter aims to familiarize the fundamentals of MAX and MXenes linked with energy storage, particularly in supercapacitors using their mechanism, progress, and, most importantly, the factors that affect the latest progresses in MAX and MXenes have been pronounced. The MAX crystalline materials can be selectively etched to produce equivalent MXene crystals that are predictable approaches towards the production of MXene. MXenes surface acquired through following selective etching strategy holds numerous functional groups. MXene with beneficial progress can be obtained *via* the CVD approach with results in its crystalline layers. Characteristics of MXenes have been projected *via* DFT approaches suggestively better than MXenes that have been experimentally synthesized so far. Thus, developing novel approaches toward the synthesis of MXene is required. Multifunctional proficiencies of MXenes display good progress in energy-storage applications that shield the areas from supercapacitors. Monovalent and multivalent cations, along with polar organic molecules, intercalate MXenes, thus permitting the tailored controller of d-spacing and allowing its entire utilization in energy storage systems. A combination of higher electronic conductivity and an oxide- or hydroxide-like surface containing exposed redox-active transition metal atoms makes MXenes very attractive for fabricating electrodes. Moreover, numerous parameters of MXenes are still unidentified that require intention. These parameters comprise the number of its family affiliates, theoretical capacity for Li storage, and other cation storage. Special consideration of multivalent intercalation (*i.e.*, Mg^{2+}, Zn^{2+}, and Al^{3+}) and larger organic ions would offer strong recommendations for the growth of electrode materials to acquire next-generation supercapacitors.

Further advancement and understanding of MXenes' characteristics were offered by theoretical predictions that will continue to lead to the extension of the MXene family *via* various elemental arrangements. Furthermore, photocatalytic responses can be personalized by developing MXene composites with TiO_2 (extensive bandgap semiconductor) and phosphorene (tunable bandgap semiconductors). The potential of MXenes in energy applications is enormous. Higher energy along with power density is complementary to achieve for supercapacitors. MXene composite materials (as an electrode) can attain higher volumetric capacitance. This is because long polymer molecules can associate with MXene layers, and the chains can tolerate repeated stretching and compression; MXene/polymer composites display outstanding mechanical properties. Consistent with steady insertion species and extended chain conductive polymers, supercapacitors can achieve decent electrochemical progress and exceptional mechanical features that enable the advances of all-solid flexible supercapacitors. Finally, MXene-based composites can meaningfully lift the progress of supercapacitors.

REFERENCES

[1]   M. Naguib, and Y. Gogotsi, "Synthesis of two-dimensional materials by selective extraction", *Acc. Chem. Res.*, vol. 48, no. 1, pp. 128-135, 2015.
[http://dx.doi.org/10.1021/ar500346b] [PMID: 25489991]

[2]   M.W. Barsoum, "The MN+1AXN phases: A new class of solids", *Prog. Solid State Chem.*, vol. 28, no. 1-4, pp. 201-281, 2000.
[http://dx.doi.org/10.1016/S0079-6786(00)00006-6]

[3]   S. Panda, K. Deshmukh, S.K. Khadheer Pasha, J. Theerthagiri, S. Manickam, and M.Y. Choi, "MXene based emerging materials for supercapacitor applications: Recent advances, challenges, and future perspectives", *Coord. Chem. Rev.*, vol. 462, p. 214518, 2022.
[http://dx.doi.org/10.1016/j.ccr.2022.214518]

[4]   T. Schultz, N.C. Frey, K. Hantanasirisakul, S. Park, S.J. May, V.B. Shenoy, Y. Gogotsi, and N. Koch, "Surface termination dependent work function and electronic properties of Ti 3 C 2 Tx MXene", *Chem. Mater.*, vol. 31, no. 17, pp. 6590-6597, 2019.
[http://dx.doi.org/10.1021/acs.chemmater.9b00414]

[5]   J. Peng, X. Chen, W.J. Ong, X. Zhao, and N. Li, "Surface and heterointerface engineering of 2D MXenes and their nanocomposites: Insights into electro- and photocatalysis", *Chem,* vol. 5, no. 1, pp. 18-50, 2019.
[http://dx.doi.org/10.1016/j.chempr.2018.08.037]

[6]   A. Sinha, Dhanjai, H. Zhao, Y. Huang, X. Lu, J. Chen, and R. Jain, "MXene: An emerging material for sensing and biosensing", *Trends Analyt. Chem.*, vol. 105, pp. 424-435, 2018.

[http://dx.doi.org/10.1016/j.trac.2018.05.021]

[7]   M. Naguib, V.N. Mochalin, M.W. Barsoum, and Y. Gogotsi, "25th anniversary article: MXenes: A new family of two-dimensional materials", *Adv. Mater.,* vol. 26, no. 7, pp. 992-1005, 2014.
[http://dx.doi.org/10.1002/adma.201304138] [PMID: 24357390]

[8]   M. Naguib, M. Kurtoglu, V. Presser, J. Lu, J. Niu, M. Heon, L. Hultman, Y. Gogotsi, and M.W. Barsoum, "Two-dimensional nanocrystals produced by exfoliation of Ti3 AlC2", *Adv. Mater.,* vol. 23, no. 37, pp. 4248-4253, 2011.
[http://dx.doi.org/10.1002/adma.201102306] [PMID: 21861270]

[9]   Y. Gogotsi, and B. Anasori, "The rise of MXenes", *ACS Nano,* vol. 13, no. 8, pp. 8491-8494, 2019.
[http://dx.doi.org/10.1021/acsnano.9b06394] [PMID: 31454866]

[10]   Z. Liu, L. Zheng, L. Sun, Y. Qian, J. Wang, and M. Li, "(Cr2/3 Ti1/3) 3 AlC2 and (Cr5/8 Ti3/8) 4 AlC3 : New MAX -phase compounds in Ti – Cr – Al – C system", *J. Am. Ceram. Soc.,* vol. 97, no. 1, pp. 67-69, 2014.
[http://dx.doi.org/10.1111/jace.12731]

[11]   M. Ghidiu, M.R. Lukatskaya, M.Q. Zhao, Y. Gogotsi, and M.W. Barsoum, "Conductive two-dimensional titanium carbide 'clay' with high volumetric capacitance", *Nature,* vol. 516, no. 7529, pp. 78-81, 2014.
[http://dx.doi.org/10.1038/nature13970] [PMID: 25470044]

[12]   A. Champagne, and J.C. Charlier, "Physical properties of 2D MXenes: From a theoretical perspective", *Journal of Physics: Materials,* vol. 3, no. 3, p. 032006, 2021.
[http://dx.doi.org/10.1088/2515-7639/ab97ee]

[13]   H. Zhang, L. Wang, Q. Chen, P. Li, A. Zhou, X. Cao, and Q. Hu, "Preparation, mechanical and anti-friction performance of MXene/polymer composites", *Mater. Des.,* vol. 92, pp. 682-689, 2016.
[http://dx.doi.org/10.1016/j.matdes.2015.12.084]

[14]   X. Jiang, A.V. Kuklin, A. Baev, Y. Ge, H. Ågren, H. Zhang, and P.N. Prasad, "Two-dimensional MXenes: From morphological to optical, electric, and magnetic properties and applications", *Phys. Rep.,* vol. 848, pp. 1-58, 2020.
[http://dx.doi.org/10.1016/j.physrep.2019.12.006]

[15]   V. Shukla, "The tunable electric and magnetic properties of 2D MXenes and their potential applications", *Materials Advances,* vol. 1, no. 9, pp. 3104-3121, 2020.
[http://dx.doi.org/10.1039/D0MA00548G]

[16]   X. Li, F. Ran, F. Yang, J. Long, and L. Shao, "Advances in MXene films: Synthesis, assembly, and applications", *Trans. Tianjin Univ.,* vol. 27, no. 3, pp. 217-247, 2021.
[http://dx.doi.org/10.1007/s12209-021-00282-y]

[17]   H. Jing, H. Yeo, B. Lyu, J. Ryou, S. Choi, J.H. Park, B.H. Lee, Y.H. Kim, and S. Lee, "Modulation of the electronic properties of MXene (Ti 3 C 2 Tx) via surface-covalent functionalization with diazonium", *ACS Nano,* vol. 15, no. 1, pp. 1388-1396, 2021.
[http://dx.doi.org/10.1021/acsnano.0c08664] [PMID: 33400488]

[18]   O. Mashtalir, M. Naguib, V.N. Mochalin, Y. Dall'Agnese, M. Heon, M.W. Barsoum, and Y. Gogotsi, "Intercalation and delamination of layered carbides and carbonitrides", *Nat. Commun.,* vol. 4, no. 1, p. 1716, 2013.
[http://dx.doi.org/10.1038/ncomms2664] [PMID: 23591883]

[19]   N.C. Osti, M. Naguib, A. Ostadhossein, Y. Xie, P.R.C. Kent, B. Dyatkin, G. Rother, W.T. Heller, A.C.T. van Duin, Y. Gogotsi, and E. Mamontov, "Effect of metal ion intercalation on the structure of mxene and water dynamics on its internal surfaces", *ACS Appl. Mater. Interfaces,* vol. 8, no. 14, pp. 8859-8863, 2016.
[http://dx.doi.org/10.1021/acsami.6b01490] [PMID: 27010763]

[20]   L. Wang, L. Yuan, K. Chen, Y. Zhang, Q. Deng, S. Du, Q. Huang, L. Zheng, J. Zhang, Z. Chai, M.W.

Barsoum, X. Wang, and W. Shi, "Loading actinides in multilayered structures for nuclear waste treatment: The first case study of uranium capture with vanadium carbide MXene", *ACS Appl. Mater. Interfaces,* vol. 8, no. 25, pp. 16396-16403, 2016.
[http://dx.doi.org/10.1021/acsami.6b02989] [PMID: 27267649]

[21]   K. Rasool, M. Helal, A. Ali, C.E. Ren, Y. Gogotsi, and K.A. Mahmoud, "Antibacterial activity of Ti 3 C 2 Tx MXene", *ACS Nano,* vol. 10, no. 3, pp. 3674-3684, 2016.
[http://dx.doi.org/10.1021/acsnano.6b00181] [PMID: 26909865]

[22]   K. Deshmukh, T. Kovářík, and S.K. Khadheer Pasha, "State of the art recent progress in two dimensional MXenes based gas sensors and biosensors: A comprehensive review", *Coord. Chem. Rev.,* vol. 424, p. 213514, 2020.
[http://dx.doi.org/10.1016/j.ccr.2020.213514]

[23]   X. Liu, T. Chen, Y. Xue, J. Fan, S. Shen, M.S.A. Hossain, M.A. Amin, L. Pan, X. Xu, and Y. Yamauchi, "Nanoarchitectonics of MXene/semiconductor heterojunctions toward artificial photosynthesis via photocatalytic CO2 reduction", *Coord. Chem. Rev.,* vol. 459, p. 214440, 2022.
[http://dx.doi.org/10.1016/j.ccr.2022.214440]

[24]   M. Marian, S. Tremmel, S. Wartzack, G. Song, B. Wang, J. Yu, and A. Rosenkranz, "Mxene nanosheets as an emerging solid lubricant for machine elements – Towards increased energy efficiency and service life", *Appl. Surf. Sci.,* vol. 523, p. 146503, 2020.
[http://dx.doi.org/10.1016/j.apsusc.2020.146503]

[25]   R. Bian, G. He, W. Zhi, S. Xiang, T. Wang, and D. Cai, "Ultralight MXene-based aerogels with high electromagnetic interference shielding performance", *J. Mater. Chem. C Mater. Opt. Electron. Devices,* vol. 7, no. 3, pp. 474-478, 2019.
[http://dx.doi.org/10.1039/C8TC04795B]

[26]   C. Yang, H. Huang, H. He, L. Yang, Q. Jiang, and W. Li, "Recent advances in MXene-based nanoarchitectures as electrode materials for future energy generation and conversion applications", *Coord. Chem. Rev.,* vol. 435, p. 213806, 2021.
[http://dx.doi.org/10.1016/j.ccr.2021.213806]

[27]   S. Niu, Z. Wang, M. Yu, M. Yu, L. Xiu, S. Wang, X. Wu, and J. Qiu, "MXene-based electrode with enhanced pseudocapacitance and volumetric capacity for power-type and ultra-long life lithium storage", *ACS Nano,* vol. 12, no. 4, pp. 3928-3937, 2018.
[http://dx.doi.org/10.1021/acsnano.8b01459] [PMID: 29589911]

[28]   X. Zheng, J. Shen, Q. Hu, W. Nie, Z. Wang, L. Zou, and C. Li, "Vapor phase polymerized conducting polymer/MXene textiles for wearable electronics", *Nanoscale,* vol. 13, no. 3, pp. 1832-1841, 2021.
[http://dx.doi.org/10.1039/D0NR07433K] [PMID: 33434252]

[29]   D. Jiang, J. Zhang, S. Qin, Z. Wang, K.A.S. Usman, D. Hegh, J. Liu, W. Lei, and J.M. Razal, "Superelastic Ti 3 C 2 Tx MXene-based hybrid aerogels for compression-resilient devices", *ACS Nano,* vol. 15, no. 3, pp. 5000-5010, 2021.
[http://dx.doi.org/10.1021/acsnano.0c09959] [PMID: 33635074]

[30]   J. Zhao, L. Zhang, X-Y. Xie, X. Li, Y. Ma, Q. Liu, W-H. Fang, X. Shi, G. Cui, and X. Sun, "Ti 3 C 2 T x (T = F, OH) MXene nanosheets: Conductive 2D catalysts for ambient electrohydrogenation of N 2 to NH 3", *J. Mater. Chem. A Mater. Energy Sustain.,* vol. 6, no. 47, pp. 24031-24035, 2018.
[http://dx.doi.org/10.1039/C8TA09840A]

[31]   Z.W. Seh, K.D. Fredrickson, B. Anasori, J. Kibsgaard, A.L. Strickler, M.R. Lukatskaya, Y. Gogotsi, T.F. Jaramillo, and A. Vojvodic, "Two-dimensional molybdenum carbide (MXene) as an efficient electrocatalyst for hydrogen evolution", *ACS Energy Lett.,* vol. 1, no. 3, pp. 589-594, 2016.
[http://dx.doi.org/10.1021/acsenergylett.6b00247]

[32]   N. Zhang, Y. Hong, S. Yazdanparast, and M. Asle Zaeem, "Superior structural, elastic and electronic properties of 2D titanium nitride MXenes over carbide MXenes: A comprehensive first principles study", *2D Materials,* vol. 5, no. 4, p. 045004, 2018.

[http://dx.doi.org/10.1088/2053-1583/aacfb3]

[33] Y. Ying, W. Ying, Q. Li, D. Meng, G. Ren, R. Yan, and X. Peng, "Recent advances of nanomaterial-based membrane for water purification", *Appl. Mater. Today,* vol. 7, pp. 144-158, 2017.
[http://dx.doi.org/10.1016/j.apmt.2017.02.010]

[34] D. Ma, Y. Li, J. Yang, H. Mi, S. Luo, L. Deng, C. Yan, M. Rauf, P. Zhang, X. Sun, X. Ren, J. Li, and H. Zhang, "New strategy for polysulfide protection based on atomic layer deposition of TiO 2 onto ferroelectric-encapsulated cathode: Toward ultrastable free-standing room temperature sodium–sulfur batteries", *Adv. Funct. Mater.,* vol. 28, no. 11, p. 1705537, 2018.
[http://dx.doi.org/10.1002/adfm.201705537]

[35] X. Chen, G. Xu, X. Ren, Z. Li, X. Qi, K. Huang, H. Zhang, Z. Huang, and J. Zhong, "A black/red phosphorus hybrid as an electrode material for high-performance Li-ion batteries and supercapacitors", *J. Mater. Chem. A Mater. Energy Sustain.,* vol. 5, no. 14, pp. 6581-6588, 2017.
[http://dx.doi.org/10.1039/C7TA00455A]

[36] D. Ma, Y. Li, H. Mi, S. Luo, P. Zhang, Z. Lin, J. Li, and H. Zhang, "Robust SnO 2− x nanoparticle-impregnated carbon nanofibers with outstanding electrochemical performance for advanced sodium-ion batteries", *Angew. Chem. Int. Ed.,* vol. 57, no. 29, pp. 8901-8905, 2018.
[http://dx.doi.org/10.1002/anie.201802672] [PMID: 29684238]

[37] Z. Wang, Z. Xu, H. Huang, X. Chu, Y. Xie, D. Xiong, C. Yan, H. Zhao, H. Zhang, and W. Yang, "Unraveling and regulating self-discharge behavior of Ti 3 C 2 Tx MXene-based supercapacitors", *ACS Nano,* vol. 14, no. 4, pp. 4916-4924, 2020.
[http://dx.doi.org/10.1021/acsnano.0c01056] [PMID: 32186846]

[38] M. Hu, H. Zhang, T. Hu, B. Fan, X. Wang, and Z. Li, "Emerging 2D MXenes for supercapacitors: Status, challenges and prospects", *Chem. Soc. Rev.,* vol. 49, no. 18, pp. 6666-6693, 2020.
[http://dx.doi.org/10.1039/D0CS00175A] [PMID: 32781463]

[39] Z. Wen, M.H. Yeh, H. Guo, J. Wang, Y. Zi, W. Xu, J. Deng, L. Zhu, X. Wang, C. Hu, L. Zhu, X. Sun, and Z.L. Wang, "Self-powered textile for wearable electronics by hybridizing fiber-shaped nanogenerators, solar cells, and supercapacitors", *Sci. Adv.,* vol. 2, no. 10, p. e1600097, 2016.
[http://dx.doi.org/10.1126/sciadv.1600097] [PMID: 27819039]

[40] Q. Xue, J. Sun, Y. Huang, M. Zhu, Z. Pei, H. Li, Y. Wang, N. Li, H. Zhang, and C. Zhi, "Recent progress on flexible and wearable supercapacitors", *Small,* vol. 13, no. 45, p. 1701827, 2017.
[http://dx.doi.org/10.1002/smll.201701827] [PMID: 28941073]

[41] T. Kar, S. Godavarthi, S.K. Pasha, K. Deshmukh, L. Martínez-Gómez, and M.K. Kesarla, "Layered materials and their heterojunctions for supercapacitor applications: A review", *Crit. Rev. Solid State Mater. Sci.,* vol. 47, no. 3, pp. 357-388, 2022.
[http://dx.doi.org/10.1080/10408436.2021.1886048]

[42] Y. Shao, M.F. El-Kady, J. Sun, Y. Li, Q. Zhang, M. Zhu, H. Wang, B. Dunn, and R.B. Kaner, "Design and mechanisms of asymmetric supercapacitors", *Chem. Rev.,* vol. 118, no. 18, pp. 9233-9280, 2018.
[http://dx.doi.org/10.1021/acs.chemrev.8b00252] [PMID: 30204424]

[43] J.L. Hart, K. Hantanasirisakul, A.C. Lang, B. Anasori, D. Pinto, Y. Pivak, J.T. van Omme, S.J. May, Y. Gogotsi, and M.L. Taheri, "Control of MXenes' electronic properties through termination and intercalation", *Nat. Commun.,* vol. 10, no. 1, p. 522, 2019.
[http://dx.doi.org/10.1038/s41467-018-08169-8] [PMID: 30705273]

[44] M.R. Lukatskaya, O. Mashtalir, C.E. Ren, Y. Dall'Agnese, P. Rozier, P.L. Taberna, M. Naguib, P. Simon, M.W. Barsoum, and Y. Gogotsi, "Cation intercalation and high volumetric capacitance of two-dimensional titanium carbide", *Science,* vol. 341, no. 6153, pp. 1502-1505, 2013.
[http://dx.doi.org/10.1126/science.1241488] [PMID: 24072919]

[45] M.D. Levi, M.R. Lukatskaya, S. Sigalov, M. Beidaghi, N. Shpigel, L. Daikhin, D. Aurbach, M.W. Barsoum, and Y. Gogotsi, "Solving the capacitive paradox of 2D MXene using electrochemical quartz-crystal admittance and in situ electronic conductance measurements", *Adv. Energy Mater.,* vol.

5, no. 1, p. 1400815, 2015.
[http://dx.doi.org/10.1002/aenm.201400815]

[46]   N. Shpigel, M.D. Levi, S. Sigalov, T.S. Mathis, Y. Gogotsi, and D. Aurbach, "Direct assessment of nanoconfined water in 2D Ti 3 C 2 electrode interspaces by a surface acoustic technique", *J. Am. Chem. Soc.,* vol. 140, no. 28, pp. 8910-8917, 2018.
[http://dx.doi.org/10.1021/jacs.8b04862] [PMID: 29928793]

[47]   J. Come, J.M. Black, M.R. Lukatskaya, M. Naguib, M. Beidaghi, A.J. Rondinone, S.V. Kalinin, D.J. Wesolowski, Y. Gogotsi, and N. Balke, "Controlling the actuation properties of MXene paper electrodes upon cation intercalation", *Nano Energy,* vol. 17, pp. 27-35, 2015.
[http://dx.doi.org/10.1016/j.nanoen.2015.07.028]

[48]   N. Shpigel, M.R. Lukatskaya, S. Sigalov, C.E. Ren, P. Nayak, M.D. Levi, L. Daikhin, D. Aurbach, and Y. Gogotsi, "In situ monitoring of gravimetric and viscoelastic changes in 2D intercalation electrodes", *ACS Energy Lett.,* vol. 2, no. 6, pp. 1407-1415, 2017.
[http://dx.doi.org/10.1021/acsenergylett.7b00133]

[49]   M. Hu, "Surface functional groups and interlayer water determine the electrochemical capacitance of Ti3C2Tx MXene", *ACS Nano,* vol. 12, no. 4, pp. 3578-3586, 2018.
[http://dx.doi.org/10.1021/acsnano.8b00676]

[50]   M.R. Lukatskaya, S.M. Bak, X. Yu, X.Q. Yang, M.W. Barsoum, and Y. Gogotsi, "Probing the mechanism of high capacitance in 2D titanium carbide using *in situ* X-ray absorption spectroscopy", *Adv. Energy Mater.,* vol. 5, no. 15, p. 1500589, 2015.
[http://dx.doi.org/10.1002/aenm.201500589]

[51]   M. Hu, Z. Li, T. Hu, S. Zhu, C. Zhang, and X. Wang, "High-capacitance mechanism for Ti 3 C 2Tx MXene by in situ electrochemical raman spectroscopy investigation", *ACS Nano,* vol. 10, no. 12, pp. 11344-11350, 2016.
[http://dx.doi.org/10.1021/acsnano.6b06597] [PMID: 28024328]

[52]   C. Zhan, M. Naguib, M. Lukatskaya, P.R.C. Kent, Y. Gogotsi, and D. Jiang, "Understanding the MXene pseudocapacitance", *J. Phys. Chem. Lett.,* vol. 9, no. 6, pp. 1223-1228, 2018.
[http://dx.doi.org/10.1021/acs.jpclett.8b00200] [PMID: 29461062]

[53]   Y. Gogotsi, and R.M. Penner, "Energy storage in nanomaterials – Capacitive, pseudocapacitive, or battery-like?", *ACS Nano,* vol. 12, no. 3, pp. 2081-2083, 2018.
[http://dx.doi.org/10.1021/acsnano.8b01914] [PMID: 29580061]

[54]   Q.X. Xia, N.M. Shinde, T. Zhang, J.M. Yun, A. Zhou, R.S. Mane, S. Mathur, and K.H. Kim, "Seawater electrolyte-mediated high volumetric MXene-based electrochemical symmetric supercapacitors", *Dalton Trans.,* vol. 47, no. 26, pp. 8676-8682, 2018.
[http://dx.doi.org/10.1039/C8DT01375F] [PMID: 29897071]

[55]   X. Zhao, C. Dall'Agnese, X.F. Chu, S. Zhao, G. Chen, Y. Gogotsi, Y. Gao, and Y. Dall'Agnese, "Electrochemical behavior of Ti 3 C 2 Tx MXene in environmentally friendly methanesulfonic acid electrolyte", *ChemSusChem,* vol. 12, no. 19, pp. 4480-4486, 2019.
[http://dx.doi.org/10.1002/cssc.201901746] [PMID: 31397541]

[56]   Z. Lin, P. Rozier, B. Duployer, P-L. Taberna, B. Anasori, Y. Gogotsi, and P. Simon, "Electrochemical and in-situ X-ray diffraction studies of Ti 3 C 2 T x MXene in ionic liquid electrolyte", *Electrochem. Commun.,* vol. 72, pp. 50-53, 2016.
[http://dx.doi.org/10.1016/j.elecom.2016.08.023]

[57]   Y. Dall'Agnese, P. Rozier, P.L. Taberna, Y. Gogotsi, and P. Simon, "Capacitance of two-dimensional titanium carbide (MXene) and MXene/carbon nanotube composites in organic electrolytes", *J. Power Sources,* vol. 306, pp. 510-515, 2016.
[http://dx.doi.org/10.1016/j.jpowsour.2015.12.036]

[58]   M. Okubo, A. Sugahara, S. Kajiyama, and A. Yamada, "MXene as a charge storage host", *Acc. Chem. Res.,* vol. 51, no. 3, pp. 591-599, 2018.

[http://dx.doi.org/10.1021/acs.accounts.7b00481] [PMID: 29469564]

[59] Y. Li, H. Shao, Z. Lin, J. Lu, L. Liu, B. Duployer, P.O.Å. Persson, P. Eklund, L. Hultman, M. Li, K. Chen, X.H. Zha, S. Du, P. Rozier, Z. Chai, E. Raymundo-Piñero, P.L. Taberna, P. Simon, and Q. Huang, "A general Lewis acidic etching route for preparing MXenes with enhanced electrochemical performance in non-aqueous electrolyte", *Nat. Mater.,* vol. 19, no. 8, pp. 894-899, 2020.
[http://dx.doi.org/10.1038/s41563-020-0657-0] [PMID: 32284597]

[60] X. Wang, T.S. Mathis, K. Li, Z. Lin, L. Vlcek, T. Torita, N.C. Osti, C. Hatter, P. Urbankowski, A. Sarycheva, M. Tyagi, E. Mamontov, P. Simon, and Y. Gogotsi, "Influences from solvents on charge storage in titanium carbide MXenes", *Nat. Energy,* vol. 4, no. 3, pp. 241-248, 2019.
[http://dx.doi.org/10.1038/s41560-019-0339-9]

[61] M.A. Hope, A.C. Forse, K.J. Griffith, M.R. Lukatskaya, M. Ghidiu, Y. Gogotsi, and C.P. Grey, "NMR reveals the surface functionalisation of Ti 3 C 2 MXene", *Phys. Chem. Chem. Phys.,* vol. 18, no. 7, pp. 5099-5102, 2016.
[http://dx.doi.org/10.1039/C6CP00330C] [PMID: 26818187]

[62] X. Yu, X. Cai, H. Cui, S.W. Lee, X.F. Yu, and B. Liu, "Fluorine-free preparation of titanium carbide MXene quantum dots with high near-infrared photothermal performances for cancer therapy", *Nanoscale,* vol. 9, no. 45, pp. 17859-17864, 2017.
[http://dx.doi.org/10.1039/C7NR05997C] [PMID: 29119157]

[63] T. Li, L. Yao, Q. Liu, J. Gu, R. Luo, J. Li, X. Yan, W. Wang, P. Liu, B. Chen, W. Zhang, W. Abbas, R. Naz, and D. Zhang, "Fluorine-free synthesis of high-purity Ti $_3$ C $_2$ T$_x$ (T=OH, O) *via* alkali treatment", *Angew. Chem. Int. Ed.,* vol. 57, no. 21, pp. 6115-6119, 2018.
[http://dx.doi.org/10.1002/anie.201800887] [PMID: 29633442]

[64] Y. Dall'Agnese, M.R. Lukatskaya, K.M. Cook, P.L. Taberna, Y. Gogotsi, and P. Simon, "High capacitance of surface-modified 2D titanium carbide in acidic electrolyte", *Electrochem. Commun.,* vol. 48, pp. 118-122, 2014.
[http://dx.doi.org/10.1016/j.elecom.2014.09.002]

[65] J. Li, X. Yuan, C. Lin, Y. Yang, L. Xu, X. Du, J. Xie, J. Lin, and J. Sun, "Achieving high pseudocapacitance of 2D titanium carbide (MXene) by cation intercalation and surface modification", *Adv. Energy Mater.,* vol. 7, no. 15, p. 1602725, 2017.
[http://dx.doi.org/10.1002/aenm.201602725]

[66] R.B. Rakhi, B. Ahmed, M.N. Hedhili, D.H. Anjum, and H.N. Alshareef, "Effect of postetch annealing gas composition on the structural and electrochemical properties of Ti 2 CTx MXene electrodes for supercapacitor applications", *Chem. Mater.,* vol. 27, no. 15, pp. 5314-5323, 2015.
[http://dx.doi.org/10.1021/acs.chemmater.5b01623]

[67] C. Yang, Y. Tang, Y. Tian, Y. Luo, M. Faraz Ud Din, X. Yin, and W. Que, "Flexible nitrogen-doped 2D titanium carbides (MXene) films constructed by an *ex situ* solvothermal method with extraordinary volumetric capacitance", *Adv. Energy Mater.,* vol. 8, no. 31, p. 1802087, 2018.
[http://dx.doi.org/10.1002/aenm.201802087]

[68] Y. Wen, T.E. Rufford, X. Chen, N. Li, M. Lyu, L. Dai, and L. Wang, "Nitrogen-doped Ti 3 C 2 T x MXene electrodes for high-performance supercapacitors", *Nano Energy,* vol. 38, pp. 368-376, 2017.
[http://dx.doi.org/10.1016/j.nanoen.2017.06.009]

[69] Y. Tian, W. Que, Y. Luo, C. Yang, X. Yin, and L.B. Kong, "Surface nitrogen-modified 2D titanium carbide (MXene) with high energy density for aqueous supercapacitor applications", *J. Mater. Chem. A Mater. Energy Sustain.,* vol. 7, no. 10, pp. 5416-5425, 2019.
[http://dx.doi.org/10.1039/C9TA00076C]

[70] X. Wang, Y. Yu, C. Yang, C. Shao, K. Shi, L. Shang, F. Ye, and Y. Zhao, "Microfluidic 3D printing responsive scaffolds with biomimetic enrichment channels for bone regeneration", *Adv. Funct. Mater.,* vol. 31, no. 40, p. 2105190, 2021.
[http://dx.doi.org/10.1002/adfm.202105190]

[71]   J. Zhu, and U. Schwingenschlögl, "P and Si functionalized MXenes for metal-ion battery applications", *2D Materials,* vol. 4, no. 2, p. 025073, 2017.
[http://dx.doi.org/10.1088/2053-1583/aa69fe]

[72]   Q. Tao, M. Dahlqvist, J. Lu, S. Kota, R. Meshkian, J. Halim, J. Palisaitis, L. Hultman, M.W. Barsoum, P.O.Å. Persson, and J. Rosen, "Two-dimensional Mo1.33C MXene with divacancy ordering prepared from parent 3D laminate with in-plane chemical ordering", *Nat. Commun.,* vol. 8, no. 1, p. 14949, 2017.
[http://dx.doi.org/10.1038/ncomms14949] [PMID: 28440271]

[73]   O. Mashtalir, M.R. Lukatskaya, A.I. Kolesnikov, E. Raymundo-Piñero, M. Naguib, M.W. Barsoum, and Y. Gogotsi, "The effect of hydrazine intercalation on the structure and capacitance of 2D titanium carbide (MXene)", *Nanoscale,* vol. 8, no. 17, pp. 9128-9133, 2016.
[http://dx.doi.org/10.1039/C6NR01462C] [PMID: 27088300]

[74]   X. Su, J. Zhang, H. Mu, J. Zhao, Z. Wang, Z. Zhao, C. Han, and Z. Ye, "Effects of etching temperature and ball milling on the preparation and capacitance of Ti3C2 MXene", *J. Alloys Compd.,* vol. 752, pp. 32-39, 2018.
[http://dx.doi.org/10.1016/j.jallcom.2018.04.152]

[75]   S. Kajiyama, L. Szabova, H. Iinuma, A. Sugahara, K. Gotoh, K. Sodeyama, Y. Tateyama, M. Okubo, and A. Yamada, "Enhanced li-ion accessibility in MXene titanium carbide by steric chloride termination", *Adv. Energy Mater.,* vol. 7, no. 9, p. 1601873, 2017.
[http://dx.doi.org/10.1002/aenm.201601873]

[76]   C. Zhang, L. Wang, W. Lei, Y. Wu, C. Li, M.A. Khan, Y. Ouyang, X. Jiao, H. Ye, S. Mutahir, and Q. Hao, "Achieving quick charge/discharge rate of 3.0 V s−1 by 2D titanium carbide (MXene) via N-doped carbon intercalation", *Mater. Lett.,* vol. 234, pp. 21-25, 2019.
[http://dx.doi.org/10.1016/j.matlet.2018.08.124]

[77]   M.Q. Zhao, C.E. Ren, Z. Ling, M.R. Lukatskaya, C. Zhang, K.L. Van Aken, M.W. Barsoum, and Y. Gogotsi, "Flexible MXene/carbon nanotube composite paper with high volumetric capacitance", *Adv. Mater.,* vol. 27, no. 2, pp. 339-345, 2015.
[http://dx.doi.org/10.1002/adma.201404140] [PMID: 25405330]

[78]   M. Ghidiu, S. Kota, J. Halim, A.W. Sherwood, N. Nedfors, J. Rosen, V.N. Mochalin, and M.W. Barsoum, "Alkylammonium cation intercalation into Ti 3 C 2 (MXene): Effects on properties and ion-exchange capacity estimation", *Chem. Mater.,* vol. 29, no. 3, pp. 1099-1106, 2017.
[http://dx.doi.org/10.1021/acs.chemmater.6b04234]

[79]   J. Tang, T.S. Mathis, N. Kurra, A. Sarycheva, X. Xiao, M.N. Hedhili, Q. Jiang, H.N. Alshareef, B. Xu, F. Pan, and Y. Gogotsi, "Tuning the electrochemical performance of titanium carbide mxene by controllable in situ anodic oxidation", *Angew. Chem. Int. Ed.,* vol. 58, no. 49, pp. 17849-17855, 2019.
[http://dx.doi.org/10.1002/anie.201911604] [PMID: 31574196]

[80]   E. Kayali, A. VahidMohammadi, J. Orangi, and M. Beidaghi, "controlling the dimensions of 2D MXenes for ultrahigh-rate pseudocapacitive energy storage", *ACS Appl. Mater. Interfaces,* vol. 10, no. 31, pp. 25949-25954, 2018.
[http://dx.doi.org/10.1021/acsami.8b07397] [PMID: 30044609]

[81]   M.R. Lukatskaya, S. Kota, Z. Lin, M-Q. Zhao, N. Shpigel, M.D. Levi, J. Halim, P-L. Taberna, M.W. Barsoum, P. Simon, and Y. Gogotsi, "Ultra-high-rate pseudocapacitive energy storage in two-dimensional transition metal carbides", *Nat. Energy,* vol. 2, no. 8, p. 17105, 2017.
[http://dx.doi.org/10.1038/nenergy.2017.105]

[82]   Y. Zhu, K. Rajouâ, S. Le Vot, O. Fontaine, P. Simon, and F. Favier, "Modifications of MXene layers for supercapacitors", *Nano Energy,* vol. 73, p. 104734, 2020.
[http://dx.doi.org/10.1016/j.nanoen.2020.104734]

[83]   Y. Xia, T.S. Mathis, M.Q. Zhao, B. Anasori, A. Dang, Z. Zhou, H. Cho, Y. Gogotsi, and S. Yang, "Thickness-independent capacitance of vertically aligned liquid-crystalline MXenes", *Nature,* vol.

557, no. 7705, pp. 409-412, 2018.
[http://dx.doi.org/10.1038/s41586-018-0109-z] [PMID: 29769673]

[84] C. Yang, Y. Tang, Y. Tian, Y. Luo, Y. He, X. Yin, and W. Que, "Achieving of flexible, free-standing, ultracompact delaminated titanium carbide films for high volumetric performance and heat-resistant symmetric supercapacitors", *Adv. Funct. Mater.*, vol. 28, no. 15, p. 1705487, 2018.
[http://dx.doi.org/10.1002/adfm.201705487]

[85] X. Wang, C. Garnero, G. Rochard, D. Magne, S. Morisset, S. Hurand, P. Chartier, J. Rousseau, T. Cabioc'h, C. Coutanceau, V. Mauchamp, and S. Célérier, "A new etching environment (FeF 3 /HCl) for the synthesis of two-dimensional titanium carbide MXenes: A route towards selective reactivity vs. water", *J. Mater. Chem. A Mater. Energy Sustain.*, vol. 5, no. 41, pp. 22012-22023, 2017.
[http://dx.doi.org/10.1039/C7TA01082F]

[86] M.R. Lukatskaya, O. Mashtalir, C.E. Ren, Y. Dall'Agnese, P. Rozier, P.L. Taberna, M. Naguib, P. Simon, M.W. Barsoum, and Y. Gogotsi, "Cation intercalation and high volumetric capacitance of two-dimensional titanium carbide", *Science,* vol. 341, no. 6153, pp. 1502-1505, 2013.
[http://dx.doi.org/10.1126/science.1241488] [PMID: 24072919]

[87] M. Lu, W. Han, H. Li, W. Shi, J. Wang, B. Zhang, Y. Zhou, H. Li, W. Zhang, and W. Zheng, "Tent-pitching-inspired high-valence period 3-cation pre-intercalation excels for anode of 2D titanium carbide (MXene) with high Li storage capacity", *Energy Storage Mater.,* vol. 16, pp. 163-168, 2019.
[http://dx.doi.org/10.1016/j.ensm.2018.04.029]

[88] C. Eames, and M.S. Islam, "Ion intercalation into two-dimensional transition-metal carbides: global screening for new high-capacity battery materials", *J. Am. Chem. Soc.,* vol. 136, no. 46, pp. 16270-16276, 2014.
[http://dx.doi.org/10.1021/ja508154e] [PMID: 25310601]

[89] K. Hantanasirisakul, M-Q. Zhao, P. Urbankowski, J. Halim, B. Anasori, S. Kota, C.E. Ren, M.W. Barsoum, and Y. Gogotsi, "Fabrication of Ti 3 C 2 Tx MXene transparent thin films with tunable optoelectronic properties", *Adv. Electron. Mater.,* vol. 2, no. 6, p. 1600050, 2016.
[http://dx.doi.org/10.1002/aelm.201600050]

[90] Z. Zhang, S. Yang, P. Zhang, J. Zhang, G. Chen, and X. Feng, "Mechanically strong MXene/Kevlar nanofiber composite membranes as high-performance nanofluidic osmotic power generators", *Nat. Commun.,* vol. 10, no. 1, p. 2920, 2019.
[http://dx.doi.org/10.1038/s41467-019-10885-8] [PMID: 31266937]

[91] Y. Fan, Z. Li, and J. Wei, "Application of aramid nanofibers in nanocomposites: A brief review", *Polymers,* vol. 13, no. 18, p. 3071, 2021.
[http://dx.doi.org/10.3390/polym13183071] [PMID: 34577972]

[92] P. Xu, H. Xiao, X. Liang, T. Zhang, F. Zhang, C. Liu, B. Lang, and Q. Gao, "A MXene-based EDA-Ti3C2Tx intercalation compound with expanded interlayer spacing as high performance supercapacitor electrode material", *Carbon,* vol. 173, pp. 135-144, 2021.
[http://dx.doi.org/10.1016/j.carbon.2020.11.010]

[93] T. Najam, S.S.A. Shah, L. Peng, M.S. Javed, M. Imran, M-Q. Zhao, and P. Tsiakaras, "Synthesis and nano-engineering of MXenes for energy conversion and storage applications: Recent advances and perspectives", *Coord. Chem. Rev.,* vol. 454, p. 214339, 2022.
[http://dx.doi.org/10.1016/j.ccr.2021.214339]

[94] J. Tang, W. Yi, X. Zhong, C.J. Zhang, X. Xiao, F. Pan, and B. Xu, "Laser writing of the restacked titanium carbide MXene for high performance supercapacitors", *Energy Storage Mater.,* vol. 32, pp. 418-424, 2020.
[http://dx.doi.org/10.1016/j.ensm.2020.07.028]

[95] L. Shen, X. Zhou, X. Zhang, Y. Zhang, Y. Liu, W. Wang, W. Si, and X. Dong, "Carbon-intercalated Ti 3 C 2 T x MXene for high-performance electrochemical energy storage", *J. Mater. Chem. A Mater. Energy Sustain.,* vol. 6, no. 46, pp. 23513-23520, 2018.

[http://dx.doi.org/10.1039/C8TA09600G]

[96]   M. Hu, R. Cheng, Z. Li, T. Hu, H. Zhang, C. Shi, J. Yang, C. Cui, C. Zhang, H. Wang, B. Fan, X. Wang, and Q.H. Yang, "Interlayer engineering of Ti 3 C 2 T x MXenes towards high capacitance supercapacitors", *Nanoscale,* vol. 12, no. 2, pp. 763-771, 2020.
[http://dx.doi.org/10.1039/C9NR08960H] [PMID: 31830197]

[97]   M. Mariano, O. Mashtalir, F.Q. Antonio, W.H. Ryu, B. Deng, F. Xia, Y. Gogotsi, and A.D. Taylor, "Solution-processed titanium carbide MXene films examined as highly transparent conductors", *Nanoscale,* vol. 8, no. 36, pp. 16371-16378, 2016.
[http://dx.doi.org/10.1039/C6NR03682A] [PMID: 27722443]

[98]   C. Cui, M. Hu, C. Zhang, R. Cheng, J. Yang, and X. Wang, "High-capacitance Ti 3 C 2Tx MXene obtained by etching submicron Ti 3 AlC 2 grains grown in molten salt", *Chem. Commun.,* vol. 54, no. 58, pp. 8132-8135, 2018.
[http://dx.doi.org/10.1039/C8CC04350G] [PMID: 29975377]

[99]   S. Yang, P. Zhang, F. Wang, A.G. Ricciardulli, M.R. Lohe, P.W.M. Blom, and X. Feng, "Fluoride-free synthesis of two-dimensional titanium carbide (MXene) using a binary aqueous system", *Angew. Chem. Int. Ed.,* vol. 57, no. 47, pp. 15491-15495, 2018.
[http://dx.doi.org/10.1002/anie.201809662] [PMID: 30289581]

[100]   A. Qian, J.Y. Seo, H. Shi, J.Y. Lee, and C.H. Chung, "Surface functional groups and electrochemical behavior in dimethyl sulfoxide-delaminated Ti 3 C 2 Tx MXene", *ChemSusChem,* vol. 11, no. 21, pp. 3719-3723, 2018.
[http://dx.doi.org/10.1002/cssc.201801759] [PMID: 30180299]

[101]   X. Huang, J. Huang, D. Yang, and P. Wu, "A multi-scale structural engineering strategy for high-performance mxene hydrogel supercapacitor electrode", *Adv. Sci. (Weinh.),* vol. 8, no. 18, p. 2101664, 2021.
[http://dx.doi.org/10.1002/advs.202101664] [PMID: 34338445]

[102]   Y. Tang, J. Zhu, C. Yang, and F. Wang, "Enhanced capacitive performance based on diverse layered structure of two-dimensional Ti 3 C 2 MXene with long etching time", *J. Electrochem. Soc.,* vol. 163, no. 9, pp. A1975-A1982, 2016.
[http://dx.doi.org/10.1149/2.0921609jes]

[103]   A.E. Ghazaly, W. Zheng, J. Halim, E.N. Tseng, P.O.Å. Persson, B. Ahmed, and J. Rosen, "Enhanced supercapacitive performance of Mo1.33C MXene based asymmetric supercapacitors in lithium chloride electrolyte", *Energy Storage Mater.,* vol. 41, pp. 203-208, 2021.
[http://dx.doi.org/10.1016/j.ensm.2021.05.006]

[104]   J.B. Lee, G.H. Choi, and P.J. Yoo, "Oxidized-co-crumpled multiscale porous architectures of MXene for high performance supercapacitors", *J. Alloys Compd.,* vol. 887, p. 161304, 2021.
[http://dx.doi.org/10.1016/j.jallcom.2021.161304]

[105]   Z. Wu, X. Liu, T. Shang, Y. Deng, N. Wang, X. Dong, J. Zhao, D. Chen, Y. Tao, and Q-H. Yang, "Reassembly of MXene hydrogels into flexible films towards compact and ultrafast supercapacitors", *Adv. Funct. Mater.,* vol. 31, no. 41, p. 2102874, 2021.
[http://dx.doi.org/10.1002/adfm.202102874]

[106]   S.Y. Lin, and X. Zhang, "Two-dimensional titanium carbide electrode with large mass loading for supercapacitor", *J. Power Sources,* vol. 294, pp. 354-359, 2015.
[http://dx.doi.org/10.1016/j.jpowsour.2015.06.082]

[107]   S. Xu, G. Wei, J. Li, Y. Ji, N. Klyui, V. Izotov, and W. Han, "Binder-free Ti 3 C 2 T x MXene electrode film for supercapacitor produced by electrophoretic deposition method", *Chem. Eng. J.,* vol. 317, pp. 1026-1036, 2017.
[http://dx.doi.org/10.1016/j.cej.2017.02.144]

[108]   C.J. Zhang, B. Anasori, A. Seral-Ascaso, S.H. Park, N. McEvoy, A. Shmeliov, G.S. Duesberg, J.N. Coleman, Y. Gogotsi, and V. Nicolosi, "Transparent, flexible, and conductive 2D titanium carbide

(MXene) films with high volumetric capacitance", *Adv. Mater.,* vol. 29, no. 36, p. 1702678, 2017.
[http://dx.doi.org/10.1002/adma.201702678] [PMID: 28741695]

[109] Q. Fu, J. Wen, N. Zhang, L. Wu, M. Zhang, S. Lin, H. Gao, and X. Zhang, "Free-standing Ti 3 C 2 T x electrode with ultrahigh volumetric capacitance", *RSC Advances,* vol. 7, no. 20, pp. 11998-12005, 2017.
[http://dx.doi.org/10.1039/C7RA00126F]

[110] Z. Ling, C.E. Ren, M.Q. Zhao, J. Yang, J.M. Giammarco, J. Qiu, M.W. Barsoum, and Y. Gogotsi, "Flexible and conductive MXene films and nanocomposites with high capacitance", *Proc. Natl. Acad. Sci. USA,* vol. 111, no. 47, pp. 16676-16681, 2014.
[http://dx.doi.org/10.1073/pnas.1414215111] [PMID: 25389310]

[111] M. Boota, B. Anasori, C. Voigt, M.Q. Zhao, M.W. Barsoum, and Y. Gogotsi, "Pseudocapacitive electrodes produced by oxidant-free polymerization of pyrrole between the layers of 2D titanium carbide (MXene)", *Adv. Mater.,* vol. 28, no. 7, pp. 1517-1522, 2016.
[http://dx.doi.org/10.1002/adma.201504705] [PMID: 26660424]

[112] M. Boota, M. Pasini, F. Galeotti, W. Porzio, M-Q. Zhao, J. Halim, and Y. Gogotsi, "Interaction of polar and nonpolar polyfluorenes with layers of two-dimensional titanium carbide (MXene): Intercalation and pseudocapacitance", *Chem. Mater.,* vol. 29, no. 7, pp. 2731-2738, 2017.
[http://dx.doi.org/10.1021/acs.chemmater.6b03933]

[113] J. Yan, C.E. Ren, K. Maleski, C.B. Hatter, B. Anasori, P. Urbankowski, A. Sarycheva, and Y. Gogotsi, "Flexible MXene/graphene films for ultrafast supercapacitors with outstanding volumetric capacitance", *Adv. Funct. Mater.,* vol. 27, no. 30, p. 1701264, 2017.
[http://dx.doi.org/10.1002/adfm.201701264]

[114] Y. Wang, H. Dou, J. Wang, B. Ding, Y. Xu, Z. Chang, and X. Hao, "Three-dimensional porous MXene/layered double hydroxide composite for high performance supercapacitors", *J. Power Sources,* vol. 327, pp. 221-228, 2016.
[http://dx.doi.org/10.1016/j.jpowsour.2016.07.062]

[115] N. Kurra, B. Ahmed, Y. Gogotsi, and H.N. Alshareef, "MXene-on-paper coplanar microsupercapacitors", *Adv. Energy Mater.,* vol. 6, no. 24, p. 1601372, 2016.
[http://dx.doi.org/10.1002/aenm.201601372]

[116] Z. Lin, D. Barbara, P-L. Taberna, K.L. Van Aken, B. Anasori, Y. Gogotsi, and P. Simon, "Capacitance of Ti3C2Tx MXene in ionic liquid electrolyte", *J. Power Sources,* vol. 326, pp. 575-579, 2016.
[http://dx.doi.org/10.1016/j.jpowsour.2016.04.035]

[117] Y.Y. Peng, B. Akuzum, N. Kurra, M-Q. Zhao, M. Alhabeb, B. Anasori, E.C. Kumbur, H.N. Alshareef, M-D. Ger, and Y. Gogotsi, "All-MXene (2D titanium carbide) solid-state microsupercapacitors for on-chip energy storage", *Energy Environ. Sci.,* vol. 9, no. 9, pp. 2847-2854, 2016.
[http://dx.doi.org/10.1039/C6EE01717G]

[118] H. Li, Y. Hou, F. Wang, M.R. Lohe, X. Zhuang, L. Niu, and X. Feng, "Flexible all-solid-state supercapacitors with high volumetric capacitances boosted by solution processable mxene and electrochemically exfoliated graphene", *Adv. Energy. Mater.,* vol. 7, no. 4, p. 1601847, 2017.
[http://dx.doi.org/10.1002/aenm.201601847]

[119] K. Krishnamoorthy, P. Pazhamalai, S. Sahoo, and S.J. Kim, "Titanium carbide sheet based high performance wire type solid state supercapacitors", *J. Mater. Chem. A Mater. Energy Sustain.,* vol. 5, no. 12, pp. 5726-5736, 2017.
[http://dx.doi.org/10.1039/C6TA11198J]

[120] X. Ji, K. Xu, C. Chen, B. Zhang, Y. Ruan, J. Liu, L. Miao, and J. Jiang, "Probing the electrochemical capacitance of MXene nanosheets for high-performance pseudocapacitors", *Phys. Chem. Chem. Phys.,* vol. 18, no. 6, pp. 4460-4467, 2016.
[http://dx.doi.org/10.1039/C5CP07311A] [PMID: 26790481]

[121] Y. Dall'Agnese, P.L. Taberna, Y. Gogotsi, and P. Simon, "Two-dimensional vanadium carbide (MXene) as positive electrode for sodium-ion capacitors", *J. Phys. Chem. Lett.,* vol. 6, no. 12, pp. 2305-2309, 2015.
[http://dx.doi.org/10.1021/acs.jpclett.5b00868] [PMID: 26266609]

CHAPTER 13

Recent Developments in the Field of Supercapacitor Materials

Mani Jayakumar[1] and **Venkatesa Prabhu S.**[2,*]

[1] *Department of Chemical Engineering, Haramaya Institute of Technology, Haramaya University, Haramaya, Dire Dawa, Ethiopia*

[2] *Center of Excellence for Bioprocess and Biotechnology, Department of Chemical Engineering, College of Biological and Chemical Engineering, Addis Ababa Science and Technology University, Addis Ababa, Ethiopia*

Abstract: Energy storage is one of the crucial requirements for today's life to store energy for later use. Energy storage critically reduces the dependence on backup power supplies. In recent times, fascinatingly, energy storage systems have been developed in compactable sizes and shapes with sustainable and appreciable backup power. However, due to some challenges in conventional storage systems, supercapacitors are gaining a huge interest which satisfies the increasing demands of energy storage devices. Supercapacitors are well-recognized for their long cycle life, high power density, and the ability for less charge and discharge time. Accordingly, in supercapacitor electrodes, activated carbon is extensively used as a base material. Current research documented that the amorphous mixed metal oxides and nanostructured oxides are also used in supercapacitor devices to exhibit high performance. Keeping this in view, this chapter is reviewed to provide information on the different types of recent developments in supercapacitors and their performance. In addition, recent developments in green approaches to supercapacitor applications in MnO_2-based electrodes and composites, supercapacitors performance, power capability, and cycle life are also discussed.

Keywords: Activated carbon, And capacitary performance, Composites, Metal oxides, Power capability, Supercapacitors.

INTRODUCTION

Nowadays, environmental pollution significantly influences climate change resulting in different impacts on the global economy, which triggers the utilization of fossil fuels like coal and natural gas [1]. To address these issues and appropriate exploitation of existing energy, developing a strategy for utilizing

---

* **Corresponding author Venkatesa Prabhu S.:** Center of Excellence for Bioprocess and Biotechnology, Department of Chemical Engineering, College of Biological and Chemical Engineering, Addis Ababa Science and Technology University, Addis Ababa, Ethiopia; E-mail: venkatesa.prabhu@aastu.edu.et

Sanjeev Verma, Shivani Verma, Saurabh Kumar & Bhawna Verma (Eds.)

renewable forms of energy (wind, tidal, and solar) is much required. However, some concerns hinder the use of resources for renewable energy, hence, the production of electricity seems to be inconsistent. Accordingly, there is an urgent need for developing sustainable, and effective energy storage systems [2]. Undeniably, there may be an unexpected imbalance between the demand and supply of electricity that may create unwanted temporal gaps and nonexistent spatial grid systems for the end users. To address this issue, there is a crucial requirement for energy storage systems. In recent times, supercapacitors (SCs), are a new form of energy storage devices that are gaining substantial attention since they exhibit a fast rate of discharge, comparatively long cycle life, and high-power density. Studies show that supercapacitors have prominent potential applications in hybrid power automobiles, electronic devices on portable scale, and even in renewable energy systems. By definition, supercapacitors are energy source devices that can be positioned between capacitors and batteries [3]. Compared to normal capacitors, SCs have a good capacity to hold larger energy and can deliver it at larger power units. In addition to this, SCs have well-appreciable cyclability and long-lasting stability. The application of SCs is still being utilized in different energy sectors and industries. In general, the SCs are classified as pseudo capacitors and electrical double-layer capacitors. Normally, electrical double-layer capacitors accumulate the charge on the interface between the electrode and electrolyte. Differently, for the pseudo capacitors, the energy is stored *via* the process, called, reversible faradaic redox [4]. The two key factors, operative voltage, and capacitive voltage, are significantly influencing the energy density. In such a way, energy density could be promoted towards the improvement of the capacitance performance of the electrode and raising the potential window. So far, several investigations have been carried out toward ameliorating the electrochemical activity of SCs which was found to be dependent on the design and synthesis procedures of materials. In this line, the materials for preparing electrodes, specifically for SCs, show a substantial impact on their performance [5].

In general, carbon materials, conductive polymers, and transition metal oxides are the three different materials used for electrode preparation. Carbon-based materials are generally used in electrical double-layer capacitors as electrodes [6]. They have high specific surface area and well-distributed pore size that can be adjusted to attain better electrical conductivity. In realization, studies revealed that carbon-based materials can provide high power density with deprived energy density. This is because of the method followed for storage in electrical double-layer capacitors that restricts the overall outcomes [7]. Accordingly, graphene, carbon nanotubes, and carbon nanofibers have been well-explored carbon-based materials as electrodes. Further investigations explicated that carbon-based materials have significantly affected the capacity of electrical double-layer capacitors. Additionally, the high cost of such materials limits the usage of these materials. The conductive polymers are known to be good, but they exhibit low stability. Recently, transition metal oxide-based electrode materials have been found to have much better energy density, chemical stability, and specific capacity than conductive polymers. Up to now, various transition metal oxides, such as MnO_2, ZnO, and Co_3O_4 are being studied for achieving extraordinary capacitance materials [8].

However, some metal oxides, like Co_2O_4, Ni_2O_4, Mn_2O_4, and Mo_2O_4 show comparatively low electrical conductivity due to their synergistic effects of elements and metal ions. Interestingly, spinel cobaltates ($CuCo_2O_4$, $MnCo_2O_4$, $NiCo_2O_4$, *etc.*) and metal molybdates are gaining potential interest because of easy availability, low cost, and enhanced electrochemical activity [9]. Numerous classes of supercapacitors are illustrated in Fig. (**1**).

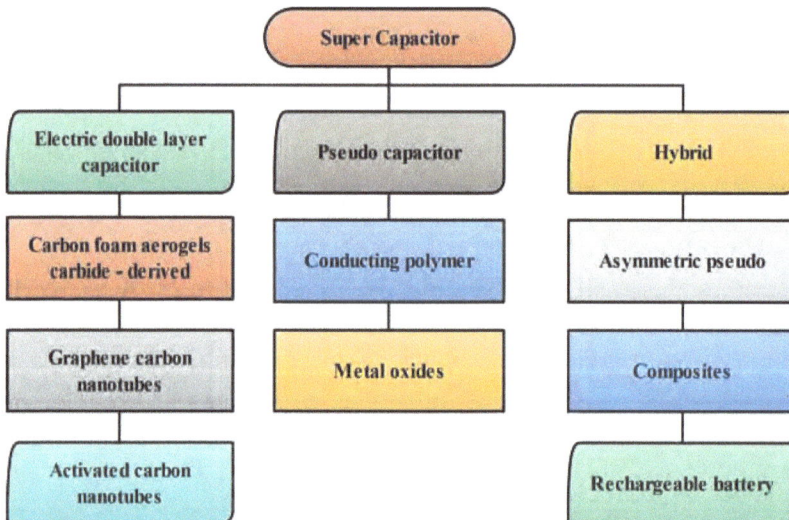

Fig. (1). Different types of Supercapacitors.

Carbon composite-based Supercapacitor Electrode Materials

Kim and co-researchers [10] have fabricated stretchable SC electrodes using a network of graphene/CNT. The fabricated electrodes have structures like a vertically aligned honeycomb. They adopted a crystallization process followed by radial compression to prepare reentrant structures. Fascinatingly, such a porous network provides better stretchability. Moreover, the resulting fabrication shows excellent conductivities. The studies revealed that such vertically aligned structures can overcome some issues based on the ion-accessible and tensile properties of the electrode. The stretchable property of the electrode is essential for designing compactable epidermal electronics. Li and his co-workers [11] followed a simple method by preparing self-standing binder-free film using reduced carbon nanotube/graphene oxide. They prepared the film using hydrothermal treatment (483 K) to develop a self-assembled composite with high-density carbon loading [12]. Table **1** presents different carbon-based SC materials and their capacities for power and energy densities.

Table 1. The values of power and energy density for selected carbon-based materials.

Sl No	Carbon Based-Material	Power Density (Wh Kg^{-1})	Energy Density (Wh Kg^{-1})	References
1	CNT/Graphene/ MnO$_2$	1200	29	[13]
2	Carbon cloth/CNT	44	5.73	[14]
3	MnO$_2$/CNT/Graphene	1280	3.2	[15]
4	CNT/ Active Carbon	14.4	0.016	[16]
5	CNT/ Carbon foam	3700	28	[17]
6	CNT/Nickel foams/ Graphene	5398	19.24	[18]

Recent developments in Capacitor Materials of Metal-oxide and its Composites

Metal-oxides (MOs) are proven as one of the key factors for developing high-performance energy storage devices. MOs exhibit high specific capacitance through the process, called faradaic charge transfer. Hence, recently, they have receiving quite significant attention for the use of SCs. However, SCs prepared by MOs have low energy density which is a drawback to be addressed. So far, several MOs, such as manganese oxide (MnO$_2$), vanadium pentoxide (V$_2$O$_5$), nickel oxide (NiO), ruthenium oxide (RuO$_2$), and cobalt oxide (Co$_3$O$_4$) have been extensively investigated as electrode materials to address the aforementioned issue. They have been observed as potential candidates to prepare the electrode with better energy storage because of their capability of multiple oxidation states and larger surface area. In addition, they can be synthesized by different component morphologies.

Besides, MOs can be exploited by managing their flaws towards modifying their interfaces/surfaces on a nanoscale. Accordingly, the capacitance can be significantly ameliorated. However, still there is room for research towards the improvement of MOs electrodes because they have low electrical conductivity, slow ions transport, and unexpected volume expansion. The most common MOs and their application as SC are elaborated as follows based on the recent literature.

In recent times, RuO_2 has been found to be an excellent material as a supercapacitor because of its very good capacitive performance, such as excellent conductivity, larger charge-carrying properties, stability at low temperatures, superior cycling stability, and higher electrical conductivity. Even though Ru has a high cost that hinders the wide commercialization of RuO_2 to be used as the electrode material in SCs, various investigations have suggested developing an effective pseudocapacitive electrode using aqueous electrolytes. While compared with other MOs, RuO_2 has more chemical stability that can be accomplished through reversible redox reactions. Hence, RuO_2 is considered for its greater specific capacitance. Ates and Yildirim have synthesized RuO_2/rGO/PANI and PANI /RuO_2 nanocomposites that showed a very good specific capacitance of 723.09 F g^{-1} and 40.2 F g^{-1}, respectively [19]. In another study, an electrode was prepared using RuO_2/PANI with the highest specific capacity (816 Fg^{-1}), as reported by Zhang *et al.* [20]. Thin-film composites using RuO_2- PANI prepared by Deshmukh *et al.* showed a very good capacitance of 830 Fg^{-1} [21]. In their study, they observed comparable values of 4.16 kW kg^{-1} and 216 W h kg^{-1} for specific power and specific energy densities, respectively. In the investigation demonstrated elsewhere by Liu *et al.* the use of the composite PEDOT/RuO_2 showed better specific capacitance of 1217 F g^{-1} [22].

Another MO, MnO_2 attracts as one of the significant materials for oxidation catalysts which is extensively applied in SCs. Because of its environmental friendliness, low toxicity, high reserve capacity, and ease of fabrication, MnO_2 gains a promising interest as an electrode material, specific to SC. Moreover, its performances are observed with higher cost-effectiveness and more operational capacity. For this MO, chemical corrosion is found to be low which provides an outstanding electrochemical stuff in the presence of neutral electrolyte. However, MnO_2 has weak conductivity which makes it to have less actual specific capacitance compared with theoretical specific capacitance. As yet another drawback, MnO_2 can be easily dissolved in the electrolyte due to its structural instability; in a nutshell, it shows a lower cycling ability. It has some limitations to use in energy storage devices since it doesn't have greater electron-transport capacity. In a study elsewhere, using MnO_2 and PANI, a composite was prepared by Zhao and Wang with a very good specific capacitance of 497 F g^{-1} [23]. They found that the composite showed better cycle stability to attain 88.2% at 10 A g^{-1} even after 5000 cycles. Grover *et al.* examined the MnO_2/MWCNT/PANI supercapacitor while MWCNT and MnO_2 were executed as negative and positive electrodes, respectively [24]. In this study, the specific capacitance of the prepared electrode was observed to be 324 F g^{-1} with a capacitance retention of 78%.

NiO is one of the notable SC electrode materials since it has very good theoretical capacitance (2584 F g^{-1}). It is widely applied in SCs for its low toxicity, cost-effectiveness in production, less impact on the environment, and larger surface area. In addition, Ni occurs in various oxidation states, which seems this metal is fascinating. But then, experimental studies showed that Ni has comparatively lower electrical conductivity and poor reversibility resulting in the restricted capacitance of Ni materials. Some investigations showed that the hollow/porous design of electrode materials using Ni aiming to improve active surface area can solve this issue quite significantly. Accordingly, the nanocomposites using NiO/PANI were fabricated by Singu *et al.* which showed an attractive specific capacitance ($514 \ Fg^{-1}$) [25]. A ternary composite, ACNF/NiO/PANI has been fabricated by Zhang *et al.* which exhibited remarkable specific capacitance having the value of $1157 \ Fg^{-1}$. They found 93.89% of capacitance retention after 5000 cycles [26]. In yet another investigation, an asymmetric supercapacitor was prepared by Han *et al.* using PPy-6/NiO as a positive electrode and activated carbon (AC) as a negative electrode [27]. They found that the AC/NiO/PPy-6 attained a specific capacitance and energy density of $937.5 \ F \ g^{-1}$ and $333.3 \ W \ h \ kg^{-1}$, respectively.

Co_3O_4 was found to be an emerging supercapacitor material because it is proven to be a great redox characteristic, feasible to synthesize, and has high theoretical capacitance. Studies suggested that it can be a substitute for non-environmentally friendly RuO_2. It has two different oxidation states (Co^{2+} and Co^{3+}) which offer the development of various types of nanocomposites aiming for higher storage capability. In this context, Vanadium pentoxide (V_2O_5), Ferrous oxide (Fe_3O_4), Tungsten trioxide (WO_3), and Tin oxide (SnO_2) are extensively applied to enable the different types of fabrication of SC electrodes. Recently, nanostructured SnO_2 has been gaining huge interest among the different MOs because of its greater broadband gap (3.6 eV) that significantly improves the capacitance. Additionally, it is environmentally friendly and inexpensive, which makes SnO_2 one of the potential candidates for SC materials. Different researchers have carried out investigations on SC performance using the combination of conducting polymers (CPs) and MOs. In general, CPs have good conductivity and are easy to synthesize in low cast. Based on the electrolyte/CP chains interface, there were several studies carried out using MO-CP combinations as summarized in Table **2**.

Table. 2. Summary of MO-CP-based supercapacitors and their performance.

Metal Oxide	Conducting Polymer	Power Density (W/kg)	Energy Density (Wh/kg)	Specific Capacitance (F/g)	Electrolyte	References
MnO_2	PANI	500	35.97	259	$1 \ M \ H_2SO_4$	[28]
MnO_2	PANI	875	11.4	417	$1 \ M \ H_2SO_4$	[29]

(Table 2) cont.....

Metal Oxide	Conducting Polymer	Power Density (W/kg)	Energy Density (Wh/kg)	Specific Capacitance (F/g)	Electrolyte	References
MnO_2	PANI	2.40 mW cm^{-2}	47.25 µWh cm^{-2}	21.1	H2SO4/PVA	[30]
$Ag@MnO_2$	PANI	1599.75	49.77	1028.66	2 M KOH	[31]
MnO_2	PANI	322	119	289	1 M H_2SO_4	[32]
NiO	PANI	800	109.8	308.8	6 M KOH	[33]
TiO_2/RuO_2	PANI	60	3.37	67.4	0.1 M H_2SO_4	[34]
RuO_2	PANI	2400	26.7	428	0.5 M H_2SO_4	[35]
Mn_3O_4	PANI	600	23	460	1 M H_2SO_4	[36]
Co_3O_4	PANI	160	58.84	3105.46	6 M KOH	[37]
Co_3O_4	PANI	751.51	52.81	1407	KOH	[38]
SnO_2	PANI	200	27	540	1 M H_2SO_4	[39]
ZnO	PANI	~403	~5.61	~40	1 M H_2SO_4	[40]
MoO_3	PANI	155	31	45	1 M H_2SO_4	[41]
CeO_2	PANI	830	100.8	504	0.1 M Na_2SO_4	[42]
Sm_2O_3	PANI	-	141	881	1 M H_2SO_4	[43]
CeO_2	PANI	850	46.27	684	1 M H_2SO_4	[44]

Capacitors using Conducting Polymers-carbon

Similar to how carbon-based materials are employed, conductive polymers, in particular, have been used for micro-supercapacitors (MSCs). The majority of conducting polymers are false capacitors. A fast-redox reaction provides the majority of capacitation [45]. Conducting polymers outperform carbon-based MSCs in terms of electrochemical performance and have a shorter life cycle compared to carbon-related materials. Numerous materials can be used to create MSC electrodes. In this regard, transition MOs, such as carbon-based materials and conducting polymers are extensively used. Recent developments in the power storage mechanisms and their studies on the desolation of ions have aided in the creation of higher capacitance electrodes, commonly, it is known as electric double-layer capacitors (EDLCs), such as porous carbon-based electrodes. Combining graphene, metal oxides, and polymers as electrode-active materials will improve the performance of SCs. In such a way, carbon-based materials are known to be a prevalent type of MSC material due to their availability and distinguishing characteristics. The ability of charge storage at the electrode-electrolyte interface requires a carbon-based MSC to have a large specific surface area, just like an EDLC does. Carbon nanotubes (CNTs), graphene, and mesoporous carbon are the most extensive materials to make MSCs [46].

Composite Graphene Capacitors

Recent developments in graphene and polymer composites have shown the very promising properties of these materials for SC applications. Given its vast surface area and strong electrical conductivity, graphene is a proven material to create electrodes. There are three different varieties of graphene: reduced graphene oxide, graphene oxide, and graphene oxide [47]. There has been a lot of research done on graphene. Because of its unique qualities, single-layer graphite is a crucial component of electrodes used in energy applications, particularly electrochemical energy storage [48, 49]. The most of the surface area of any carbon material is found in graphene, which also has several other qualities that are advantageous for making supercapacitors, such as high electrical and thermal conductivity, lightweight, and excellent chemical stability. Graphene-based materials are more desirable for energy storage due to their mix of chemical, physical, and mechanical capabilities [50, 51].

Activated carbons (ACs) are still used as electrode materials in commercial EDLCs due to their high electrochemical performance and ease of implementation for large-scale production. Due to the requirement for energy, ACs are made from organic biomaterials and polymers. However, as the charge/discharge current density increases, the electrochemical performance of ordinary activated carbons demonstrates a quick degradation. When utilized in SCs with organic electrolytes, the specific capacitance of ACs significantly decreases [52]. A new carbon material called chemically modified graphene, which has a mesoporous structure and great electrical conductivity, is a perfect filler for microporous carbon [53]. High electrochemical activity, conductivity, flexibility, strength, and low weight are only a few of the benefits that graphene fibers exhibit collectively. A potential option for wire-shaped electrochemical SCs is graphene composite fibers [54]. Due to its higher specific surface area and more substantial high specific surface area, which improve capacitance and boost the device's ability to store electrostatic charge, the material is frequently suggested as an alternative to activated carbon in electrochemical capacitors. Additionally, these gadgets make use of the material's inherent lightness, elasticity, and mechanical strength [55].

Composite Capacitors with CNTs

In comparison to activated carbon, more sophisticated carbons are intensively researched for their ability to control structures and improve characteristics. CNTs are cylindrical nanostructures with a structure that is very close to being one-dimensional [56]. Carbon nanotubes come in two basic varieties: single-walled and multi-walled. A single graphite sheet that has been coiled into a cylindrical shape forms the single-walled carbon nanotube. Many single-walled carbon nanotubes with various diameters but the same core make up multi-walled carbon nanotubes. Composite electrodes are made using metal oxides or polymers like carbon nanotubes and polypyrrole conducting polymer [57]. The substantial pseudo-capacitance of conducting polymers combined with the conductivity and mechanical strength of the CNT make the composites of CNT and conducting polymer even more attractive and promising [58]. Carbon-based materials make it possible for a capacitive double layer of charge to form. They also have a large surface area, which improves the interaction of the electrolyte and the deposited pseudo-capacitive material [59].

Increased capacitance is the outcome of faradic reactions. Thus, through faradaic processes, pseudo-capacitive materials can increase the electrode's capacitance [60]. Because of its large specific surface area, superior electrical conductivity, and tunable regular porosity network, CNTs are extensively subjected to research for application in SCs. Because of their appreciable mechanical flexibility and good aspect ratio, CNTs are proven to be very flexible in making films for different processes, such as vacuum filtering, dry drawing, self-assembly, blade coating, and so forth. The developed freestanding membrane/paper-like CNT-based electrodes are suggested as a suitable material for flexible SCs since it has outstanding flexibility, high conductivity, and is lightweight [61]. In addition to the polymeric film materials, flexible SCs have also been utilized. CNTs that were drop-dry placed onto office sheets as electrodes [62, 63]. The characteristics of the CNT-based SCs were further enhanced by the incorporation of granulated CNTs and activated carbon into CNT papers [64, 65].

Advancements in Micro-supercapacitors

Due to their sufficient power density and ability to maintain a quick frequency response, MSCs are the preferred option for advanced miniaturized energy storage systems. MSCs are typically sandwiched structures that range in size from a few microns to a few centimeters [66]. EDLCs and pseudo-capacitors are two different types of MSCs. Energy is primarily stored through faradaic processes in pseudo-capacitors as opposed to EDLCs, which store energy on the electrode materials. Recently, several techniques developed, including, 3D printing, screen printing, and laser writing, have been improved to create MSCs. Due to their high efficiency, printing, and laser writing are the current trends in this industry. In comparison to metal oxide, which has a high redox performance but low electrical conductivity and stability, carbon-based active materials, particularly graphene, perform better in electric double-layer capacitors. Additionally, MXenes offer a wide range of possible applications in MSCs due to their chemical and mechanical stability, adjustable thickness, and high electrical conductivity. There have been many approaches to MSC preparation developed up to this point, including photolithography, laser direct writing, printing, and focused ion beam (FIB) [67]. The creation of wearable electronics has grown quickly in response to the rising demand for dependable stretchable and flexible electronics. Recent studies have described the creation of numerous stretchable devices, including soft surgical instruments [68, 69]. Epidermal electronics, wearable photovoltaics, sensitive robotic skin, and organic or inorganic light-emitting diodes (LEDs). Since MSCs offer a higher power density, a longer cycle life, and superior safety than batteries, they have a significant potential to be used as energy storage devices for LEDs and sensors [70 - 74, 73, 75].

1. Supercapacitors' Benefits

❖ SCs have a large energy storage capacity compared to electrolyte capacitors and batteries because the use of activated carbon material improves capacitance value.

❖ More durable than batteries. Batteries use a chemical reaction inside the electrode material to store and release energy, which results in electrode material degradation.

❖ SCs can fast recharge and provide surges of power when needed.

❖ They also have a high power density and can deliver large power bursts for a brief period.

2. Standard uses for Supercapacitor Materials

SCs are used widely in the consumer, public, and industrial sectors. They are also essential in energy recovery and renewable energy technologies, as well as in the fields of medicine, aviation, and the military. SCs are also used in hybrid electric vehicles, trains, buses, light rail systems, trams, aerial lifts, forklifts, tractors, and even motor racing cars [76, 77]. Various standard uses of SC materials are illustrated in the below Fig. (**2**).

Electric and Hybrid Vehicles

The largest market for SCs is transportation. The E-bus project aims to expand and lower the cost of zero-emission public transportation [78]. SCs can also be used to complement batteries in the locomotive starter system. The technology that combines SCs and batteries, however, is more advantageous for buses as a whole. SCs are typically employed in situations where a high number of charge/discharge cycles or a longer lifespan of an appliance is required. SCs are most frequently employed for the voltage-leveling activities offered by rechargeable batteries, such as lead-acid batteries in electric vehicle-driven systems. SCs can also be used to complement batteries in the locomotive starter system [79].

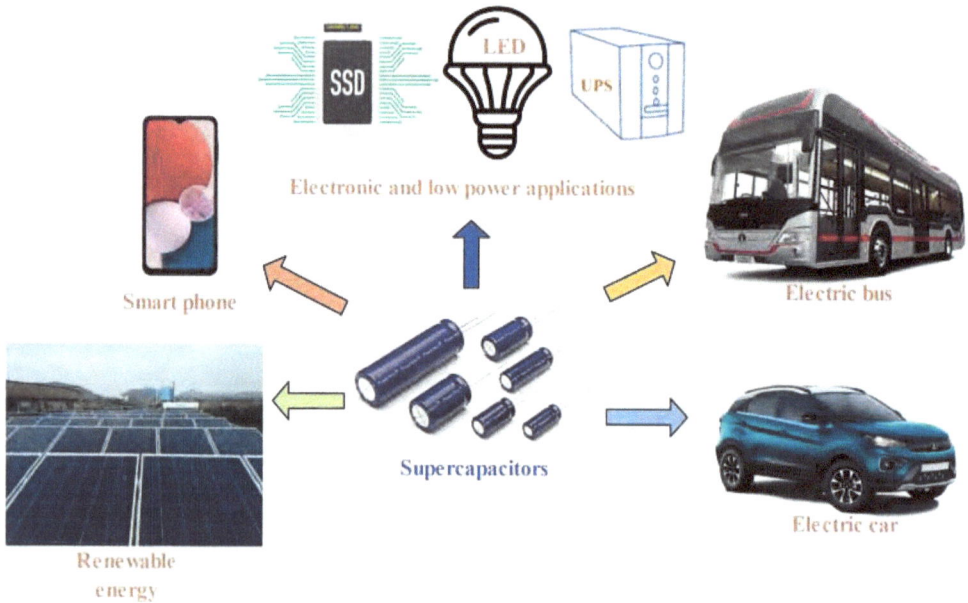

Fig. (2). Various standard applications of SC materials.

Electronic and Low-power Applications

The exceptional storage capacity of SCs has led to their widespread use in a variety of electronic and low-power applications utilized in radio tuners, mobile phones, UPS, laptop memory, street lights, energy storage in tools such as electric screwdrivers, and other electronic devices. They are utilized in situations like LED flash units where a brief power surge is necessary, in addition, they are more commonly utilized in applications that require quick charge and discharge cycles.

Military and Defense Applications

An untapped field for hybrid supercapacitor applications is instruments that rely on batteries, such as navigational, sensing, and communication devices. The radar system, electromagnetic pulse weapons, torpedoes, *etc.*, can all be operated with the appropriate hybrid supercapacitor assembly. Applications requiring high particular power include phased array radar antennas, avionics display systems, airbag exploitation power, GPS, and rockets.

Renewable Energy

Among the improvements in electrochemical energy storage and conversion, SCs are regarded as the most practical auxiliary devices. Compared to existing energy storage devices, they are a new breed of technology. They provide more power than batteries and can store more energy than traditional capacitors.

Industrial and Biomedical Applications

In the industrial sector, there are uses for the combination of batteries and hybrid SCs. The partnership limits the benefits of both energy storage device types and may be useful in many applications involving power. For uninterrupted operation, sophisticated factories need clean, uninterrupted power. For these circumstances, SCs provide a long-lasting, maintenance-free solution. Applications exist for ride-through, energy recapture, peak load cutting, and other UPS setups. The hybrid SCs have uses for high-voltage pulse delivery in the medical field. Patients who have experienced mental trauma are treated using similar techniques. A novel use of hybrid SCs, together with dental technology, *etc.*, is the ventilator backup. This industry has the broadest range of uses for hybrid SCs with good backup power supplies to prevent any catastrophic failures until power is restored.

Traction

The three different phases that the traction vehicle must go through are the acceleration phase, the cruising phase (driving at a constant speed), and the deceleration phase. The power can be both immensely positive and negative throughout these stages. Supercapacitors are integrated with a DC-DC converter as a short-term storage device to solve the aforementioned issue.

SUMMARY AND OUTLOOK

SCs have demonstrated significant potential in a wide range of sectors and fields as a new kind of green and effective energy storage device. SC development will have countless opportunities thanks to the enormous potential market. A detailed discussion on SCs, carbon-based supercapacitor electrode materials, and recent developments in capacitor materials such as metal-oxide composites and capacitors using conducting polymers—carbon is provided in this chapter. Graphene, metal oxide, and carbon are three electrode materials for SCs that are frequently utilized. Although metal oxides have a higher specific capacity, their use in SCs is limited due to their low conductivity, high cost, and negative environmental impact. Graphene is one of the ideal materials for SCs because of its excellent electrochemical properties (such as high conductivity and big specific surface area). The pieces of literature enlighten us with the advancements in micro-supercapacitors. Furthermore, the benefits of and standard uses for SCs' materials were also briefly discussed. In addition, numerous applications of SCs were also addressed in detail.

Future research will be required to create structural electrodes with considerably higher specific energy and power without compromising strength or modulus. One can anticipate active research efforts on SC materials with higher performance attributes in terms of recyclability, prospective window, material fabrication capabilities, and cost considerations, as is the case with any scientific activity. Higher-performing SC materials in terms of recyclability, potential windows, material fabrication possibilities, and cost considerations. Future research should put a heavy emphasis on integrating the mechanical toughness of carbon fiber SCs with cutting-edge methods to balance energy, power, and mechanical properties through the design of the electrode's form, density, and porosity. Therefore, tremendous efforts are required to enhance the current fabrication strategies and create a completely new, efficient way with few limitations. There is a reason to anticipate that enhanced micro-SCs will soon be able to resolve the energy issue for nanoscale devices, given the significant advancements already made and the increased interest and input in the field.

REFERENCES:

[1]   J. Li, Y. Peng, and P. Wang, "Environmental pollution and migrant settlement decision: Evidence from China", *Zhongguo Renkou Ziyuan Yu Huanjing,* vol. 20, no. 4, pp. 357-368, 2022.
[http://dx.doi.org/10.1016/j.cjpre.2022.11.006]

[2]   K. Yousefipour, R. Sarraf-Mamoory, and A. Chaychi Maleki, "A new strategy for the preparation of multi-walled carbon nanotubes/NiMoO4 nanostructures for high-performance asymmetric supercapacitors", *J. Energy Storage,* vol. 59, p. 106438, 2023.
[http://dx.doi.org/10.1016/j.est.2022.106438]

[3]   M. Şahin, F. Blaabjerg, and A. Sangwongwanich, "A comprehensive review on supercapacitor applications and developments", *Energies,* vol. 15, no. 3, p. 674, 2022.
[http://dx.doi.org/10.3390/en15030674]

[4]   R. Brooke, J. Åhlin, K. Hübscher, O. Hagel, J. Strandberg, A. Sawatdee, and J. Edberg, "Large-scale paper supercapacitors on demand", *J. Energy Storage,* vol. 50, p. 104191, 2022.
[http://dx.doi.org/10.1016/j.est.2022.104191]

[5]   K. Thiyagarajan, W-J. Song, H. Park, V. Selvaraj, S. Moon, J. Oh, M-J. Kwak, G. Park, M. Kong, M. Pal, J. Kwak, A. Giri, J-H. Jang, S. Park, and U. Jeong, "Electroactive 1T-MoS2 fluoroelastomer ink

for intrinsically stretchable solid-state in-plane supercapacitors", *ACS Appl. Mater. Interfaces,* vol. 13, no. 23, pp. 26870-26878, 2021.
[http://dx.doi.org/10.1021/acsami.1c01463] [PMID: 34085807]

[6]   R. Brooke, J. Edberg, M.G. Say, A. Sawatdee, A. Grimoldi, J. Åhlin, G. Gustafsson, M. Berggren, and I. Engquist, "Supercapacitors on demand: all-printed energy storage devices with adaptable design", *Flexibl. Print. Electron.,* vol. 4, no. 1, p. 015006, 2019.
[http://dx.doi.org/10.1088/2058-8585/aafc4f]

[7]   Poonam, K. Sharma, A. Arora, and S.K. Tripathi, "Review of supercapacitors: Materials and devices", *J. Energy Storage,* vol. 21, pp. 801-825, 2019.
[http://dx.doi.org/10.1016/j.est.2019.01.010]

[8]   S. Koohi-Fayegh, and M.A. Rosen, "A review of energy storage types, applications and recent developments", *J. Energy Storage,* vol. 27, p. 101047, 2020.
[http://dx.doi.org/10.1016/j.est.2019.101047]

[9]   S. Hajiaghasi, A. Salemnia, and M. Hamzeh, "Hybrid energy storage system for microgrids applications: A review", *J. Energy Storage,* vol. 21, pp. 543-570, 2019.
[http://dx.doi.org/10.1016/j.est.2018.12.017]

[10]   B.S. Kim, K. Lee, S. Kang, S. Lee, J.B. Pyo, I.S. Choi, K. Char, J.H. Park, S-S. Lee, J. Lee, and J.G. Son, "2D reentrant auxetic structures of graphene/CNT networks for omnidirectionally stretchable supercapacitors", *Nanoscale,* vol. 9, no. 35, pp. 13272-13280, 2017.
[http://dx.doi.org/10.1039/C7NR02869E] [PMID: 28858356]

[11]   Y. Li, Z. Kang, X. Yan, S. Cao, M. Li, Y. Guo, Y. Huan, X. Wen, and Y. Zhang, "A three-dimensional reticulate CNT-aerogel for a high mechanical flexibility fiber supercapacitor", *Nanoscale,* vol. 10, no. 19, pp. 9360-9368, 2018.
[http://dx.doi.org/10.1039/C8NR01991F] [PMID: 29737983]

[12]   G. Zhu, Z. He, J. Chen, J. Zhao, X. Feng, Y. Ma, Q. Fan, L. Wang, and W. Huang, "Highly conductive three-dimensional MnO2-carbon nanotube-graphene-Ni hybrid foam as a binder-free supercapacitor electrode", *Nanoscale,* vol. 6, no. 2, pp. 1079-1085, 2014.
[http://dx.doi.org/10.1039/C3NR04495E] [PMID: 24296659]

[13]   Z. Li, Y. Li, L. Wang, L. Cao, X. Liu, Z. Chen, D. Pan, and M. Wu, "Assembling nitrogen and oxygen co-doped graphene quantum dots onto hierarchical carbon networks for all-solid-state flexible supercapacitors", *Electrochim. Acta,* vol. 235, pp. 561-569, 2017.
[http://dx.doi.org/10.1016/j.electacta.2017.03.147]

[14]   Y. Cheng, S. Lu, H. Zhang, C.V. Varanasi, and J. Liu, "Synergistic effects from graphene and carbon nanotubes enable flexible and robust electrodes for high-performance supercapacitors", *Nano Lett.,* vol. 12, no. 8, pp. 4206-4211, 2012.
[http://dx.doi.org/10.1021/nl301804c] [PMID: 22823066]

[15]   K. Shi, M. Ren, and I. Zhitomirsky, "M. Ren. L. Zhitomirsky. Activated carbon-coated carbon nanotubes for energy storage in supercapacitors and capacitive water purification", *ACS Sustain. Chem.& Eng.,* vol. 2, no. 5, pp. 1289-1298, 2014.
[http://dx.doi.org/10.1021/sc500118r]

[16]   Z. Li, K. Xu, and Y. Pan, "Recent development of supercapacitor electrode based on carbon materials", *Nanotechnol. Rev.,* vol. 8, no. 1, pp. 35-49, 2019.
[http://dx.doi.org/10.1515/ntrev-2019-0004]

[17]   C. Xiong, T. Li, T. Zhao, Y. Shang, A. Dang, X. Ji, H. Li, and J. Wang, "Two – step approach of fabrication of three – dimensional reduced graphene oxide – carbon nanotubes – nickel foams hybrid as a binder – free supercapacitor electrode", *Electrochim. Acta,* vol. 217, pp. 9-15, 2016.
[http://dx.doi.org/10.1016/j.electacta.2016.09.068]

[18]   M. Ates, and M. Yildirim, "The synthesis of rGO/RuO2, rGO/PANI, RuO2/PANI and rGO/RuO2/PANI nanocomposites and their supercapacitors", *Polym. Bull.,* vol. 77, no. 5, pp. 2285-

2307, 2020.
[http://dx.doi.org/10.1007/s00289-019-02850-8]

[19]   G. Zhang, L. Deng, J. Liu, J. Zhang, J. Wang, W. Li, and X. Li, "Controllable intercalated polyaniline nanofibers highly enhancing the utilization of delaminated RuO 2 nanosheets for high-performance hybrid supercapacitors", *ChemElectroChem*, vol. 9, no. 9, p. e202200039, 2022.
[http://dx.doi.org/10.1002/celc.202200039]

[20]   P.R. Deshmukh, R.N. Bulakhe, S.N. Pusawale, S.D. Sartale, and C.D. Lokhande, "Polyaniline–RuO 2 composite for high performance supercapacitors: Chemical synthesis and properties", *RSC Adv.,* vol. 5, no. 36, pp. 28687-28695, 2015.
[http://dx.doi.org/10.1039/C4RA16969G]

[21]   R. Liu, J. Duay, T. Lane, and S. Bok Lee, "Synthesis and characterization of RuO(2)/poly(3,4-ethylenedioxythiophene) composite nanotubes for supercapacitors", *Phys. Chem. Chem. Phys.,* vol. 12, no. 17, pp. 4309-4316, 2010.
[http://dx.doi.org/10.1039/b918589p] [PMID: 20407700]

[22]   Y. Zhao, and C.A. Wang, "Nano-network MnO2/polyaniline composites with enhanced electrochemical properties for supercapacitors", *Mater. Des.,* vol. 97, pp. 512-518, 2016.
[http://dx.doi.org/10.1016/j.matdes.2016.02.120]

[23]   S. Grover, V. Sahu, G. Singh, and R.K. Sharma, "High specific energy ternary nanocmposite polyaniline: Manganeese dioxide@ MWCNT electrode for asymmetric supercapacitor", *J. Energy Storage,* vol. 29, p. 101411, 2020.
[http://dx.doi.org/10.1016/j.est.2020.101411]

[24]   B.S. Singu, S. Palaniappan, and K.R. Yoon, "Polyaniline-nickel oxide nanocomposites for supercapacitor", *J. Appl. Electrochem.,* vol. 46, no. 10, pp. 1039-1047, 2016.
[http://dx.doi.org/10.1007/s10800-016-0988-3]

[25]   J. Zhang, L. Su, L. Ma, D. Zhao, C. Qin, Z. Jin, and K. Zhao, "Preparation of inflorescence-like ACNF/PANI/NiO composite with three-dimension nanostructure for high performance supercapacitors", *J. Electroanal. Chem.,* vol. 790, pp. 40-49, 2017.
[http://dx.doi.org/10.1016/j.jelechem.2017.02.047]

[26]   K. Han, Y. Liu, H. Huang, Q. Gong, Z. Zhang, and G. Zhou, "Tremella-like NiO microspheres embedded with fish-scale-like polypyrrole for high-performance asymmetric supercapacitor", *RSC Advances,* vol. 9, no. 38, pp. 21608-21615, 2019.
[http://dx.doi.org/10.1039/C9RA03046H] [PMID: 35518896]

[27]   H. Heydari, M. Abdouss, S. Mazinani, A.M. Bazargan, and F. Fatemi, "Electrochemical study of ternary polyaniline/MoS2−MnO2 for supercapacitor applications", *J. Energy Storage,* vol. 40, p. 102738, 2021.
[http://dx.doi.org/10.1016/j.est.2021.102738]

[28]   S.A. Jadhav, S.D. Dhas, K.T. Patil, A.V. Moholkar, and P.S. Patil, "Polyaniline (PANI)-manganese dioxide (MnO2) nanocomposites as efficient electrode materials for supercapacitors", *Chem. Phys. Lett.,* vol. 778, p. 138764, 2021.
[http://dx.doi.org/10.1016/j.cplett.2021.138764]

[29]   Y. Wei, W. Luo, X. Li, Z. Lin, C. Hou, M. Ma, J. Ding, T. Li, and Y. Ma, "PANI-MnO2 and Ti3C2Tx (MXene) as electrodes for high-performance flexible asymmetric supercapacitors", *Electrochim. Acta,* vol. 406, p. 139874, 2022.
[http://dx.doi.org/10.1016/j.electacta.2022.139874]

[30]   M.B. Poudel, M. Shin, and H.J. Kim, "Polyaniline-silver-manganese dioxide nanorod ternary composite for asymmetric supercapacitor with remarkable electrochemical performance", *Int. J. Hydrogen Energy,* vol. 46, no. 1, pp. 474-485, 2021.
[http://dx.doi.org/10.1016/j.ijhydene.2020.09.213]

[31]   P. Dirican, "Polyaniline/MnO2/porous carbon nanofiber electrodes for supercapacitors", *J.*

Electroanal. Chem., vol. 861, p. 113995, 2020.
[http://dx.doi.org/10.1016/j.jelechem.2020.113995]

[32] C. Huang, C. Hao, W. Zheng, S. Zhou, L. Yang, X. Wang, C. Jiang, and L. Zhu, "Synthesis of polyaniline/nickel oxide/sulfonated graphene ternary composite for all-solid-state asymmetric supercapacitor", *Appl. Surf. Sci.,* vol. 505, p. 144589, 2020.
[http://dx.doi.org/10.1016/j.apsusc.2019.144589]

[33] M.A. Arvizu, F.J. González, A. Romero-Galarza, F.J. Rodríguez-Varela, C.R. Garcia, and M.A. Garcia-Lobato, "Symmetric Supercapacitors of PANI Coated RuO 2 /TiO 2 Macroporous Structures Prepared by Electrostatic Spray Deposition", *J. Electrochem. Soc.,* vol. 169, no. 2, p. 020564, 2022.
[http://dx.doi.org/10.1149/1945-7111/ac5482]

[34] C.W. Kuo, J.C. Chang, B.W. Wu, and T.Y. Wu, "Electrochemical characterization of RuO2-Ta2O5/polyaniline composites as potential redox electrodes for supercapacitors and hydrogen evolution reaction", *Int. J. Hydrogen Energy,* vol. 45, no. 42, pp. 22223-22231, 2020.
[http://dx.doi.org/10.1016/j.ijhydene.2019.08.059]

[35] R. Boddula, R. Bolagam, and P. Srinivasan, "Incorporation of graphene-Mn3O4 core into polyaniline shell: supercapacitor electrode material", *Ionics,* vol. 24, no. 5, pp. 1467-1474, 2018.
[http://dx.doi.org/10.1007/s11581-017-2300-x]

[36] Y. Fan, H. Chen, Y. Li, D. Cui, Z. Fan, and C. Xue, "PANI-Co3O4 with excellent specific capacitance as an electrode for supercapacitors", *Ceram. Int.,* vol. 47, no. 6, pp. 8433-8440, 2021.
[http://dx.doi.org/10.1016/j.ceramint.2020.11.208]

[37] K. Chhetri, A.P. Tiwari, B. Dahal, G.P. Ojha, T. Mukhiya, M. Lee, T. Kim, S-H. Chae, A. Muthurasu, and H.Y. Kim, "A ZIF-8-derived nanoporous carbon nanocomposite wrapped with Co3O4-polyaniline as an efficient electrode material for an asymmetric supercapacitor", *J. Electroanal. Chem.,* vol. 856, p. 113670, 2020.
[http://dx.doi.org/10.1016/j.jelechem.2019.113670]

[38] R. Bolagam, R. Boddula, and P. Srinivasan, "Hybrid material of PANi with TiO2-SnO2: pseudocapacitor electrode for higher performance supercapacitors", *ChemistrySelect,* vol. 2, no. 1, pp. 65-73, 2017.
[http://dx.doi.org/10.1002/slct.201601421]

[39] S. Palsaniya, H.B. Nemade, and A.K. Dasmahapatra, "Hierarchical PANI-RGO-ZnO ternary nanocomposites for symmetric tandem supercapacitor", *J. Phys. Chem. Solids,* vol. 154, p. 110081, 2021.
[http://dx.doi.org/10.1016/j.jpcs.2021.110081]

[40] R. Gottam, and P. Srinivasan, "Composite electrode material of MoO3-MC-SiO2-PANI : Aqueous supercapacitor cell with high energy density, 1 V and 250,000 CD cycles", *Polym. Adv. Technol.,* vol. 32, no. 6, pp. 2465-2475, 2021.
[http://dx.doi.org/10.1002/pat.5276]

[41] R. Bortamuly, G. Konwar, P.K. Boruah, M.R. Das, D. Mahanta, and P. Saikia, "CeO2-PANI-HCl and CeO2-PANI-PTSA composites: synthesis, characterization, and utilization as supercapacitor electrode materials", *Ionics,* vol. 26, no. 11, pp. 5747-5756, 2020.
[http://dx.doi.org/10.1007/s11581-020-03690-7]

[42] Z.H. Mahmoud, R.A. AL-Bayati, and A.A. Khadom, "Synthesis and supercapacitor performance of polyaniline-titanium dioxide-samarium oxide (PANI/TiO2-Sm2O3) nanocomposite", *Chem. Zvesti,* vol. 76, no. 3, pp. 1401-1412, 2022.
[http://dx.doi.org/10.1007/s11696-021-01948-6]

[43] A. Jeyaranjan, T.S. Sakthivel, C.J. Neal, and S. Seal, "Scalable ternary hierarchical microspheres composed of PANI/ rGO/CeO2 for high performance supercapacitor applications", *Carbon,* vol. 151, pp. 192-202, 2019.
[http://dx.doi.org/10.1016/j.carbon.2019.05.043]

[44]   M.R. Khawar, N.A. Shad, S. Hussain, Y. Javed, M.M. Sajid, A. Jilani, M. Faheem, and A. Asghar, "Cerium oxide nanosheets-based tertiary composites (CeO2/ZnO/ZnWO4) for supercapattery application and evaluation of faradic & non-faradic capacitive distribution by using Donn's model", *J. Energy Storage,* vol. 55, p. 105778, 2022.
[http://dx.doi.org/10.1016/j.est.2022.105778]

[45]   H. Fang, D. Yang, Z. Su, X. Sun, J. Ren, L. Li, and K. Wang, "Preparation and application of graphene and derived carbon materials in supercapacitors: A review", *Coatings,* vol. 12, no. 9, p. 1312, 2022.
[http://dx.doi.org/10.3390/coatings12091312]

[46]   K. De, S. Majumder, and P. Kumar, "Is E-mobility a panacea for emission mitigation? A case study of an Indian city", *Environ. Prog. Sustain. Energy,* vol. 40, no. 2, p. e13500, 2021.
[http://dx.doi.org/10.1002/ep.13500]

[47]   A. Yu, V. Chabot, and J. Zhang, *Electrochemical supercapacitors for energy storage and delivery: Fundamentals and applications.* Taylor & Francis, 2013.

[48]   P. Sharma, and V. Kumar, "Current technology of supercapacitors: A review", *J. Electron. Mater.,* vol. 49, no. 6, pp. 3520-3532, 2020.
[http://dx.doi.org/10.1007/s11664-020-07992-4]

[49]   S. Kumar, G. Saeed, L. Zhu, K.N. Hui, N.H. Kim, and J.H. Lee, "0D to 3D carbon-based networks combined with pseudocapacitive electrode material for high energy density supercapacitor: A review", *Chem. Eng. J.,* vol. 403, p. 126352, 2021.
[http://dx.doi.org/10.1016/j.cej.2020.126352]

[50]   N.I. Jalal, R.I. Ibrahim, and M.K. Oudah, "A review on Supercapacitors: Types and components", *J. Phys. Conf. Ser.,* vol. 1973, no. 1, p. 012015, 2021.
[http://dx.doi.org/10.1088/1742-6596/1973/1/012015]

[51]   L.L. Zhang, R. Zhou, and X.S. Zhao, "Graphene-based materials as supercapacitor electrodes", *J. Mater. Chem.,* vol. 20, no. 29, pp. 5983-5992, 2010.
[http://dx.doi.org/10.1039/c000417k]

[52]   C. Li, W. Yang, W. He, X. Zhang, and J. Zhu, "Multifunctional surfactants for synthesizing high-performance energy storage materials", *Energy Storage Mater.,* vol. 43, pp. 1-19, 2021.
[http://dx.doi.org/10.1016/j.ensm.2021.08.033]

[53]   L. Ji, P. Meduri, V. Agubra, X. Xiao, and M. Alcoutlabi, "Graphene-based nanocomposites for energy storage", *Adv. Energy Mater.,* vol. 6, no. 16, p. 1502159, 2016.
[http://dx.doi.org/10.1002/aenm.201502159]

[54]   K. Mohan, D. Narsimhaswamy, and V. Ravi, "Graphene/carbon-based materials for advanced energy conversion applications", *Adv. Nat. Sci.: Nanosci. Nanotechnol.,* vol. 13, no. 3, p. 033005, 2022.
[http://dx.doi.org/10.1088/2043-6262/ac8672]

[55]   D. Ghosh, and S.O. Kim, "Chemically modified graphene based supercapacitors for flexible and miniature devices", *Electron. Mater. Lett.,* vol. 11, no. 5, pp. 719-734, 2015.
[http://dx.doi.org/10.1007/s13391-015-9999-1]

[56]   S. Prasad, S. Srivastava, R. Nagar, and N. Bhagat, "Graphene-based Supercapacitors and their Future", *J. Nanostructures,* vol. 12, no. 2, pp. 375-388, 2022.

[57]   Z. Niu, W. Zhou, J. Chen, G. Feng, H. Li, Y. Hu, W. Ma, H. Dong, J. Li, and S. Xie, "A repeated halving approach to fabricate ultrathin single-walled carbon nanotube films for transparent supercapacitors", *Small,* vol. 9, no. 4, pp. 518-524, 2013.
[http://dx.doi.org/10.1002/smll.201201587] [PMID: 23117974]

[58]   L. SS, "Kim BY Kim JH Chung H", *Lee SY Kim W. ACS Nano,* vol. 6, pp. 6400-6406, 2012.

[59]   C. Zheng, W. Qian, C. Cui, Q. Zhang, Y. Jin, M. Zhao, P. Tan, and F. Wei, "Hierarchical carbon nanotube membrane with high packing density and tunable porous structure for high voltage

supercapacitors", *Carbon,* vol. 50, no. 14, pp. 5167-5175, 2012.
[http://dx.doi.org/10.1016/j.carbon.2012.06.058]

[60]　G. Xu, C. Zheng, Q. Zhang, J. Huang, M. Zhao, J. Nie, X. Wang, and F. Wei, "Binder-free activated carbon/carbon nanotube paper electrodes for use in supercapacitors", *Nano Res.,* vol. 4, no. 9, pp. 870-881, 2011.
[http://dx.doi.org/10.1007/s12274-011-0143-8]

[61]　P.C. Okonkwo, E. Collins, and E. Okonkwo, "Application of biopolymer composites in super capacitor", In: *Biopolymer composites in electronics.* Elsevier, 2017, pp. 487-503.
[http://dx.doi.org/10.1016/B978-0-12-809261-3.00018-8]

[62]　M. Conte, "Supercapacitors technical requirements for new applications", *Fuel Cells,* vol. 10, no. 5, pp. 806-818, 2010.
[http://dx.doi.org/10.1002/fuce.201000087]

[63]　A. Muzaffar, M.B. Ahamed, K. Deshmukh, and J. Thirumalai, "A review on recent advances in hybrid supercapacitors: Design, fabrication and applications", *Renew. Sustain. Energy Rev.,* vol. 101, pp. 123-145, 2019.
[http://dx.doi.org/10.1016/j.rser.2018.10.026]

[64]　A. Berrueta Irigoyen, A. Ursúa Rubio, I. San Martín Biurrun, A. Eftekhari, and P. Sanchis Gúrpide, "Supercapacitors: electrical characteristics, modeling, applications, and future trends", *IEEE Access,* vol. 7, p. 2019, 2019.

[65]　A.M. van Voorden, L.M.R. Elizondo, G.C. Paap, J. Verboomen, and L. van der Sluis, "The application of supercapacitors to relieve battery-storage systems in autonomous renewable energy systems", *2007 IEEE Lausanne Power Tech,* Lausanne, Switzerland, 01-05 July, 2007, pp.479-484

[66]　M.F. Elmorshedy, M.R. Elkadeem, K.M. Kotb, I.B.M. Taha, and D. Mazzeo, "Optimal design and energy management of an isolated fully renewable energy system integrating batteries and supercapacitors", *Energy Convers. Manage.,* vol. 245, p. 114584, 2021.
[http://dx.doi.org/10.1016/j.enconman.2021.114584]

[67]　W. Lhomme, P. Delarue, P. Barrade, A. Bouscayrol, and A. Rufer, "Design and control of a supercapacitor storage system for traction applications", *Fourtieth IAS Annual Meeting. Conference Record of the 2005 Industry Applications Conference,* Hong Kong, China, 02-06 Oct, 2005.
[http://dx.doi.org/10.1109/IAS.2005.1518724]

[68]　X. Sun, K. Chen, F. Liang, C. Zhi, and D. Xue, "Perspective on micro-supercapacitors", *Front Chem.,* vol. 9, p. 807500, 2022.
[PMID: 35087793]

[69]　C. Peng, S. Zhang, D. Jewell, and G.Z. Chen, "Carbon nanotube and conducting polymer composites for supercapacitors", *Prog. Nat. Sci.,* vol. 18, no. 7, pp. 777-788, 2008.
[http://dx.doi.org/10.1016/j.pnsc.2008.03.002]

[70]　F. Bu, W. Zhou, Y. Xu, Y. Du, C. Guan, and W. Huang, "Recent developments of advanced micro-supercapacitors: Design, fabrication and applications", *npj Flexible Elect.,* vol. 4, p. 31, 2020.

[71]　C. Zhang, J. Hu, J. Cong, Y. Zhao, W. Shen, H. Toyoda, M. Nagatsu, and Y. Meng, "Pulsed plasma-polymerized alkaline anion-exchange membranes for potential application in direct alcohol fuel cells", *J. Power Sources,* vol. 196, no. 13, pp. 5386-5393, 2011.
[http://dx.doi.org/10.1016/j.jpowsour.2011.02.073]

[72]　D.H. Kim, N. Lu, R. Ghaffari, Y-S. Kim, S.P. Lee, L. Xu, J. Wu, R-H. Kim, J. Song, Z. Liu, J. Viventi, B. de Graff, B. Elolampi, M. Mansour, M.J. Slepian, S. Hwang, J.D. Moss, S-M. Won, Y. Huang, B. Litt, and J.A. Rogers, "Materials for multifunctional balloon catheters with capabilities in cardiac electrophysiological mapping and ablation therapy", *Nat. Mater.,* vol. 10, no. 4, pp. 316-323, 2011.
[http://dx.doi.org/10.1038/nmat2971] [PMID: 21378969]

[73]   L. Xu, S.R. Gutbrod, Y. Ma, A. Petrossians, Y. Liu, R.C. Webb, J.A. Fan, Z. Yang, R. Xu, J.J. Whalen III, J.D. Weiland, Y. Huang, I.R. Efimov, and J.A. Rogers, "Materials and fractal designs for 3D multifunctional integumentary membranes with capabilities in cardiac electrotherapy", *Adv. Mater.,* vol. 27, no. 10, pp. 1731-1737, 2015.
[http://dx.doi.org/10.1002/adma.201405017] [PMID: 25641076]

[74]   D. H. Kim, "Epidermal electronics", *Science,* vol. 333, pp. 838-843, 2011.

[75]   H. JC, "Fearing RS Javey A", *Nat. Mater.,* vol. 9, pp. 821-826, 2010.

[76]   J. Yoon, A.J. Baca, S-I. Park, P. Elvikis, J.B. Geddes III, L. Li, R.H. Kim, J. Xiao, S. Wang, T-H. Kim, M.J. Motala, B.Y. Ahn, E.B. Duoss, J.A. Lewis, R.G. Nuzzo, P.M. Ferreira, Y. Huang, A. Rockett, and J.A. Rogers, "Ultrathin silicon solar microcells for semitransparent, mechanically flexible and microconcentrator module designs", *Nat. Mater.,* vol. 7, no. 11, pp. 907-915, 2008.
[http://dx.doi.org/10.1038/nmat2287] [PMID: 18836435]

[77]   T. Sekitani, H. Nakajima, H. Maeda, T. Fukushima, T. Aida, K. Hata, and T. Someya, "Stretchable active-matrix organic light-emitting diode display using printable elastic conductors", *Nat. Mater.,* vol. 8, no. 6, pp. 494-499, 2009.
[http://dx.doi.org/10.1038/nmat2459] [PMID: 19430465]

[78]   X. Hu, P. Krull, B. de Graff, K. Dowling, J.A. Rogers, and W.J. Arora, "Stretchable inorganic-semiconductor electronic systems", *Adv. Mater.,* vol. 23, no. 26, pp. 2933-2936, 2011.
[http://dx.doi.org/10.1002/adma.201100144] [PMID: 21538588]

[79]   K.O. Oyedotun, and N. Manyala, "Graphene foam-based electrochemical capacitors", *Curr. Opin. Electrochem.,* vol. 21, pp. 125-131, 2020.
[http://dx.doi.org/10.1016/j.coelec.2019.12.010]

CHAPTER 14

Supercapacitor Materials: From Research to the Real World

Ahmad Nawaz[1,2], Vikas Kumar Pandey[2] and Pradeep Kumar[2,*]

[1] *Center for Refining & Advanced Chemicals, King Fahd University of Petroleum and Minerals, Dhahran, 31261, Saudi Arabia*

[2] *Department of Chemical Engineering & Technology, Indian Institute of Technology (Banaras Hindu University), Varanasi-221005, India*

Abstract: Supercapacitors are gaining prominence in the realm of energy storage devices due to their high power density, extended cycle stability, and fast charge/discharge rates. Supercapacitors are widely used in industries such as service grids, transportation, consumer electronics, wearable and flexible systems, energy harvesting, *etc*. Due to their remarkable high-power performance, high reliability, and extended lifetime, they are a key electrochemical device for energy storage; as a result, the worldwide supercapacitor market is rapidly developing. Supercapacitors have a straightforward basic construction, but different products for various applications require cells in various configurations. The application of supercapacitors from the perspective of the industry is the subject of this chapter.

Keywords: Electrolytes, Energy storage, Electrochemical supercapacitors.

INTRODUCTION

Supercapacitors (SCs) first became commercially available in 1957 when General Electric obtained a patent for a specific type of electrolytic capacitor (EC) which included a porous carbon electrode, known as a double-layer capacitor. However, the Standard Oil Company (SOC) obtained a patent for a capacitor that resembled a disc and was made from carbon paste that had been electrolyte-soaked. Nippon Electric Corporation (NEC), who later produced the first commercially effective supercapacitor, received a license to use the patent in 1971. The primary use of these early supercapacitors was generally as memory backup for electronic applications. Several businesses signified by Panasonic have underway to make supercapacitors with a focus on memory backup devices. These early results generated interest in additional uses, likewise in electric vehicles of the hybrid

---

[*] **Corresponding author Pradeep Kumar:** Department of Chemical Engineering & Technology, Indian Institute of Technology (Banaras Hindu University), Varanasi-221005, India; E-mail: pkumar.che@itbhu.ac.in

Sanjeev Verma, Shivani Verma, Saurabh Kumar & Bhawna Verma (Eds.)

type. After that, supercapacitors' dependability and lifecycle have substantially improved [1].

Supercapacitors are classified based on their charge storage techniques such as ELDCs (electric double-layer capacitors), pseudocapacitors, and hybrids. EDLCs use adsorption to accumulate charges on the electrolyte-electrode interface; on the other hand, pseudocapacitors utilize reversible redox (Faradaic) reactions. The hybrid supercapacitors employ a combination of EDLC and pseudocapacitance to store electrochemical energy [2 - 4]. Currently, supercapacitors are highly advanced amongst the pivotal patents in the middle of the 20th century, because of their higher performance in terms of power, outstanding reliability, and lifecycle. They have played an immensely and continuously expanding role in the development of modern society. It has evolved into a substantial category of electrochemical energy-storage devices on its own; they are commonplace in daily life and come in a variety of shapes depending on the particular application scenarios. This chapter will provide an introduction to the process of making supercapacitors, focusing mostly on the viewpoint of the industry including their device designs, manufacturing procedures, and problems to take into account when attempting to improve performance.

A supercapacitor's fundamental design consists of two porous dielectric films that operate as a separator. Current collectors covered with active electrode materials are typically constructed of metals and are found between separators. The sandwiched structure is then infused with an electrolyte, which can be an organic or aqueous solution containing a significant amount of ions. The size of the basic unit is controlled by the needed capacitance and voltage, and it can be fixed to form a flat pouch cell or rolled to form a radical structure. Carbon electrodes and organic electrolytes are now the industry's typical material options. The device design may be briefly divided into small, medium, and large cells and is significantly dependent on the desired applications. Supercapacitors are made in the following ways: electrode fabrication, cell assembly, separator location, electrolyte impregnation, external connection, and system sealing. The major objectives of supercapacitor manufacturing's industrialization efforts are to boost device performance and decrease ESR (equivalent series resistance), increase resilience, and lower prices as well as the lifecycle of the entire cell. Supercapacitor applications can be broadly classified into 2 categories: (i) High capacitance supercapacitors (HCS) utilized in elevators, uninterruptible power supplies (UPS), and other applications. Supercapacitors are typically built into modules for these purposes and coupled within an electrical balancing circuit. (2) Low capacitance supercapacitors (LCS) are utilized for a variety of electrical functions such as voltage stability and backup. In these applications, supercapacitors are commonly directly welded into the circuit board and are typically the same size as other electronic components such as electrolytic and dielectric capacitors.

COMPONENTS

Electrodes

Current Collector

Most commercial supercapacitors need the deposition of active materials on a metallic current collector in order to lower the resistance, with the exception of a few assemblies that use self-support electrodes instead of current collectors.

An electrode that can stand on its own is made up of conductive carbon fabric. The equivalent series resistance (ESR) is frequently too high in the latter scenario for the effectiveness of high power density. Yet, the self-supporting electrode method is more preferential when manufacturing flexible supercapacitors. The electrode is connected to the external circuit through the current collector. The increase in ESR during cycling is mostly caused by the active material being detached from the current collector. The supercapacitors (SCs) used for commercial purposes are crucial to optimize the mechanical and chemical adherence of electrode material to current collectors. The most extensively used technique in the sector involves coating the current collector with an aqueous or organic slurry that contains activated carbon. Additionally, activated carbon can be directly laminated onto the current collector [5]. The fabrication of thick electrodes is not appropriate for this technique. The different functional groups attached to activated carbon can be used in chemical processes to increase adhesion [6]. Physical methods such as plasma can also be employed among activated carbon and aluminium current collectors to increase adhesion.

Stability in electrolytes is one of the factors to consider when selecting a current collector. Processability, density, and cost are essential to select appropriate current collectors. In an organic electrolyte, aluminium is a good choice for the collector as it is inexpensive, has a low density, and in typical organic electrolytes, and good electrochemical stability. Furthermore, its surface can be given a particular treatment to improve the adhesion between active substances. For such applications, there are two main types of Al: (1) Standard Low-cost aluminium foil is used. It is challenging to preserve material and current collector using traditional binders like polyvinylidene fluoride and polytetrafluoroethylene.

Activated Carbon-based Materials

Activated carbons (ACs) are the major active electrode components of commercial SCs. It serves as a wonderful illustration of how to improve device performance by material modification. The pore density of early activated carbons is low. The necessity to expand the ion-accessible surface in order to improve capacitance quickly gained widespread acceptance. The sugar industry was used to produce the first commercially available activated carbons for supercapacitors. These carbons have low purity, an unbalanced particle size distribution, a significant amount of surface functional groups, and short lifetimes. After that, scientists are experimenting with ACs from various feedstocks to get rid of these drawbacks. It is well-known that only around twenty percent of the porosity is available to ions, which contributes to an application for energy storage. Different studies concentrated on adjusting the porosity in order to enhance the number of pore structures that are accessible to ions [7 - 9].

Precursors and activation methods are the two basic categories used to classify activated carbons. Although several activated carbons have been created in the lab, it is important to note that scaling them for commercial application is an additional matter [10]. The performance and cost balance causes the activated carbons to be extensively adopted by supercapacitor makers. In the recent past, various economically *via*ble carbon sources have been investigated, including, coconuts [11, 12], petroleum waste [13], wood [14], and carbohydrates [14].

Binders

Supercapacitors use binders to link particles and improve electrode adhesion to the current collector which reduces equivalent series resistance and minimizes capacitance degradation [15]. Although, as most binders are insulators and a high amount of them might raise the ESR, the electrode must have a low binder content. Higher binder content can reduce gravimetric capacitance because it is a nonactive substance. Additionally, carbon and binder might adhere to one another more strongly because of electrolytes' capacity to moisten carbon [16, 17]. PTFE (Polytetrafluoroethylene) is widely used as a binder in the SC business due to its great electrochemical stability (ECS) and processability in aqueous environments [18]. A current collector was coated with a PTFE-carbon combination in one seminal patent to manufacture an EDLC [19]. Polyvinylidene fluoride, which is spread in organic solvents including NMP and DMSO as well as THF (tetrahydrofuran), is also widely used as tetrahydrofuran [20]. Additional common binders such as water-soluble polymers like CMC (carboxymethylcellulose) and PVA (polyvinyl alcohol) [21]. Due to their propensity to deteriorate, they cannot function at high voltage. As a result, these binders frequently require chemical modification [22]. Polyimide, another crucial binder, is especially helpful in high-temperature applications.

Additives used for Conductivity

A conductive addition is very necessary for enhancing the electrode's electrical conductivity and reducing the ESR. The idea to include carbon black in order to enhance the conductivity of EDLC carbon electrodes was initially put forth in 1972. The use of conductive additives in the composition of the electrode is first documented in this patent. Since then, a variety of conductive additives have been created, most notably acetylene black, carbon black, Ketjenblack, carbon nanotubes (CNTs), carbon fibers, graphite, and particles metallic fibers, *etc.* The electrode resistance of the conductive additives is influenced by their particle size [23]. As of now, carbon blacks are primarily used by manufacturers due to their excellent all-around performance, which includes electrical conductivity, purity, affordability, stability, lightweight, and processability. The majority of the particles that make up carbon blacks are spherical, and they can aggregate to create particles with a diameter ranging from dozens to hundreds of nm. An efficient 3D conductive network can then be created by the aggregates. Carbon blacks have a microstructure that is similar to graphite. Due to their substantial mesoporosity, two kinds of carbon black such as acetylene and Ketjenblack are frequently preferred [24, 25].

Electrolyte

The electrochemical stability (ECS) opening of electrolytes causes a significant impact on how well SCs function. Additionally, the electrolyte may affect capacitance and ESR. Thermal stability, conductivity, ESC, toxicity, ion mobility, ion concentration, solvent or solvent mixes, temperature, and conductivity of an electrolyte are all crucial electrolyte properties. The electrolyte's salt selection is the main crucial footstep in achieving higher conductivity. Due to their high conductivity and solubility, tetraalkylammonium ion is frequently utilized in commercial SCs. These cations have a high degree of stability due to their shallow reduction potential. Tetraalkylammonium ions can be decreased on a variety of electrodes, according to numerous studies [26 - 28]. RN^{4+} is frequently converted into trialkyl amines, alkanes, and alkenes. The common anions are SO_3CF_3, BF_4, ClO_4, and PF_6. Et_4NBF_4 (tetraethylammonium tetrafluoroborate, $TEABF_4$) and Et_3MeNBF_4 are the most widely used ions (triethylmethylammonium tetrafluoroborate, $TEMABF_4$). The usage of Et_4NBF_4 in ACN has grown quickly in popularity due to its higher conductivity. However, due to environmental concerns, some businesses switch from ACN to PC. However, compared to ACN, the lifespan of SCs built on PC is often less.

According to Zheng *et al.* [29], salt concentration has a direct impact on the voltage at which supercapacitors operate. In the workplace, compromises are frequently necessary because of the cost of the electrolyte solution, particularly the salt component. Therefore, the majority of businesses use 1 M of $TEABF_4$ in CAN or PC. The major factors to take into account while choosing a solvent are its polarizability, dielectric constant, and receptiveness to electrode's electro-active kinds.

When these conditions are combined, the most widely used commercial solvents in SCs are ethylene carbonate (EC), DME (dimethoxyethane), PC, THF, DMF (dimethyl formamide), CAN, *etc*. A key concern for industrialization is the ECS of the solvent, which is closely related to contaminants and potentials of the anode or cathode. Numerous organic electrochemical systems can be severely harmed by oxygen and trace water in particular. Because of this, SC solvent in businesses must have much higher impurity levels and electrochemical potentials that are wider as compared to the potential opening of the device.

Electrolyte Degradation

Even an extremely stable pure electrolyte can deteriorate due to interactions with other supercapacitor parts. One of the most significant methods of degradation is functional groups attached to activated carbon, particularly acidic [30] and trace amounts of water, especially when SC voltage is more than 2.3 V [31]. Gases produced by the degradation reactions clog the pores of the separator and activated carbon, raising the ESR. Gases like carbon monoxide, propene, CO_2, and H_2 were out in a PC electrolyte (1 M $TEABF_4$/PC) type SC up to a voltage of 2.6 [32].

These investigations into the electrolyte degradation processes point to a possible mode of supercapacitor failure, namely the development of pressure as a result of ongoing gas production. Therefore, to withstand enhanced internal pressure and avoid dripping, an organic supercapacitor needs to be mechanically strong. The foremost patented solutions include: condensing gas inside the component, evacuating gas from the inner side of the device *via* valve and membrane, strengthening the supercapacitor with thicker covers and cans, opening device with membrane/venting, and reducing the quantity of gas produced employing chemicals.

Thermal Stability

Supercapacitors have a broad operating temperature range (at least between 30 and 70 °C), which is a considerable benefit. Therefore, when building cells, the consideration of conductivity of various electrolytes at high temperatures. Different strategies can be used to boost conductivity at very high temperatures is very crucial. For instance, by employing $EMPyrBF_4$ (ethyl methyl pyrrolidinium tetrafluoroborate) salt or adding fluoroborenzene [33], the temperature range and conductivity of PC-based electrolyte may be expanded. Supercapacitors are kept running smoothly at deep space temperatures by using a specialized thermal management unit [34].

The key difficulty is to build an electrolyte mixture with a low melting point and appropriate ionic conductivity while reducing the ESR enhancement at lesser temperatures because commercial cells typically utilize high freezing point solvents like CAN and PC. A larger temperature range can be achieved than ACN may be attained by adding solvent [35]. High conductivity is necessary for high-power applications at high temperatures, but high temperatures also hasten to age. ACN may be utilized up to 80°C in real life. PC-type electrolytes are utilized above this temperature, however, aging can be very significant. Sulfolane or EC are two other solvents that are practical at high temperatures. Ionic liquids can be employed if an even greater temperature range is required.

Non-conventional Electrolytes

A conventional solvent with a low flash point and hazardous properties is ACN. As a result, the majority of supercapacitor producers must create cells that prevent electrolyte leakage. Additionally, cells include a cap on the amount of electrolyte to prevent liquid leakage. Given the tight environmental restrictions, ACN is still not allowed in several nations. Despite the possibility of increasing PC-based supercapacitor performance, finding substitutes for ACN in terms of thermal stability, conductivity, and electrochemical stability remains difficult. The low flash point problem, however, may be somewhat reduced by employing flame retardant [36].

By employing pure ionic liquids as the electrolyte in SCs, solvents can be eliminated. Ionic liquids appear like the perfect choice due to their lack of solvent, stability over a broad temperature range, and lack of flammability. Additionally, their purification is challenging, which, as was previously mentioned, may jeopardize the lifespan of supercapacitors. Ionic liquids are unsuited for industry, save for a few specialized uses like at high temperatures [37] and high voltage, due to this feature's high cost. Commercially accessible items based on $DEME-BF_4$ have already demonstrated superior performance to PC-based electrolytes [38].

Separators

To prevent the two electrodes from making contact with one another while allowing ions to travel through, separators are required. Less thickness (as thin as feasible, nonetheless not so thin that the carbon constituent gives rise to shorting among the electrodes), good porosity, high stability (electrochemical, chemical, and thermal), and little cost are thus required. Supercapacitor separators frequently use paper produced from cellulose fibers. They have a low density of less than 0.85 and are prepared of high-grade cellulose fibers (thickness 15–50 µm). To get rid of any H_2O presence in SCs, such separators require robust dehydrating *via* acetone washing thermal or outgassing [39, 40].

Paper separators have also a significant disadvantage, such as oxidative degradation. This results in a significant loss of strength and perhaps the disintegration of the paper, which occurs once open to high voltages of above 3 V. In order to maximize power and energy density, researchers are aiming to raise the operating voltage, which is frequently reached or approached in practice. Consequently, the industry has developed polymeric separators; some such examples are porous polypropylene films (PPF) [41 - 43], Glass fiber-based separators (GFBS) [44, 45] multi-layer separators (MLS), with 2nd layer of electrospun ultra-fine fibers placed on one normal polymeric separator. These types of assembly can be benefitted in the reduction of shorting among electrodes, and separators made of pulp fibers and thermoplastic resin. These also have outstanding mechanical strength, liquid permeability, and ion, and at the same time cost is too high.

Supercapacitor Applications

Power Electronics

There are numerous industrial, commercial, home, aerospace, and military applications for contemporary portable electronics. Power electronics are currently receiving a lot of attention in the field of energy efficiency due to the increased need for energy conservation. Energy storage systems are required by the majority of power electronics systems to offer energy backup and decoupling between power conversion stages.

The most popular storage devices are standard electrolytic capacitors and batteries because of their low cost, however, they have limitations in terms of size, performance, and cycle life. Alternatives to conventional energy storage devices currently seem to include a number of novel energy storage technologies. Electrochemical supercapacitors (ESs) are excellent choices for applications, especially where high power densities, quick transient responses, and small volumes are necessary. ESs are excellent for backup energy storage in power electronics systems that need to be smaller due to their enormous capacitances.

Memory Protection

Numerous battery technologies, including lithium varieties, have been employed in memory preservation and backup power applications for many years. Batteries aren't always the best option, though. Some batteries require complex charging circuits in order to avoid temperature runaways and explosions. They must undergo a number of conditioning processes before being put into operation due to their short lifecycles and frequent replacement. Finally, they might not live long enough to save data during a prolonged power loss when you need them. Even though data centres all over the world primarily rely on batteries to protect sensitive data on their servers and memory controllers, long lifecycles and reliability are still highly desirable attributes. ESs are used on electronic circuit boards to regulate voltages that can fluctuate because of component power consumption, distances, and conductance between the devices. In order to function, they store a fluctuating incoming electronic current and send out a consistent level. The amount of electricity an ES can store increases with size. Similar to a battery, an ES can give a circuit the power it needs to run for a specific amount of time. For example, ESs can supply enough power to maintain a mobile computing device running while batteries are replaced. ESs have been utilised in such applications for decades.

Battery Enhancement

The number of portable and mobile devices is skyrocketing in the modern world. These technologies keep improving as a result of the rising demands for computing, communications, and transportation. Batteries must perform better in terms of power and energy to support all this advancement. Actually, since 1986, computer processing power requirements have been rising significantly. Unfortunately, the growing demands of the new devices have prevented battery producers from keeping up. The kinetics and properties of the chemical reactions that batteries produce continue to be a limitation. The amount of time a battery-powered portable system, such as a laptop or other mobile application, can operate before running out of power is referred to as the battery run-time. For instance, depending on the kind, current battery technologies often have energy densities of 200 W/kg. For instance, a typical Li-ion laptop battery, which typically weighs 0.4 kg, can only power the computer load (25 Wh) for about 3 hours, which is insufficient for a full day of work or uninterrupted autonomy throughout a typical airplane travel. A Li-ion battery with a rated capacity of 4 Ah is the most practical laptop power available today. Despite significant advancements in Li-ion battery technology, the gadgets are still unable to keep up with new features for notebook usage providing digital entertainment, multimedia, and wireless connectivity. The expanded usage demands higher power and currents with higher slew rates.

The conventional approach to extending a battery-powered device's runtime is to simply increase the number of cells that make up the battery pack. However, this approach is unacceptable since it limits the portability and market acceptance of the device and merely raises the size, weight, and maintenance requirements of a portable system.

Portable Energy Sources

ESs may function as rechargeable stand-alone power sources for some portable electronic devices with moderate energy requirements. The most practical power sources right now are batteries. They must, however, be charged overnight and take a long time to recharge. This is seen as a contemporary technological constraint. With the help of ESs, it is possible to design devices that can be charged quickly and repeatedly charged/discharged without suffering major efficiency losses. The powering of light-emitting diodes (LEDs), which produce very effective and quickly rechargeable safety lights, is a typical use of ESs. Choosing the voltage regulators, capacities, and output voltages presented new difficulties due to the ES-based architecture. This new design gives the intended application power for the intended timeframe.

According to Jordan and Spyker, a complete ES power supply package included a DC converter circuitry on a 50 F, 2.5 V ELNA Dynacap [46]. ESs can circumvent the energy delivery restrictions of traditional capacitors and power delivery limitations of batteries for portable devices. After managing high-power activities like wireless transmission, GPS, audio, LED flash, video, and battery hot swaps, they are then recharged from a battery at an average power rate. A common cell phone, for instance, has two Class D amplifiers that are powered by a normal 3.6 V battery and power a pair of 8 speakers. This kind of design only produces 2.25 W of peak power, indicating that it can only support music with weak bass beats and a thin sound. Standard camera phones have trouble producing a strong enough flash to take clear images in dim light. High-current LEDs require up to 400% more power than a battery can offer in order to produce full light intensity.

Adjustable Speed Drives (ASDs)

Due to their effectiveness, adjustable speed drives (ASDs) are frequently employed in industrial applications; yet, they are frequently vulnerable to power interruptions and fluctuations. Industrial disruptions are very unwanted, and equipment downtime during an ongoing process can result in large financial losses. Frequency variations, waveform distortions, swells (over-voltages), sags (under-voltages), voltage fluctuations, transients, and interruptions are the main causes of ASD shutdown. These issues may result in ADS symptoms such as frequent repairs and replacements, inconsistent process parameter control, inexplicable component failures and/or fuse blowing, continuous running of motor cooling systems, and/or frequent motor overheating trips.

An energy storage device must function as a backup power supply for an ASD to ride through disruptions at full power. There are several choices available. Up to an hour of ride-through can be provided using flywheel and battery systems. Although batteries are more cost-effective solutions, their sizes and maintenance needs are the main drawbacks of flywheel systems. Although they can store a lot of energy, fuel cells cannot react rapidly. Also, they need sophisticated cooling systems, and superconducting magnetic energy storage systems can offer reasonable ride-through capabilities.

ESs offer special benefits such as extended lifetimes, quick responses to voltage variations, and less maintenance as compared to these devices, and this can also be easily observed as their charge status is voltage-dependent [47]. However, the selection of an energy storage method is mostly determined by the amount of power needed and the intended ride-through duration. Most power fluctuations typically last no longer than 0.8 seconds. With a 100 kVA rating, an ES can offer ride-through times of up to 5 seconds. As a result, in the majority of applications, ESs are expected to offer a sufficient ride-through.

High Power Sensors and Actuators

A system or mechanism can be moved or controlled by an actuator, a sort of motor. It operates on electric power and transforms that power into motion in some way. For instance, mechanical actuators can change rotary motion into linear motion or the other way around. The majority of actuation systems require pulsed currents with high peak power demands and relatively low average power demands [48]. While an ES bank by itself is probably unable to store enough energy, a battery and ES can be used to create a system that can handle both average and peak load requirements. If only a battery is used to satisfy both demands, an enormous configuration is necessary. For applications with limited space where weight must be kept to a minimum, this is undesirable. When compared to a conventional power source, a hybrid power source made up of a battery and an ES bank can save weight by up to 60% when compared to the use of a battery alone.

Hybrid Electric Vehicles

The potential application of ESs in electric vehicles has generated considerable interest in the technology. Due to its energy efficiency and the potential for recovering energy lost during braking, this technology appeals to those who are concerned about the environment. Several power sources now being considered for installation in electric vehicles (EVs) are insufficient for accelerating. Due to their high energy densities, fuel cells are promising, but their power specs are still constrained. Fuel cell and ES technology can be used in tandem to meet the power and energy needs of an EV. In general, a mixed power source arrangement enables the average load requirements to be met by a high-energy-density device, such as a fuel cell. Fig. (**1a**) demonstrates supercapacitor-based hybrid electric vehicles, while Fig. (**1b**) shows an electric drive train scheme for supercapacitor-fuel cell-based vehicles.

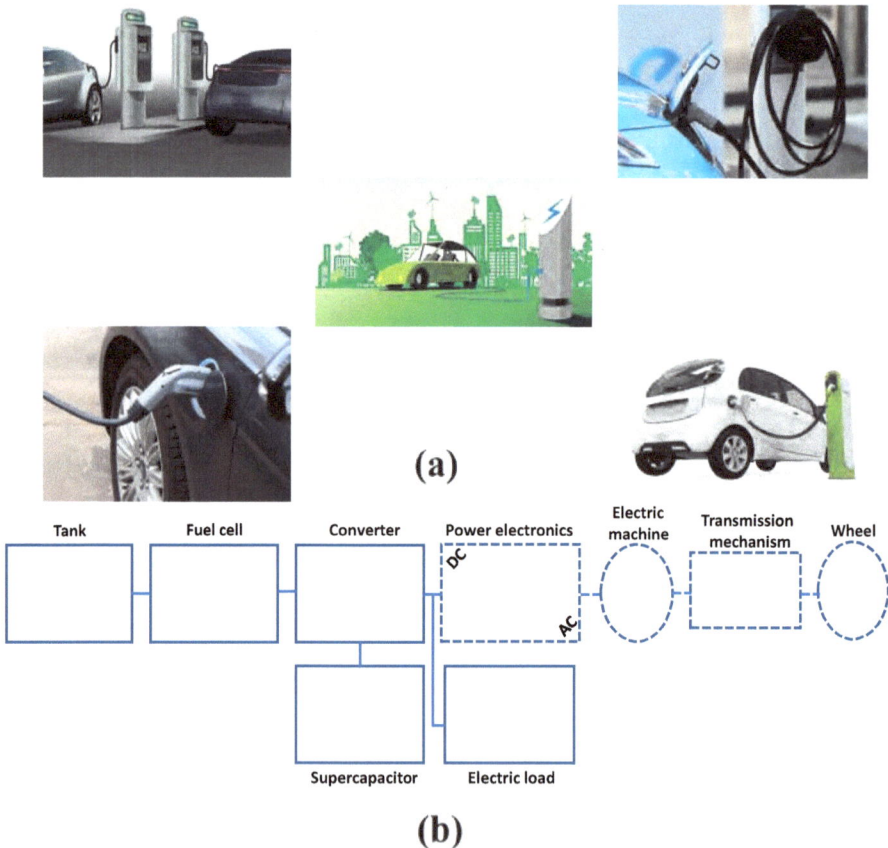

(a)

(b)

Fig. (1). (a) Supercapacitor-based hybrid electric vehicles, **(b)** Electric drive train for vehicles using a combination of supercapacitors with fuel cells.

ES banks are high-power devices that can handle the peak load demands brought on by either accelerating or climbing hills. Regenerative braking is also made possible by the use of ES. It is feasible to store some of the energy from an already moving vehicle and so improve fuel efficiency because the ES bank can be refilled. ESs can be utilized in hybrid vehicles to maximize the efficiency of internal combustion engines (ICEs), in addition to their use in hybrid electric vehicles. To meet the increasing power needs of luxury vehicles, 42 V electrical systems, for example, were proposed, with starter motor substitutes [49]. Greater generating capacity and start-stop engine operation may be made possible by the introduction of an ISA. When the vehicle must halt for an extended amount of time, the ISA can start an ICE fast and easily, allowing the ICE to be turned off instead of burning gasoline needlessly. In this setup, a battery maintains the charge on an ES bank, which powers the engine cranking. The battery is just used to charge the ES; it does not supply power to start the ICE. Battery life can be considerably increased with this method.

Military and Aerospace Applications

As ES technology has advanced, the military market sector has shown a growing demand for ES devices to address a range of power constraints based on ES's special properties. They produce power quickly, perform well at low temperatures, and can withstand up to one million cycles. The common uses of ES in military applications include providing backup power for electronics in armored vehicles, fire control systems in tanks, airbag deployments, and black boxes on helicopters, as well as backup power and memory devices for portable emergency radios.

In peak power applications, ES-based modules are also suitable for cold engine starts, active suspension for vehicle stabilization, GPS-guided missiles and projectile systems, dependable communication transmission on land-based vehicles, and bus voltage hold-up during peak currents. In marine applications, ES-based modules are employed for data retention when switching from ground to onboard aircraft subsystems, high power discharge of unmanned boats in naval warfare, all-electric launches on aircraft carriers, power sources for magnetic drivers that power munitions elevators on aircraft carriers, and all-electric launches. Unmanned aerial vehicles and pulsed laser radar systems both frequently use ES components (UAVs).

CONCLUSION

Electrochemical supercapacitors (ESs) offer significant benefits that make them suitable for a variety of applications. This chapter provides a quick overview of a number of significant ES use cases. Examples include power electronics, off-peak energy storage from renewable power, hybrid electric vehicles, memory protection, portable energy sources, battery enhancements, power quality improvement, high-power actuators, and military and aerospace applications.

ESs are known for their high power densities, lengthy cycle durations, low ESRs, and compact and light weights. They dramatically increase battery cycle life when utilized with batteries. The extension can be made possible by lowering internal battery losses caused by the reduced impedance brought on by the battery-ES connection. Additionally, the hybrid system's capacity to provide power has improved along with the loss decrease, enhancing its performance for pulsating loads like microprocessors. Compared to a parallel battery capacitor connection, the system's running time can be increased by adding an inductor and connecting it in series with the battery.

In a real-world application intended to improve power quality, the utilization of an ES bank to offer ride-through in adjustable speed drives proved useful in correcting for deep voltage sags and transient voltage interruptions while maintaining a constant DC link voltage level. It is important to note that DC-DC converters are necessary to control cell voltage because of the current-dependent ES cell voltage. This will increase system costs and result in some parasitic power loss.

REFERENCES

[1] W. Raza, F. Ali, N. Raza, Y. Luo, K-H. Kim, J. Yang, S. Kumar, A. Mehmood, and E.E. Kwon, "Recent advancements in supercapacitor technology", *Nano Energy,* vol. 52, pp. 441-473, 2018.
[http://dx.doi.org/10.1016/j.nanoen.2018.08.013]

[2] S. Verma, T. Das, V.K. Pandey, and B. Verma, "Facile and scalable synthesis of reduced-graphene oxide using different green reducing agents and its characterizations", *Diamond Related Mater.,* vol. 129, p. 109361, 2022.
[http://dx.doi.org/10.1016/j.diamond.2022.109361]

[3] V.K. Pandey, S. Verma, T. Das, and B. Verma, "Supercapacitive behavior of polyaniline-waste derived carbon-copper cobaltite based ternary composite", *Bioresour. Technol. Rep.,* vol. 20, p. 101255, 2022.
[http://dx.doi.org/10.1016/j.biteb.2022.101255]

[4] S. Verma, V.K. Pandey, and B. Verma, "Synthesis and supercapacitor performance studies of graphene oxide based ternary composite", *Mater. Technol.,* vol. 37, no. 14, pp. 2915-2931, 2022.
[http://dx.doi.org/10.1080/10667857.2022.2086767]

[5] C. Portet, P.L. Taberna, P. Simon, and E. Flahaut, "Modification of al current collector/active material interface for power improvement of electrochemical capacitor electrodes", *J. Electrochem. Soc.,* vol. 153, no. 4, p. A649, 2006.
[http://dx.doi.org/10.1149/1.2168298]

[6] K. Babel, and K. Jurewicz, "KOH activated carbon fabrics as supercapacitor material", *J. Phys. Chem. Solids,* vol. 65, no. 2-3, pp. 275-280, 2004.
[http://dx.doi.org/10.1016/j.jpcs.2003.08.023]

[7] J. Zhao, H. Lai, Z. Lyu, Y. Jiang, K. Xie, X. Wang, Q. Wu, L. Yang, Z. Jin, Y. Ma, J. Liu, and Z. Hu,

"Hydrophilic hierarchical nitrogen-doped carbon nanocages for ultrahigh supercapacitive performance", *Adv. Mater.,* vol. 27, no. 23, pp. 3541-3545, 2015.
[http://dx.doi.org/10.1002/adma.201500945] [PMID: 25931030]

[8]   K. Xie, X. Qin, X. Wang, Y. Wang, H. Tao, Q. Wu, L. Yang, and Z. Hu, "Carbon nanocages as supercapacitor electrode materials", *Adv. Mater.,* vol. 24, no. 3, pp. 347-352, 2012.
[http://dx.doi.org/10.1002/adma.201103872] [PMID: 22139896]

[9]   J. Zhao, Y. Jiang, H. Fan, M. Liu, O. Zhuo, X. Wang, Q. Wu, L. Yang, Y. Ma, and Z. Hu, "Porous 3D few-layer graphene-like carbon for ultrahigh-power supercapacitors with well-defined structure-performance relationship", *Adv. Mater.,* vol. 29, no. 11, p. 1604569, 2017.
[http://dx.doi.org/10.1002/adma.201604569] [PMID: 28044378]

[10] V.V.N. Obreja, "On the performance of supercapacitors with electrodes based on carbon nanotubes and carbon activated material-A review", *Physica E,* vol. 40, no. 7, pp. 2596-2605, 2008.
[http://dx.doi.org/10.1016/j.physe.2007.09.044]

[11] K. Yang, J. Peng, C. Srinivasakannan, L. Zhang, H. Xia, and X. Duan, "Preparation of high surface area activated carbon from coconut shells using microwave heating", *Bioresour. Technol.,* vol. 101, no. 15, pp. 6163-6169, 2010.
[http://dx.doi.org/10.1016/j.biortech.2010.03.001] [PMID: 20303745]

[12] V.K. Pandey, S. Verma, and B. Verma, "Polyaniline/activated carbon/copper ferrite (PANI/AC/CuF) based ternary composite as an efficient electrode material for supercapacitor", *Chem. Phys. Lett.,* vol. 802, p. 139780, 2022.
[http://dx.doi.org/10.1016/j.cplett.2022.139780]

[13] S. Yao, Z. Zhang, Y. Wang, Z. Liu, and Z. Li, "Simple one-pot strategy for converting biowaste into valuable graphitized hierarchically porous biochar for high-efficiency capacitive storage", *J. Energy Storage,* vol. 44, p. 103259, 2021.
[http://dx.doi.org/10.1016/j.est.2021.103259]

[14] H. Benaddi, T.J. Bandosz, J. Jagiello, J.A. Schwarz, J.N. Rouzaud, D. Legras, and F. Béguin, "Surface functionality and porosity of activated carbons obtained from chemical activation of wood", *Carbon,* vol. 38, no. 5, pp. 669-674, 2000.
[http://dx.doi.org/10.1016/S0008-6223(99)00134-7]

[15] Z. Zhu, S. Tang, J. Yuan, X. Qin, Y. Deng, R. Qu, and G.M. Haarberg, "Effects of various binders on supercapacitor performances", *Int. J. Electrochem. Sci.,* vol. 11, no. 10, pp. 8270-8279, 2016.
[http://dx.doi.org/10.20964/2016.10.04]

[16] W. Qiao, Y. Korai, I. Mochida, Y. Hori, and T. Maeda, "Preparation of an activated carbon artifact: oxidative modification of coconut shell-based carbon to improve the strength", *Carbon,* vol. 40, no. 3, pp. 351-358, 2002.
[http://dx.doi.org/10.1016/S0008-6223(01)00110-5]

[17] D. Qu, "Studies of the activated carbons used in double-layer supercapacitors", *J. Power Sources,* vol. 109, no. 2, pp. 403-411, 2002.
[http://dx.doi.org/10.1016/S0378-7753(02)00108-8]

[18] L. Bonnefoi, P. Simon, J.F. Fauvarque, C. Sarrazin, J.F. Sarrau, and A. Dugast, "Electrode compositions for carbon power supercapacitors", *J. Power Sources,* vol. 80, no. 1-2, pp. 149-155, 1999.
[http://dx.doi.org/10.1016/S0378-7753(99)00069-5]

[19] M. Vijayakumar, A. Bharathi Sankar, D. Sri Rohita, T.N. Rao, and M. Karthik, "Conversion of biomass waste into high performance supercapacitor electrodes for real-time supercapacitor applications", *ACS Sustain. Chem. Eng.,* vol. 7, no. 20, pp. 17175-17185, 2019.
[http://dx.doi.org/10.1021/acssuschemeng.9b03568]

[20] Y. Gao, "Graphene and polymer composites for supercapacitor applications: A review", *Nanoscale Res. Lett.,* vol. 12, no. 1, p. 387, 2017.

[http://dx.doi.org/10.1186/s11671-017-2150-5] [PMID: 28582964]

[21]   F. Markoulidis, C. Lei, C. Lekakou, D. Duff, S. Khalil, B. Martorana, and I. Cannavaro, "A method to increase the energy density of supercapacitor cells by the addition of multiwall carbon nanotubes into activated carbon electrodes", *Carbon,* vol. 68, pp. 58-66, 2014.
       [http://dx.doi.org/10.1016/j.carbon.2013.08.040]

[22]   C. Benoit, D. Demeter, D. Bélanger, and C. Cougnon, "A redox-active binder for electrochemical capacitor electrodes", *Angew. Chem. Int. Ed.,* vol. 55, no. 17, pp. 5318-5321, 2016.
       [http://dx.doi.org/10.1002/anie.201601395] [PMID: 26997572]

[23]   A.G. Pandolfo, G.J. Wilson, T.D. Huynh, and A.F. Hollenkamp, "The influence of conductive additives and inter-particle voids in carbon EDLC electrodes", *Fuel Cells,* vol. 10, no. 5, pp. 856-864, 2010.
       [http://dx.doi.org/10.1002/fuce.201000027]

[24]   D. Tashima, H. Yoshitama, M. Otsubo, S. Maeno, and Y. Nagasawa, "Evaluation of electric double layer capacitor using Ketjenblack as conductive nanofiller", *Electrochim. Acta,* vol. 56, no. 24, pp. 8941-8946, 2011.
       [http://dx.doi.org/10.1016/j.electacta.2011.07.124]

[25]   D. Tashima, K. Kurosawatsu, M. Taniguchi, M. Uota, and M. Otsubo, "Basic characteristics of electric double layer capacitor mixing ketjen black as conductive filler", *Electr. Eng. Jpn.,* vol. 165, no. 1, pp. 1-8, 2008.
       [http://dx.doi.org/10.1002/eej.20772]

[26]   S.D. Ross, M. Finkelstein, and R.C. Petersen, "Mechanism of the electroreduction of benzyltriethylammonium nitrate in dimethylformamide at aluminum and platinum cathodes", *J. Am. Chem. Soc.,* vol. 92, no. 20, pp. 6003-6006, 1970.
       [http://dx.doi.org/10.1021/ja00723a032]

[27]   Y. Matsuda, M. Morita, M. Ishikawa, and M. Ihara, "New electric double-layer capacitors using polymer solid electrolytes containing tetraalkylammonium salts", *J. Electrochem. Soc.,* vol. 140, no. 7, pp. L109-L110, 1993.
       [http://dx.doi.org/10.1149/1.2220779]

[28]   E.K. Humphreys, P.K. Allan, R.J.L. Welbourn, T.G.A. Youngs, A.K. Soper, C.P. Grey, and S.M. Clarke, "A neutron diffraction study of the electrochemical double layer capacitor electrolyte tetrapropylammonium bromide in acetonitrile", *J. Phys. Chem. B,* vol. 119, no. 49, pp. 15320-15333, 2015.
       [http://dx.doi.org/10.1021/acs.jpcb.5b08248] [PMID: 26513141]

[29]   J.P. Zheng, and T.R. Jow, "The effect of salt concentration in electrolytes on the maximum energy storage for double layer capacitors", *J. Electrochem. Soc.,* vol. 144, no. 7, pp. 2417-2420, 1997.
       [http://dx.doi.org/10.1149/1.1837829]

[30]   V. Ruiz, T. Huynh, S.R. Sivakkumar, and A.G. Pandolfo, "Ionic liquid–solvent mixtures as supercapacitor electrolytes for extreme temperature operation", *RSC Advances,* vol. 2, no. 13, p. 5591, 2012.
       [http://dx.doi.org/10.1039/c2ra20177a]

[31]   Y. Liu, B. Soucaze-Guillous, P.L. Taberna, and P. Simon, "Understanding of carbon-based supercapacitors ageing mechanisms by electrochemical and analytical methods", *J. Power Sources,* vol. 366, pp. 123-130, 2017.
       [http://dx.doi.org/10.1016/j.jpowsour.2017.08.104]

[32]   D. Hulicova, M. Kodama, and H. Hatori, "Electrochemical performance of nitrogen-enriched carbons in aqueous and non-aqueous supercapacitors", *Chem. Mater.,* vol. 18, no. 9, pp. 2318-2326, 2006.
       [http://dx.doi.org/10.1021/cm060146i]

[33]   R.S. Ball, "Electrolytes for lithium and lithium-ion batteries", *Johnson Matth. Technol. Rev.,* vol. 59, no. 1, pp. 30-33, 2015.

[http://dx.doi.org/10.1595/205651315X685517]

[34]   M. Ayadi, O. Briat, A. Eddahech, R. German, G. Coquery, and J.M. Vinassa, "Thermal cycling impacts on supercapacitor performances during calendar ageing", *Microelectron. Reliab.,* vol. 53, no. 9-11, pp. 1628-1631, 2013.
[http://dx.doi.org/10.1016/j.microrel.2013.07.079]

[35]   E. Perricone, M. Chamas, J-C. Leprêtre, P. Judeinstein, P. Azais, E. Raymundo-Pinero, F. Béguin, and F. Alloin, "Safe and performant electrolytes for supercapacitor. Investigation of esters/carbonate mixtures", *J. Power Sources,* vol. 239, pp. 217-224, 2013.
[http://dx.doi.org/10.1016/j.jpowsour.2013.03.123]

[36]   T. Ye, D. Li, H. Liu, X. She, Y. Xia, S. Zhang, H. Zhang, and D. Yang, "Seaweed biomass-derived flame-retardant gel electrolyte membrane for safe solid-state supercapacitors", *Macromolecules,* vol. 51, no. 22, pp. 9360-9367, 2018.
[http://dx.doi.org/10.1021/acs.macromol.8b01955]

[37]   A.J.R. Rennie, N. Sanchez-Ramirez, R.M. Torresi, and P.J. Hall, "Ether-bond-containing ionic liquids as supercapacitor electrolytes", *J. Phys. Chem. Lett.,* vol. 4, no. 17, pp. 2970-2974, 2013.
[http://dx.doi.org/10.1021/jz4016553] [PMID: 24920995]

[38]   S. Zhang, S. Brahim, and S. Maat, "High-voltage operation of binder-free CNT supercapacitors using ionic liquid electrolytes", *J. Mater. Res.,* vol. 33, no. 9, pp. 1179-1188, 2018.
[http://dx.doi.org/10.1557/jmr.2017.455]

[39]   Y.M. Shulga, S.A. Baskakov, Y.V. Baskakova, Y.M. Volfkovich, N.Y. Shulga, E.A. Skryleva, Y.N. Parkhomenko, K.G. Belay, G.L. Gutsev, A.Y. Rychagov, V.E. Sosenkin, and I.D. Kovalev, "Supercapacitors with graphene oxide separators and reduced graphite oxide electrodes", *J. Power Sources,* vol. 279, pp. 722-730, 2015.
[http://dx.doi.org/10.1016/j.jpowsour.2015.01.032]

[40]   M. Wu, C. Yang, H. Xia, and J. Xu, "Comparative analysis of different separators for the electrochemical performances and long-term stability of high-power lithium-ion batteries", *Ionics,* vol. 27, no. 4, pp. 1551-1558, 2021.
[http://dx.doi.org/10.1007/s11581-021-03943-z]

[41]   N. Wang, G. Han, Y. Xiao, Y. Li, H. Song, and Y. Zhang, "Polypyrrole/graphene oxide deposited on two metalized surfaces of porous polypropylene films as all-in-one flexible supercapacitors", *Electrochim. Acta,* vol. 270, pp. 490-500, 2018.
[http://dx.doi.org/10.1016/j.electacta.2018.03.090]

[42]   N. Wang, G. Han, H. Song, Y. Xiao, Y. Li, Y. Zhang, and H. Wang, "Integrated flexible supercapacitor based on poly (3, 4-ethylene dioxythiophene) deposited on Au/porous polypropylene film/Au", *J. Power Sources,* vol. 395, pp. 228-236, 2018.
[http://dx.doi.org/10.1016/j.jpowsour.2018.05.074]

[43]   N. Wang, X. Wang, Y. Zhang, W. Hou, Y. Chang, H. Song, Y. Zhao, and G. Han, "All-in-one flexible asymmetric supercapacitor based on composite of polypyrrole-graphene oxide and poly(3,4-ethylenedioxythiophene)", *J. Alloys Compd.,* vol. 835, p. 155299, 2020.
[http://dx.doi.org/10.1016/j.jallcom.2020.155299]

[44]   B.K. Deka, A. Hazarika, O.B. Kwon, D. Kim, Y.B. Park, and H.W. Park, "Multifunctional enhancement of woven carbon fiber/ZnO nanotube-based structural supercapacitor and polyester resin-domain solid-polymer electrolytes", *Chem. Eng. J.,* vol. 325, pp. 672-680, 2017.
[http://dx.doi.org/10.1016/j.cej.2017.05.093]

[45]   D. Xu, G. Teng, Y. Heng, Z. Chen, and D. Hu, "Eco-friendly and thermally stable cellulose film prepared by phase inversion as supercapacitor separator", *Mater. Chem. Phys.,* vol. 249, p. 122979, 2020.
[http://dx.doi.org/10.1016/j.matchemphys.2020.122979]

[46]   B.A. Jordan, and R.L. Spyker, "Integrated capacitor and converter package", *APEC 2000. Fifteenth*

Annual IEEE Applied Power Electronics Conference and Exposition (Cat. No.00CH37058), New Orleans, LA, USA, 06-10 Feb, 2000.
[http://dx.doi.org/10.1109/APEC.2000.822576]

[47]   A. von Jouanne, P.N. Enjeti, and B. Banerjee, "Assessment of ride-through alternatives for adjustable-speed drives", *IEEE Trans. Ind. Appl.,* vol. 35, no. 4, pp. 908-916, 1999.
[http://dx.doi.org/10.1109/28.777200]

[48]   S.A. Merryman, "Chemical double-layer capacitor power sources for electrical actuation applications", *Proceedings of the 31st Intersociety Energy Conversion Engineering Conference,* Washington, DC, USA, 11-16 Aug, 1996, pp. 251-254
[http://dx.doi.org/10.1109/IECEC.1996.552879]

[49]   D. Spillane, D. O' Sullivan, M.G. Egan, and J.G. Hayes, "Supervisory control of a HV integrated starter-alternator with ultracapacitor support within the 42 V automotive electrical system", *Eighteenth Annual IEEE Applied Power Electronics Conference and Exposition,* Miami Beach, FL, USA, 09-13 Feb, 2003, pp. 1111-1117
[http://dx.doi.org/10.1109/APEC.2003.1179356]

CHAPTER 15

Future Outlook and Challenges for Supercapacitors

Vikas Kumar Pandey[1] and **Bhawna Verma**[1,*]

[1] Department of Chemical Engineering & Technology, Indian Institute of Technology (Banaras Hindu University), Varanasi-221005, India

Abstract: The contemporary research environment calls for developing next-generation devices using cutting-edge technologies for energy storage applications. Supercapacitors are becoming burgeoning contenders in the energy sector due to their increased durability and quicker charge storage capacity. In contrast to batteries and fuel cells, supercapacitors are a less realistic solution for practical applications. Additionally, there is a pressing need for fabrication techniques that must be addressed to deliver an appropriate supercapacitor electrode. The book chapter will better describe the difficulties encountered during various supercapacitor research, development, and commercial application phases. Finally, a conclusive prognosis has been given on how the discussion above will deliver essential insights and create chances to expand the possible application of new-generation supercapacitors.

Keywords: EDLCs, Energy storage, Pseudocapacitors, Supercapacitors.

INTRODUCTION

The most active research area for the current generation, second only to battery development, is electrochemical supercapacitors, also known as ultracapacitors. Between conventional capacitors and batteries, supercapacitors act as a bridge. There are different types of supercapacitor devices such as wearable supercapacitors, flexible supercapacitors, hybrid supercapacitors, and conventional supercapacitors as illustrated in Fig. (**1**). The in-depth research and development into new materials for supercapacitor component components have enhanced the electrochemical performance and feasibility of the technology in recent years. Significant technical breakthroughs in supercapacitors have been made in the last ten years thanks to significant advancements in material development. The supercapacitor electrode and electrolyte materials are two important domains that require attention. Higher energy storage capacities will be

---

* **Corresponding author Bhawna Verma:** Department of Chemical Engineering & Technology, Indian Institute of Technology (Banaras Hindu University), Varanasi-221005, India; E-mail: bverma.che@itbhu.ac.in

Sanjeev Verma, Shivani Verma, Saurabh Kumar & Bhawna Verma (Eds.)

made possible by increasing the capacitances of active electrode materials. However, cost and cyclability difficulties must be considered while creating these high-capacitance materials. Important parts of supercapacitor devices include electrolyte materials (solvents and ionic species). Aqueous electrolyte solutions have typically been employed as they are easy to handle and non-hazardous. On the contrary, the electrochemically stable operating potential windows provided by aqueous electrolytes are very small (~1.0 V).

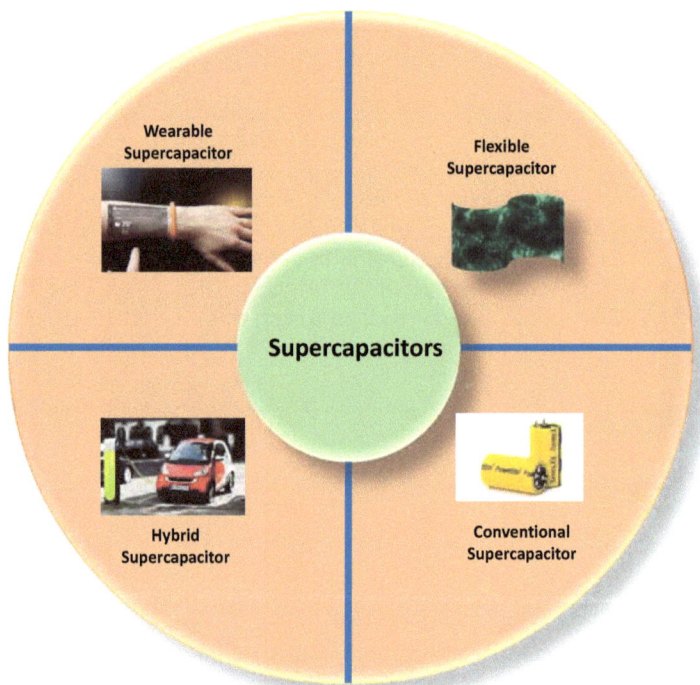

Fig. (1). Different types of supercapacitor devices.

Alternative electrolyte solutions with significantly larger operating voltage stability windows can significantly increase the energy storage capacity of supercapacitors since the energy storage capacity is directly proportional to the square of the operating voltage. Researchers are examining organic and ionic liquid electrolytes on that front. Organic electrolytes are used in the majority of commercially marketed supercapacitor systems. However, there are several material handling and technological challenges with organic and ionic liquid electrolytes that must be resolved.

Several difficulties persist despite significant advancements in the supercapacitor electrode and electrolyte material fields in recent years. This chapter will first cover the market obstacles to supercapacitor development efforts, then it will go into great depth about the development of electrode and electrolyte materials, including the advances made and the difficulties encountered. Also computational techniques will be covered that can be used to support efforts at material development. Finally, a few viewpoints and future directions for study and development will be examined.

MARKET CHALLENGES

Although a small number of supercapacitor devices are currently available, they continue to have several drawbacks, including high cost, a lack of performance capabilities, and issues with long-term endurance. Future energy systems will undoubtedly include supercapacitors as integral components, but to get around the challenges and increase this emerging technology's economic viability, considerable efforts are needed to build new system components. A comparison between specific power density *vs.* specific energy density (Ragone plots) for different electrochemical energy storage and conversion conversion devices has been illustrated in Fig. (**2**).

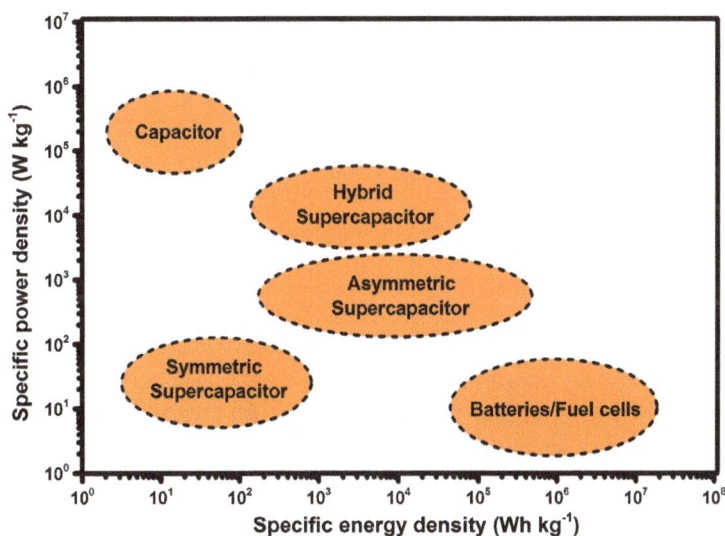

Fig. (2). Ragone plot showing energy density *vs.* power density for various electrochemical energy storage and conversion devices.

If supercapacitor devices are to preserve their competitive advantages over current battery technology, energy density must be significantly increased. Recent developments in this area, including using organic and alternative electrolytes with higher operational voltages, and significant advancements have been achieved owing to the development of new active electrode materials with higher capacitance capabilities. Although these improvements are fairly encouraging, the gap in energy density between supercapacitors and conventional batteries has proven to be extremely challenging to close without sacrificing cyclability. The culprits behind these issues with long-term durability are the electrolyte and electrode components. Charging-discharging, especially with pseudocapacitive materials, may lead to compositional/volumetric changes that jeopardize the electrode's structures and stability and degrade its performance over time [1]. At higher working voltages, electrolyte breakdown is a significant problem that is more frequent and may arise on the surfaces of EDLC-based carbon electrodes, which could lead to higher electrode resistances on account of pore blockages. Another significant issue is the corrosion of aluminum current collectors, which can be solved by using surface treatments to increase stability.

Challenges Based On Electrode Materials

Current collector challenges

An active electric double layer (EDL) based carbon or pseudocapacitive component is frequently put over a conductive metallic current collector (usually made of aluminum) to make a supercapacitor electrode. The interface between the collector and the active material is the main technological issue when employing a metallic foil current collector. The internal resistance of the entire system would certainly rise due to the charge transfer resistance produced by this interface, which will lower the performance of the overall supercapacitor. Increasing the active material-current collector contact area and altering the interface will both help to overcome this problem by lowering the charge transfer resistance. A two-step process was used to modify aluminium current collectors, according to several studies. In order to create surface roughness, a chemical or electrochemical *etc*hing method is used in the initial phase. Due to this, the internal resistance of the current collector is decreased, allowing for greater capacitance during operation and enlarging the area at which the current collector and active material interact. After electrochemically *etc*hing a commercial current collector, it has been shown that these etching techniques can lower a supercapacitor's cell's internal resistance from 50 to 5 Ω cm^2 [2]. Following *etc*hing, the carbon-based coatings may be applied using carbonaceous sol-gel deposition or chemical vapour deposition techniques which are considered capable of further lowering the internal resistances of supercapacitor devices to as low as 0.4 Ω cm^2 [2, 3]. Another plausible strategy for enhancing supercapacitors' performance is the creation of innovative current collector architectures. For instance, the use of a current collector made of carbon nanotubes (CNTs) has reportedly resulted in extremely high capacitance values and cyclability [4].

Better electronic conductivities, higher surface areas, durability, and compatibility with pseudocapacitive materials like conductive polymers [5, 6] and transition metal oxides [7 - 9] are just a few of the advantages that using highly graphitic nanostructured materials like CNTs can provide. These stand-alone composites of active material and current collector are very promising and possess the enormous potential required to greatly reduce the weight of electrode frameworks. Nevertheless, more research and development work is needed to assess whether they are viable. The present collectors' electrochemical stability continues to be a key technical problem that must be considered in all facets of research and development efforts. For instance, material corrosion may cause increased resistance, active material separation, and unavoidable considerable performance loss over extended periods of operation. Two viable strategies for addressing this issue are researching special modification methods for conventional current collectors and creating new current collector materials. The unavoidable objective is to develop current collector structures with lower internal resistance and cost without compromising operational robustness.

Electrical Double-layer-based Materials

Due to their high porosity, electrochemical stability, and conductivity, carbonaceous materials are nearly always used as active components of double-layer electrodes. The most feasible active carbon-based electrode materials are currently activated carbons. They may be synthesized with the help of various easily accessible precursors having high surface areas and are cheap to develop.

By expanding surface areas and porosities, active carbon materials with greater capacitances are ideally developed. There was a widespread misconception that expanding pore volumes relative to solvated ion species' size would immediately enhance capacitance, but recent reports suggested that these materials' capacitances might be increased by the inclusion of micropores (2 nm) [10]. This suggests that some form of molecular-level ion desolvation mechanism facilitates adsorption and ion transport in these micropores. Ion size has also been demonstrated to significantly correlate with ion transport which suggests that it will be challenging to create active carbon materials with customisable pores for specific types of ion species. By using these techniques along with molecular scale computational techniques, it would be possible to prepare high-performance electrode materials through hit and trial methodology. A variety of electrolyte and ion transport issues could arise and limit the performance of electrodes manufactured from activated carbon materials, despite the fact that these materials are frequently utilized and have high capacitance values. Constructing ordered electrode architectures appears to hold promise for getting around this restriction.

For instance, CNTs have received much attention because of their one-dimensional architectures, which provide electrodes with porous networks.

The limited overall surface area that is accessible to ion adsorption is the main issue with CNT-based electrodes; however, the use of the enlarged surface areas of graphene sheets might get over this restriction and help produce better capacitance capabilities [11]. Additionally, CNTs offer rigid conductive routes in the active layer and prevent the accumulation of graphene sheets, which prevents electrolyte penetration and results in electrode blockages. In reality, employing electrodes made entirely of graphene presents a problem because of the tendency for graphene sheets to aggregate. Increased exploitation of the carbon surface area requires features such as distinct in-plane designs [12].

The inherent difficulties of CNTs and graphene can be solved by combining pseudocapacitive materials with them. High-performance pseudocapacitive materials can be used to make up for the CNT's lower active surfaces that are accessible for the development of double layers while still allowing the CNTs to maintain an electrode structure that allows for easy access to the electrolyte. The pseudocapacitive materials with graphene add to the total capacitance and function as spacers to prevent agglomeration. CNTs and graphene are just two examples of the materials under investigation for electric double-layer electrodes. Other examples of carbonaceous materials with use in supercapacitor technologies include graphene [13], activated carbons [14], carbon onion [15], nanostructures, carbon nanofibers [16], and specific template mesoporous carbons [17]. It will take further research to create active carbon-based materials with greater capacitances and electrode configurations that are good for mass transfer. Moreover, low-cost and up-scalable production methods are needed to make these materials commercially viable.

Pseudocapacitive Materials

Conducting Polymers

Conductive polymers can achieve excellent capacitance storage capacity and good electronic conductivity [18]. Additionally, they may store energy across the bulk material, making nanoarchitecture tuning a useful tool needed to give ion access and mass transit to boost internal redox center utilization. Consistent improvement has been made in this area, as shown by developing aligned polyaniline nanorods or nanowires [19] to improve reactant availability. However, the inherent instabilities brought on by their unique redox charge storage methods essentially restrict the use of conductive polymers as pseudocapacitive materials. The long-term operational stability of conductive polymer-based electrodes will also be hampered by the volumetric changes and irreversibilities that are brought about by cycling. The literature suggests that techniques like ultrasonic irradiation and substituent group insertion during polymer processing may help attenuate the effects of these inherent problems [20 - 23].

Transition Metal Oxides

Ruthenium oxide-based materials have undergone extensive development as pseudocapacitive materials because of their excellent operational stability and extraordinarily high values of theoretical capacitance [24] Due to the limited supply and high ruthenium price, these materials' long-term usefulness is constrained. Therefore, research into non-precious metal oxides is crucial to take advantage of their unique cost advantages. A number of metal oxides, such as iron oxide, tin oxide, manganese oxide, titanium oxide, and molybdenum oxide have been synthesized with various phase structures and nanostructures. The two main problems with transition metal oxide development attempts are (1) electrochemical instability and inferior cyclic ability as a consequence of redox nature and (2) their appalling theoretical capacitance and limited electronic conductivity in contrast to ruthenium oxide. Some metal oxides have varied electrochemically stable potential windows that entirely rule out their use in supercapacitor devices.

Numerous researchers examined novel strategies to get over the poor electrical conductivity of metal oxides and their limitations in energy storage. To get around these problems, deliberate nanostructure control is a frequent strategy. The crystallographic structure offers the maximum energy storage capacity with the precise crystal structures, specifically with manganese oxide, having been found to exert considerable impacts on specific capacitance [25, 26]. Additionally, the influence of morphology can customize the particular capacitances of these materials. To achieve this, many manganese oxide nanostructures have been investigated. Performance enhancements with needle-like morphologies [27], nanoplates [28], and nanorods [29] have shown promising results.

Utilizing systems with two or more metal species is a typical method for enhancing the performance of transition metal oxides. These intricate transition metal oxide systems have advantageous electrical characteristics and have a lot of potential as pseudo-capacitive materials. These include $MnFe_2O_4$ [30], $FeTiO_3$ [31], and high-surface-area microporous $NiCo_2O_4$ spinel material [32, 33]. These materials' complementing redox and electrical characteristics make their use in supercapacitor devices particularly promising. The introduction of nitride or sulfide species may also significantly improve performance [34 - 37]. More effort is needed to fully comprehend these materials' performances and create new formulas and architectures. The stability of operations must also be taken into account. A device's overall performance will be jeopardized by problems like electrochemical instability and volumetric variations during cycling.

Composite Based Materials

As stated in the previous section, several strategies have been used to creatively handle the technological difficulties connected with synthesizing pseudocapacitive materials and electric double-layer carbon. Although these materials' ability to store energy has barely improved in recent years, innovations in energy storage technologies are still required. It has become clear that combining carbons with pseudocapacitive materials can close the energy storage gap between batteries and conventional capacitor materials and achieve much better energy storage capacities. This approach uses particular charge storage systems that can work well together in cutting-edge supercapacitor designs.

Compared to other microporous carbon blacks, CNTs have smaller surface areas but maintain strong, durable mechanical electrode structures that can lead to advantageous 3D architectural arrangements. This kind of network configuration can offer highly electronic conductive paths to the redox centers of the active materials with porous structures and can easily promote the transport of electrolyte species. This scaffold-like architecture's structure, porosity, and pore size distribution can also be adjusted by utilizing CNTs with different wall thicknesses, surface characteristics, and diameters. The redox centers found in the majority of conductive polymer materials can be more efficiently used by using these distinctive electrode architectures. Additionally, the volumetric fluctuations caused by charging and discharging processes can be accommodated by a flexible CNT network configuration which can keep the active material layer's mechanical integrity intact, enhancing the materials' cyclabilities [38]. Numerous conjugated conductive polymer architectures can interact positively with the surfaces of CNTs, and the resulting composites can easily form charge transfer complexes [39]. Further research should be done to study the polymer-CNT interface and determine the accurate type of these interactions. Addressing metal's poor electrical conductivity and irreversibility, the carbon-metal oxide composite can be developed to produce thin films containing manganese oxide and carbon nanotubes as composites [40]. Manganese oxide nanoparticles supported on graphene nanoplatelets [41], carbon nanofoams covered in manganese oxide [42], nickel oxide nanoparticles supported on graphene oxide [43], and functionalized graphene with scattered MnO_2 nanosheets [44]. Metal oxide incorporation into CNT-based electrodes is a frequently researched strategy. Metal oxide particles are directly deposited on the surfaces of CNTs by using procedures like electrodeposition [45, 46], hydrothermal processes [47], wet chemistry [48], and microwave-assisted deposition therapy [49]. An effective contact between the two is created by direct deposition, which, while preserving the underlying benefits, enables the CNTs to serve as electronic charge carriers with negligible interfacial resistance. These types of 1D nanostructures can be synthesized to aligned electrodes using CNTs as backbone structures, as was previously mentioned, to get morphologies with improved capacity for storing and transporting electrolytes. The CNT layouts can reduce volumetric changes that occur significantly with charging-discharging cycles and increase the cyclic stabilities of metal oxides-based materials.

Graphene-based composite materials can be developed to circumvent the issue of the low surface areas of CNTs by offering adequate surface areas and mechanical and electrical qualities. Direct deposition of metal oxide species can create complementary composite materials on graphene surfaces. For instance, graphene was coated with different morphologies of manganese oxides using an aqueous precipitation technique [50]. These composite materials' specific capacitances were greater than those of their constituent parts, demonstrating the effects of the synthesis circumstances and composite morphologies. The use of graphene in electrode structures can be used to create electronically conductive channels and function as a buffer to decrease the effects of volumetric changes that take place during operation [51]. To stop graphene agglomerations, metal oxide species might function as spacers and mold into electrode structures; these composite materials' porous characteristics are beneficial for allowing electrolyte access. The anticipated decline in CNT and graphene prices and the improvement of production techniques may make these composites a financially feasible alternative for supercapacitor electrode materials.

Composite materials have amazing potential for use in supercapacitor devices [52, 53]. Researchers can make advancements in the values of capacitances and simultaneously address the difficulties in inferior cyclability by combining the various charge storage methods of carbons and pseudocapacitive materials [54]. To achieve this, researchers must deepen their understanding of the underlying characteristics that this large class of composite materials exhibits throughout design and manufacture. Future supercapacitors research and development may benefit from the discovery of innovative composite arrangements with distinctive morphologies and electrode layouts. Additionally, the physical characteristics of the materials used for the electrodes need to be adapted to a particular kind of electrolyte used as electrode materials operate substantially differently in organic and aqueous electrolytes with different ions.

Electrolytes

Energy storage density can be improved quite practically by raising the operating voltages of supercapacitor devices. In terms of material handling and processing, aqueous electrolytes are favored, but they are prone to breakdown above approx. 1.0 V and gas evolution takes place. It is possible to utilize organic and nonaqueous electrolytes, which will expand the stable potential window to 2.5 V. Acetonitrile and propylene carbonate are the most popular organic electrolytes, with the former being employed more frequently as they offer better ionic conductivity. The effectiveness of these materials has increased as our understanding of ion desolvation, and transport mechanisms has evolved. Due to the safety issues, environmental toxicity of organic electrolytes, and limited operational potential windows, the focus has recently shifted to developing and testing ionic electrolytes for future supercapacitors. When employing ionic liquid electrolytes, stable operational voltages can be increased to 3.5 V or higher without experiencing instability or breakdown issues.

Additionally, ionic liquids possess well-defined ion sizes and do not have ion salvation or desolvation mechanisms in contrast to aqueous and organic electrolyte systems. Because of the huge sizes and shapes of the constituent ion species, their use faces a major technological difficulty due to the limited ionic conductivity. Because of these things, it was formerly thought that ionic liquids were not helpful for operations at low temperatures. With the proper design and material selections, however, recent research has demonstrated that supercapacitors based on the ionic liquid can function reasonably across various operating temperatures. Ionic liquids' temperature-dependent behaviors and physical characteristics, for instance, can be greatly altered by using carefully chosen mixes of anionic and cationic species [55].

Although the development of ionic liquid electrolyte-based supercapacitors is still in its early stages, this is a promising strategy. Additional studies on carefully selected ionic liquid mixtures and their compatibility investigations with various electrode active materials are expected to result in a significant improvement for the supercapacitors industry.

Current Collectors

An appropriate current collector is another essential requirement, in addition to the electrode and electrolyte, for the effective operation of supercapacitors. In practical terms, the thickness of the substrates has an impact on the electrochemical performance. In the lab, substrates are often loaded with only 2-3 mg cm^{-2} of mass, which results in high specific capacitance. However, as the substrates are employed on an industrial scale, the value decreases and negatively affects the supercapacitor device's overall rate performance.

A current collector must offer reasonable electrical conductivity with a low charge transfer and contact resistance. This electrocatalyst is uniformly deposited on the surface of a CC once more [56]. The CC must have strong stability, mechanical resistance, and flexibility while being light and thin. Despite the fact that aluminium makes up the majority of the current collectors used in supercapacitor technology, aluminium has a high charge transfer resistance. Alternative materials like nickel foam, copper foil, and carbon cloth must be employed to reduce contact resistance without impacting the overall cost and performance effectiveness of supercapacitors.

Cu wire with a flexible textile substrate can make an extremely useful substrate for wearable supercapacitors [57]. It is expected that this study would lessen the need for carbon-based materials for negative electrodes as they generally require liquid electrolytes. Further, research on fabric substrates with ink prints will be helpful for the development of flexible substrates and provide some fresh future materials [58].

Computational Aspects

Fundamentally, the precise mechanics of energy storage in electrochemical devices are still not well understood. It is crucial to use in-depth computational methods, such as molecular-level modeling, to analyze the physical and chemical processes of charging, storage, and discharging, especially for composites and electrolytes (organic and ionic liquids). The effects of pore diameters on double-layer capacitance were the main focus of the most recent computer investigations. According to empirical data, capacitance can be influenced by pores that are smaller than the diameters of solvated ion species. This proves that ion desolvation, transport, and adsorption in micropores are not well understood. In order to gain an understanding of the precise mechanisms at play in these systems, density functional theory has been employed. For different size regimes, significantly different charge storage processes for different pore diameters were discovered (microporous, mesoporous, and macroporous). Studies of this kind can provide crucial insights essential for developing novel electrode materials. Additionally, the choice of electrolyte and ion species, as well as their compatibility with various electrode materials, will have a significant impact on how well the supercapacitors perform.

A more profound comprehension of the electrolyte dynamics of a system will enable the design of novel electrolytes with superior ion transport properties and molecular interactions at the electrode interfaces, resulting in noticeably better performance.

Protocols and strategies

There are specific guidelines and practices for all industrial uses. Supercapacitor technology should adhere to the fundamental workings of the component and its materials, life expectancy, performance during deterioration, reliability, physical phenomena like voltage resistance and temperature resistance, device engineering, the performance of all components, electronics components, and failure mechanisms. The device must undergo thorough testing before being used in its final form. Supercapacitors' lifetimes need to be researched and estimated. The viability of supercapacitors for practical applications and the perspective element of the future depends on the safety guidelines, attention being paid to circuit design, and adherence to safety guidelines.

The large e-buses/ electric vehicles developed by Aowei Technology Co., Ltd. (Shanghai, China) and "Capabus" inspired researchers to work on supercapacitors on a global scale and symbolized its promising future. Several phrases that are frequently used in connection with supercapacitor technology, such as high voltage, packaging, self-discharge, temperature resistance, and low-cost materials for energy storage, have generated substantial interest. To address the practical energy problem from a future perspective, supercapacitor technology development requires a well-organized plan. For future commercialization to be possible, wearable supercapacitors, photo supercapacitors, and hybrid supercapacitors should all be created using the same supercapacitor technology. The creation of energy storage systems must benefit from endless natural resources, including water, solar energy, and air. The wearable, self-healing, micro supercapacitor, a small electrochemical storage device, has a higher power density and may be easier to travel. Flexible/wearable devices should be created using industry standards and computational tools. Researchers should understand the underlying concepts and the charge storage mechanism to prevent self-discharge, short circuit failure, and series resistance.

The data analysis should be flawless and thoroughly researched before moving forward. To prevent numerous chaotic processes, some cost-effective material synthesis techniques, such as microwave, coprecipitation, hydrothermal, and electrospinning, should be adopted or developed. This can help MXene and MOFs even more by eliminating conductivity and thwarting problems with aggregation, volume expansion, structural scaffolds, *etc*. The reliable electrochemical analysis data for the commercialization point should come from the two electrodes or asymmetric device system with greater mass loading and flexible electrodes to support the large-scale industrial production of supercapacitors without capacitance loss. Aqueous binders or solid-state electrolytes will be more advantageous for making supercapacitors. The development of non-toxic and non-flammable electrolytes will be helpful for supercapacitor technology.

For the right supercapacitor technology, it's also important to find a solution to issues with supercapacitors including resistance, packaging, sustainability, self-discharge, voltage loss or low voltage, electrolyte, adaptability, and stability. Self-discharge at greater temperatures is the major flaw with supercapacitors which results in the loss of energy storage capability. At the interface between the electrode and electrolyte, parasitic redox processes take place, and a short circuit might result in self-discharge. Utilizing some lithium-ion battery material will help to solve this issue. A higher density will be provided by the supercapacitor's connection, reducing self-discharge. Even so, the coupling is still difficult, although self-discharge can be managed with this [59]. As the square of the applied voltage and capacitance is directly correlated with energy, innovative strategies are required to increase the capacitance and voltage. The equivalent series resistances (ESR) of a cell or device should be taken care of by regulating the surface of current collectors and the active material, thereby minimizing the ohmic drop. In the case of very high operational voltages, particularly in aqueous electrolytes, the electrolyte quickly breaks down, ultimately resulting in self-discharge. The migration of charges in the hindered electric double layer can lead to limited self-discharge. EDL between two electrodes is impeded, and self-discharge is reduced [60]. Numerous approaches have been examined to assess the capacitance and resistance of devices by getting better energy and power densities. However, now is the time to standardize the evaluation procedures regarding commercial use.

Low or high-temperature supercapacitors might prove to be more beneficial in terms of safe operation. Because of the sluggish ion and charge motion, the performance of traditional capacitors frequently declines with changing temperatures. The AVX Corporation's commercial supercapacitors can withstand temperatures of up to 175 °C. However, the ionic liquid electrolytes can remain stable up to 200 °C by improving ionic mobility [61].

CONCLUSION

With numerous applications, including portable electronic devices, hybrid electric vehicles, industrial-scale power production, and energy management, the exciting development of energy storage technologies, such as supercapacitors and batteries, has captured the interest of energy storage research over the past two decades. This chapter discusses the development of improved energy storage materials to create commercially viable supercapacitor electrodes.

A discussion of the development and workings of supercapacitors follows a brief overview of energy storage systems and devices. The ongoing research on various supercapacitors is then highlighted, along with its potential applications in the future. Future reasonable, ultra-flexible, compact, and sustainable energy storage systems will significantly rely on portable and flexible cutting-edge electronics. Before further processing, the MXene, MOFs, COFs, and metal oxides/sulfides require extra consideration and care. Data interpretation should have a crucial component that is accurate and considers theoretical calculations. Additionally, before expansion, computational tools should be acquired. In Fig. (**3**), we have outlined the potential of supercapacitors in the future. For the foundational requirements of supercapacitor technology development, knowledge of the mechanism, supercapacitor kinds, and device fabrication is crucial. The industry standards must be adhered to using some original tactics. Supercapacitors that are small, hybrid, intelligent, and adaptable can be created using the energy resources and technology available today. Consequently, we may anticipate a suitable plan for developing energy substances and electrolytes, and by adhering to industrial protocol, the recently produced supercapacitors will be put into use.

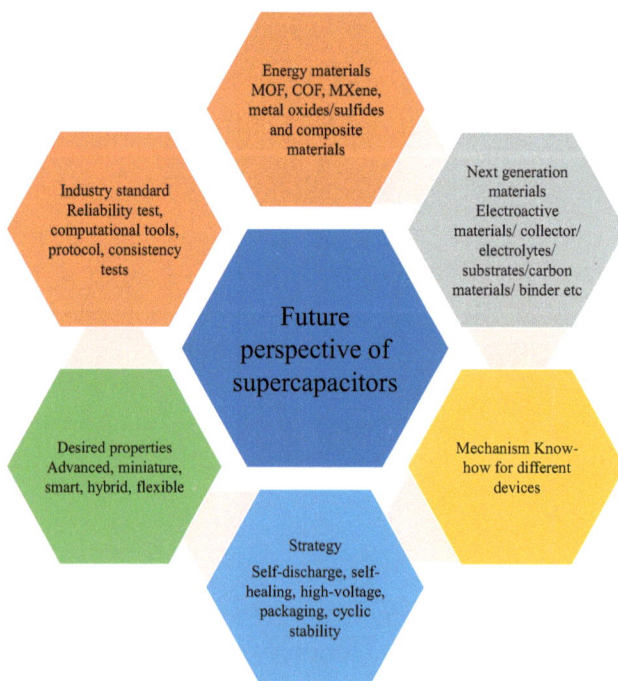

Fig. (3). Future perspective of supercapacitors.

REFERENCES

[1]   X. Zhao, B.M. Sánchez, P.J. Dobson, and P.S. Grant, "The role of nanomaterials in redox-based supercapacitors for next generation energy storage devices", *Nanoscale,* vol. 3, no. 3, pp. 839-855, 2011.
[http://dx.doi.org/10.1039/c0nr00594k] [PMID: 21253650]

[2]   C. Portet, P.L. Taberna, P. Simon, and C. Laberty-Robert, "Modification of Al current collector surface by sol–gel deposit for carbon–carbon supercapacitor applications", *Electrochim. Acta,* vol. 49, no. 6, pp. 905-912, 2004.
[http://dx.doi.org/10.1016/j.electacta.2003.09.043]

[3]   C. Portet, P.L. Taberna, P. Simon, and E. Flahaut, "Modification of al current collector/active material interface for power improvement of electrochemical capacitor electrodes", *J. Electrochem. Soc.,* vol. 153, no. 4, p. A649, 2006.
[http://dx.doi.org/10.1149/1.2168298]

[4]   R. Zhou, C. Meng, F. Zhu, Q. Li, C. Liu, S. Fan, and K. Jiang, "High-performance supercapacitors using a nanoporous current collector made from super-aligned carbon nanotubes", *Nanotechnology,* vol. 21, no. 34, p. 345701, 2010.
[http://dx.doi.org/10.1088/0957-4484/21/34/345701] [PMID: 20683140]

[5]   S.R. Sivakkumar, J.M. Ko, D.Y. Kim, B.C. Kim, and G.G. Wallace, "Performance evaluation of CNT/polypyrrole/MnO2 composite electrodes for electrochemical capacitors", *Electrochim. Acta,* vol. 52, no. 25, pp. 7377-7385, 2007.
[http://dx.doi.org/10.1016/j.electacta.2007.06.023]

[6]   C. Peng, J. Jin, and G.Z. Chen, "A comparative study on electrochemical co-deposition and capacitance of composite films of conducting polymers and carbon nanotubes", *Electrochim. Acta,* vol. 53, no. 2, pp. 525-537, 2007.
[http://dx.doi.org/10.1016/j.electacta.2007.07.004]

[7]   M. Wu, G.A. Snook, G.Z. Chen, and D.J. Fray, "Redox deposition of manganese oxide on graphite for supercapacitors", *Electrochem. Commun.,* vol. 6, no. 5, pp. 499-504, 2004.
[http://dx.doi.org/10.1016/j.elecom.2004.03.011]

[8]   S.B. Ma, K.Y. Ahn, E.S. Lee, K.H. Oh, and K.B. Kim, "Synthesis and characterization of manganese dioxide spontaneously coated on carbon nanotubes", *Carbon,* vol. 45, no. 2, pp. 375-382, 2007.
[http://dx.doi.org/10.1016/j.carbon.2006.09.006]

[9]   J.H. Jang, K. Machida, Y. Kim, and K. Naoi, "Electrophoretic deposition (EPD) of hydrous ruthenium oxides with PTFE and their supercapacitor performances", *Electrochim. Acta,* vol. 52, no. 4, pp. 1733-1741, 2006.
[http://dx.doi.org/10.1016/j.electacta.2006.01.075]

[10]   J. Chmiola, G. Yushin, Y. Gogotsi, C. Portet, P. Simon, and P. L. Taberna, "Anomalous increase in carbon capacitance at pore sizes less than 1 nanometer", *Science,* vol. 313, no. 5794, pp. 1760-1763, 2006.
[http://dx.doi.org/10.1126/science.1132195]

[11]   A. Yu, I. Roes, A. Davies, and Z. Chen, "Ultrathin, transparent, and flexible graphene films for supercapacitor application", *Appl. Phys. Lett.,* vol. 96, no. 25, p. 253105, 2010.
[http://dx.doi.org/10.1063/1.3455879]

[12]   J.J. Yoo, K. Balakrishnan, J. Huang, V. Meunier, B.G. Sumpter, A. Srivastava, M. Conway, A.L. Reddy, J. Yu, R. Vajtai, and P.M. Ajayan, "Ultrathin planar graphene supercapacitors", *Nano Lett.,* vol. 11, no. 4, pp. 1423-1427, 2011.
[http://dx.doi.org/10.1021/nl200225j] [PMID: 21381713]

[13]   S. Verma, and B. Verma, "Graphene-based nanomaterial for supercapacitor application", In: *Nanostructured Materials for Supercapacitors.* Springer, 2022, pp. 221-244.
[http://dx.doi.org/10.1007/978-3-030-99302-3_11]

[14]   V.K. Pandey, S. Verma, and B. Verma, "Polyaniline/activated carbon/copper ferrite (PANI/AC/CuF) based ternary composite as an efficient electrode material for supercapacitor", *Chem. Phys. Lett.,* vol. 802, p. 139780, 2022.
[http://dx.doi.org/10.1016/j.cplett.2022.139780]

[15]   D. Pech, M. Brunet, H. Durou, P. Huang, V. Mochalin, Y. Gogotsi, P-L. Taberna, and P. Simon, "Ultrahigh-power micrometre-sized supercapacitors based on onion-like carbon", *Nat. Nanotechnol.,* vol. 5, no. 9, pp. 651-654, 2010.
[http://dx.doi.org/10.1038/nnano.2010.162] [PMID: 20711179]

[16]   Q. Li, F. Liu, L. Zhang, B.J. Nelson, S. Zhang, C. Ma, X. Tao, J. Cheng, and X. Zhang, "In situ construction of potato starch based carbon nanofiber/activated carbon hybrid structure for high-performance electrical double layer capacitor", *J. Power Sources,* vol. 207, pp. 199-204, 2012.
[http://dx.doi.org/10.1016/j.jpowsour.2012.01.142]

[17]   J. Monk, R. Singh, and F.R. Hung, "Effects of pore size and pore loading on the properties of ionic liquids confined inside nanoporous CMK-3 carbon materials", *J. Phys. Chem. C,* vol. 115, no. 7, pp. 3034-3042, 2011.
[http://dx.doi.org/10.1021/jp1089189]

[18]   T. Das, V.K. Pandey, S. Verma, S.K. Pandey, and B. Verma, "Optimization of the ratio of aniline, ammonium persulfate, para-toluenesulfonic acid for the synthesis of conducting polyaniline and its use in energy storage devices", *Int. J. Energy Res.,* vol. 46, no. 14, pp. 19914-19928, 2022.
[http://dx.doi.org/10.1002/er.8690]

[19]   K. Wang, J. Huang, and Z. Wei, "Conducting Polyaniline Nanowire Arrays for High Performance Supercapacitors", *J. Phys. Chem. C,* vol. 114, no. 17, pp. 8062-8067, 2010.
[http://dx.doi.org/10.1021/jp9113255]

[20]   W. Li, J. Chen, J. Zhao, J. Zhang, and J. Zhu, "Application of ultrasonic irradiation in preparing conducting polymer as active materials for supercapacitor", *Mater. Lett.,* vol. 59, no. 7, pp. 800-803, 2005.
[http://dx.doi.org/10.1016/j.matlet.2004.11.024]

[21]   A.M.P. Hussain, A. Kumar, F. Singh, and D.K. Avasthi, "Effects of 160 MeV Ni 12+ ion irradiation on HCl doped polyaniline electrode", *J. Phys. D Appl. Phys.,* vol. 39, no. 4, pp. 750-755, 2006.
[http://dx.doi.org/10.1088/0022-3727/39/4/023]

[22]   R. Sivakumar, and R. Saraswathi, "Redox properties of poly(N-methylaniline)", *Synth. Met.,* vol. 138, no. 3, pp. 381-390, 2003.
[http://dx.doi.org/10.1016/S0379-6779(03)00023-7]

[23]   C. Arbizzani, M. Catellani, M. Mastragostino, and C. Mingazzini, "N- and P-doped Polydithieno[3,4-B:3',4'-D] thiophene: A narrow band gap polymer for redox supercapacitors", *Electrochim. Acta,* vol. 40, no. 12, pp. 1871-1876, 1995.
[http://dx.doi.org/10.1016/0013-4686(95)00096-W]

[24]   C.C. Hu, W.C. Chen, and K.H. Chang, "How to achieve maximum utilization of hydrous ruthenium oxide for supercapacitors", *J. Electrochem. Soc.,* vol. 151, no. 2, p. A281, 2004.
[http://dx.doi.org/10.1149/1.1639020]

[25]   Y. Yang, and C. Huang, "Effect of synthetical conditions, morphology, and crystallographic structure of MnO2 on its electrochemical behavior", *J. Solid State Electrochem.,* vol. 14, no. 7, pp. 1293-1301, 2010.
[http://dx.doi.org/10.1007/s10008-009-0938-7]

[26]   S. Devaraj, and N. Munichandraiah, "Effect of crystallographic structure of mno 2 on its electrochemical capacitance properties", *J. Phys. Chem. C,* vol. 112, no. 11, pp. 4406-4417, 2008.
[http://dx.doi.org/10.1021/jp7108785]

[27]   S. Verma., S. Verma, S. Kumar and B. Verma, "Nanowires for Supercapacitors", In, *Nanowires, Boca*

Raton: CRC Press, vol. 22, no. 5, pp. 135-140, 2023.

[28]   Z. Ai, L. Zhang, F. Kong, H. Liu, W. Xing, and J. Qiu, "Microwave-assisted green synthesis of MnO2 nanoplates with environmental catalytic activity", *Mater. Chem. Phys.,* vol. 111, no. 1, pp. 162-167, 2008.
       [http://dx.doi.org/10.1016/j.matchemphys.2008.03.043]

[29]   Y. Li, H. Xie, J. Wang, and L. Chen, "Preparation and electrochemical performances of α-MnO2 nanorod for supercapacitor", *Mater. Lett.,* vol. 65, no. 2, pp. 403-405, 2011.
       [http://dx.doi.org/10.1016/j.matlet.2010.10.048]

[30]   S.L. Kuo, and N.L. Wu, "Electrochemical Capacitor of MnFe[sub 2]O[sub 4] with Organic Li-Ion Electrolyte", *Electrochem. Solid-State Lett.,* vol. 10, no. 7, p. A171, 2007.
       [http://dx.doi.org/10.1149/1.2737541]

[31]   T. Tao, A.M. Glushenkov, H. Liu, Z. Liu, X.J. Dai, H. Chen, S.P. Ringer, and Y. Chen, "Ilmenite FeTiO 3 nanoflowers and their pseudocapacitance", *J. Phys. Chem. C,* vol. 115, no. 35, pp. 17297-17302, 2011.
       [http://dx.doi.org/10.1021/jp203345s]

[32]   J. Xiao, and S. Yang, "Sequential crystallization of sea urchin-like bimetallic (Ni, Co) carbonate hydroxide and its morphology conserved conversion to porous NiCo2O4 spinel for pseudocapacitors", *RSC Advances,* vol. 1, no. 4, p. 588, 2011.
       [http://dx.doi.org/10.1039/c1ra00342a]

[33]   T.Y. Wei, C.H. Chen, H.C. Chien, S.Y. Lu, and C.C. Hu, "A cost-effective supercapacitor material of ultrahigh specific capacitances: spinel nickel cobaltite aerogels from an epoxide-driven sol-gel process", *Adv. Mater.,* vol. 22, no. 3, pp. 347-351, 2010.
       [http://dx.doi.org/10.1002/adma.200902175] [PMID: 20217716]

[34]   F. Tao, Y.Q. Zhao, G.Q. Zhang, and H.L. Li, "Electrochemical characterization on cobalt sulfide for electrochemical supercapacitors", *Electrochem. Commun.,* vol. 9, no. 6, pp. 1282-1287, 2007.
       [http://dx.doi.org/10.1016/j.elecom.2006.11.022]

[35]   A.M. Glushenkov, D. Hulicova-Jurcakova, D. Llewellyn, G.Q. Lu, and Y. Chen, "Structure and capacitive properties of porous nanocrystalline VN prepared by temperature-programmed ammonia reduction of V 2 O 5", *Chem. Mater.,* vol. 22, no. 3, pp. 914-921, 2010.
       [http://dx.doi.org/10.1021/cm901729x]

[36]   D. Choi, G.E. Blomgren, and P.N. Kumta, "Fast and reversible surface redox reaction in nanocrystalline vanadium nitride supercapacitors", *Adv. Mater.,* vol. 18, no. 9, pp. 1178-1182, 2006.
       [http://dx.doi.org/10.1002/adma.200502471]

[37]   S.J. Bao, C.M. Li, C.X. Guo, and Y. Qiao, "Biomolecule-assisted synthesis of cobalt sulfide nanowires for application in supercapacitors", *J. Power Sources,* vol. 180, no. 1, pp. 676-681, 2008.
       [http://dx.doi.org/10.1016/j.jpowsour.2008.01.085]

[38]   C. Meng, C. Liu, and S. Fan, "Flexible carbon nanotube/polyaniline paper-like films and their enhanced electrochemical properties", *Electrochem. Commun.,* vol. 11, no. 1, pp. 186-189, 2009.
       [http://dx.doi.org/10.1016/j.elecom.2008.11.005]

[39]   W. Feng, X.D. Bai, Y.Q. Lian, J. Liang, X.G. Wang, and K. Yoshino, "Well-aligned polyaniline/carbon-nanotube composite films grown by in-situ aniline polymerization", *Carbon,* vol. 41, no. 8, pp. 1551-1557, 2003.
       [http://dx.doi.org/10.1016/S0008-6223(03)00078-2]

[40]   S.W. Lee, J. Kim, S. Chen, P.T. Hammond, and Y. Shao-Horn, "Carbon nanotube/manganese oxide ultrathin film electrodes for electrochemical capacitors", *ACS Nano,* vol. 4, no. 7, pp. 3889-3896, 2010.
       [http://dx.doi.org/10.1021/nn100681d] [PMID: 20552996]

[41]   S. Verma, and B. Verma, "Synergistic optimization of nanostructured graphene oxide based ternary

composite for boosting the performance of supercapacitor electrode material *via* response surface methodology", *Colloids and Surfaces A: Physicochemical and Engineering Aspects,* vol. 682, p. 132893, 2024.
[http://dx.doi.org/0.1016/j.colsurfa.2023.132893]

[42] J.W. Long, M.B. Sassin, A.E. Fischer, D.R. Rolison, A.N. Mansour, V.S. Johnson, P.E. Stallworth, and S.G. Greenbaum, "Multifunctional MnO 2 −Carbon Nanoarchitectures Exhibit Battery and Capacitor Characteristics in Alkaline Electrolytes", *J. Phys. Chem. C,* vol. 113, no. 41, pp. 17595-17598, 2009.
[http://dx.doi.org/10.1021/jp9070696]

[43] M.S. Wu, Y.P. Lin, C.H. Lin, and J.T. Lee, "Formation of nano-scaled crevices and spacers in NiO-attached graphene oxidenanosheets for supercapacitors", *J. Mater. Chem.,* vol. 22, no. 6, pp. 2442-2448, 2012.
[http://dx.doi.org/10.1039/C1JM13818A]

[44] J. Zhang, J. Jiang, and X.S. Zhao, "Synthesis and capacitive properties of manganese oxide nanosheets dispersed on functionalized graphene sheets", *J. Phys. Chem. C,* vol. 115, no. 14, pp. 6448-6454, 2011.
[http://dx.doi.org/10.1021/jp200724h]

[45] H. Zhang, G. Cao, Z. Wang, Y. Yang, Z. Shi, and Z. Gu, "Growth of manganese oxide nanoflowers on vertically-aligned carbon nanotube arrays for high-rate electrochemical capacitive energy storage", *Nano Lett.,* vol. 8, no. 9, pp. 2664-2668, 2008.
[http://dx.doi.org/10.1021/nl800925j] [PMID: 18715042]

[46] S.L. Chou, J.Z. Wang, S.Y. Chew, H.K. Liu, and S.X. Dou, "Electrodeposition of MnO2 nanowires on carbon nanotube paper as free-standing, flexible electrode for supercapacitors", *Electrochem. Commun.,* vol. 10, no. 11, pp. 1724-1727, 2008.
[http://dx.doi.org/10.1016/j.elecom.2008.08.051]

[47] S. Chen, J. Zhu, X. Wu, Q. Han, and X. Wang, "Graphene oxide--MnO2 nanocomposites for supercapacitors", *ACS Nano,* vol. 4, no. 5, pp. 2822-2830, 2010.
[http://dx.doi.org/10.1021/nn901311t] [PMID: 20384318]

[48] J. Yan, T. Wei, B. Shao, Z. Fan, W. Qian, M. Zhang, and F. Wei, "Preparation of a graphene nanosheet/polyaniline composite with high specific capacitance", *Carbon,* vol. 48, no. 2, pp. 487-493, 2010.
[http://dx.doi.org/10.1016/j.carbon.2009.09.066]

[49] A. Yu, A. Sy, and A. Davies, "Graphene nanoplatelets supported MnO2 nanoparticles for electrochemical supercapacitor", *Synth. Met.,* vol. 161, no. 17-18, pp. 2049-2054, 2011.
[http://dx.doi.org/10.1016/j.synthmet.2011.04.034]

[50] L. Mao, K. Zhang, H.S. On Chan, and J. Wu, "Nanostructured MnO 2 /graphene composites for supercapacitor electrodes: the effect of morphology, crystallinity and composition", *J. Mater. Chem.,* vol. 22, no. 5, pp. 1845-1851, 2012.
[http://dx.doi.org/10.1039/C1JM14503G]

[51] S. Verma, V.K. Pandey, and B. Verma, "Synthesis and supercapacitor performance studies of graphene oxide based ternary composite", *Mater. Technol.,* vol. 37, no. 14, pp. 2915-2931, 2022.
[http://dx.doi.org/10.1080/10667857.2022.2086767]

[52] V.K. Pandey, S. Verma, T. Das, and B. Verma, "Supercapacitive behavior of polyaniline-waste derived carbon-copper cobaltite based ternary composite", *Bioresour. Technol. Rep.,* vol. 20, p. 101255, 2022.
[http://dx.doi.org/10.1016/j.biteb.2022.101255]

[53] S. Verma, V.K. Pandey, and B. Verma, "Facile synthesis of graphene oxide-polyaniline-copper cobaltite (GO/PANI/CuCo2O4) hybrid nanocomposite for supercapacitor applications", *Synth. Met.,* vol. 286, p. 117036, 2022.
[http://dx.doi.org/10.1016/j.synthmet.2022.117036]

[54]   S. Verma, T. Das, V.K. Pandey, and B. Verma, "Nanoarchitectonics of GO/PANI/CoFe2O4 (Graphene Oxide/polyaniline/Cobalt Ferrite) based hybrid composite and its use in fabricating symmetric supercapacitor devices", *J. Mol. Struct.,* vol. 1266, p. 133515, 2022.
[http://dx.doi.org/10.1016/j.molstruc.2022.133515]

[55]   R. Lin, P-L. Taberna, S. Fantini, V. Presser, C.R. Pérez, F. Malbosc, N.L. Rupesinghe, K.B.K. Teo, Y. Gogotsi, and P. Simon, "Capacitive Energy Storage from −50 to 100 °C Using an Ionic Liquid Electrolyte", *J. Phys. Chem. Lett.,* vol. 2, no. 19, pp. 2396-2401, 2011.
[http://dx.doi.org/10.1021/jz201065t]

[56]   G.F. Yang, K.Y. Song, and S.K. Joo, "A metal foam as a current collector for high power and high capacity lithium iron phosphate batteries", *J. Mater. Chem. A Mater. Energy Sustain.,* vol. 2, no. 46, pp. 19648-19652, 2014.
[http://dx.doi.org/10.1039/C4TA03890H]

[57]   T. Purkait, G. Singh, D. Kumar, M. Singh, and R.S. Dey, "High-performance flexible supercapacitors based on electrochemically tailored three-dimensional reduced graphene oxide networks", *Sci. Rep.,* vol. 8, no. 1, p. 640, 2018.
[http://dx.doi.org/10.1038/s41598-017-18593-3] [PMID: 29330476]

[58]   P. Sundriyal, and S. Bhattacharya, "Textile-based supercapacitors for flexible and wearable electronic applications", *Sci. Rep.,* vol. 10, no. 1, p. 13259, 2020.
[http://dx.doi.org/10.1038/s41598-020-70182-z] [PMID: 32764660]

[59]   A. González, E. Goikolea, J.A. Barrena, and R. Mysyk, "Review on supercapacitors: Technologies and materials", *Renew. Sustain. Energy Rev.,* vol. 58, pp. 1189-1206, 2016.
[http://dx.doi.org/10.1016/j.rser.2015.12.249]

[60]   A. Borenstein, O. Hanna, R. Attias, S. Luski, T. Brousse, and D. Aurbach, "Carbon-based composite materials for supercapacitor electrodes: A review", *J. Mater. Chem. A Mater. Energy Sustain.,* vol. 5, no. 25, pp. 12653-12672, 2017.
[http://dx.doi.org/10.1039/C7TA00863E]

[61]   R.S. Borges, A.L.M. Reddy, M-T.F. Rodrigues, H. Gullapalli, K. Balakrishnan, G.G. Silva, and P.M. Ajayan, "Supercapacitor operating at 200 degrees celsius", *Sci. Rep.,* vol. 3, no. 1, p. 2572, 2013.
[http://dx.doi.org/10.1038/srep02572] [PMID: 23999206]

SUBJECT INDEX

A

Absorption spectroscopy 155, 249
  electrochemical X-ray 249
AC electrodes, heteroatoms-based 180
Acid 40, 51, 75, 76, 122, 124, 127, 175, 182, 203, 245, 251, 266
  chlorosulfonic 203
  electrolytes 175, 182
  glutaric 75
  hydrochloric 124
  hydrofluoric 122, 245
  methanesulfonic 51, 251
  oleic 76
  phenylsulfonic 127
Activities, electrocatalytic 112
Adsorption 35, 93, 106
  dye 93
  electrochemical 35
  energy 106
Aerospace applications 315, 316
Applications 4, 49, 54, 194, 201, 272, 303
  electronic 4, 49, 54, 194, 201, 272, 303
  energy-storage 272
  flexible electronics 194, 201
  photocatalytic 49
Atomic layer deposition (ALD) 128
Augur electron spectroscopy (AES) 146, 155

B

Ball milling method 264
Behavior 74, 254, 256, 262, 264
  electro-chemiluminescent 74
  electrochemical 256, 262, 264
  robust optical absorption 254
Biomass-derived activated carbon 33
Building blocks 89
  organic 89
  supra-molecular 89

C

Capacitors 16, 30, 32, 35, 101, 112, 120, 135, 213, 225, 226, 258, 285, 286, 290, 291, 296, 303, 305
  dielectric 32, 305
  electrochemical 16, 30, 258, 291
  electrochemical double layer 226
  potassium-ion-based 101
Carbide-derived carbons (CDC) 160
Carbon 13, 36, 37, 38, 60, 61, 64, 73, 74, 75, 76, 78, 79, 81, 82, 100, 103, 105, 107, 179, 180, 194, 196, 213, 216, 220, 221, 226, 237, 255, 270, 285, 287, 290, 291, 292, 307, 324, 326, 328, 329
  materials synthesis technique 226
  nanofibers 13, 36, 38, 237, 285, 326
  nanomaterials 36, 37, 38, 179
  nanotubes (CNTs) 38, 60, 61, 64, 100, 103, 105, 107, 179, 194, 196, 270, 287, 291, 292, 307, 324, 326, 328, 329
  polymorphs 61
  quantum dots (CQDs) 73, 74, 75, 76, 78, 79, 81, 82, 213, 216, 220, 221
  related materials 290
  source materials 180
  vacancies 255
Carbon-based 17, 38, 292
  active materials 292
  composite materials 17
  electrode materials 38
Carbon fiber(s) 206, 214
  materials 206
  oxidizing 214
Catalytic transformations 93
Cellulose fibers 310
Charge carrier(s) 121, 329
  mobility 121
  electronic 329
Charge-discharge 33, 146, 151, 217, 218, 264
  cycles 151, 217, 218, 264
  mechanism 33

process 146

Charge storage 6, 29, 98, 128, 135, 147, 158, 163, 169, 213, 247, 304, 329, 331, 332
  mechanisms 6, 29, 98, 135, 147, 332
  methodologies 158
  methods 329
  pathways 247
  process 158, 163, 169, 213, 331
  properties 128
  techniques 304

Charge-storing 29, 247, 251
  method 29
  pathways 247, 251

Chemical 2, 14, 59, 72, 91, 107, 108, 112, 128, 179, 194
  properties 2, 72, 108, 112
  reduction methods 59
  techniques 14
  transformations 91
  vapor deposition (CVD) 107, 128, 179, 194

Cobalt phenylphosphonate (CPs) 15, 289

Composite materials 17, 18, 39, 49, 107, 110, 175, 273, 329
  complementary 329
  developed polymer-based 17

Computational techniques 323, 325
  molecular scale 325

Computational tools 332, 334

Conductance-based sensor 55

Conducting polymers 7, 36, 37, 39, 77, 78, 81, 233, 234, 237, 289, 290, 292
  electrical 39
  polypyrrole 292

Conductive polymer materials 328

Corrosion 12, 17, 288, 324
  chemical 288
  resistive 12

Crystal structure topology 91

CVD growth method 60

Cycling 134, 146, 153, 170, 175, 201, 266, 305, 327, 328
  electrochemical 153
  galvanostatic 266
  stability 134, 146, 170, 175, 201

D

Defects 104, 124, 128, 129, 262, 265, 266
  atomic 124
  metal site 104

sonication-induced 262

Degradation 17, 93, 182, 194, 205, 268, 308, 310
  catalytic 17
  oxidative 310
  photocatalytic 17

Depletion of electrolytes 40

Detection, colorimetric 93

Detoxifying agents 109

Development 2, 195
  industrial 2
  wearable electronic device 195

Devices 2, 18, 21, 56, 63, 81, 153, 191, 192, 213, 225, 285, 295, 303, 304, 311, 315, 331
  communication 295
  efficient energy 213
  electrochemical 153, 225, 303, 331
  electrochemical energy-storage 304
  energy source 285
  fabricated 81
  memory 191, 315
  mobile 311
  optoelectronic 2, 21, 63
  renewable energy 56
  solar energy 18
  wearable electrical 192

Dispersion, mechano-chemical 19

Dry polymer electrolyte 177

E

Electric vehicles (EVs) 29, 97, 132, 221, 303, 314, 332

Electrical conductivity 55, 60, 88, 95, 129, 134, 136, 194, 203, 206, 219, 226, 307

Electrocatalysis 15, 93, 107

Electrocatalysts 63, 331

Electrochemical 15, 34, 61, 73, 78, 92, 98, 99, 110, 121, 126, 146, 148, 149, 151, 152, 161, 163, 169, 170, 171, 173, 175, 176, 178, 179, 181, 201, 212, 216, 247, 251, 254, 255, 258, 260, 262, 264, 265, 268, 270, 271, 273, 310, 311, 312, 313, 314, 315, 316
  activity 78, 110, 179, 216, 271
  applications 15, 92, 99
  behaviour 251, 268
  comparison 152
  efficiency 258, 264, 265

impedance spectroscopy (EIS) 146, 151, 152, 163, 181, 201
method 34, 73, 126
microscopy 146, 148
progress 247, 254, 255, 260, 270, 273
properties 61, 98, 149, 212, 262, 271
quartz crystal microbalance (EQCM) 146, 161, 163
storage 121, 251
supercapacitors (ESs) 169, 170, 171, 173, 175, 176, 178, 310, 311, 312, 313, 314, 315, 316
Electrochemical energy 95, 108, 133
storage performances 108, 133
storage technologies 95
Electrode 102, 238, 247, 304
fabrication 247, 304
kinetics 102
mechanism 238
Electrodeposition 328
Electrolyte(s) 5, 38, 40, 98, 152, 175, 177, 215, 225, 238, 263, 264, 288, 309
carbonate-based 263
ceramic 177
corrosive 225, 238
gel polymeric 40
ion diffusion 215
leakage 98, 152, 309
liquid-based 40
neutral 5, 175, 288
transport 38, 264
Electron 99, 121, 134, 136, 181, 236, 249
transfer 99, 134, 136, 181, 236, 249
transportation 121
Electron transport 97, 113, 264
pathway 113
Electronic(s) 2, 15, 16, 21, 30, 36, 52, 60, 61, 108, 129, 134, 135, 177, 191, 192, 201, 206, 246, 247, 285, 293, 294, 315
conductivity 134, 135
devices 30, 36, 52, 177, 191, 206, 285, 294
properties 16, 129
wearable 192, 201, 247, 293
Elemental 155, 246
distribution profiles 155
stoichiometries 246
Energy resources 120, 334
renewable 120
Energy storage 5, 6, 17, 29, 30, 62, 77, 78, 88, 89, 95, 102, 107, 112, 120, 150, 195,

205, 206, 213, 226, 244, 246, 247, 284, 285, 288, 291, 293, 295, 313, 323, 328, 332, 333, 334
devices (ESDs) 62, 77, 78, 88, 89, 120, 244, 246, 247, 284, 285, 288, 293, 295
electrochemical 95, 213, 291, 295, 323
mechanisms 5, 6
method 313
of devices 205
properties 17
systems 29, 30, 95, 102, 112, 150, 195, 284, 285, 332, 334
technologies 30, 107, 226, 328, 333
wearable 206
Energy technologies 30, 102, 293
renewable 30, 293
Entropy, thermodynamic 72
Equivalent series resistance (ESR) 169, 170, 171, 174, 197, 304, 305, 306, 307, 308, 333
Etching 125, 324
method, electrochemical 324
techniques 125, 324

F

Fabrication 130, 238, 271, 321
methods 130, 271
techniques 238, 321
Faradaic system 148
Field effect transistors (FET) 54, 64
FT-IR 160
absorption 160
spectroscopy 160
Fuels, conventional 213

G

Galvanostatic charge-discharge (GCD) 81, 146, 150, 199, 267, 268
Gas 2, 4, 21, 93, 94, 308
adsorption 93, 94
production 308
sensors 2, 4, 21
sorption 93
Gel 177, 195, 196, 197, 198, 251
ionic-liquid 251
polymer electrolytes 177, 195, 196, 197, 198

Gel electrolytes 196, 197, 215
  sandwiching 215
  water-based 197
Glass fiber-based separators (GFBS) 310
Graphene quantum dots (GQDs) 13, 59, 60, 64, 75, 78, 79, 82, 213, 214, 216, 219, 220

H

High 88, 105, 291
  electrochemical activity 291
  energy storage performances 88
  resolution transmission electron microscopy (HRTEM) 105
Hydrogen evolution reaction, photocatalytic 92
Hydrothermal 54, 79, 91, 126, 133, 135, 150, 214, 287
  conditions 54, 214
  interactions 150
  method 79, 214
  reactions 91, 126, 133, 135, 214
  treatment 287

I

Imaging 13, 18, 59, 77, 89, 153
  biological 18, 77
  biomedical 59, 89
  fluorescence 13
  magnetic resonance 13
Internal 315, 324
  combustion engines (ICE) 315
  resistance 324
Ion(s) 136, 159
  electrolytic 159
  diffusion pathway 136
Ionic 40, 129, 169, 171, 175, 176, 177, 178, 182, 201, 309
  concentration 201
  conductivity 129, 169, 171, 175, 176, 177, 178, 182, 309
  electrolytes 40
Ionic liquid(s) 40, 169, 172, 174, 176, 177, 178, 180, 182, 183, 194, 197, 271, 309, 322, 330, 331, 333
  and organic electrolytes 174

electrolytes 40, 182, 183, 197, 271, 322, 330, 333

L

Laser ablation technique 73
Layer double hydroxides (LDHs) 111, 112, 220, 259
Light-emitting diodes, inorganic 293
Liquid 169, 171, 173, 177, 178, 192, 201, 236, 331
  anisotropic 201
  electrolytes 169, 171, 173, 177, 178, 192, 236, 331
Low capacitance supercapacitors (LCS) 304
Luminescence properties 95

M

Manganese oxide 5, 328
  electrode material 5
  nanoparticles 328
Mechanical flexibility 106
Mechanisms 29, 30, 31, 35, 36, 100, 146, 153, 159, 163, 173, 181, 183, 331, 334
  electrical 100
  electrochemical 159
  energy-storing 35, 36
Metal-organic 89
  coordination network (MOCN) 89
  materials (MOMs) 89
Methods 107, 215
  solvothermal 107
  ultra-sonication 215
Microwave 1, 73, 75, 76, 91, 328
  absorbing materials 1
  assisted deposition therapy 328
  synthesis 73, 75, 76, 91
Military and defense applications 295
MOF(s) 92, 93, 106, 109, 111, 238
  and carbonaceous materials 109
  and graphene combination 106
  composites, oxide-based 111
  derivatives 238
  metals-based 92, 93
Multi-layer 158, 310
  graphene (MLG) 158
  separators (MLS) 310
Multifaceted topologies 94

MXene 129, 215, 257, 258, 259, 260, 261, 262, 263, 264, 265, 269
  electrodes 129, 257, 259, 260, 261, 262, 263, 264, 265, 269
  films, conductor 258
  hydrogel supercapacitor electrode 265
  quantum dot (MQDs) 215
MXene-based 120, 130, 244, 247, 252, 266, 270, 271, 273
  composites 244, 270, 273
  electrochemical energy storage applications 244
  electrochemical energy storage systems 247
  High-performance Supercapacitors 266
  nanocomposites 120
  organic devices 252
  solid-state supercapacitors 271
  transparent films 130
  two-dimensional materials 266

N

Nanoparticles 7, 18, 107, 155
  encapsulate 107
  inorganic 18
  magnetic metal oxide 7
  metal hydroxide 155
Network 89, 292
  metal-organic coordination 89
  tunable regular porosity 292
Nippon electric corporation (NEC) 303
NMR spectroscopy 159
Nuclear magnetic resonance (NMR) 130, 146, 159, 163, 253
  spectroscopy 159

O

Ohm's law 151
Optoelectronic properties 130
Organic electrolyte systems 330
Oxidation 7, 34, 51, 54, 56, 63, 73, 74, 75, 79, 111, 131, 149, 155, 182, 245
  chemical 54, 75, 79
  electrochemical 73, 74
  fuel 63
Oxides 7, 35, 111, 327
  crystalline niobium 35

iron 327
magnetic metal 7
multimetallic 111
Oxygen 61, 97, 98, 110, 152
  adsorbed 152
  dissolved 152
  evolution reaction (OER) 61, 97, 110
  reduction reaction (ORR) 61, 97, 98, 110

P

Panasonic commercialize 190
Photoelectron effect theory 153
Photolithography 293
PMMA microsphere 257
Polymer electrolytes 171, 177, 197
  gel-based 171
Polymeric film materials 292
Porous 89, 310
  coordination polymers (PCPs) 89
  polypropylene films (PPF) 310
Power fluctuations 313
Process 8, 53, 74, 101, 102, 328
  electrochemical-mediated 102
  faradaic charge-transfer 101
  hydrogenation 53
  hydrothermal 328
  natural 8
  ultrasonic 74
Properties 13, 19, 60, 105, 247
  electrochemical capacitive 247
  electromagnetic 19
  fluorescent 13
  photochemical 105
  photoelectric 60
Pyrolysis 76, 108, 111

Q

QENS 162
  method 162
  technique 162
Quantum dots function 220

R

Raman spectroscopy (RS) 105, 146, 156, 158, 159, 267
Reaction kinetics 78, 133

electrochemical 133
Reactions 31, 32, 36, 41, 54, 61, 78, 91, 95,
    97, 100, 101, 107, 111, 154, 156, 158,
    270, 293, 304, 311
    chemical 32, 41, 100, 101, 158, 293, 311
    electrochemical 31, 95, 107, 111, 156
    oxidation-reduction 270
    oxygen reduction 61, 97
Redox processes, reversible electrochemical
    267
Redox reactions 7, 31, 34, 35, 37, 38, 78, 150,
    152, 156, 236, 262
    electrochemical 34
Resistance 17, 39, 40, 63, 151, 152, 197, 201,
    220, 250, 305, 333
    charge transportation 197
    corrosion 63
    electrolyte's 39
Resonance, nuclear magnetic 146, 253

S

Sauebrey's equation 161
Scanning 146, 148, 153
    electrochemical microscopy 148
    electron microscopy 146, 153
Sensors 15, 18, 52, 55, 56, 64, 77, 89, 93, 111,
    237, 246
    electrical 64
    electrochemical 18
Solar 19, 56, 64, 108
    cells, photoelectrochemical 56, 64
    energy harvesting 108
    energy management 19
Solid-state supercapacitors (SSS) 40, 62, 82,
    235, 268
Solvothermal 91, 108
    ion-assisted 108
Sonication and ion intercalation methods 127
Spectroscopy 146, 151, 161, 181, 201
    electrochemical impedance 146, 151, 181,
    201
    hydrodynamic 161
Storage 15, 52, 89
    energy-based 52
    gas 15, 89
Storage applications 104, 108, 110, 247
    durable energy 104
    flexible electrochemical energy 247

Storage devices 93, 96, 120, 135, 196, 205,
    212, 221, 287, 310
    conventional energy 310
    developing high-performance energy 287
    emerging energy 96
    fiber-shape energy 196
    next-generation energy 120, 135
    pseudocapacitive energy 205
Supercapacitor 4, 5, 6, 8, 34, 36, 38, 214, 215,
    225, 226, 231, 232, 236, 238, 270, 271,
    286, 321, 322, 323, 324, 327, 329, 330,
    331
    devices 4, 6, 34, 321, 322, 323, 324, 327,
    329, 330, 331
    electrode materials 36, 225, 238, 286, 329
    electrodes 5, 8, 214, 215, 226, 231, 232,
    236, 270, 271, 321, 322, 324
    fabrication 38
Synthesis 74, 77, 79, 91, 214, 229, 231
    economical 77
    electrochemical 74, 91, 229
    hydrothermal 214, 231
    microwave-assisted 91, 229
    sonochemical 229
    wet-chemical 79
Systems 62, 225
    electrochemical energy depository 62
    energy accumulator 62
    hybrid transportation 225

T

Techniques 55, 73, 74, 75, 91, 145, 146, 148,
    161, 162, 181, 192, 214, 270, 271, 292,
    295, 305, 324, 325, 327
    chemical-oxidative polymerization 214
    chemical polymerization 214
    chemical vapour deposition 324
    economical microwave pyrolysis 75
    electrochemical 145, 146, 148
    hydrothermal 192
    laser-reduction 192
    sonication 271
    thermal decomposition 73
    vapor-phase 73
    wet-chemical 73
Technology, dental 295
Thermal 72, 76, 131, 135, 291
    conductivity 131, 135, 291
    decomposition 76

energy 72
Thermogravimetric analyses 156
Toxicity, environmental 330
Transmission 97, 146, 153, 315
   communication 315
   electron microscopy 146, 153
Transport mechanisms 330

V

Vacuum filtration 133

W

Wet 123, 190, 201, 202
   chemical etching 123
   spinning technique 190, 201, 202

X

X-ray photoelectron spectroscopy 146, 153

www.ingramcontent.com/pod-product-compliance
Lightning Source LLC
Chambersburg PA
CBHW050804220326
41598CB00006B/113